Structural Dynamics

Vibrations & Systems

Structural Dynamics
Vibrations & Systems

Madhujit Mukhopadhyay
Indian Institute of Technology,
Kharagpur

CRC Taylor & Francis
Taylor & Francis Group
Boca Raton London New York

CRC is an imprint of the Taylor & Francis Group,
an informa business

Ane Books India

Structural Dynamics

First Published in 2008 by

Ane Books India

4821 Parwana Bhawan, 1st Floor
24 Ansari Road, Darya Ganj, New Delhi -110 002, India
Tel: +91 (011) 2327 6843-44, 2324 6385
Fax: +91 (011) 2327 6863
e-mail: anebooks@vsnl.com
Website: www.anebooks.com

For

CRC Press

Taylor & Francis Group
6000 Broken Sound Parkway, NW, Suite 300
Boca Raton, FL 33487 U.S.A.
Tel : 561 998 2541
Fax : 561 997 7249 or 561 998 2559
Web : www.taylorandfrancis.com

For distribution in rest of the world other than the Indian sub-continent

ISBN-10 : 1 42007 066 5
ISBN-13 : 978 1 42007 066 8

British Library Cataloguing in Publication Data
A catalogue record for this book is available from the British Library

Printed at Gopsons Paper Ltd., Noida

The book is dedicated

to the memory of

The Late Professor Bidhu Ranjan Sen

Preface

During last three decades the subject "Vibrations and Dynamics of Structural and Mechanical Systems" has undergone significant development. There has been an urge and urgency of designing present day modern and complex structures and systems in their proper perspective, a task which could not be conceived even three decades back, This has been possible due to the advent of the electronic digital computer. With the exception of the self weight of the structures, strictly speaking, no other load can be treated as static load. In order to truly analyse these systems, a knowledge of treatment the dynamic load is of paramount importance.

The existing book on the subject mainly deal with one of the following aspects:

1. Vibrations of Mechanical Systems.
2. Structural Dynamics.

Though basic principles and solution techniques are same for both the above topics, they primarily differ in their application areas. The main objective of the book is to present these aspects in a unified and integrated manner. The chapters have been arranged in such a way that starting from the basics, the reader is acquainted with more advanced topics in a gradual manner.

The material of the book has been developed while teaching the students of Indian Institute of Technology, Kharagpur. A similar book was written by the present author which is no longer availaible in the market and also it is withdrawn. As such a necessity has arisen for the publication of the present book.

The author is indebted to a number of persons, without whose help, the preparation of the manuscript would not have been possible. To name the most important persons amongst them are - Ashoke Das, Prabir Sinha, Parimal Kumar Roy, Chinmoy Mukherjee and Ratna, the wife of the author.

Lastly, the author expresses his sincere thanks to Mr. Sunil Saxena, Jai Raj Kapoor and Sudipta Ghosh - all of Ane Books, for their help and the excellent cooperation received.

Madhujit Mukhopadhyay

Contents

Chapter 1

Introduction

1.1 INTRODUCTION

Majority of today's structures is subjected to load which varies with time. In fact, with the possible exception of dead load, no structural load can really be considered as static. However, in many cases the variation of the force is slow enough, which allows the structures to be treated as static. For highrise buildings subjected to wind and earthquake, offshore platforms surrounded by waves, aeroplanes flying through storms, vehicles moving on the road, reciprocating engines, rotors placed on the floor and many other categories of loading, the dynamic effect associated with the load must be accounted for in the proper evaluation of safety, performance and reliability of these systems.

Many of the structures such as buildings, slab-beam bridges, ship etc., are designed on the basis of static or pseudo-static analysis. The main reason for so doing lies in the simplicity of these analyses. A dynamic analysis of structures is much more involved and time consuming than an equivalent static analysis. Structures thus designed have been found to be safe, though no idea could be obtained from this as to the extent of its safety as also to its actual behaviour.

The analysis of vibration problems in machines follows a similar approach. The torsional vibrations of shafting systems with its gearing arrangement, the vibration of turbine blades, the whirling of rotating shafts and various other related problems have assumed added significance. The effective working condition of these mechanical systems, keeping them free from critical conditions associated with vibration, can only be achieved from a thorough understanding of the vibration analytical procedures.

With high speed digital computers at the disposal of the analyst, and the tremendous advancement made in the analytical and numerical procedures, more accurate representation of the structural behaviour due to dynamic loads, as also the study of more complex problems related to machine vibrations have now become possible. As such, many of the structures which could not have been conceived in terms of its length and breadth even four decades ago, have become a reality today. Large systems having internal intricacies and huge dimensions are now daringly attempted. The increasing interest in the vibrations and dynamics of structures and machines is thus only natural.

1.2 BRIEF HISTORY OF VIBRATIONS

During the period of development of basic science in Greece, the birth of vibration theory and its subsequent growth was a natural process. The Ionion School of natural philosophy introduced the scientific methods of dealing with natural phenomena. Vibration theory was initiated by

Pythagoras in the fifth century B.C. By establishing a rational method of measuring sound frequencies, Pythagoras quantified the theory of acoustics. He conducted several experiments with hammers, strings, pipes and shells for determining fundamental frequency. He proved with his hammer experiments that natural frequencies are system properties and they do not depend on the magnitude of excitation. [1.1], [1.2].

Aristotle and his Peripatetic School founded mechanics and developed an understanding between statics and dynamics. The basics were attempted to be implanted on fundamental philosophical ideas, instead of empirically founded axioms. [1.3] In the Hellenistic period there were substantial engineering developments in the field of mechanical design and vibration.

The pendulum as a time and vibration measuring device, was known in antiquity to the extent that by the end of first millenium A.D., it was used by astronomers as a vibrating and perhaps, time measuring device.

Galileo experimented with pendulum in the seventeenth century. He observed that there exists a relationship between the length and the frequency of a pendulum. He further noticed that the density, tension, frequency and length of a vibrating string are related to each other. The relationship between sound and vibration of mechanical system was understood for a long time, but Galileo was the first to prove that pitch is determined from the frequency of vibration.

In the development of vibration theory, Taylor, Bernoulli, D'Alembert, Euler, Lagrange, Duhamel, Hertz and Fourier are the names to conjure with. Thomas Young was the first to show how important dynamical effect of load can be [1.4]. Poncelet demonstrated analytically that a suddenly applied load produces twice the stress the same load would produce if applied gradually. From experiments, Savart noted the mode shapes from the study of nodal lines in aelotropic circular plates. He also coined the term fundamental for the lowest natural frequency and harmonics for the other. The concept of the principle of linear superposition of harmonics was put forward by Bernoulli.

Coulomb carried out theoretical and experimental studies of torsional oscillation of a metal cylinder suspended by a wire in 1784.

By covering a plate with fine sand, Chladni proved the existence of nodal lines for various modes of vibration. The French Emperor Napoleon was very impressed by Chladni's experiments. At his suggestion, the French Academy announced in 1811 a prize for the problem of deriving a mathematical theory of plate vibrations. In response to it, Sophie Germain, a French lady proposed a solution, but was denied the prize due to an incorrect computation of the variation of a particular integral.

The Academy proposed the subject again in 1813. Though the error previously mentioned was rectified by now, but Germain failed to give a satisfactory explanation of an assumption. So the prize again eluded her. When the Academy proposed the subject once more, Germain did finally win the prize in the third attempt in 1816, though the judges comprising of Legendre, Laplace and Poisson were still far from being happy. Poisson treated her as an inferior novice in the company of giants. Though the differential equation of plate vibration was correctly derived by Germain, but she made mistakes in boundary conditions which were corrected by Kirchhoff in 1850.

Lord Rayleigh wrote his classic book on the theory of sound [1.5] which is still being treated with great respect. Rayleigh introduced the concepts of generalised forces and generalised coordinates, the treatment of which proved advantageous to engineers. In his method an approximation is made of the suitable form of the type of motion for obtaining natural frequencies of complicated systems. The famous Rayleigh's method is based on the principle of conservation of energy for conservative systems. This idea of calculating natural frequencies directly from the energy consideration, without taking recourse to the differential equations of vibration was later utilised by Ritz. The so called Rayleigh-Ritz method is now a very popular approach for studying not only vibrations, but in solving problems in elasticity, theory of structures, nonlinear mechanics and other branches of physics.

Euler and Bernoulli derived the differential equation of a vibrating beam undergoing small deflections. Kirchhoff investigated the vibration of bars of variable cross-section and found 'exact' solution for certain cases. Duhamel proposed a general method for analysing the forced vibration of elastic plates. This method was also used by Saint Venant in studying lateral forced vibration of beams.

One of the first problems in which the importance of studying vibrations was recognised by engineers was that of torsional vibration of propeller shafts of steamships. Frahm was the first to investigate the problem theoretically and experimentally. Approximate methods have been introduced for studying beam problems of non-prismatic sections. One such idea of successively approximating the integration of differential equations was introduced by Vianello, who used it for calculating buckling load of struts. Its extension to vibration problems was made by Stodola.

The whirling of a shaft carrying a disc was first investigated by Foppl. Investigation on two-span beam had been carried out theoretically and experimentally by Ayre, Ford and Jacobsen. Timoshenko included the shear deformation and rotary inertia in his beam vibration problem to yield an improved theory of beam vibration. Similarly, Mindlin contributed greatly to propose an improved theory of plate vibrations.

Mathematical theory of nonlinear vibration was developed towards the end of last century [1.6]. Duffing and Van der Pol were the first to propose definite solutions in nonlinear vibrations.

Taylor introduced the concept of correlation function in 1920 and Wiener and Khinchin the spectral density in 1930s [1.7,1.8]. These opened new vistas for laying down the theory of random vibrations [1.9-1.12]. A number of monographs on the topic are now available.

Vibrations and structural dynamics have made rapid strides in last few decades. The advent of computer has brought in a revolution in this area like many others. As a result of which new numerical techniques have been developed, out of which the finite element method has proved to be most versatile [1.13]. These newly developed methods of analysis permits the structural analyst to formulate the increasingly complex problems [1.13].

1.3 COMPARISON BETWEEN STATIC AND DYNAMIC ANALYSES

The prime difference between static and dynamic analyses lies in the interpretation of the load considered. Dynamic means time-varying. As such, time is the additional parameter in the dynamic analysis.

What is the difference between the two regarding the interpretation of forces? A block resting on the ground is shown in Fig.1.1. The ground reactions can be evaluated for the following conditions from elementary knowledge of statics:

(a) When the block is at rest,
(b) When the block is on the verge of incipient motion,
(c) When it is moving with a uniform velocity.

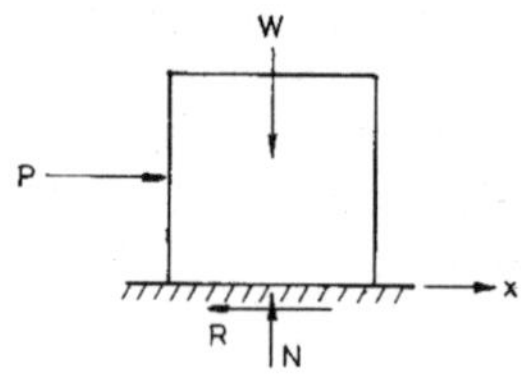

Fig. 1.1 A block resting on the ground.

The forces that may arise in the above three conditions have been indicated in Fig.1.1. If the block now starts moving with acceleration a, then it will experience an inertia force which is equal to mass times acceleration. By applying Newton's second law, one obtains

$$P - R = ma \tag{1.1}$$

where m is the mass of the block.

It is the acceleration of the body which converts the problem into a dynamic one.

1.4 D'ALEMBERT'S PRINCIPLE

The above example of the accelerating body can be considered as an equilibrium problem. The state of equilibrium is referred to as dynamic equilibrium. The principle that is adopted for its solution is known as D'Alembert's principle.

Equation (1.1) is rewritten in the following form

$$P - R - ma = 0 \tag{1.2}$$

Equation (1.2) can be conceived as an equation of equilibrium, i.e. $\sum F_x = 0$ which means that in addition to the real forces acting on the body, a fictitious force $-ma$ acting opposite to the line and direction of motion is considered (Fig. 1.2).

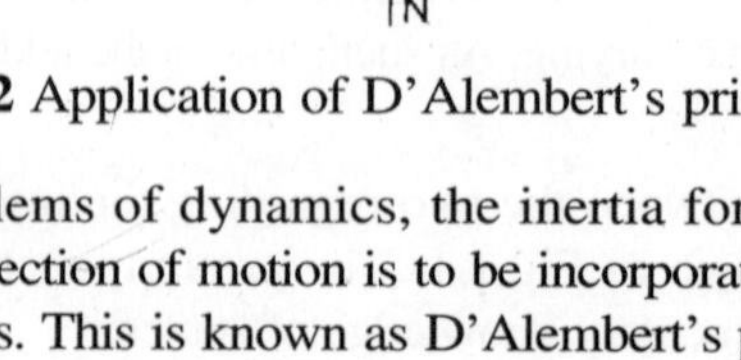

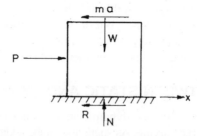

Fig. 1.2 Application of D'Alembert's principle.

For the solution of problems of dynamics, the inertia force, which is mass times the acceleration opposite to the direction of motion is to be incorporated in the freebody diagram in addition to all impressed forces. This is known as D'Alembert's principle.

Structural dynamics and machine vibrations have their basic input to the fundamental theory of vibration. To start with, let us look into some basic definitions.

1.5 SOME BASIC DEFINITIONS

When an elastic system is acted upon by an externally applied disturbance, it will respond by oscillating to and fro.

1.5.1 Vibration and Oscillation

If the motion of the body is oscillating or reciprocating in character, it is called vibration if it involves deformation of the body. In case the reciprocating motion involves only the rigid body movement without involving its deformation, then it is called oscillation. The motion of a pendulum is simple harmonic oscillation and so is the case of a ship if considered as a rigid body moving over waves.

1.5.2 Free Vibration

Vibration of an elastic system may be initiated by the application of the external force, which may subsequently be withdrawn. The resulting vibration, where no external force is present is referred to as free vibration.

1.5.3 Forced Vibration

If the external force is considered to be operative during the vibratory motion of the body, such a motion is referred to as forced vibration.

1.5.4 Damping

For all practical structures, resistance is encountered during vibration. As a result of which, the amplitude will not remain constant, but will gradually go on reducing with time in the case of free vibration, till a time comes when all the motion will be damped out.

Damping forces may depend on the vibrating system, as also on factors outside it. Damping in a system arises due to the internal friction of the material, or due to drag effects of surrounding air or other fluids, in which the structure is immersed. The exact representation of them involves difficulties. However, as to its nature, damping is of the following four types:

(a) Structural damping,
(b) Viscous damping,
(c) Coulomb damping, and
(d) Negative damping.

Structural damping is due to the internal molecular friction of the material of the structure, or due to the connections inherent in a structural system.

Viscous damping occurs in a system vibrating in a fluid. The damping force in this case is proportional to the velocity and is given by

$$F_0 = c\dot{x} \tag{1.3}$$

where $\dot{x}$ = velocity = $\dfrac{dx}{dt}$ and c is the damping constant.

Coulomb damping, also known as dry friction, occurs when the motion of the body is on a dry surface. The damping force is given by

$$F_0 = \mu N \tag{1.4}$$

where μ is the coefficient of kinematic friction and N is the normal pressure of the moving body on the surface.

Negative damping is a very special case, when the system is such that damping adds energy to the system instead of it being dissipated. As such, amplitude increases progressively in such cases. It occurs in the wires of transmission line tower.

In this book, damping will be considered throughout to be of viscous type unless otherwise specified.

1.5.5 **Degrees of Freedom**

Any mass can undergo six possible displacements in space - three translations and three rotations (about an orthogonal axis system). However, it happens with most systems that they are allowed to move in certain directions, while their movements in other directions are constrained. The minimum number of coordinate systems required to indicate the position of the mass at any instant of time is referred to as the degrees of freedom. A cantilever column with a concentrated weight at the top is shown in Fig. 1.3. If one is interested in studying the axial vibration, then one coordinate in the y-direction is sufficient to describe the motion. The system then has a single degree of freedom. If one is interested only in the flexural vibration of the structure, then the x-coordinate displacement describes the motion and the system still is of single degree of freedom. But if one intends to study the axial and flexural vibration in a combined manner, then the problem at hand becomes a two-degree of freedom system (the x and y displacements of the mass).

Depending on the independent coordinates required to describe the motion, the vibratory system is divided into the following categories:

(a) Single degree of freedom system (SDF system),
(b) Multiple degrees of freedom system (MDF system),
(c) Continuous system.

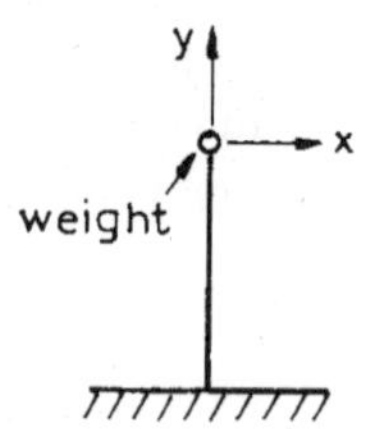

Fig. 1.3 A cantilever column with a concentrated weight at the top.

If a single coordinate is sufficient to define at any instant of time the position of the mass of the system, it is referred to as a single degree of freedom system (Fig. 1.4).

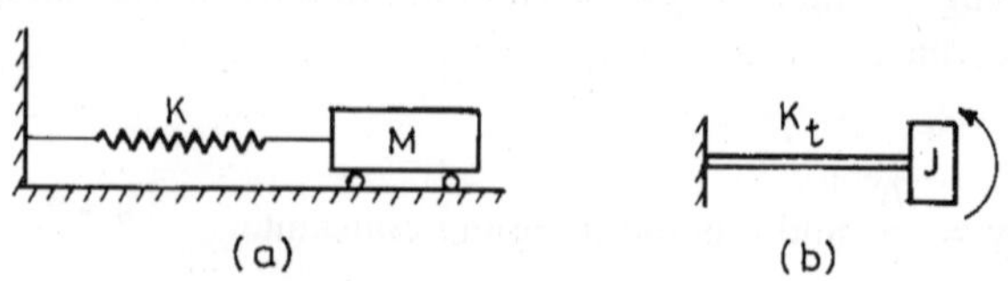

Fig. 1.4 Single degree of freedom system.

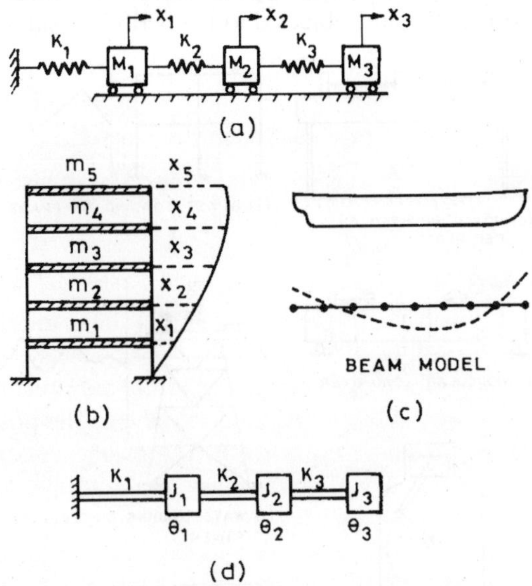

Fig. 1.5 Multiple degrees of freedom system

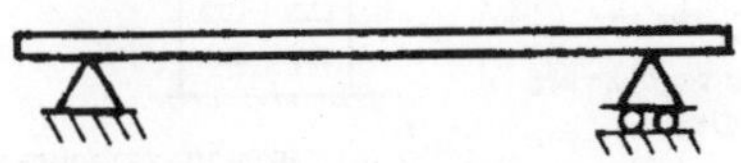

Fig. 1.6 A continuous system

If more than one independent coordinate is required to define at any instant of time the position of different masses of the system, it is referred to as multiple degrees of freedom system. A few examples of multiple degrees of freedom system have been shown in Fig. 1.5. If a system requires three independent coordinates as shown in Fig. 1.5(a) and 1.5(d), it is called three degrees of freedom system.

The mass of a system, such as the beam of Fig. 1.6 may be considered to be distributed over its length, in which case the mass is considered to have infinite degrees of freedom. Such a system is referred to as a continuous system.

1.6 DYNAMIC LOADING

Dynamic load is that in which its magnitude, direction or position vary with time. Almost all structures are subjected to dynamic load sometime or other during its lifespan. They are the result of various manmade and natural causes. The particulars of the load vary with time for different sources from which they are generated. Some common types of influences resulting in the disturbance are due to initial conditions, applied actions and support motions.

A few cases of dynamic loads acting on different systems are shown in Fig. 1.7. In general the deterministic loading that act on different systems can be divided into two basic categories: periodic and nonperiodic. In periodic loading the variation of the load is same in each cycle. The simplest periodic loading is a simple harmonic motion. For the rotor, the reciprocating engine and the ship of Fig. 1.7, the dynamic load is periodic. The form of periodic loading varies from simple sine or cosine curves to more complex shapes. The bridge, the frame, the offshore tower and the building of Fig. 1.7 are subjected to nonperiodic loading.

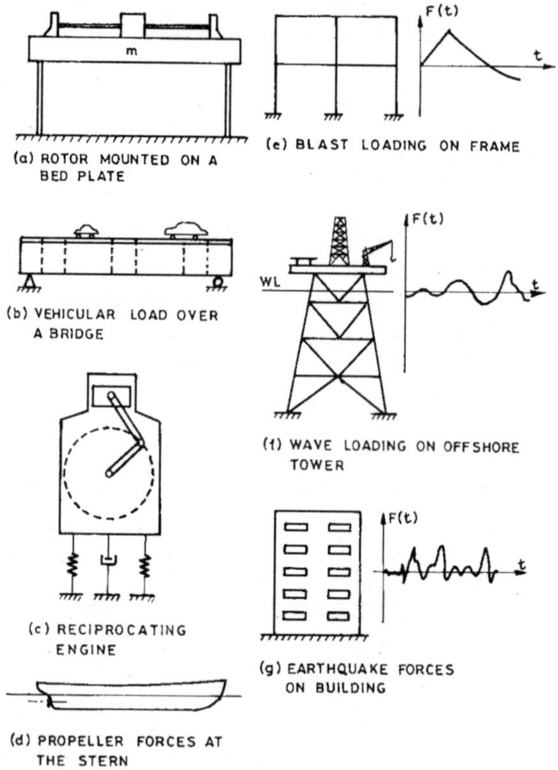

Fig. 1.7 Dynamic loads on different systems

Again, the form of the loading for these cases varies. The blast loading of Fig. 1.7(e) is of short duration, whereas the wave and earthquake loading of Fig. 1.7(f) and 1.7(g) are having a much longer tenure. Due to the characteristics of these loadings, the mathematical treatment is different for different cases.

1.7 FINITE ELEMENT DISCRETIZATION

One of the initial steps to be followed in the finite element analysis is the discretization of the model. For framed structure the discretization is obvious, as one member is considered at a time which is the discrete element. However, for the sake of analysis, the joint may be artificially created by considering a joint at any point of the member. The tapered beam of Fig.1.8 is divided into four elements. Fig. 1.8 shows the beam consisting of four elements and 5 joints. A joint or a node is defined as the intersection of two or more elements, a support point or a free end. For the analysis of one-dimensional structures, the choice of the element is obvious in many cases. For example, the portal frame of Fig.1.9 has 3 elements and 4 joints/nodes. In this case the choice of the element is obvious, as each member (naturally occurring) is considered as an element.

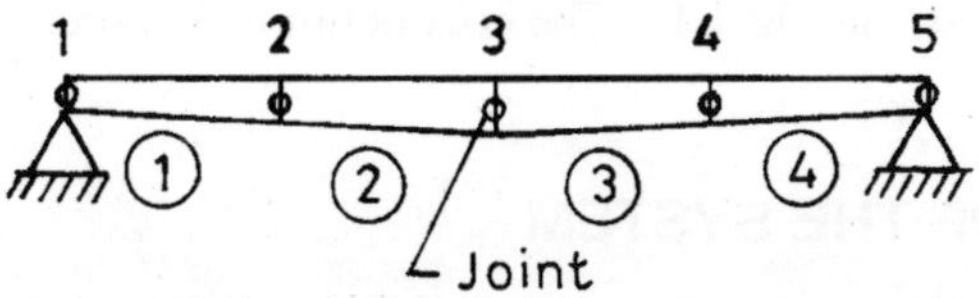

Fig. 1.8 A tapered beam with elements

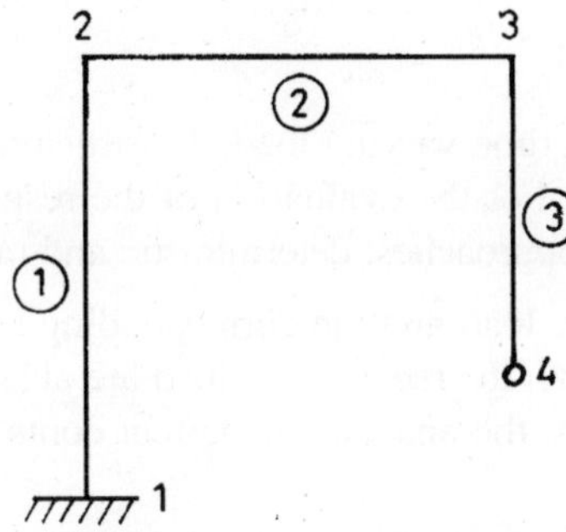

Fig. 1.9 A portal frame

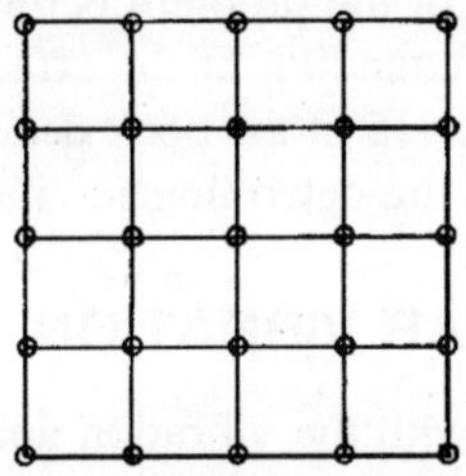

Fig. 1.10 (a) A discretized plate

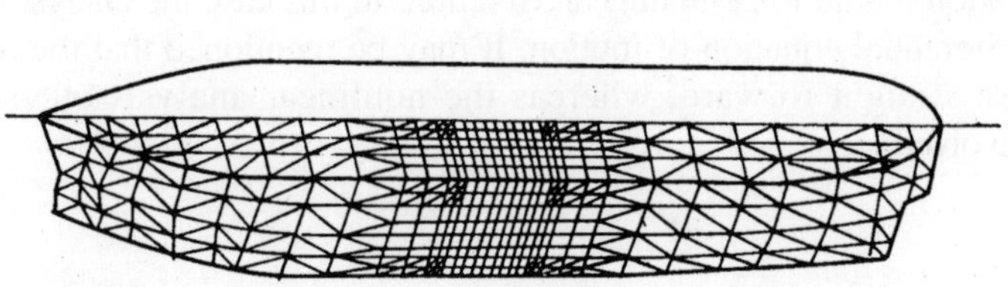

Fig. 1.10 (b) Discretized ship

When we focus our attention to two and three-dimensional structures such as plates and shells we immediately face difficulty with the choice of discretization, as no obvious choice exists. In such cases, like conventional one-dimensional structures which are considered as an assemblage of structural elements interconnected at their intersections, the continuum is assumed to consist of imaginary divisions. Thus the real continuum is divided into a finite number of discrete elements as shown in Fig 1.10. The continuum is considered as an assemblage of these 'finite elements'.

1.8 RESPONSE OF THE SYSTEM

The dynamic load applied to the system not only produces certain motions, but also cause certain associated stresses, strains, reactions, etc. The term response of a system indicates any of the above effects produced by the dynamic load. More commonly, however, the dynamic response indicates the displacement time history.

1.9 TYPES OF ANALYSIS

Dynamic load being defined as a time varying load, the structural response to dynamic loading is also time varying or dynamic. For the evaluation of the response due to the dynamic load, there are basically two different approaches: deterministic and random.

If the records of the dynamic load and the corresponding responses are taken many times under identical conditions, and both the records obtained are alike in all cases, then the vibration is said to be deterministic. In this, the analyst has perfect control over all the variables related to the problem.

If the records of the dynamic load and the corresponding responses are taken many times, when all other conditions under the control are maintained the same, but the records are found to differ continually from each other, then the vibration is said to be random. This unpredictability associated with the input and output of the problem is referred to as randomness. The degree of randomness depends on the understanding of the effects of different parameters involved and the ability to control them. Chapter 15 of the book deals with random vibration, whereas the treatment in other remaining chapters are deterministic vibration.

1.10 LINEAR AND NONLINEAR VIBRATION

If the basic components associated with the vibration analysis, such as the spring, mass and the damper behave linearly, then the emanated vibration is referred to as linear vibration. In this case the differential equation of motion is linear.

On the other hand if one or more of the basic components behave in a nonlinear manner, the resulting vibration is said to be nonlinear vibration. In this case the analyst will have to deal with nonlinear differential equation of motion. It may be mentioned that the linear analysis of vibration is more straight forward, whereas the nonlinear analysis of vibration is more mathematically involved.

References

1.1 M. R. Cohen and I. E. Drabkin, A Source Book on Greek Science, Harvard University Press, Cambridge, 1958

1.2 A. D. Dimarogonas, The origins of vibration theory, Journal of Sound and Vibration, V. 140, N`o. 2, 1993, pp. 181-189

1.3 M. G. Evans, The Physical Philosophy of Aristotle, The University of New Mexico Press, Albuquerque, 1964

1.4 S. P. Timoshenko, History of Strength of Materials, McGraw-Hill, New York 1953

1.5 J. W. Strutt [Lord Rayleigh], The Theory of Sound, Dover, 1945.

1.6 F. Dinca and C. Theodosin, Nonlinear and Random Vibration, Academic Press Inc., 1973.

1.7 S. H. Crandall and W. D. Mark, Random Vibration in Mechanical Systems, Academic Press, New York, 1963.

1.8 J. D. Robson, Random Vibration, Edinburgh University Press, Edinburgh, U. K., 1964.

1.9 D. E. Newland, An Introduction to Random Vibrations and Spectral Analysis, Longman, London, 1975.

1.10 R.W.Clough and J. Penzien, Dynamics of Structures, McGraw-Hill Inc., 1993

1.11 S.O. Rice, Mathematical Analysis of Random Noise, Dover Publications, New York, 1954

1.12 J.S. Bendat, Principles and Applications for Random Noise Theory, John Wiley & Sons, New York, 1966

1.13 O. C. Zienkiewicz and R.L. Taylor, The Finite Element Method, Vol I & Vol II, Fourth Edition, McGraw-Hill Book Company, UK, 1989

1.14 L. Meirovitch, Computational Methods in Structural Dynamics, Sijthoff and Nordhoff, The Netherlands, 1980

Chapter 2

Free Vibration of Single Degree of Freedom System

2.1 INTRODUCTION

The simplest physical system is one having single degree of freedom. Almost all practical systems are much more complex than this simple model. However, for obtaining an approximate idea about vibration characteristics, some of the systems are at times reduced to that of a single degree of freedom, such as the water tower of Fig. 2.1.

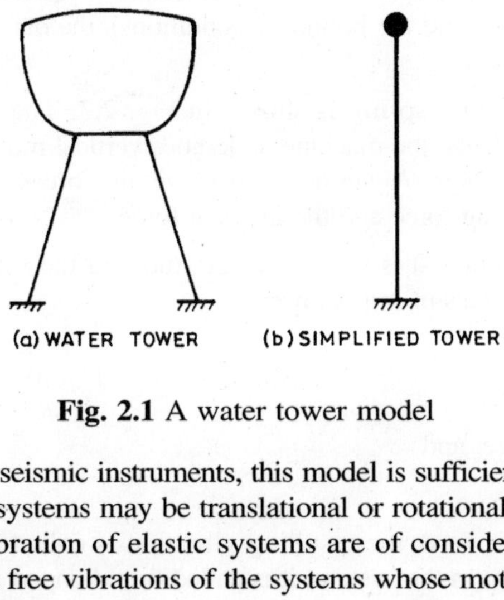

Fig. 2.1 A water tower model

In some cases such as seismic instruments, this model is sufficient for the necessary study. Single degree of freedom systems may be translational or rotational. The rotational problems related to the torsional vibration of elastic systems are of considerable importance. In this chapter, we shall deal with free vibrations of the systems whose motion can be described by a single coordinate [2.1 - 2.10].

2.2 EQUATION OF MOTION OF SINGLE DEGREE OF FREEDOM (SDF) SYSTEM

Consider the single bay portal frame of Fig. 2.2. The masses of the columns are considered to be small in comparison to the mass of the girder, and they are

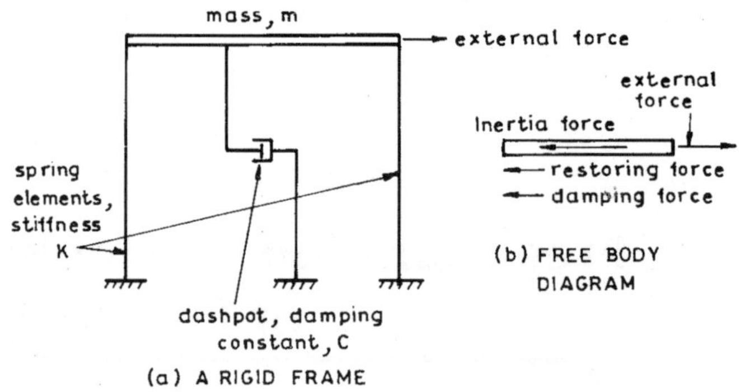

Fig. 2.2 A portal frame and its free body diagram

neglected in the analysis. Further, the girder is assumed to be infinitely rigid, so that the stiffness of the system is provided only by columns. The entire motion may be determined once the mass centre of the girder is known. When the frame vibrates in the horizontal direction, the forces associated with the motion are the inertia forces connected with the mass of the body, the restoring force provided by spring elements (the spring constant can be determined from the properties of the column and the boundary conditions), the damping force provided by the damper and the external force.

A machine mounted on a spring is shown in Fig. 2.3. The damping of the system is indicated by the dashpot. When the machine undergoes vertical motion, the forces acting on its mass are indicated in Fig. 2.3(b). Again in the machine, the forces that act are the inertia force, the spring force, the damping force and the external force.

From the above examples it is seen, that for studying the vibration of any system, the following are the four elements of importance:

(a) The inertia force,

(b) The damping force,

(c) The restoring force, and

(d) The exciting force.

If the external force is zero, the resulting vibration is called the free vibration. In such cases, if damping is present in the system, it is called free damped vibration and if damping is absent in the system, it is called free undamped vibration.

The two systems mentioned above, can be idealised in the form of springmass system as shown in Fig. 2.4. All structures can be reduced to systems consisting of combinations of springs, masses and dashpots and is an idealisation in a convenient form of the actual structure.

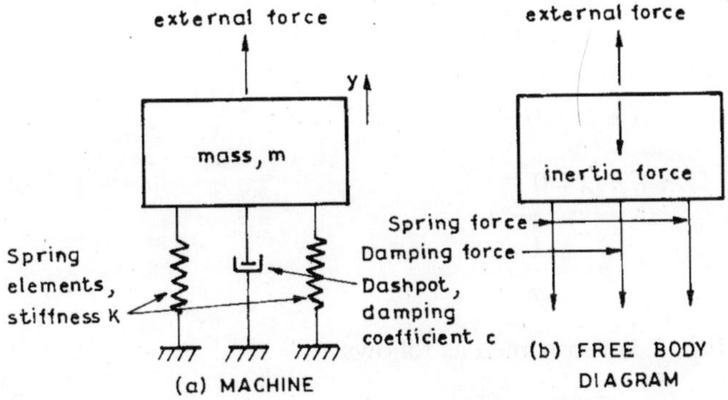

Fig. 2.3 A spring mounted machine and its freebody

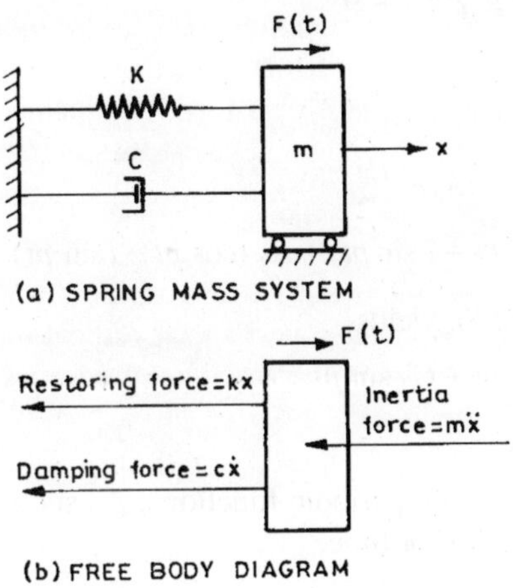

Fig. 2.4 A spring-mass-dashpot system

The only independent coordinate of the spring-mass system of Fig. 2.4 is the x-directional translation and the displacement will simply be given by x, velocity $\dot{x}\left(=\dfrac{dx}{dt}\right)$ and acceleration $\ddot{x}\left(=\dfrac{d^2x}{dt^2}\right)$. The inertia force of the mass is $m\ddot{x}$, the damping force is $c\dot{x}$ and the spring force is *kx*. The external force acting on the system is *F*(*t*).

Applying D'Alembert's principle and considering equilibrium of all the forces in the x-direction, the general equation of motion for the SDF system is

$$m\ddot{x} + c\dot{x} + kx = F(t) \tag{2.1}$$

The solution of Eq. (2.1) gives the response of the mass to the applied force in the SDF system.

2.3 FREE UNDAMPED VIBRATION OF THE SDF SYSTEM

When the case of free undamped vibration is considered, $F(t) = 0$ and $c = 0$ in Eq. (2. 1) and the resulting equation becomes

$$m\ddot{x} + kx = 0 \qquad (2.2)$$

or

$$\ddot{x} + p^2 x = 0 \qquad (2.3)$$

where

$$p^2 = \frac{k}{m}$$

The solution of Eq. (2.3) is assumed as follows

$$x = Ae^{\lambda t} \qquad (2.4)$$

Substituting x and its derivatives from Eq. (2.4) to Eq. (2.3), yields

$$A\lambda^2 e^{\lambda t} + p^2 Ae^{\lambda t} = 0$$

or

$$\lambda = \pm ip$$

The solution of Eq. (2.3) then is

$$x = A_1 e^{ipt} + A_2 e^{-ipt}$$

or

$$x = A_1 (\cos pt + i \sin pt) + A_2 (\cos pt - i \sin pt) \qquad (2.5)$$

Rearranging the terms of Eq. (2.5), yields

$$x = C_1 \cos pt + C_2 \sin pt \qquad (2.6)$$

where $C_1$and C_2 are constants.

The factors cos pt and sin pt are periodic functions. As such the motion is periodic and it repeats itself after a certain interval of time.

Since $pT = 2\pi$, then

$$T = \frac{2\pi}{p} = 2\pi\sqrt{\frac{m}{k}} \qquad (2.7)$$

This time interval T is called the period of undamped free vibration. The number of times that the motion repeats itself in one second is called natural frequency of vibration f.

$$f = \frac{1}{T} = \frac{1}{2\pi}\sqrt{\frac{k}{m}} \qquad (2.8)$$

The parameter p can now be given a physical significance, since

$$p = \frac{2\pi}{T} = 2\pi f \qquad (2.9)$$

p is called the circular or angular frequency of vibration.

The vibratory motion represented by Eq. (2.6) is a harmonic motion. The constants C_1 and C_2 can be determined from the initial conditions of the motion. If at $t = 0$, $x = x_0$ and $\dot{x} = \dot{x}_0$, then applying the initial conditions

$$C_1 = x_0 \text{ and } C_2 = \frac{\dot{x}_0}{p} \tag{2.10}$$

Therefore, the equation of motion becomes

$$x = x_0 \cos pt + \frac{\dot{x}_0}{p} \sin pt \tag{2.11}$$

Assuming, $x_0 = A\cos\in$ and $\frac{\dot{x}_0}{p} = A\sin\in$, Eq. (2.11) becomes

$$x = A\cos\in \cos pt + A\sin\in \sin pt$$

$$x = A\cos(pt - \in) \tag{2.12}$$

By squaring and adding the assumed values of x_0 and $\dot{x}_0$ it can be shown that

$$A = \sqrt{x_0^2 + \frac{\dot{x}_0^2}{p^2}} \tag{2.13}$$

Thus, the resulting motion is simple harmonic, having an amplitude A and phase difference ε. It has been shown graphically in Fig. 2.5.

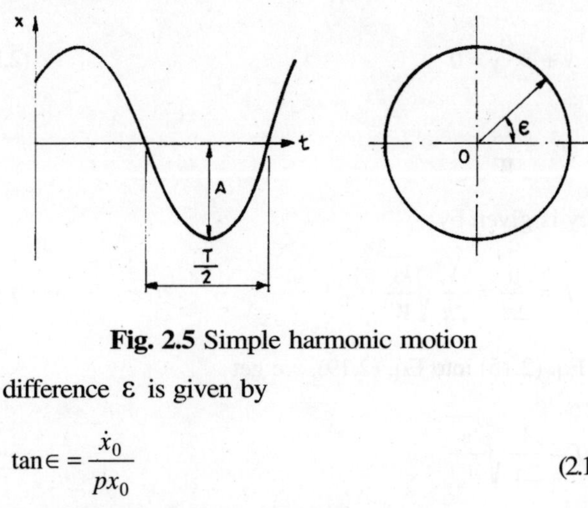

Fig. 2.5 Simple harmonic motion

Similarly, the phase difference ε is given by

$$\tan\in = \frac{\dot{x}_0}{px_0} \tag{2.14}$$

Example 2.1

A weight $W = 15$ N is vertically suspended by a spring of stiffness $k = 2$ N/mm. Determine the natural frequency of free vibration of the weight.

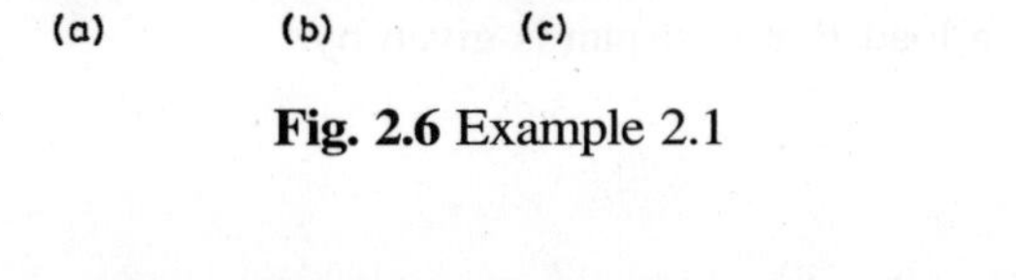

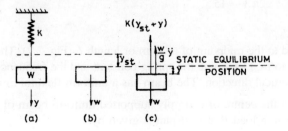

Fig. 2.6 Example 2.1

A vertically suspended weight W attached to a spring of stiffness k is shown in Fig. 2.6. Due to the application of the weight W, the spring will have vertical deflection, which is given by

$$y_{st} = \frac{W}{k} \tag{2.15}$$

The spring will vibrate about its mean position, which is the static equilibrium position. The freebody diagram of the weight, when it is displaced by y during free vibration is shown in Fig. 2.6(c). Applying D'Alembert's principle, one gets

$$\frac{W}{g}\ddot{y} + k(y_{st} + y) - W = 0 \tag{2.16}$$

Substituting y_{st} from Eq. (2.15) to Eq. (2.16), we get

$$\frac{W}{g}\ddot{y} + W + ky - W = 0$$

or
$$\frac{W}{g}\ddot{y} + ky = 0 \tag{2.17}$$

or
$$\ddot{y} + p^2 y = 0 \tag{2.18}$$

where
$$p^2 = \frac{kg}{W}$$

The natural frequency is given by

$$f = \frac{p}{2\pi} = \frac{1}{2\pi}\sqrt{\frac{kg}{W}} \tag{2.19}$$

Substituting y_{st} from Eq. (2.15) into Eq. (2.19), we get

$$f = \frac{1}{2\pi}\sqrt{\frac{g}{y_{st}}} \tag{2.20}$$

Putting the numerical values of W and k, we get

$$f = \frac{1}{2\pi}\sqrt{\frac{2 \times 9810}{15}} = 5.756 \text{ Hz (cycles/s)}$$

Example 2.2

A mass m is attached to the midpoint of a beam of length L (Fig. 2.7). The mass of the beam is small in comparison to m. Determine the spring constant and the frequency of the free vibration of the beam in the vertical direction. The beam has a uniform flexural rigidity EI.

The deflection at the centre of a simply supported uniform beam of length L and flexural rigidity EI subjected to a load P at midspan is given by

$$\delta = \frac{PL^3}{48\,EI}$$

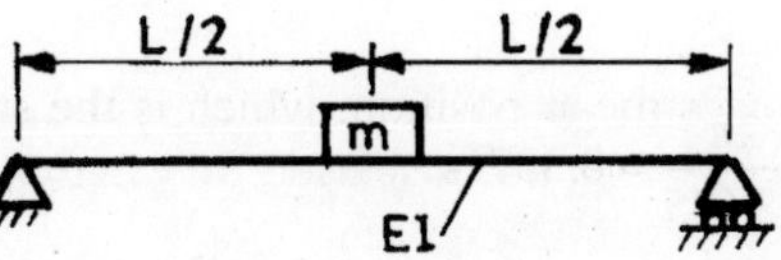

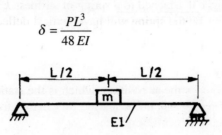

Fig. 2.7 Example 2.2

Therefore, from definition, the stiffness which is the force required to produce unit displacement, becomes

$$k = \frac{P}{\delta} = \frac{48\,EI}{L^3}$$

The natural frequency of the massless beam is given by

$$f = \frac{1}{2\pi}\sqrt{\frac{k}{m}} = \frac{1}{2\pi}\sqrt{\frac{48EI}{mL^3}} = 1.10\sqrt{\frac{EI}{mL^3}}$$

Example 2.3

Calculate the natural angular frequency in sidesway for the frame of Fig. 2.8 and also the natural period of vibration. If the initial displacement is 25 mm and the initial velocity is 25 mm/s, what is the amplitude and displacement at $t = 1\text{s}$?

If the horizontal deflection is at the top of a member fixed at both ends, then the moments and forces developed at the supports is shown in Fig. 2.8(b). For such a member, the stiffness is $12EI/L^3$ for a unit displacement at the top support. For the frame of Fig. 2.8(a), there are two vertical columns connected by a rigid beam. The freebody diagram of the beam with restoring forces is shown in Fig. 2.8(c).

Total restoring force in the columns = $(K_{AB} + K_{CD})\,x$. The stiffnesses of two columns *AB* and *CD* in this case can be conceived as two springs connected in parallel.

The equivalent stiffness of the columns therefore, is

$$k_{eq} = \frac{12(EI)_{AB}}{L_{AB}^3} + \frac{12(EI)_{CD}}{L_{CD}^3}$$

where L_{AB} and L_{CD} are the lengths of the members *AB* and *CD*.

Putting the numerical values

$$k_{eq} = \frac{12\times 30\times 10^{12}}{(1000)^3} + \frac{12\times 30\times 10^{12}}{(800)^3} = 1{,}063{,}125 \text{ N/mm}$$

Therefore, the natural angular frequency is given by

$$p = \sqrt{\frac{k_{eq}}{m}} = \sqrt{\frac{1063125 \times 9810}{30 \times 10^6}} = 18.645 \text{ rad/s}$$

The time period is given by

$$T = \frac{2\pi}{p} = \frac{2\pi}{18.645} = 0.337 \text{ s}$$

The system can be idealised into a spring-mass system and the equation of motion is given by Eq. (2.11).

$$x = x_0 \cos pt + \frac{\dot{x}_0}{p} \sin pt$$

$$= 25 \cos 18.645\, t + \frac{25}{18.645} \sin 18.645t$$

$$= A \cos(18.645t - \in)$$

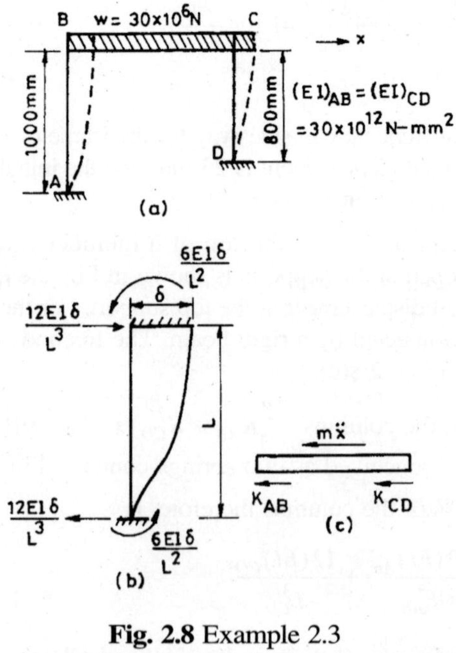

Fig. 2.8 Example 2.3

where $$A = \sqrt{(25)^2 + \left(\frac{25}{18.645}\right)^2} = 25.04 \text{ mm}$$

The amplitude, of the motion is 25.04 mm

$$\in = \tan^{-1} \frac{\dot{x}_0}{p x_0} \qquad \text{[Eq. (2.14)]}$$

$$= \tan^{-1}\frac{25}{18.645 \times 25} = 0.054 \text{ rad}$$

At t = 1s, the displacement x is given by

$$x = 25.04 \cos(18.645 - 0.054) = 24.21 \text{ mm}$$

Example 2.4

A multibay bent is shown in Fig. 2.9, where the girder is considered to be infinitely stiff and the uniform columns are assumed to have negligible mass in comparison to total mass M of the girder. Determine the period of free horizontal vibration.

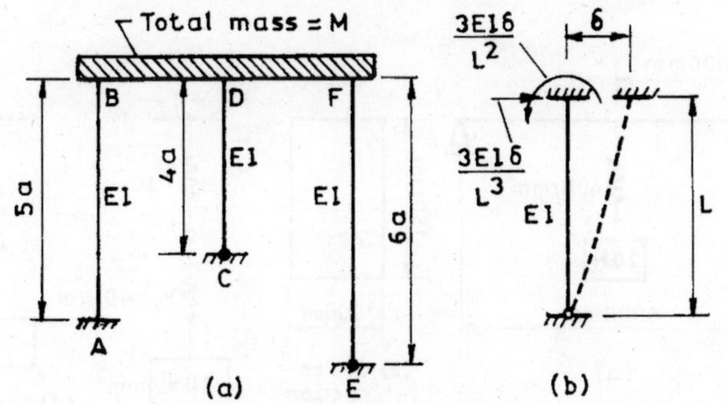

Fig. 2.9 Example 2.4

Like the previous example, the equivalent stiffness of the columns is to be determined first. The columns CD and EF have one end hinged and the other end fixed. For such cases, the stiffness for a unit displacement at the top is $3EI / L^3$, where L is the length of the member. Therefore, the equivalent stiffness of the columns is

$$k_{eq} = \frac{12EI}{(5a)^3} + \frac{3EI}{(4a)^3} + \frac{3EI}{(6a)^3} = 0.157\,\frac{EI}{a^3}$$

The period of free horizontal vibration then is

$$T = 2\pi\sqrt{\frac{M}{k_{eq}}} = 2\pi\sqrt{\frac{Ma^3}{0.157EI}} = 15.86\sqrt{\frac{Ma^3}{EI}}$$

Example 2.5

Find the natural frequency of the system shown in Fig. 2.10. The mass of the beam is negligible in comparison to the suspended mass.

$$E = 2.1 * 10^5 \text{ N/mm}^2$$

In addition to the flexible beam which provides a spring action, there is a spring attached to the mass. In this the deformation of the central mass is equal to the sum of the deformations of the beam at the centre and the attached spring.

The beam can be replaced by a spring k_1 of stiffness $48EI/L^3$, where EI is the flexural rigidity of the beam and L, its length. Two springs in this case is said to be connected in series (Fig. 2.10(c)), which can be replaced by an equivalent spring having stiffness k_{eq}. Due to a unit force, the deformation in both the cases will be identical. For the equivalent spring. the deformation due to unit force is $1/k_{eq}$ and for the two springs, they are $1/k_1$ and $1/k_2$.

Therefore $$\frac{1}{k_{eq}} = \frac{1}{k_1} + \frac{1}{k_2}$$

or $$k_{eq} = \frac{k_1 k_2}{k_1 + k_2}$$

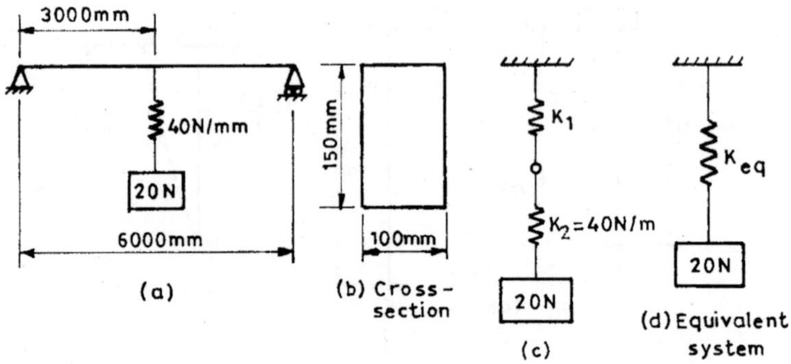

Fig. 2.10 Example 2.5

Substituting numerical values for the problem

$$I = \frac{1}{12} \times (100)\ (150)^3 = 28{,}125{,}000\ \text{mm}^4$$

$$k_1 = \frac{48 \times 2.1 \times 10^5 \times 28125000}{(6000)^3} = 1312.5\ \text{N/mm}$$

$$k_{eq} = \frac{1312.5 \times 40}{1312.5 + 40} = 38.82\ \text{N/mm}$$

The natural frequency of the system is

$$f = \frac{1}{2\pi}\sqrt{\frac{k_{eq}}{m}} = \frac{1}{2\pi}\sqrt{\frac{38.82 \times 9810}{20}} = 21.962\ \text{Hz}$$

Example 2.6

A mass hangs from a spring having stiffness k. A second mass drops through a height h and sticks to mass m_1 without rebound (Fig. 2.11). Determine the subsequent motion.

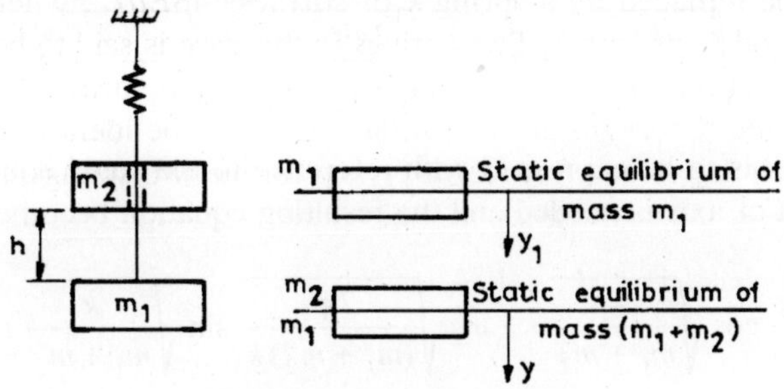

Fig. 2.11 Example 2.6

The velocity of mass m_2 just before impact is given by

$$v_2^2 = 2gh$$

Applying the principle of conservation of momentum, if $\dot{y}_0$ is the initial velocity of the combined mass $(m_1 + m_2)$, then

$$(m_1 + m_2)\,\dot{y}_0 = m_2\, v_2$$

or
$$\dot{y}_0 = \frac{m_2\sqrt{2gh}}{m_1 + m_2}$$

After impact, the combined mass $(m_1 + m_2)$ will vibrate about the static equilibrium position of $(m_1 + m_2)$. At the instant of impact, the mass m_1 is at its static equilibrium position. Therefore, the initial displacement is equal to the distance between static equilibrium position of mass $(m_1 + m_2)$ and the static equilibrium position of mass m_1 alone. Thus

$$y_0 = -\left[\frac{(m_1 + m_2)\,g}{k} - \frac{m_1 g}{k}\right] = -\frac{m_2 g}{k}$$

The minus sign has been put in the above expression, as the initial displacement is in the negative direction. Therefore, the equation of motion is given by (Eq. 2.11)

$$y = y_0 \cos pt + \frac{\dot{y}_0}{p}\sin pt$$

Now
$$p = \sqrt{\frac{k}{m_1 + m_2}}$$

Substitution of the values of y_0, $\dot{y}_0$ and p in the above equation, we get

$$y = -\frac{m_2 g}{k}\cos\sqrt{\frac{k}{m_1 + m_2}}\,t + \frac{m_2\sqrt{2gh}}{m_1 + m_2}\sqrt{\frac{m_1 + m_2}{k}}\sin\sqrt{\frac{k}{m_1 + m_2}}\,t$$

$$= -\frac{m_2 g}{k}\cos\sqrt{\frac{k}{m_1+m_2}}\,t + m_2\sqrt{\frac{2gh}{(m_1+m_2)k}}\sin\sqrt{\frac{k}{m_1+m_2}}\,t$$

If the above equation is to be expressed with reference to the static equilibrium position of mass m_1, then a shift of axis is needed and the resulting equation becomes

$$y = \frac{m_2 g}{k} - \frac{m_2 g}{k}\cos\sqrt{\frac{k}{m_1+m_2}}\,t + m_2\sqrt{\frac{2gh}{(m_1+m_2)k}}\sin\sqrt{\frac{k}{m_1+m_2}}\,t$$

2.4 FREE DAMPED VIBRATION OF SDF SYSTEM

From Eq. (2.1), it follows that the equation of motion for free vibration for a damped SDF system is

$$m\ddot{x} + c\dot{x} + kx = 0 \qquad (2.21)$$

Equation (2.21) is a linear differential equation of second order and can be solved by using standard procedure.

The general solution of the equation can be assumed in the following form

$$x = Ae^{\lambda t} \qquad (2.22)$$

Substituting x and its time derivatives from Eq.(2.22) into Eq.(2.21) and rearranging the terms, we get

$$A\lambda^2 e^{\lambda t} + \frac{c}{m}A\lambda e^{\lambda t} + \frac{k}{m}Ae^{\lambda t} = 0$$

or

$$\lambda^2 + 2n\lambda + p^2 = 0 \qquad (2.23)$$

where $n = \frac{c}{2m}$ and $p^2 = \frac{k}{m}$

Equation (2.23) is a quadratic equation and its solution is given by

$$\lambda_{1,2} = -n \pm \sqrt{n^2 - p^2} \qquad (2.24)$$

Relative values of n and p will govern the resulting solution. Three cases may arise and they are discussed below one after another.

Case I. Overdamped system ($n > p$)

Roots λ_1 and λ_2 of Eq.(2.24) are real and negative when n is greater than p. The solution of the equation is

$$x = A_1e^{\lambda_1 t} + A_2e^{\lambda_2 t} \qquad (2.25)$$

The solution is shown graphically in Fig. 2.12. The solution does not contain any periodicity; as such it does not represent a vibratory motion. The roots of Eq. (2.24) are unequal and two terms on the right hand side represent motions which decay exponentially at two different rates.

The motion associated with root λ_1 predominates only for a short time, which however goes on decreasing in an exponential manner with time. In this case, the viscous resistance of the body is so large that when it is displaced from the equilibrium position, it only creeps back to that position. The displacement finally vanishes as t approaches infinity. This system is known as overdamped system.

Case II. Critically damped system ($n = p$)

When $n = p$, roots λ_1 and λ_2 of Eq. (2.24) are real, negative and equal, that is, repeating.

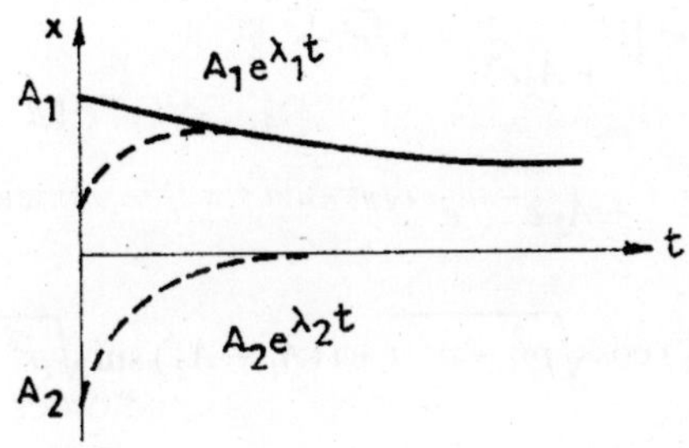

Fig. 2.12 Overdamped motion

The solution of Eq. (2.21) is

$$x = (A_1 + A_2 t)e^{\lambda t} \qquad (2.26)$$

where $\lambda_1 = \lambda_2 = \lambda$.

The displacement of the mass reaches zero asymtotically with time for this case, as shown in Fig. 2.13. The damping in the system is great enough not to set up oscillatory motion. The system is known as critically damped system.

At critical damping, damping coefficient c is given by

$$c = 2mp \qquad (2.27)$$

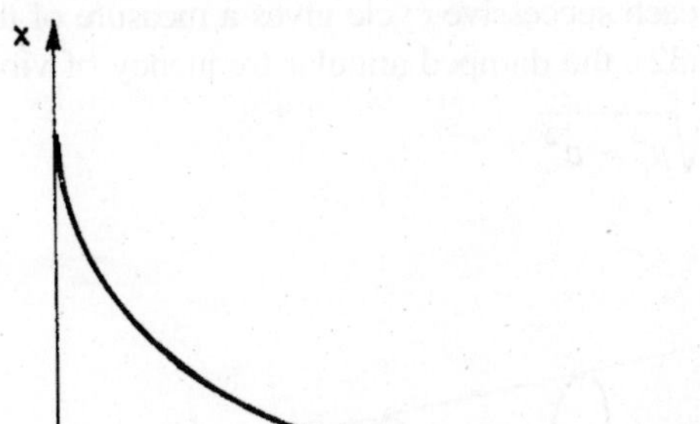

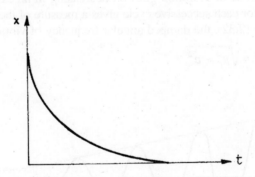

Fig. 2.13 Critically damped motion

The damping constant in other cases is expressed in terms of a certain percentage of critical damping. For a structure having 10 percent of critical damping, it will have a damping constant c given by

$$c = (0.1)\ 2mp = 0.2\ mp \tag{2.28}$$

Case III. Underdamped or Damped system ($n < p$)

Roots λ_1 and λ_2 are complex when n is less than p, and they are given by

$$\lambda_{1,2} = -n \pm i\sqrt{p^2 - n^2} \tag{2.29}$$

The solution of Eq. (2.21) will be

$$x = A_1 e^{\left(-n + i\sqrt{p^2 - n^2}\right)t} + A_2 e^{\left(-n - i\sqrt{p^2 - n^2}\right)t} \tag{2.30}$$

or $$x = A_1 e^{-nt} e^{i\sqrt{p^2 - n^2}\,t} + A_2 e^{-nt} e^{-i\sqrt{p^2 - n^2}\,t}$$

or $$x = e^{-nt}\left[(A_1 + A_2)\cos\sqrt{p^2 - n^2}\,t + i(A_1 - A_2)\sin\sqrt{p^2 - n^2}\,t\right]$$

or $$x = e^{-nt}\left(C_1 \cos\sqrt{p^2 - n^2}\,t + C_2 \sin\sqrt{p^2 - n^2}\,t\right) \tag{2.31}$$

where C_1 and C_2 are constants, which can be determined from initial conditions. This system is known as underdamped or simply damped system of vibration.

If at $t = 0$, $x = x_0$ and $\dot{x} = \dot{x}_0$, then putting these conditions in Eq. (2.31), we get

$$x = e^{-nt}\left(x_0 \cos\sqrt{p^2 - n^2}\,t + \frac{nx_0 + \dot{x}_0}{\sqrt{p^2 - n^2}}\sin\sqrt{p^2 - n^2}\,t\right) \tag{2.32}$$

The variation of the displacement of the damped system with time given by Eq. (2.32) is shown in Fig. 2.14. The amplitude of vibration goes on decreasing in an exponential manner – the reduction in magnitude for each successive cycle gives a measure of the damping in the system. As can be seen from Eq.(2.32), the damped angular frequency of vibration is given by

$$p_d = \sqrt{p^2 - n^2} \tag{2.33}$$

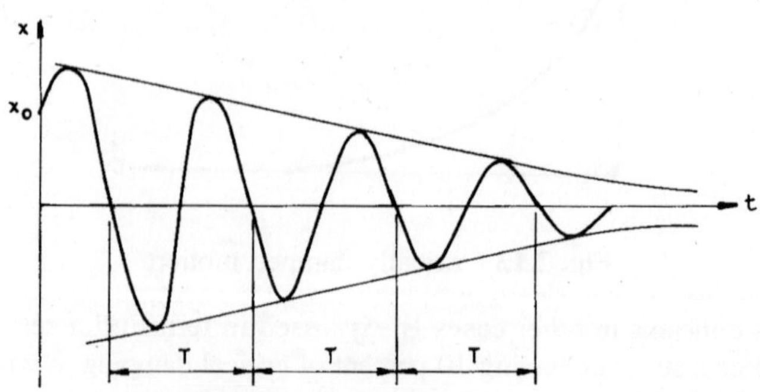

Fig. 2.14 Damped motion

Time period for damped free vibration is

$$T_d = \frac{2\pi}{\sqrt{p^2 - n^2}} = \frac{2\pi}{p} \frac{1}{\sqrt{1 - \left(\frac{n}{p}\right)^2}} \tag{2.34}$$

Equation (2.34) reveals that the damped time period is more than the undamped time period, but n is usually a small quantity in comparison to p. This increase is a small quantity of second order. Therefore, for all practical purposes, it can be assumed with sufficient accuracy that a small viscous damping does not affect the period of vibration.

Timewise variation of the displacement for different systems have been shown in Fig. 2.15, so that a relative idea of the displacements in different cases can be obtained.

Example 2.7

A mass 1 kg is attached to the end of a spring with a stiffness 0.7 KN/mm. Determine the critical damping constant.

The stiffness of the spring is $k = 0.7$ kN/mm $= 700$ N/mm. The natural angular frequency is

$$p = \sqrt{\frac{k}{m}} = \sqrt{\frac{700}{1}} = 26.46 \text{ rad/s}$$

At critical damping

$$n = p$$

or

$$\frac{c}{2m} = 26.46$$

or

$$c = 26.46 \times 2 \times 1 = 52.92 \text{ Ns/m}$$

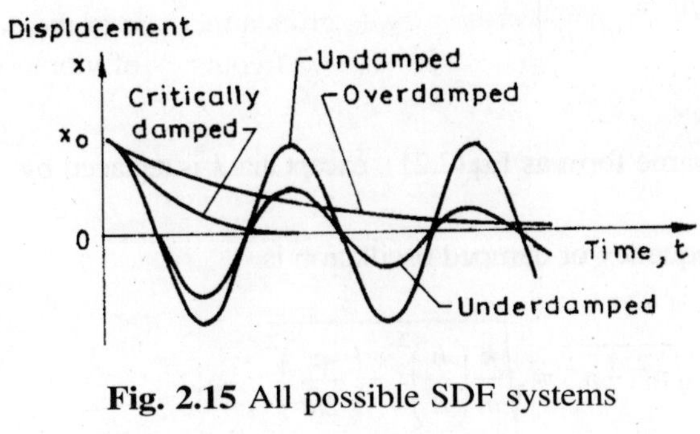

Fig. 2.15 All possible SDF systems

Example 2.8

A mass, spring and dashpot attached to a rigid bar is shown in Fig. 2.16. Write the equation of motion. Determine the natural frequency of damped free oscillation and critical damping coefficient.

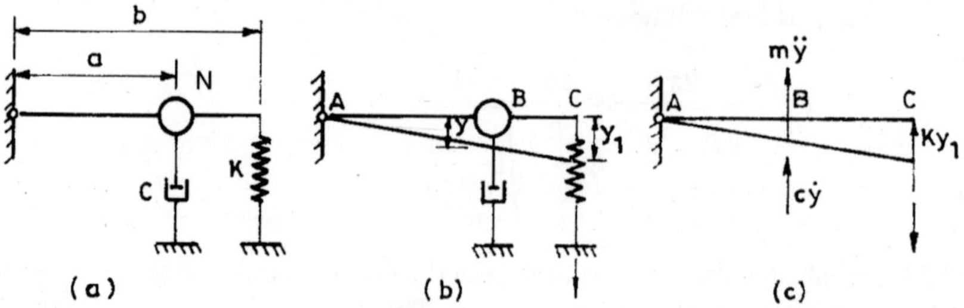

Fig. 2.16 Example 2.8

The bar *AC* being rigid, the point *B* where the mass and the dashpot are attached and the end *C* where the spring is attached will undergo different displacements and they are related. From similar triangles, it can be shown that

$$y_1 = \frac{b}{a}\,y \tag{a}$$

During oscillation, the forces acting on the system are the inertia force $m\ddot{y}$, the damping force $c\dot{y}$ and the spring force ky_1. Taking moment about A [Fig. 2.16(c)], yields

$$m\ddot{y}a + c\dot{y}a + ky_1b = 0 \tag{b}$$

Substituting y_1 from Eq. (a) into Eq. (b), we get

$$m\ddot{y} + c\dot{y} + k\left(\frac{b}{a}\right)^2 y = 0 \tag{c}$$

Equation (c) can be written as follows

$$\ddot{y} + 2n\dot{y} + p^2 y = 0 \tag{d}$$

where $n = \dfrac{c}{2m}$ and $p^2 = \dfrac{k}{m}\left(\dfrac{b}{a}\right)^2$

Equation (c) is of the same form as Eq.(2.21), except that k is replaced by $k\left(\dfrac{b}{a}\right)^2$

The natural angular frequency of damped oscillation is

$$p_d = \sqrt{p^2 - n^2} = \sqrt{\frac{k}{m}\left(\frac{b}{a}\right)^2 - \left(\frac{c}{2m}\right)^2}$$

At critical damping

$$n = p$$

or

$$\frac{c}{2m} = \frac{b}{a}\sqrt{\frac{k}{m}}$$

or $$c = \frac{2b}{a}\sqrt{km}$$

Example 2.9

A piston of mass 5 kg is travelling inside a cylinder with a velocity of 15 m/s and engages a spring and damper as shown in Fig. 2.17. Determine the maximum displacement of the piston after engaging the spring-damper.

How many seconds does it take?

The mass has an initial velocity. Referring to Eq. (2.32)

$$n = \frac{c}{2m} = \frac{175}{2 \times 5} = 17.5$$

$$p = \sqrt{\frac{k}{m}} = \sqrt{\frac{3000}{5}} = 24.295 \text{ rad/s}$$

The displacement of the spring is given by

$$x = e^{-nt}\left(C_1 \cos\sqrt{p^2 - n^2}\, t + C_2 \sin\sqrt{p^2 - n^2}\, t\right)$$

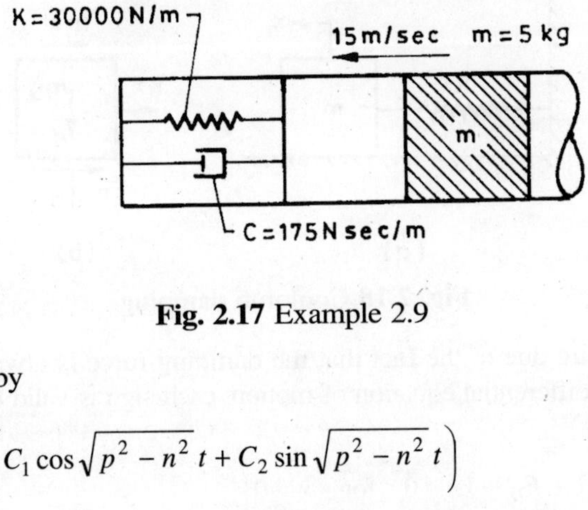

Fig. 2.17 Example 2.9

The velocity is given by

$$\dot{x} = -ne^{-nt}\left(C_1 \cos\sqrt{p^2 - n^2}\, t + C_2 \sin\sqrt{p^2 - n^2}\, t\right)$$

$$+ \exp(-nt)\sqrt{p^2 - n^2}\,(-C_1 \sin\sqrt{p^2 - n^2}\, t + C_2 \cos\sqrt{p^2 - n^2}\, t)$$

At $t = 0$, $x = 0$ and $\dot{x} = 15$

Putting these initial conditions in the above expressions for velocity and displacement, yields

$$C_1 = 0 \quad \text{and} \quad C_2 = 0.875$$

Therefore, the expression for displacement reduces to

$$x = 0.875e^{-nt} \sin\sqrt{p^2 - n^2}\, t$$

Maximum displacement will occur when $t = \frac{T}{4}$, that is

$$t = 0.045 \text{ s}$$

The maximum displacement is

$$x = 0.875\, e^{-17.5\times 0.045} \sin(17.85 \times 0.045) = 0.277 \text{ m}$$

2.5 FREE VIBRATION WITH COULOMB DAMPING

So far we have considered viscous damping. However, the damping resulting from the friction of two dry surfaces gives constant damping force and the resulting damping is referred to as Coulomb or dry friction damping. A practical example of it is the friction of two dry surfaces as occurs in riveted joints.

A SDF system with Coulomb friction is shown in Fig. 2.18. The damping force F_D is equal to the product of the normal force N and the coefficient of friction m and is independent of the velocity. For a single degree of freedom system, the equation of motion will be

$$m\ddot{x} + kx \pm F_D = 0 \tag{2.35}$$

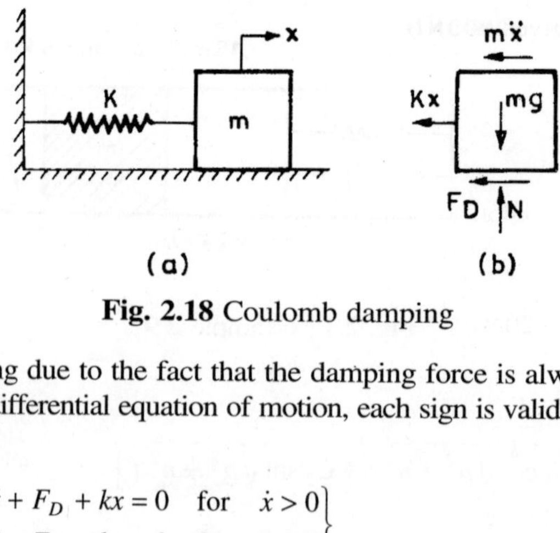

Fig. 2.18 Coulomb damping

The ± sign is arising due to the fact that the damping force is always opposite to that of the velocity and in the differential equation of motion, each sign is valid for half-cycle intervals. Or in other words

$$\left.\begin{aligned} m\ddot{x} + F_D + kx = 0 \quad \text{for} \quad \dot{x} > 0 \\ m\ddot{x} - F_D + kx = 0 \quad \text{for} \quad \dot{x} < 0 \end{aligned}\right\} \tag{2.36}$$

The solution of Eq. (2.35) is

$$x = A\cos pt + B\sin pt - F_D \operatorname{sgn}(\dot{x})/k \tag{2.37}$$

where $p^2 = \frac{k}{m}$ and A and B are constants.

If at $t = 0$, $x = x_0$ and $\dot{x} = 0$ then

$$A = x_0 - \frac{F_D}{k}\,\text{sgn}\,(\dot{x}) \text{ and } B = 0$$

On substitution of the values of A and B, Eq. (2.37) becomes

$$x = \left[x_0 - \frac{F_D}{k}\,\text{sgn}\,(\dot{x})\right]\cos pt + \frac{F_D}{k}\,\text{sgn}\,(\dot{x}) \tag{2.38}$$

For the first half-cycle ($pt = \pi$), the displacement is

$$x = -\,x_0 + 2\,\frac{F_D}{k}\,\text{sgn}\,(\dot{x}) \tag{2.39}$$

Hence, the decrease in amplitude is $2\,\frac{F_D}{k}$ per half-cycle. The amplitude change per cycle for free vibration with Coulomb damping then is $\frac{4F_D}{k}$. It may further be noted from Eq. (2.38), that the time period for the free vibration with Coulomb damping is same as the undamped case.

The resulting motion is plotted in Fig. 2.19. A characteristic feature of the response is that, the amplitude decays in a linear manner and not exponentially as in the case of viscous damping.

Example 2.10

For the SDF system of Fig. 2.18, determine the coefficient of friction if a tensile force $P(= mg)$ elongates the spring by 6 mm. The initial amplitude of x_0 = 600 mm reduces to 0.8 of its value after 20 cycles.

The spring stiffness in given units is

$$k = \frac{P}{6}$$

For a SDF system with Coulomb damping, the amplitude reduces in each cycle by $\frac{4F_D}{k}$.

Therefore $$20 \times 4\,\frac{F_D}{k} = 600 \times 0.2$$

or $$\frac{F_D}{k} = \frac{120}{80}$$

or $$F_D = \frac{120}{80}k = \frac{120}{80} \times \frac{P}{6} = \frac{P}{4} = \frac{1}{4}P \text{ (P will be the normal reaction).}$$

Therefore, the coefficient of friction is 1/4.

The variation of the displacement is shown in Fig. 2.19.

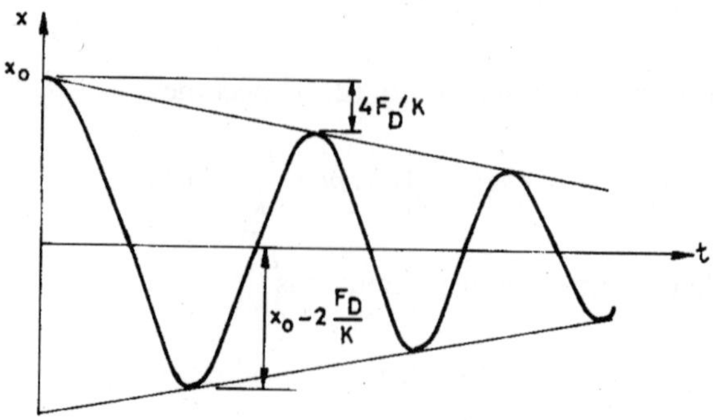

Fig. 2.19 The variation of displacement

2.6 ENERGY METHOD AND FREE TORSIONAL VIBRATION

Using the principle of conservation of energy, natural frequencies of vibrating systems can be conveniently determined. For a mechanical vibrating system, the energy is partly kinetic and partly potential. The kinetic energy T_E is associated with the velocity of the mass and the potential energy U is due to the strain energy stored in the spring. Thus

$$T_E + U = \text{Total mechanical energy} = \text{constant} \tag{2.40}$$

Therefore, the rate of change of energy is zero. Mathematically

$$\frac{d}{dt}(T_E + U) = 0 \tag{2.41}$$

The equation of motion of the spring-mass system of Fig. 1.4 can be derived from energy consideration. The kinetic energy of the mass is given by

$$T_E = \frac{1}{2} m\dot{x}^2 \tag{2.42}$$

The potential energy of the system is due to strain energy stored in the spring (Fig. 2.20) and is given by

$$U = \int_0^x kx\, dx = \frac{1}{2} kx^2 \tag{2.43}$$

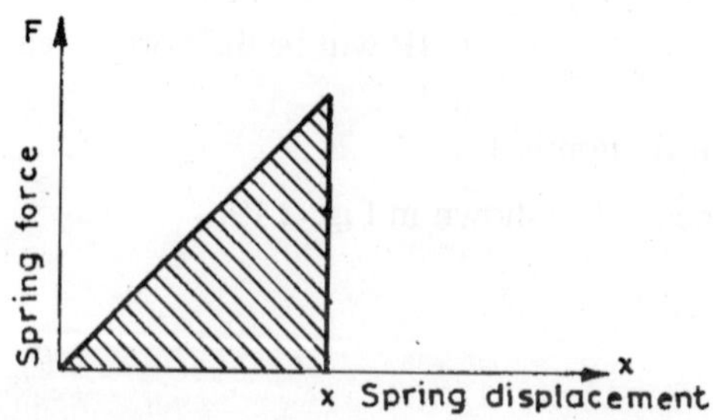

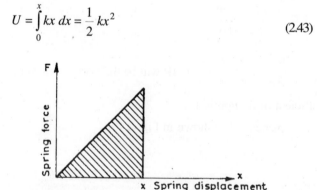

Fig. 2.20 Force – displacement relation

Substituting Eqs. (2.42) and (2.43) into Eq. (2.41), we get

$$\frac{d}{dt}\left(\frac{1}{2}m\dot{x}^2 + \frac{1}{2}kx^2\right) = 0 \quad (2.44)$$

or

$$\frac{1}{2}m2\dot{x}\ddot{x} + \frac{1}{2}k2x\dot{x} = 0$$

or

$$m\ddot{x} + kx = 0 \quad (2.45)$$

Equation (2.45) is same as Eq. (2.2).

2.6.1 Torsional Vibration of the SDF System

Applying the above energy principle, the free vibration equation for the torsion of the shaft can be derived.

A vertical shaft with a disc attached at its lower end is shown in Fig. 2.21. The mass moment of inertia of the shaft is neglected in the analysis. Let ρ be the mass density of the material, I_p be the mass moment of inertia of the disc, D be the diameter of the disc, d be the diameter of the shaft, L be the length of the shaft, and t be the thickness of the disc, then

$$\rho = \frac{4W}{\pi D^2 tg} \quad (2.46)$$

$$I_P = \rho J_D \cdot t = \frac{4W}{\pi D^2 tg} \cdot \frac{\pi D^4}{32} t = \frac{WD^2}{8g} \quad (2.47)$$

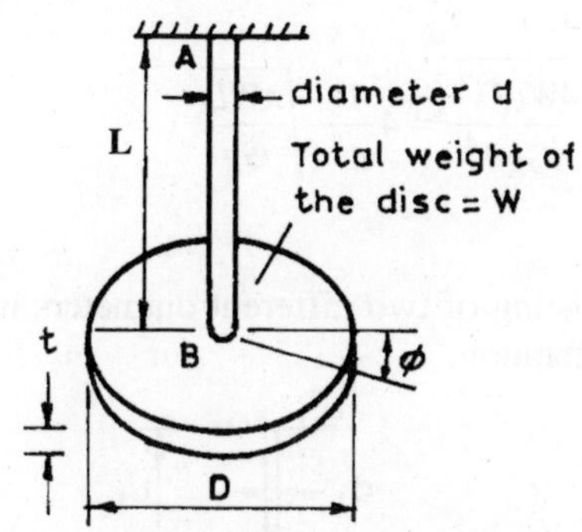

Fig. 2.21 Torsional vibration

where J_D is the polar moment of inertia of the disc. In free torsional vibration of the shaft, the spring stiffness is provided by the shaft and it is given by

$$k = \frac{\pi d^4 G}{32L} \quad (2.48)$$

where G is shear modulus of elasticity of the material of the shaft.

If ϕ be the angular displacement at any instant of time, the kinetic energy of the system is

$$T_E = \frac{1}{2} I_P \dot{\phi}^2 \tag{2.49}$$

The corresponding potential energy, is nothing but the strain energy in the spring and is given by

$$U = \int_0^{\phi} k\phi \, d\phi = \frac{1}{2} k\phi^2 \tag{2.50}$$

Substituting the values of T_E and U from Eqs. (2.49) and (2.50) respectively into Eq. (2.41), we get

$$\frac{d}{dt}\left(\frac{1}{2} I_P \dot{\phi}^2 + \frac{1}{2} k\phi^2\right) = 0$$

or

$$I_P \ddot{\phi} + k\phi = 0 \tag{2.51}$$

Equation (2.51) is the free torsional vibration equation of the shaft and is similar to Eq. (2.2), and its solution is given by

$$\phi = \phi_0 \cos pt + \frac{\dot{\phi}_0}{p} \sin pt \tag{2.52}$$

where ϕ_0 and $\dot{\phi}_0$ are the initial angular displacement and the initial angular velocity, respectively.

Angular frequency, $$p = \sqrt{\frac{k}{I_P}} = \sqrt{\frac{\pi d^4 G}{32L} \times \frac{8g}{WD^2}} = \sqrt{\frac{\pi G g d^4}{4WD^2 L}} \tag{2.53}$$

The time period is

$$T = 2\pi \sqrt{\frac{4WLD^2}{\pi G g d^4}} = 4\frac{D}{d^2}\sqrt{\frac{\pi WL}{Gg}} \tag{2.54}$$

Example 2.11

A system having a shaft consisting of two different diameters is shown in Fig. 2.22. Find the time period of free torsional vibration.

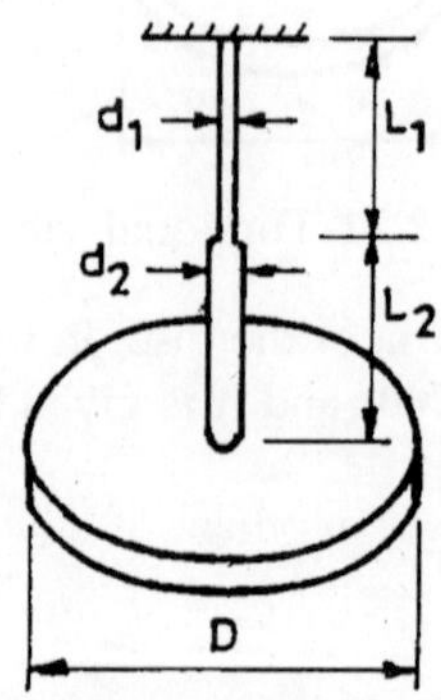

Fig. 2.22 Example 2.11

The problem is similar to that of the previous problem, except that the definition of stiffness is different in this case. If a torque T_0 is statically applied at the disc level to produce an angular displacement, then the following relationship is valid

$$\phi = \frac{32T_0L_1}{\pi d_1^4G} + \frac{32T_0L_2}{\pi d_2^4G}$$

$$= \frac{32T_0}{\pi d_1^4G}\left(L_1 + \frac{L_2 d_1^4}{d_2^4}\right)$$

$$= \frac{32T_0}{\pi d_1^4G} L$$

where $L = L_1 + L_2 \dfrac{d_1^4}{d_2^4}$ = equivalent length

$$k = \frac{T_0}{\phi} = \frac{\pi d_1^4 G}{32L}$$

$$I_P = \frac{WD^2}{8g}$$

Therefore, by definition

$$T = 2\pi\sqrt{\frac{I_P}{k}}$$

$$= 2\pi\sqrt{\frac{WD^2}{8g}\cdot\frac{32}{\pi d_1^4G}\left(L_1 + L_2\frac{d_1^4}{d_2^4}\right)}$$

Example 2.12

A disc of unknown mass moment of inertia I_P suspended at the end of a slender wire executes 40 cycles in one minute. To twist the wire 10 deg, a torque of 100 N-mm is required. Determine I_P.

The time period of the system is given by

$$T = \frac{60}{40} = 1.5 \text{ s}$$

The time period T is related to the mass moment of inertia I_P by

$$T = 2\pi\sqrt{\frac{I_P}{k}}$$

The stiffness of the system is

$$k = \frac{100}{10} \times \frac{57.3}{1000} = 0.573 \text{ Nm/rad}$$

Therefore

$$I_P = \frac{kT^2}{4\pi} = \frac{0.573 \times 1.5^2}{4\pi^2} = 0.0326 \text{ Nm s}^2$$

Example 2.13

A shaft 10 mm in diameter and 0.2 m long connects a generator to the main engine. If the mass moment of inertia of the generator rotor is 0.55 Nmm- s^2, determine the natural frequency in torsion [G = 8 × 10^6 N/cm^2].

The mass moment of inertia of the engine is very large compared to that of the generator rotor. As such the engine end is assumed to be fixed.

The stiffness of the system is given by

$$k = \frac{\pi d^4 G}{32L} = \frac{\pi (1)^4 \times 8 \times 10^6}{32 \times 20} = 3927 \text{ Ncm/rad} = 39.27 \text{ Nm/rad}$$

The natural frequency is given by

$$f = \frac{1}{2\pi}\sqrt{\frac{39.27}{0.55 \times 10^{-3}}} = 42.53 \text{ cycle/s}$$

Example 2.14

A shaft with two circular discs of uniform thickness at the ends is shown in Fig. 2.23. Determine the frequency of torsional vibration of the shaft.

The discs at the two ends are given two opposite torques, which are immediately withdrawn to start the torsional vibration. There is a particular section cd which remains immovable during vibration. The left hand and right hand portions of the shaft about cd will have the same time period. Therefore

$$\sqrt{\frac{I_{P_1}}{k_1}} = \sqrt{\frac{I_{P_2}}{k_2}} \quad \text{or} \quad \frac{k_1}{k_2} = \frac{I_{P_1}}{I_{P_2}}$$

where and are the spring stiffnesses for the left and for the right hand portions of the shaft. We know that stiffnesses are inversely proportional to their lengths

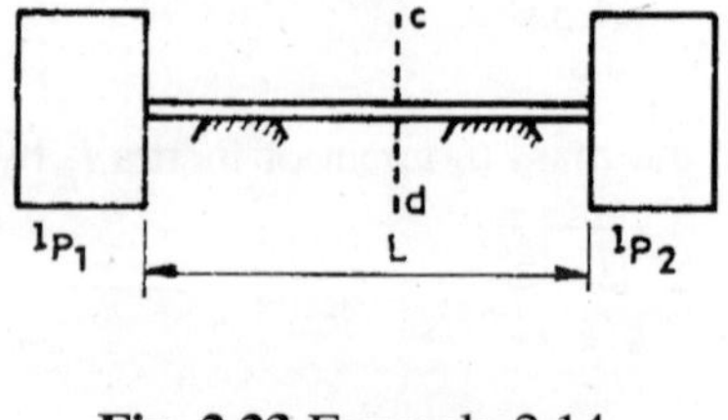

Fig. 2.23 Example 2.14

Therefore $$\frac{k_1}{k_2} = \frac{b}{a} = \frac{I_{P_1}}{I_{P_2}} \qquad \text{(a)}$$

Further $$a + b = L \qquad \text{(b)}$$

Combining Eqs. (a) and (b), gives

$$a = \frac{I_{P_2} L}{I_{P_1} + I_{P_2}} \quad \text{and} \quad b = \frac{I_{P_1} L}{I_{P_1} + I_{P_2}}$$

Therefore, frequency of torsional vibration of the shaft is

$$f = \frac{1}{2\pi}\sqrt{\frac{k_1}{I_{P_1}}} = \frac{1}{2\pi}\sqrt{\frac{\pi d^4 G (I_{P_1} + I_{P_2})}{32 L I_{P_1} I_{P_2}}}$$

It may be noted that

$$k_1 = \frac{\pi d^4 G}{32a} = \frac{\pi d^4 G}{32 L I_{P_2}} (I_{P_1} + I_{P_2})$$

2.6.2 Rayleigh's Method

A real system is rarely of single degree of freedom. It is treated either as multiple degrees of freedom system or as a system having distributed mass. The calculation of natural frequencies can be simplified by using Rayleigh's method. The vibration amplitudes are assumed a priori. On the basis of this assumed distribution, it is possible to calculate natural frequencies of the system with the help of energy principles. However, the results thus obtained are not exact.

Another interpretation of energy principle also exists. The natural frequency is a function of the rate of change of kinetic and potential energies of a system. As such, when the mass passes through the mean position, its potential energy is zero. The kinetic energy at that instant is maximum and is equal to the total mechanical energy. When the mass is at its maximum displacement, its kinetic energy is zero and the total mechanical energy consists only of the potential energy. As the total energy of the system is constant [Eq. (2.41)], we can write

$$T_{E\,max} = U_{max} \qquad (2.55)$$

Example 2.15

Determine the natural frequency of an undamped spring-mass system by incorporating the effect of mass of the spring [Fig. 2.24].

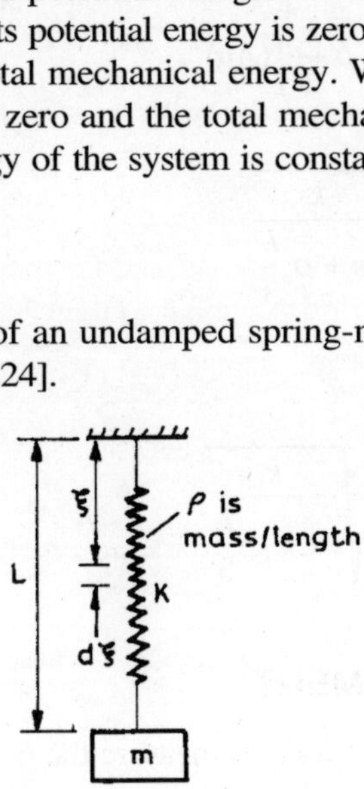

Fig. 2.24 Example 2.15

The dynamics of a helical spring is somewhat complicated. It has been simplified as an elastic rod, assuming that it is uniformly stretched. If the velocity of the lumped mass is $\dot{x}$, then the velocity of the spring element at a distance ξ is

$$\dot{x}\frac{\xi}{L}$$

If ρ is the mass per unit length of the spring, then

$$T_{E_{max}} = \frac{1}{2}m\dot{x}_{max}^2 + \int_0^L \frac{1}{2}\rho\left(\frac{\xi}{L}\dot{x}_{max}\right)^2 d\xi$$

$$= \frac{1}{2}\left(m + \rho\frac{L}{3}\right)\dot{x}_{max}^2$$

Let $\qquad x = A \sin pt$

then $\qquad \dot{x}_{max} = Ap$

Therefore $\quad T_{E_{max}} = \frac{1}{2}\left(m + \rho\frac{L}{3}\right)(Ap)^2$

The maximum potential energy of the system is

$$U_{max} = \frac{1}{2}kx_{max}^2 = \frac{1}{2}kA^2$$

Equating the maximum kinetic energy to the maximum potential energy, we get

$$\frac{1}{2}kA^2 = \frac{1}{2}\left(m + \rho\frac{L}{3}\right)(Ap)^2$$

or

$$p = \sqrt{\frac{k}{m + \rho\frac{L}{3}}}$$

The natural frequency is given by

$$f = \frac{1}{2\pi}\sqrt{\frac{k}{m + \rho\frac{L}{3}}}$$

2.7 LOGARITHMIC DECREMENT

Logarithmic decrement is a concept used to measure the damping factor of a SDF system. It is based on the measure of the decay of two successive amplitudes. Logarithmic decrement is defined as the natural logarithm of two successive amplitudes.

At any instant $t = t_1$, the amplitude from Eq.(2.31) is given by (Fig. 2.25)

$$x_1 = e^{-nt_1}(C_1 \cos\sqrt{p^2 - n^2}\, t_1 + C_2 \sin\sqrt{p^2 - n^2}\, t_1) \qquad (2.56)$$

Next positive amplitude will occur at $t = t_1 + T = t_1 + \dfrac{2\pi}{\sqrt{p^2 - n^2}}$

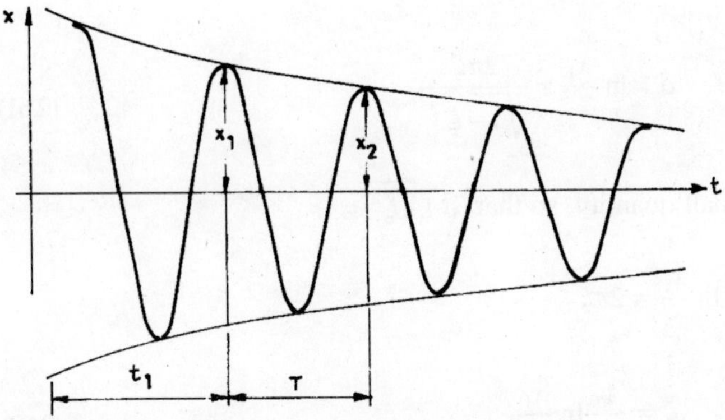

Fig. 2.25 Damped motion

The value of this amplitude at $t = t_1 + T$ is given by

$$x_2 = \exp\left(-n\left(t_1 + \frac{2\pi}{\sqrt{p^2 - n^2}}\right)\right)\left[C_1 \cos\sqrt{p^2 - n^2}\left(t_1 + \frac{2\pi}{\sqrt{p^2 - n^2}}\right)\right.$$

$$\left. + C_2 \sin\sqrt{p^2 - n^2}\left(t_1 + \frac{2\pi}{\sqrt{p^2 - n^2}}\right)\right] \qquad (2.57)$$

or

$$x_2 = e^{-nt_1} \exp\left(-\frac{2\pi n}{\sqrt{p^2 - n^2}}\right)$$

$$\left[C_1 \cos\sqrt{p^2 - n^2}\, t_1 + C_2 \sin\sqrt{p^2 - n^2}\, t_1\right] \qquad (2.58)$$

Dividing Eq. (2.58) by Eq. (2.56), we get

$$\frac{x_1}{x_2} = \exp\left(\frac{2\pi n}{\sqrt{p^2 - n^2}}\right) \qquad (2.59)$$

Taking natural logarithm on both sides, give

$$\ln \frac{x_1}{x_2} = \frac{2\pi n}{\sqrt{p^2 - n^2}} = \frac{2\pi\left(\frac{n}{p}\right)}{\sqrt{1-\left(\frac{n}{p}\right)^2}} \quad (2.60)$$

If $\zeta = \frac{n}{p}$ critical damping ratio, then the logarithmic decrement is given by

$$\delta = \ln \frac{x_1}{x_2} = \frac{2\pi\zeta}{\sqrt{1-\zeta^2}} \quad (2.61)$$

Usually ζ is a small quantity, so that $\sqrt{1-\zeta^2} \cong 1$

Then $$\ln \frac{x_1}{x_2} = 2\pi\zeta$$

or $$\zeta = \frac{1}{2\pi} \ln \frac{x_1}{x_2} \quad (2.62)$$

If a system has 5 per cent of critical damping (ζ = 0.05), the logarithmic decrement will be 0.1 π which indicates that successive peaks would be $e^{0.1\pi}$ or 1.37. Inverting this quantity, it can be said that each and every peak would have a value of 0.73 times that of the preceding. This facilitates to a great extent the visualisation of the effect of damping.

Example 2.16

For a system having mass 10 kg and spring constant 12 kN/m, the amplitude decreases to 0.2 of the initial value after six consecutive cycles. Find the damping constant of the damper.

The ratio of two successive amplitudes will remain the same.

$$\frac{x_0}{x_1} = \frac{x_1}{x_2} = \frac{x_2}{x_3} = \frac{x_3}{x_4} = \frac{x_4}{x_5} = \frac{x_5}{x_6} = e^{\delta}$$

Therefore, $$\frac{x_0}{x_1} \cdot \frac{x_1}{x_2} \cdot \frac{x_2}{x_3} \cdot \frac{x_3}{x_4} \cdot \frac{x_4}{x_5} \cdot \frac{x_5}{x_6} = \frac{x_0}{x_6} = e^{6\delta} = \frac{1}{0.2} = 5$$

$$\delta = \frac{1}{6} \ln 5 = 0.268$$

therefore $$\frac{2\pi\zeta}{\sqrt{1-\zeta^2}} = 0.268$$

or $$\zeta = 0.0427$$

The damping constant is

$$c = 2\zeta\sqrt{km} = 2\times(0.0427)\sqrt{12000\times 10} = 29.58 \text{ Ns/m}$$

Based on the above example, a general statement can be made. If the amplitude is reduced by a factor N after n cycles, then

$$\frac{x_0}{x_n} = N = e^{n\delta}$$

or

$$\delta = \frac{1}{n}\ln N \tag{2.63}$$

Example 2.17

Show that the critical damping ratio ζ can be worked out from the free vibration records, as shown in Fig. 2.26, by the expression given below

$$\zeta = \frac{1}{\pi}\ln\frac{A_1}{A_2}$$

It is assumed that the amplitude a is occurring at time instant t_1. Then, referring to Fig. 2.26

$$\frac{A_1}{A_2} = \frac{a+b}{b+c} = \frac{x_{t_1} - x_{(t_1+T/2)}}{-x_{(t_1+T/2)} + x_{(t_1+T)}} \tag{a}$$

$$x_{t_1} = e^{-nt_1}[C_1\cos p_1t_1 + C_2\sin p_1t_1)$$

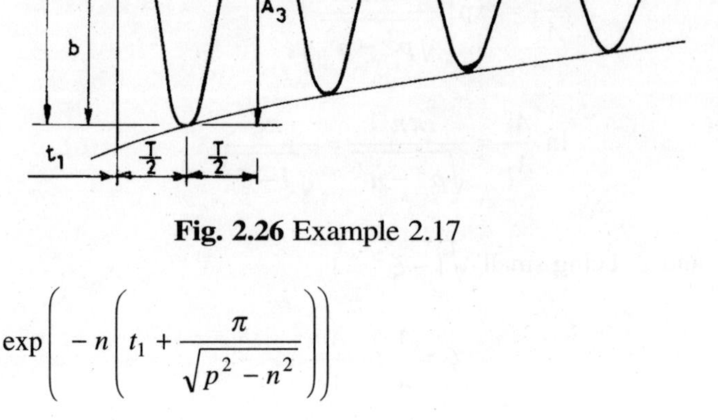

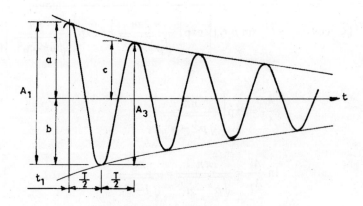

Fig. 2.26 Example 2.17

$$x_{(t_1+T/2)} = \exp\left(-n\left(t_1 + \frac{\pi}{\sqrt{p^2-n^2}}\right)\right)$$

$$\left[C_1 \cos p_1\left(t_1+\frac{\pi}{p_1}\right)+C_2 \sin p_1\left(t_1+\frac{\pi}{p_1}\right)\right]$$

$$=e^{-nt_1} \exp\left(-\frac{n\pi}{\sqrt{p^2-n^2}}\right)[-C_1 \cos p_1t_1 - C_2 \sin p_1t_1]$$

$$x(t_1+T)=\exp\left[-n\left(t_1+\frac{2\pi}{\sqrt{p^2-n^2}}\right)\right]$$

$$\times\left[C_1 \cos p_1\left(t_1+\frac{2\pi}{p_1}\right)+C_2 \sin p_1\left(t_1+\frac{2\pi}{p_1}\right)\right]$$

$$=e^{-nt_1} \exp\left(-\frac{2\pi n}{\sqrt{p^2-n^2}}\right)[C_1 \cos p_1t_1 + C_2 \sin p_1t_1]$$

In all the above equations of displacement, $p_1=\sqrt{p^2-n^2}$

Substituting the values of x_t, $x_{(t_1+T/2)}$ and $x_{(t_1+T)}$ from the above equations into Eq. (a) gives

$$\frac{A_1}{A_2}=\frac{e^{-nt_1}[C_1\cos p_1t_1 + C_2 \sin p_1t_1]\left[1+\exp\left(\frac{-n\pi}{\sqrt{p^2-n^2}}\right)\right]}{e^{-nt_1}[C_1 \cos p_1t_1 + C_2 \sin p_1t_1]\left[\exp\left(\frac{-2\pi n}{\sqrt{p^2-n^2}}\right)+\exp\left(\frac{-n\pi}{\sqrt{p^2-n^2}}\right)\right]}$$

or
$$\frac{A_1}{A_2}=\exp\left(\frac{n\pi}{\sqrt{p^2-n^2}}\right)$$

or
$$\ln\frac{A_1}{A_2}=\frac{\pi n}{\sqrt{p^2-n^2}}=\frac{\pi\zeta}{\sqrt{1-\zeta^2}}$$

where $\zeta=\dfrac{n}{p}$ and ζ being small $\sqrt{1-\zeta^2}=1$

Therefore
$$\zeta=\frac{1}{\pi}\ln\frac{A_1}{A_2}$$

References

2.1 R.A. Anderson, Fundamentals of Vibration, McMillan Co., New York, 1967.

2.2 J.M. Biggs, Introduction to Structural Dynamics, McGraw-Hill Book Inc., New York, 1964.

2.3 R.W. Clough and J. Penzien, Dynamics of Structures, 2nd Edition, McGraw-Hill Book Inc., New York, 1993.

2.4 Roy R. Craig, Jr., Structural Dynamics, John Wiley & Sons, 1981.

2.5 J.P. Den Hartog, Mechanical Vibrations, Fourth Edition, McGraw-Hill Book Inc., 1956.

2.6 D.G. Fertis, Dynamics and Vibrations of Structures, John Wiley & Sons, New York, 1973.

2.7 L.S. Jacobsen and R.S. Ayer, Engineering Vibrations, McGraw-Hill Book Inc, 1958.

2.8 L. Meirovitch, Elements of Vibration Analysis, McGraw-Hill Book Inc., New York, 1975.

2.9 W.T. Thomson, Theory of Vibration with Applications, Prentice Hall, New Jersey, 1972.

2.10 M. Paz, Structural Dynamics Theory and Computations, Second Edition, CBS Publication and Distribution, New Delhi, 1983.

EXERCISE 2

2.1 A massless beam with a concentrated mass m is shown in the figure. Determine the natural frequency of the system.

2.2 Determine the natural frequency and the time period of the uniform cantilever beam. The beam is considered as massless.

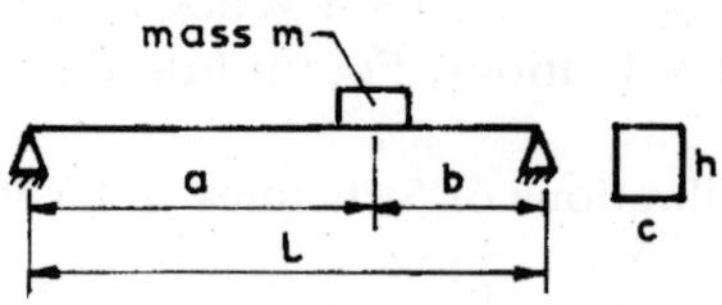

Prob. 2.1

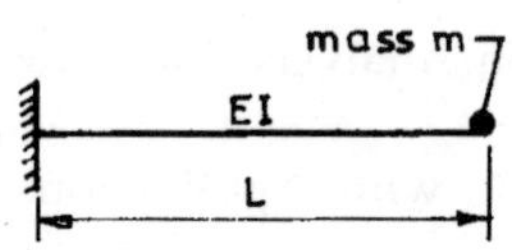

Prob. 2.2

2.3 Find the natural frequency of the system shown in the figure. The weight of the pulley is neglected.

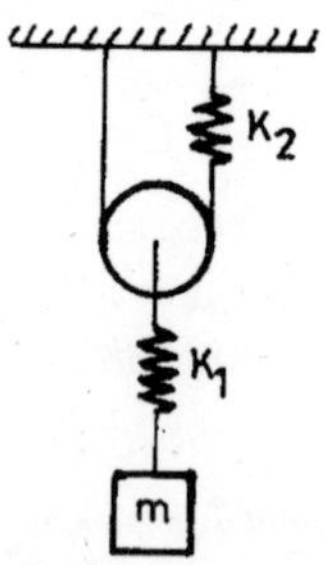

Prob. 2.3

2.4 The member *BC* of the portal frame of Fig. 2.8 is displaced horizontally by 10 mm at $t = 0$. Determine an expression for the resultant displacement of *BC*.

2.5 A mass m is suspended from a beam as shown in the figure. The beam is of negligible mass and has a uniform flexural rigidity *EI*. Find the natural frequency of the system.

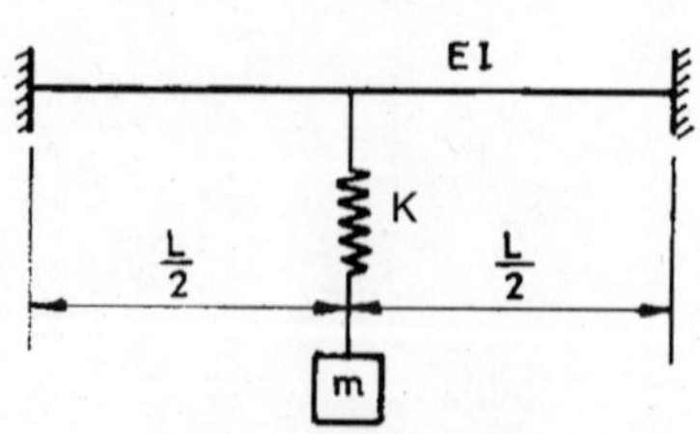

Prob. 2.5

2.6 A disc is connected to a shaft as shown in the figure. Determine the natural frequency of the system for torsional vibration.

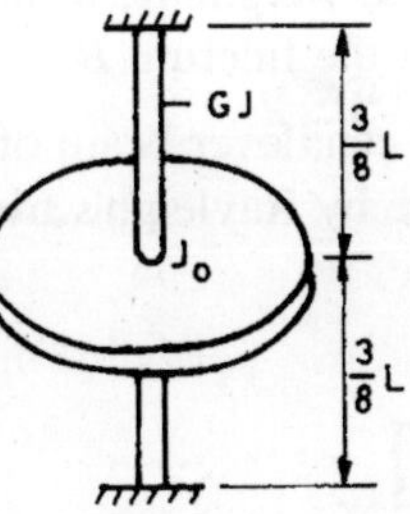

Prob. 2.6.

2.7 For a SDF system, if $m = 10$ kg, $k = 10$ N/m and $c = 4$ Ns/m, find the damping factor ζ the logarithmic decrement and ratio of any two successive amplitudes.

2.8 A SDF system has an undamped natural frequency of 7 rad/s and a damping factor of 10 percent. The initial conditions are $x_0 = 0$ and $\dot{x} = 0.5$ m/s. Determine the damped natural frequency and the equation of motion for the system.

2.9 A machine mounted on springs and having a damper is shown in the figure. Find the equation of motion in terms of initial displacements x_0 and initial velocity $\dot{x}_0$.

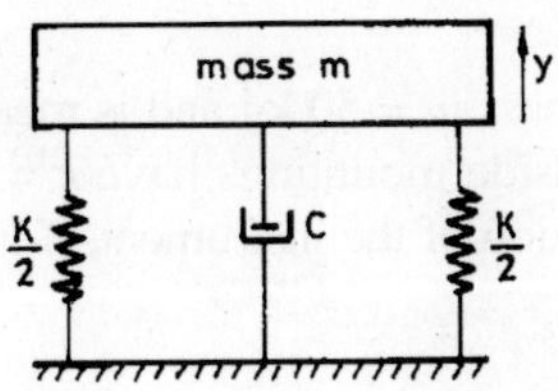

Prob. 2.9

2.10 In Prob. 2.4, it is found that for the horizontal vibration of *BC*, the maximum displacement in the positive direction is 0.8 of the preceding maximum displacement in that direction. Determine the coefficient of viscous damping, damping ratio and the logarithmic decrement.

2.11 A 0.02 N force produces a constant velocity of 3 cm/s. Find the damping factor ζ used in a system having a mass 1 kg and spring stiffness 700 N/m.

2.12 Determine effective stiffness of the springs of the figure.

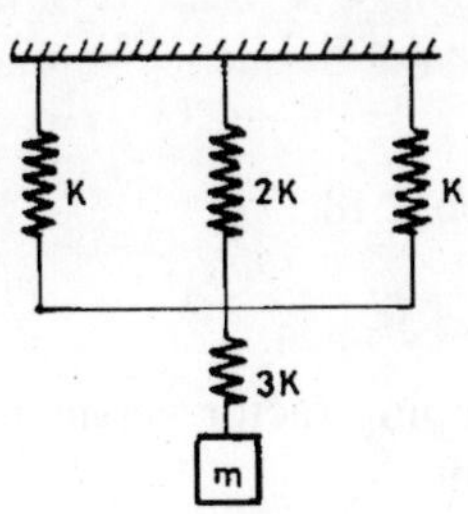

Prob. 2.12.

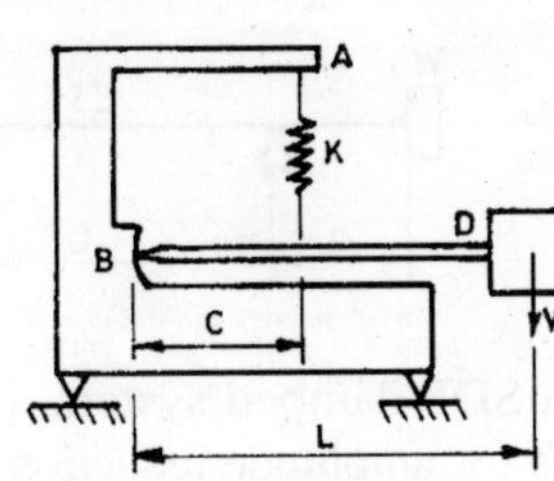

Prob. 2.13.

2.13 A device for recording ship vibration is shown in the figure. Determine the natural frequency of the vertical vibration of the weight W, if the moment of inertia I of this weight along with that of bar BD about the fulcrum B is known.

2.14 Determine the natural frequency of the cantilever beam of negligible mass having two weights attached as shown in the figure by Rayleigh's method.

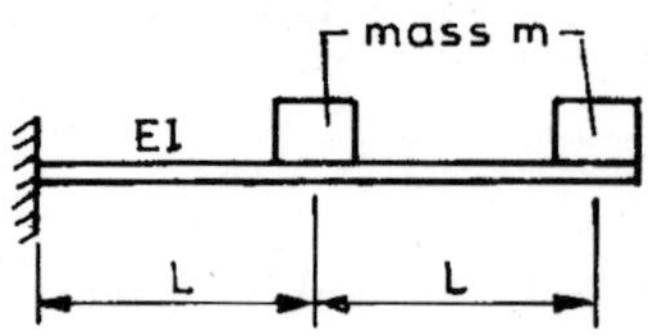

Prob. 2.14

2.15 Determine the natural frequency in torsion for the shaft carrying two discs as shown.

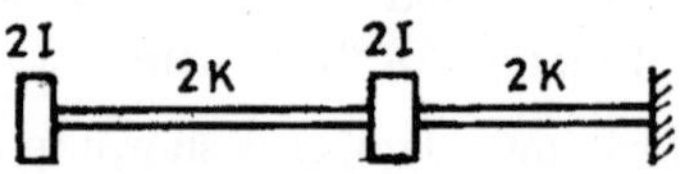

Prob. 2.15

2.16 In 5 cycles, the amplitude of a motion is observed to decay from 30 mm to 0.15 mm. Calculate the damping ratio ζ .

2.17 The instrument of the figure has mass m = 50 kg and is meant for impact on a surface. The impact velocity is 10 m/s. Inside mountings have k = 5 N/m and c = 1 Ns/mm. Determine the maximum deceleration of the instrument, if it is assumed that it will not bounce.

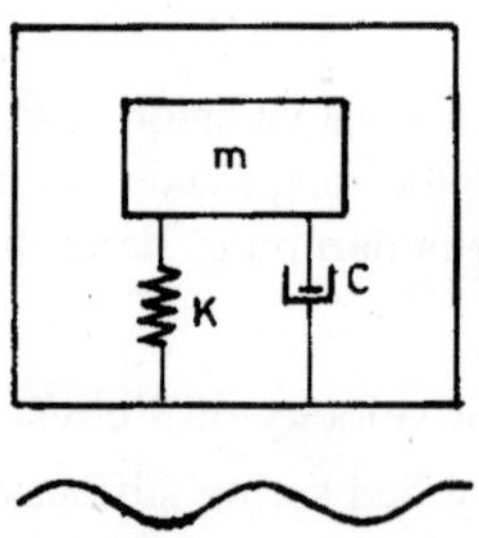

Prob. 2.17.

2.18 Applying energy principle, determine the natural frequency of the beam of the figure.

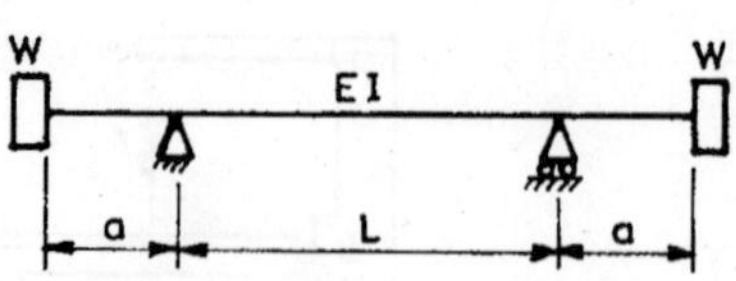

Prob. 2.18.

2.19 For a SDF damped system, plot three curves of damping factor versus number of cycles for amplitude reduction of 25, 50 and 75 per cent.

2.20 Determine the natural frequency of the horizontal vibration of the frame by Rayleigh's method, assuming the deflection of the column given by

$$x = x_{max}\ (1 - \cos \pi y / L)$$

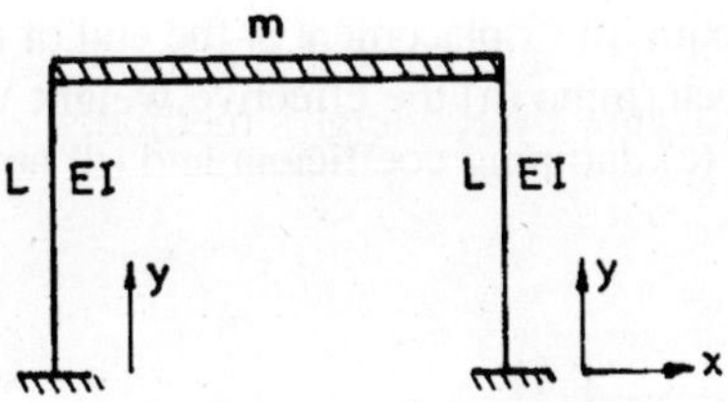

Prob. 2.20.

2.21 In a spring mass system, the parameters associated with vibration are m, c, k and x. What are the corresponding quantities for torsional vibration?

2.22 What is the effect of increase of stiffness in the natural frequency?

2.23 How is the concept of logarithmic decrement utilised?

2.24 What is the importance of critical damping?

2.25 A spring mass has a natural frequency of 12 Hz. If the spring stiffness is reduced by 1000 N/m, the frequency is altered by 50%. For the original system, determine the mass and the spring stiffness.

2.26 A spring mass system has maximum velocity 20 cm/s and time period 1s. If the initial displacement is 1 cm, determine (a) the amplitude, (b) the initial velocity, (c) the maximum acceleration and (d) the phase angle.

2.27 Find the natural frequency of the system in the figure. The bar *AB* is massless and rigid.

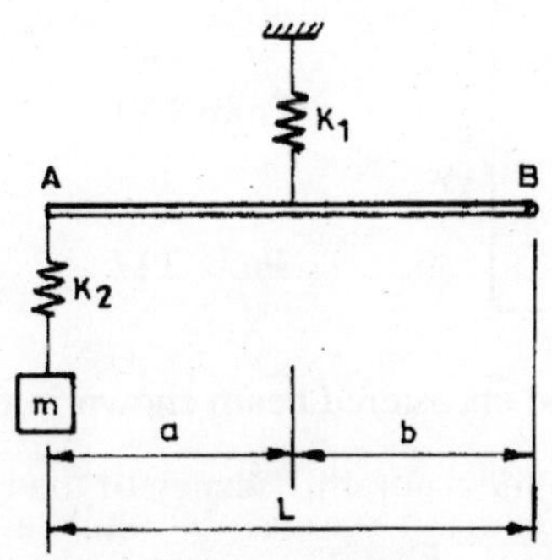

Prob. 2.27

2.28 For the system shown in the figure, the bar *AB* makes oscillations in the horizontal plane about *XY*. Determine the angular frequency of the system.

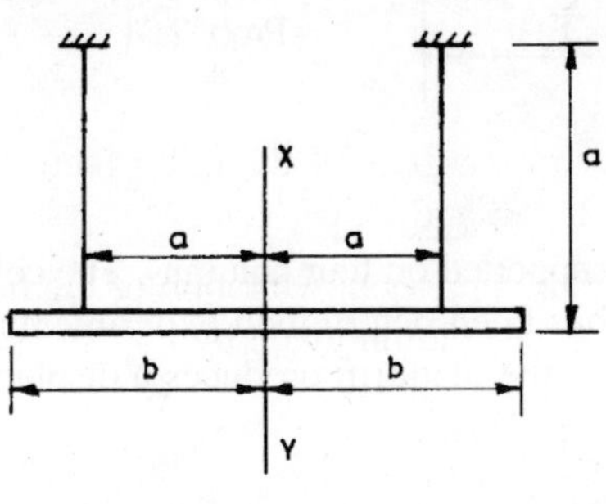

Prob. 2.28

2.29 A one-storey building is idealised as a rigid girder having total weight W and weighless columns. During the jacking operation while carrying out a free vibration test, a horizontal force of 100 kN displaces the girder laterally by 0.5 cm. After the instantaneous release of this initial displacement, the maximum displacement at the end of the first cycle is 0.4 cm. and the time period is 1.4s. Determine (a) the effective weight W of the girder, (b) undamped frequency of vibration, (c) damping coefficient and (d) amplitude after first 5 cycles.

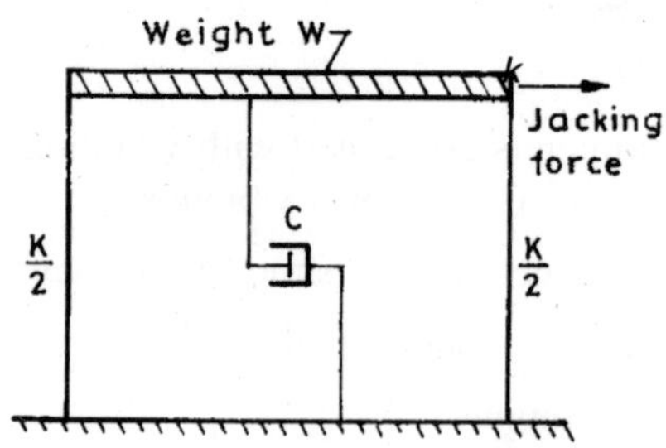

Prob. 2.29

2.30 A system shown in the Fig. having $k_1 = k_2 = 8 \times 10^4$ N/m and $k_3 = 3 \times 10^5$ N/m and m = 250 kg. Determine the natural frequency of the system.

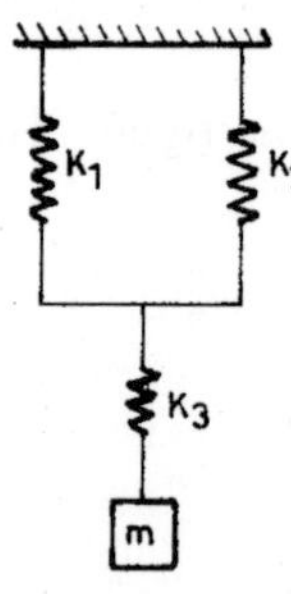

Prob. 2.30

2.31 Determine the natural frequency of the encastered beam shown in the figure.

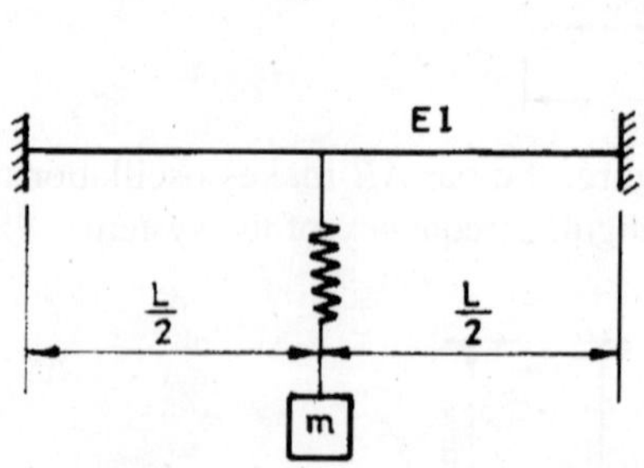

Prob. 2.31

2.32 A platform weighing 7 × 10^2 N is supported on four columns. The colunms are identical and are clamped at both ends. It has been determined experimentally that a force of 1.75 × 10^5 N horizontally applied to the platform produces a displacement of $\triangle$= 2.54

mm. Damping is 5 percent of critical damping. Determine from this structure the following: (a) undamped natural frequency (b) absolute damping coefficient, (c) logarithmic decrement and (d) the number of cycles and time required for the amplitude of motion to be reduced from an initial value of 2.54 mm to 0.254 mm.

2.33 Show that for an undamped system in free vibration the logarithmic decrement is expressed as

$$\delta = \frac{1}{k_0} \ln \frac{x_i}{x_{i+k}}$$

where k_0 is the number of cycles separating two measured peak amplitudes x_i and x_i + k.

2.34 Determine the effective torsional stiffness of the shaft and determine its natural period.

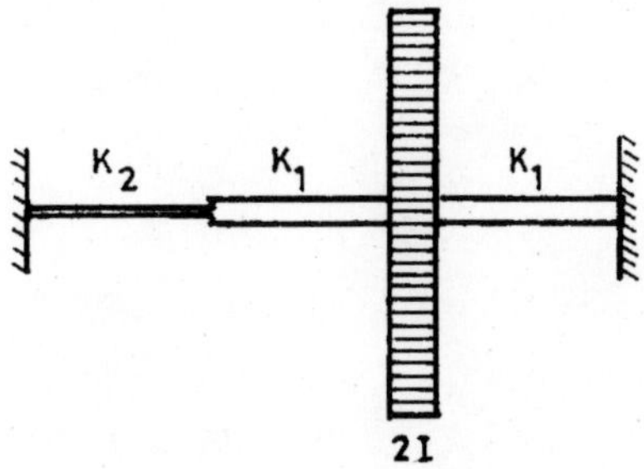

Prob. 2.34

2.35 A mass of 2 kg. is connected to a spring of stiffness 15 N/cm. Determine the critical damping constant.

2.36 A diver weighing 90 kg. stands at the end of a cantilever diving board of span 1 m. The diver oscillates at a frequency of 2 Hz. What is the flexural rigidity *EI* of the diving board?

2.37 A free vibration test is conducted on an empty elevated water tank. A cable attached to the tank applied a horizontal force of 75 kN and pulls the tank horizontally by 5 cm. The cable is suddenly cut and the resulting free vibration is recorded. At the end of 4 complete cycles, the time is 2s and the amplitude is 25 mm. From these data compute the following: (a) damping ratio, (b) natural period of undamped vibration, (c) damping coefficient and (d) number of cycles required for the displacement amplitude to decrease to 5 mm.

Chapter 3

Forced Vibration of Single Degree of Freedom System

3.1 INTRODUCTION

In the previous chapter, solutions have been obtained for the differential equation of free vibration of single degree of freedom system (SDF) and also, determination of free vibration characteristics of SDF systems by energy methods. In this chapter, attention is directed towards the study of forced vibration of SDF systems and some of its applications. The chapter is begun with the study of harmonic loading. Towards the end of the chapter the earthquake response analysis of structures has been dealt with.

3.2 RESPONSE OF DAMPED SYSTEMS TO HARMONIC LOADING

Consider a single degree of freedom damped system acted upon by force $F_0 \sin \omega t$. The amplitude of this force is F_0 and angular frequency ω. The system has a spring of constant k and a damper having coefficient c.

For the dynamical equilibrium of the system for which the freebody diagram has been shown in Fig. 3.1, the equation of motion is

$$m\ddot{x} + c\dot{x} + kx = F_0 \sin \omega t \tag{3.1}$$

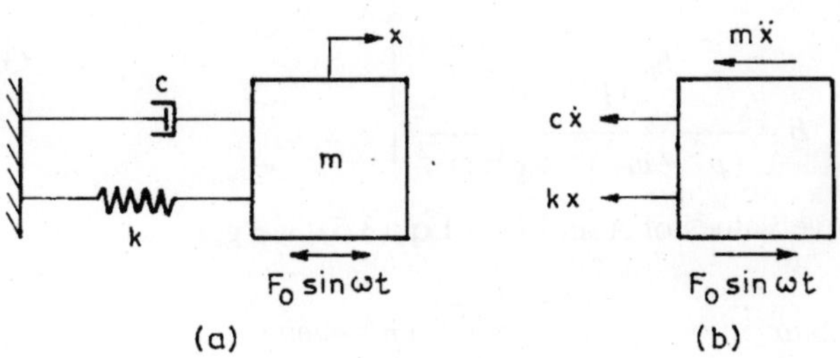

Fig. 3.1 A SDF system

Equation (3.1) is recast as

$$\ddot{x} + 2n\dot{x} + p^2 x = \frac{F_0}{m} \sin \omega t \tag{3.2}$$

where $n = c/2m$ and $p^2 = k/m$

General solution of Eq. (3.2) consists of two parts, the complementary function, which is the solution of homogeneous equation and the particular integral. Thus

$$x = x_c + x_p \tag{3.3}$$

where x_c is the complementary function and x_p is the particular integral.

For the present case, the complementary function is the solution of the damped free vibration, which has already been obtained in the previous chapter. Thus from Eq. (2.31), we have

$$x_c = e^{-nt}(C_1 \cos\sqrt{p^2 - n^2}\, t + C_2 \sin\sqrt{p^2 - n^2}\, t) \tag{3.4}$$

In order to obtain the particular integral, let the trial solution be

$$x_p = A\cos\omega t + B\sin\omega t \tag{3.5}$$

Substitution of the trial solution of Eq. (3.5) into Eq. (3.1), we get

$$-(A\omega^2 \cos\omega t + B\omega^2 \sin\omega t) + 2n(B\omega\cos\omega t - A\omega\sin\omega t)$$

$$+ p^2 A\cos\omega t + p^2 B\sin\omega t = \frac{F_0}{m}\sin\omega t \tag{3.6}$$

Equating the coefficients of $\cos\omega t$ and $\sin\omega t$ on both sides of Eq. (3.6), one obtains

$$\left.\begin{aligned} B(p^2 - \omega^2) - A\,2n\omega &= \frac{F_0}{m} \\ B\,2n\omega + A(p^2 - \omega^2) &= 0 \end{aligned}\right\} \tag{3.7}$$

Solution of simultaneous Eq. (3.7) is

$$\left.\begin{aligned} A &= \frac{-\dfrac{F_0}{m}2n\omega}{(p^2 - \omega^2)^2 + (2n\omega)^2} \\ B &= \frac{\dfrac{F_0}{m}(p^2 - \omega^2)}{(p^2 - \omega^2)^2 + (2n\omega)^2} \end{aligned}\right\} \tag{3.8}$$

Substituting the above values of A and B in Eq. (3.5), we get

$$x_p = \frac{-\dfrac{F_0}{m}\cdot 2n\omega}{(p^2 - \omega^2)^2 + (2n\omega)^2}\cos\omega t + \frac{\dfrac{F_0}{m}(p^2 - \omega^2)}{(p^2 - \omega^2)^2 + (2n\omega)^2}\sin\omega t \tag{3.9}$$

Let

$$\left.\begin{aligned} \sin\phi &= (2n\omega)\,k_1 \\ \cos\phi &= (p^2 - \omega^2)\,k_1 \end{aligned}\right\} \tag{3.10}$$

On substitution of $2n\omega$ and $(p^2 - \omega^2)$ in the numerator of the two terms on the right hand side, on the basis of Eq. (3.10), we get

$$x_p = \frac{\dfrac{F_0}{m}}{\sqrt{(p^2 - \omega^2)^2 + (2n\omega)^2}} \sin(\omega t - \phi) \qquad (3.11a)$$

where $$\phi = \tan^{-1} \frac{2n\omega}{p^2 - \omega^2} \qquad (3.11b)$$

The complete solution is given by

$$x = e^{-nt}(C_1 \cos\sqrt{p^2 - n^2}\, t + C_2 \sin\sqrt{p^2 - n^2}\, t) + x_p \qquad (3.12)$$

The first term on the right hand side having a factor e^{-nt} represents the free damped vibration. The other term having the same frequency as the disturbing force, represents forced vibration. The actual motion is a superimposition of two simple harmonic motions, having different amplitudes, different frequencies and different phases. The resulting motion is somewhat irregular and complicated in nature (Fig. 3.2). However, due to damping in the system, the free vibration part vanishes after a short time and only the forced vibration part remains, which however, is harmonic in nature.

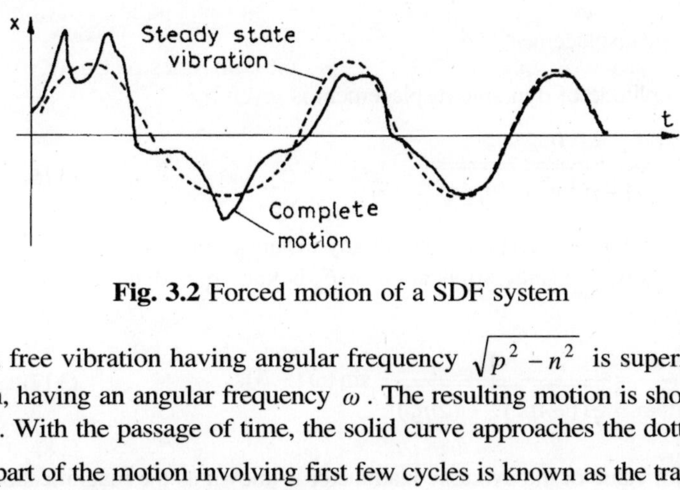

Fig. 3.2 Forced motion of a SDF system

In Fig. 3.2, free vibration having angular frequency $\sqrt{p^2 - n^2}$ is superimposed on the forced vibration, having an angular frequency ω. The resulting motion is shown by the solid curve (Fig. 3.2). With the passage of time, the solid curve approaches the dotted curve.

The initial part of the motion involving first few cycles is known as the transient vibration.

If the first term of Eq. (3.12) is neglected, i.e. the free vibration part is ignored, then the resulting motion is termed as steady state vibration.

Equation (3.12) then becomes

$$x = \frac{\dfrac{F_0}{m}}{\sqrt{(p^2 - \omega^2)^2 + (2n\omega)^2}} \sin(\omega t - \phi) \qquad (3.13)$$

Equation (3.13) can be rewritten as

$$x = \frac{\dfrac{F_0}{mp^2}}{\sqrt{\left[1-\left(\dfrac{\omega}{p}\right)^2\right]^2+\left(2\cdot\dfrac{n}{p}\cdot\dfrac{\omega}{p}\right)^2}}\sin(\omega t-\phi) \tag{3.14}$$

Introducing

$$\frac{\omega}{p} = \eta = \text{ tuning factor = resonant frequency ratio}$$

$$\frac{n}{n_c} = \frac{n}{p} = \zeta = \text{ critical damping ratio}$$

and $mp^2 = k$, Eq. (3.14) becomes

$$x = \frac{\dfrac{F_0}{k}}{\sqrt{(1-\eta^2)^2+(2\eta\zeta)^2}}\sin(\omega t-\phi) \tag{3.15}$$

$$\frac{F_0}{k} = \text{static displacement} = \delta_{st}$$

The maximum amplitude of dynamic displacement is given by

$$x_{\max} = \frac{\delta_{st}}{\sqrt{(1-\eta^2)^2+(2\eta\zeta)^2}} \tag{3.16}$$

The ratio of the dynamic displacement at any instant of time to the displacement that would have been produced by the static application of F_0 is known as dynamic load factor (DLF) and is given by

$$\text{DLF} = \frac{x}{\delta_{st}} = \frac{1}{\sqrt{(1-\eta^2)^2+(2\eta\zeta)^2}}\sin(\omega t-\phi) \tag{3.17}$$

The maximum value of the dynamic load factor is known as the magnification factor (μ), i.e. the ratio of the maximum dynamic displacement to the static displacement is defined as the magnification factor, and is given by

$$\mu = \frac{1}{\sqrt{\left(1-\eta^2\right)^2+(2\eta\zeta)^2}} \tag{3.18}$$

The magnification factor is of interest to the designer. A magnification factor of 3 immediately reveals that all maximum displacements, forces and stresses due to dynamic load will be thrice the value obtained from the static analysis.

In Fig. 3.3, the variation of the magnification factor μ with η has been indicated for different values of ζ .For small values of η, the magnification factor approaches unity, whereas for larger values of η, μ is very small. For these two extremes, the effect of damping is negligible. But the effect of damping is most pronounced in the region $0.5 < \eta < 1.5$. In order to obtain the maximum value of μ, we proceed as follows

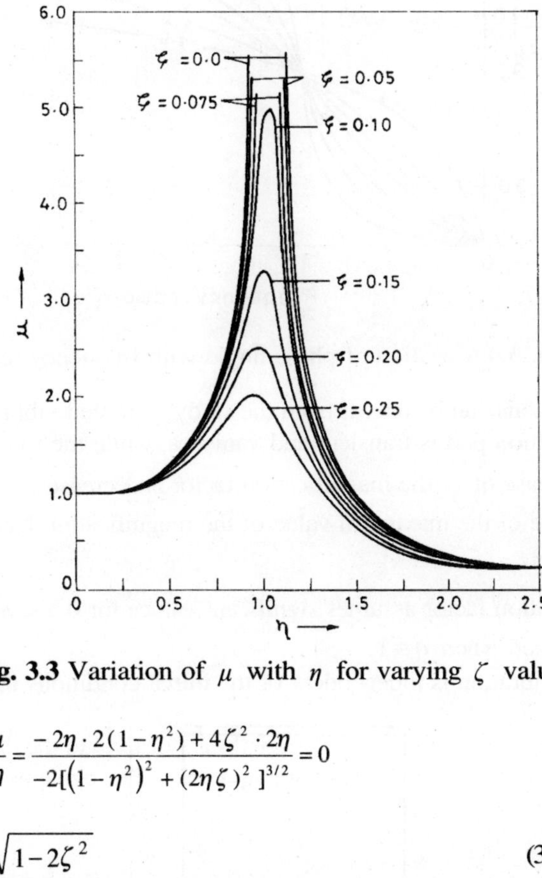

Fig. 3.3 Variation of μ with η for varying ζ values

$$\frac{d\mu}{d\eta} = \frac{-2\eta \cdot 2(1-\eta^2) + 4\zeta^2 \cdot 2\eta}{-2[(1-\eta^2)^2 + (2\eta\zeta)^2]^{3/2}} = 0$$

or,
$$\eta = \sqrt{1-2\zeta^2} \tag{3.19}$$

ζ being a small quantity, the maximum value of μ is obtained when $\eta \cong 1$. It can be seen in Fig. 3.3, that when the frequency of the external force is nearly equal to the frequency of the system in free vibration, magnification factor increases rapidly. The maximum value of the magnification factor is highly sensitive to the damping of the system. The condition of maximum amplitude is known as the condition of resonance.

The phase angle ϕ given by Eq. (3.11b) is rewritten as follows

$$\phi = \tan^{-1}\frac{2\zeta\eta}{1-\eta^2} \tag{3.20}$$

Equation (3.20) is plotted with ζ as a parameter in Fig. 3.4. For small values of η the phase angle is 90°. At larger values of η, ϕ tends to approach 180°.

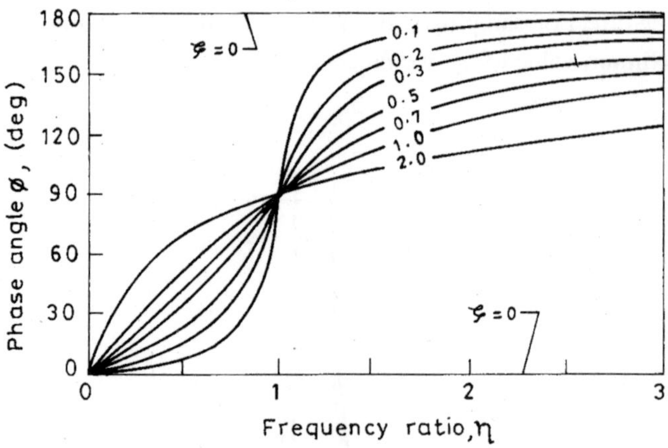

Fig. 3.4 Variation of phase angle with frequency ratio

The following conclusions can be drawn from the study of forced vibrations:

(1) The free vibration part is transient and vanishes, while the forced part persists.

(2) With the increase of ζ the magnification factor m decreases.

(3) The magnitude of the maximum value of the magnification factor is very sensitive to the value of ζ

(4) The magnification factor assumes significant values for $0.5 < \eta < 1.5$, the maximum value is obtained when $\eta \cong 1$.

(5) Steady-state vibration is independent of the initial conditions in the system.

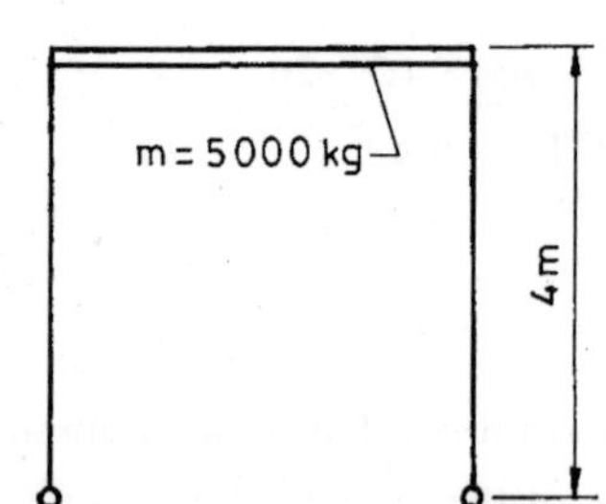

Fig. 3.5 Example 3.1

Example 3.1

A steel rigid frame of Fig. 3.5 supports a rotating machine, which exerts a horizontal force at the girder level of 50,000 sin 11*t* N. Assuming 4 percent critical damping, what is the steady-state amplitude of vibration? I for columns $= 1500 \times 10^{-7}$ m^4, $E = 21 \times 10^{10}$ N/m^2

$$\text{Stiffness for each column} = \frac{3EI}{L^3}$$

$$\text{Total stiffness} = \frac{2 \times 3 \times EI}{L^3}$$

$$= \frac{2\times 3\times 21\times 10^{10}\times 1500\times 10^{-7}}{4^3} = 2{,}953{,}125 \quad \text{N/m}$$

$$m = 5000\ \text{kg.}$$

The natural frequency is given by

$$p = \sqrt{\frac{k}{m}} = \sqrt{\frac{2953125}{5000}} = 24.3 \quad \text{rad/s}$$

$$\eta = \frac{\omega}{p} = \frac{11}{24.3} = 0.453$$

$$\zeta = 0.04$$

$$F_0 = 50{,}000\ \text{N}$$

Therefore,

$$x_{\max} = \frac{\frac{F_0}{k}}{\sqrt{(1-\eta^2) + (2\eta\zeta)^2}}$$

$$= \frac{50000}{2953125\sqrt{(1-0.453^2)^2 + (2\times 0.04\times 0.453)^2}}$$

$$= 2.13\times 10^{-2}\ \text{m} = 21.3\ mm$$

Example 3.2

Determine the magnification factor of forced vibration produced by an oscillator, fixed at the middle of the beam at a speed of 600 rpm. The weight concentrated at the middle of the beam is $W = 5000$ N and produces a statical deflection of the beam equal to $\delta_{st} = 0.025$ cm. Neglect the weight of the beam and assume that the damping is equivalent to a force acting at the middle of the beam proportional to the velocity and equal to 500 N at a velocity of 2.5 cm/s.

The frequency of forcing function is given by

$$\omega = 2\pi\times\frac{600}{60} = 20\pi \quad \text{rad/s}$$

The mass of the vibrating body is

$$m = \frac{5000}{9.8}\ \text{kg.}$$

The static deflection is

$$\delta_{st} = \frac{0.025}{100} m$$

The stiffness is given by

$$k = \frac{F_0}{\delta_{st}} = \frac{5000 \times 100}{0.025} = 20 \times 10^6 \quad \text{N/m}$$

Based on the information given, the damping constant is

$$c = 500 \times \frac{100}{2.5} = 20{,}000 \quad \text{Ns/m}$$

The natural frequency of the system is given by

$$p^2 = \frac{k}{m} = \frac{20 \times 10^6 \times 9.8}{5000} = 39{,}200$$

Therefore, $p = 198$ rad/s.

The damping constant is

$$n = \frac{c}{2m} = \frac{20000}{2 \times 5000} \times 9.8 = 19.6$$

$$\eta = \frac{\omega}{p} = \frac{20\pi}{198} = 0.32$$

$$\zeta = \frac{n}{p} = \frac{19.6}{197} = 0.099$$

The magnification factor is

$$\mu = \frac{1}{\sqrt{(1-\eta^2)^2 + (2\eta\zeta)^2}}$$

$$= \frac{1}{\sqrt{(1-0.32^2)^2 + (2 \times 0.32 \times 0.099)^2}}$$

$$= 1.111$$

Example 3.3

An automobile whose weight is 150 N is mounted on four identical springs. Due to its weight, it sags 0.23 m. Each shock absorber has a damping coefficient of 0.4 N for a velocity of 3 cm per second. The car is placed on a platform which moves vertically at resonant speed, having an amplitude of 1 cm. Find the amplitude of vibration of the car.

The centre of gravity is assumed to be at the centre of the wheel base.

The natural frequency of the car as per Eq. (2.20) is

$$p = \sqrt{\frac{9.81}{0.23}} = 6.53 \quad \text{rad/s}$$

The damping coefficient for the four shock absorbers is

$$c = \frac{0.4 \times 4}{0.03} = \frac{160}{3} \text{ Ns/m}$$

The platform vibrates with an amplitude x_0. Therefore, spring extension is $x - x_0 \sin\omega t$, where ω is the frequency of oscillation of the platform. The equation of motion is

$$m\ddot{x} + c(\dot{x} - x_0\omega \cos\omega t) + k(x - x_0 \sin\omega t) = 0$$

or

$$m\ddot{x} + c\dot{x} + kx = cx_0\omega \cos\omega t + kx_0 \sin\omega t$$

$$= \sqrt{(cx_0\omega)^2 + (kx_0)^2} \sin(\omega t + \alpha)$$

Therefore, equivalent external force is

$$F = \sqrt{(cx_0\omega)^2 + (kx_0)^2}$$

At resonance

$$\omega = p\sqrt{1 - 2\zeta^2}$$

$$\zeta = \frac{160}{3 \times 2 \times 150 \times 6.53} = 0.027$$

$$\omega = 6.53\sqrt{1 - 2(0.027)^2} = 6.53$$

$$x_0 = 0.01 \text{ m and } k = \frac{150}{0.23}$$

Therefore

$$F = \sqrt{\left(\frac{160}{3} \times 0.01 \times 6.53\right)^2 + \left(\frac{150}{0.23} \times 0.01\right)^2}$$

$$= 7.39 \text{ N}$$

$$\eta = \frac{\omega}{p} = 1$$

Therefore, the amplitude of the automobile is

$$x = \frac{F}{c\omega} = \frac{7.39 \times 3}{6.53 \times 160} = 0.021 \text{ m} = 2.1 \text{ cm}$$

3.3 ROTATING UNBALANCE

Many machines, such as a turbine or an electric motor is reciprocating in nature. Unbalance in the machines causes vibration excitation and the unbalance occurs when the mass centre of the rotor does not coincide with the axis of rotation.

A rotating machine is shown in Fig. 3.6. Total mass of the machine is M. The unbalance is represented by mass m having an eccentricity e. It is rotating with angular velocity ω. The machine is constrained to move in the vertical direction only. The displacement of the mass m from the static equilibrium position at any instant of time is $(x + e\sin\omega t)$.

The equation of motion of the system is given by

$$(M - m)\ddot{x} + m\frac{d^2}{dt^2}(x + e\sin\omega t) + c\dot{x} + kx = 0 \tag{3.21}$$

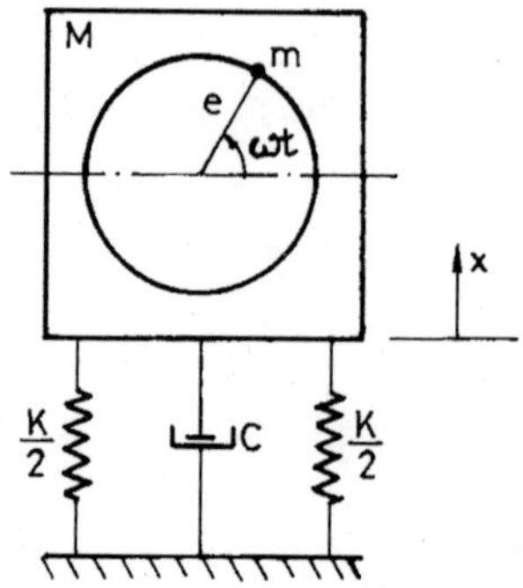

Fig. 3.6 Rotating machine

which on rearranging yields

$$M\ddot{x} + c\dot{x} + kx = me\omega^2 \sin\omega t \tag{3.22}$$

Comparing Eq. (3.22) with Eq. (3.1) reveals their identical nature. The amplitude of dynamic displacement is, therefore given by [see Eq. (3.13)]

$$x_{max} = \frac{me\omega^2}{\sqrt{(k - M\omega^2)^2 + (c\omega)^2}} \tag{3.23}$$

This can be expressed in nondimensional form as follows

$$\frac{M}{m}\frac{x_{max}}{e} = \frac{\left(\frac{\omega}{p}\right)^2}{\sqrt{\left[\left\{1 - \left(\frac{\omega}{p}\right)\right\}^2 + \left(2\zeta\frac{\omega}{p}\right)^2\right]}}$$

$$= \frac{\eta^2}{\sqrt{(1-\eta^2)^2 + (2\eta\zeta)^2}} \tag{3.24}$$

and

$$\tan\phi = \frac{2\eta\zeta}{1-\eta^2} \tag{3.25}$$

The plot of Mx_{max}/me against η for different ζ are given in Fig. 3.7.

Example 3.4

A machine of 100 kg mass has a 20 kg rotor with 0.5 mm eccentricity. The mounting springs have $k = 85$ kN/m and the damping is negligible. The operating speed is 600 rpm and the unit is constrained to move vertically. Determine the dynamic amplitude of the machine.

Assuming damping constant as zero, from Eq. (3.23), the dynamic amplitude is

$$x_{max} = \frac{me\omega^2}{k - M\omega^2}$$

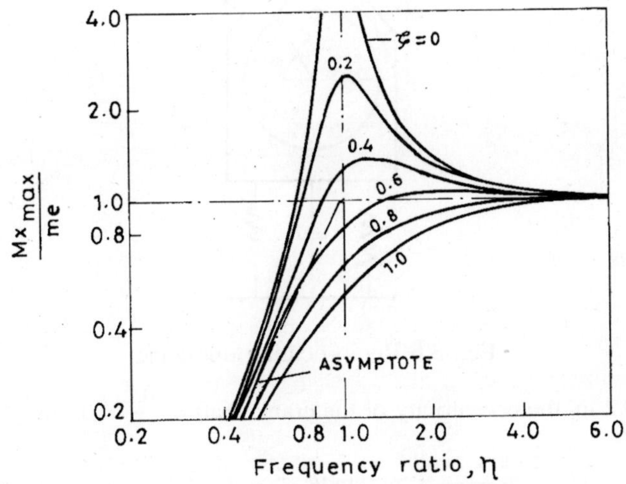

Fig. 3.7 Variation of $\frac{Mx_{max}}{me}$ with η

$e = 0.5 \times 10^{-3}$ m, $m = 20$ kg, $k = 85000$ N/m, $\omega = \frac{600 \times 2\pi}{60} = 20\pi$ rad/s and $M = 100$ kg

$$x_{max} = \frac{20 \times 0.5 \times 10^{-3} \times 20\pi \times 20\pi}{85000 - 100 \times 20\pi \times 20\pi}$$

$=1.27x10^{-4}$ m (negative sign is ignored)

3.4 RECIPROCATING UNBALANCE

The treatment of reciprocating unbalance is same as rotating unbalance. In this case, the unbalanced mass *m*, which is reciprocating consists of the mass of the piston, the wrist pin and a portion of the connecting rod (Fig. 3.8). The exciting force *F* is due to the inertia force of the reciprocating mass. This has been shown to be equal to $me\omega^2\left[\sin\omega t + \frac{e}{L}\sin 2\omega t\right]$, where e is the radius of the crank shaft and *L* is the length of the connecting rod. If *e/L* is a small quantity, the second harmonic term, $\left(\frac{e}{L}\right)\sin 2\omega t$ can be neglected. In that case the problem becomes same as that of rotating unbalance.

Example 3.5

A machine having a mass of 250 kg is supported by springs of total stiffness k = 21.5 kN/m. Assume that the damping ratio is 0.15. The machine incorporates a piston whose mass is 5 kg and it has a stroke of 200 mm. For operation at 1500 rpm, determine the dynamic amplitude.

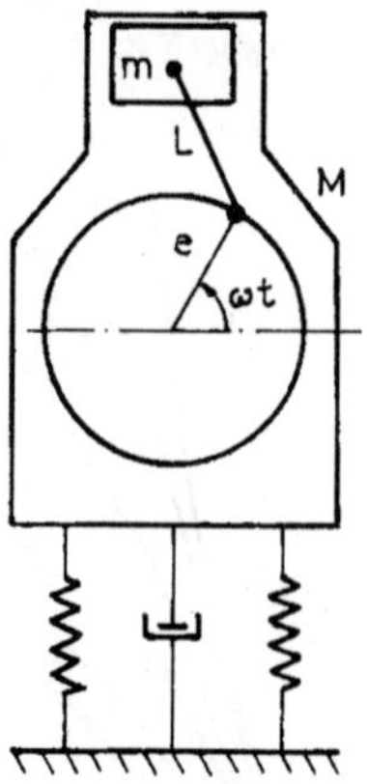

Fig. 3.8 Reciprocating unblance

The stroke being 0.2 in, the eccentricity of the crank shaft = $\frac{0.2}{2} = 0.1\,\text{m}$.

$$\omega = \frac{1500 \times 2\pi}{60} = 50\pi \text{ rad/s}$$

$$M = 250 \text{ kg}, m = 5 \text{ kg and } k = 21500 \text{ N/m}, \zeta = 0.15,$$

$$c = 0.15 \times 2 \times 250 \times \sqrt{\frac{21500}{250}} = 695.52$$

The dynamic amplitude is given by

$$x_{\max} = \frac{me\omega^2}{\sqrt{(k - M\omega^2)^2 + (c\omega)^2}}$$

$$= \frac{5 \times 0.1 \times 50\pi \times 50\pi}{\sqrt{(21500 - 250 \times 50\pi \times 50\pi)^2 + (695.52 \times 50\pi)^2}}$$

$$= 2.01 \times 10^{-3} \text{ m}$$

3.5 WHIRLING OF ROTATING SHAFTS

Rarely the geometric centre of a part of the structure or machine coincide with its centre of gravity. This is due to the defects in manufacture or non-homogeneity of the material, constituting the structure or machine. As such, there is an inherent eccentricity involved in the problem.

A disc is mounted on the shaft as shown in Fig. 3.9. The disc is having a mass, in which the mass of the shaft is negligible. The equivalent spring stiffness of the shaft is k. The distances of different points have been indicated in Fig. 3.9(b). The shaft is rotating at the rate ω. The damping in the system is neglected.

The centre of rotation in Fig. 3.9 is O, S is the geometric centre and G is the mass centre. The elastic forces set up on the shaft are $-kx$ and $-ky$. This is represented by a complex quantity $-kz$, such that $z = x + iy$. The position of S relative to G is $e_0 \, e^{i\omega t}$ and position of G relative to O is $z + re^{i\omega t}$. The equation of motion is as follows

$$m \frac{d^2}{dt^2}\left(z + re^{i\omega t}\right) = -kz$$

or

$$m\ddot{z} + kz = mr\omega^2 e^{i\omega t} \tag{3.26}$$

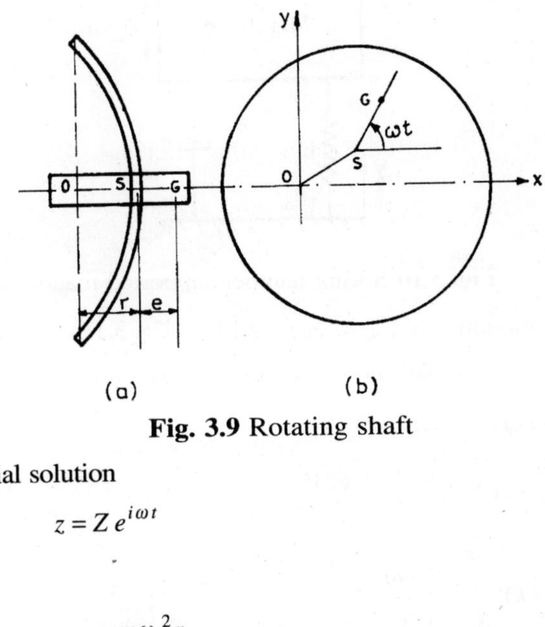

Fig. 3.9 Rotating shaft

Substitution of the trial solution

$$z = Z\, e^{i\omega t}$$

into Eq. (3.26) yields

$$Z = \frac{m\omega^2 r}{k - m\omega^2}$$

or

$$\frac{Z}{r} = \frac{\omega^2 / p^2}{1 - \dfrac{\omega^2}{p^2}} \tag{3.27}$$

With the usual notations, Eq. (3.27) becomes

$$\frac{Z}{r} = \frac{\eta^2}{1 - \eta^2} \tag{3.28}$$

Equation (3.28) reveals that the speed of the shaft becomes critical when $\omega = p$, as z tends to infinity at that frequency.

In the complex representation, both real and imaginary parts are needed to describe the motion.

3.6 VIBRATION ISOLATION AND TRANSMISSIBILITY

Machines are often connected to the foundation by springs and dampers. The main purpose behind it is to reduce the transmission of forces to the foundation.

A machine mounted on springs and dampers is shown in Fig. 3.10. Assuming the support to be unyielding, the force transmitted to the support is

$$F_T = c\dot{y} + ky \tag{3.29}$$

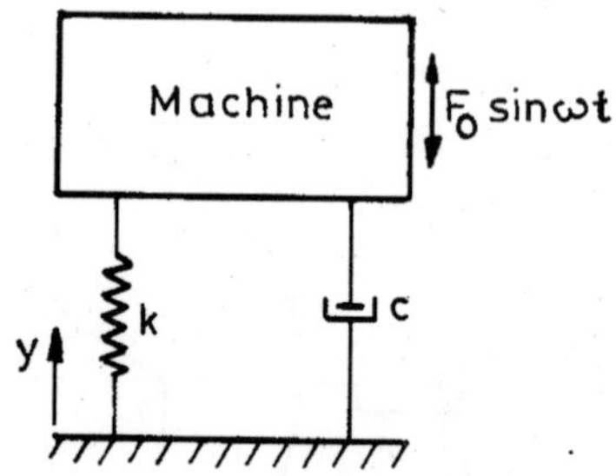

Fig. 3.10 Spring-damper supported machine

Assuming steady state motion, $y = Y\cos(\omega t - \phi)$ (see Eq. 3.14), one obtains

$$F_T = -c\omega Y \sin(\omega t - \phi) + kY\cos(\omega t - \phi) \tag{3.30}$$

The amplitude of F_T, as given by Eq. (3.30) is

$$F_{T_0} = \sqrt{(c\omega Y)^2 + (kY)^2} \tag{3.31}$$

or

$$F_{T_0} = kY\sqrt{1 + \left(\frac{c\omega}{k}\right)^2}$$

From Eq. (3.14), we get

$$Y = \frac{\dfrac{F_0}{k}}{\sqrt{(1-\eta^2)^2 + (2\eta\zeta)^2}}$$

Substituting the value of Y in Eq. (3.31), we get

$$F_{T_0} = \frac{F_0\sqrt{1 + \left(\dfrac{c\omega}{k}\right)^2}}{\sqrt{(1-\eta^2)^2 + (2\eta\zeta)^2}} \tag{3.32}$$

Now $\quad c\omega / k = c\omega / mp^2 = 2 \cdot \dfrac{c}{2m} \dfrac{\omega}{p^2} = 2 \cdot (n/p)(\omega/p) = 2\eta\zeta$

Therefore, Eq. (3.32) becomes

$$\frac{F_{T_0}}{F_0} = \frac{\sqrt{1 + (2\eta\zeta)^2}}{\sqrt{(1-\eta^2)^2 + (2\eta\zeta)^2}} \tag{3.33}$$

The ratio of the transmitted force to the applied force F_{T_0} / F_0, is defined as the transmissibility. The plot of this ratio with varying η for different values of ζ has been shown in Fig. 3.11. One of the main concerns for a designer is the reduction of the force transmitted to the foundation, i.e. to isolate the vibration. Figure 3.11 indicates that all the curves cross at $\eta = \sqrt{2}$. The transmitted force is greater than the applied force for $\eta < \sqrt{2}$ and less than the applied force for $\eta > \sqrt{2}$ Therefore, the reduction of the transmitted force may be accomplished by adjusting the stiffnesses, such that $\eta > \sqrt{2}$.

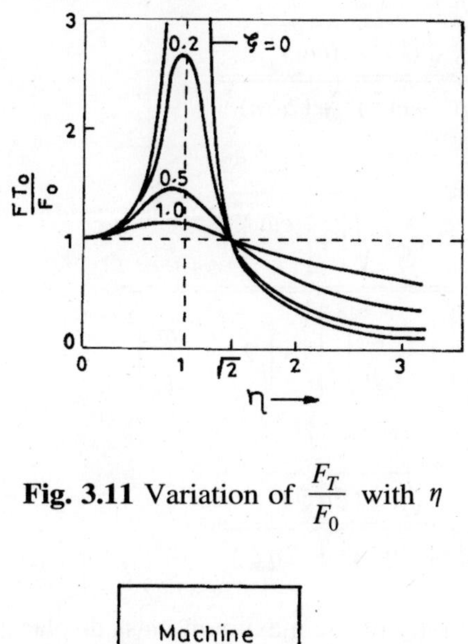

Fig. 3.11 Variation of $\dfrac{F_T}{F_0}$ with η

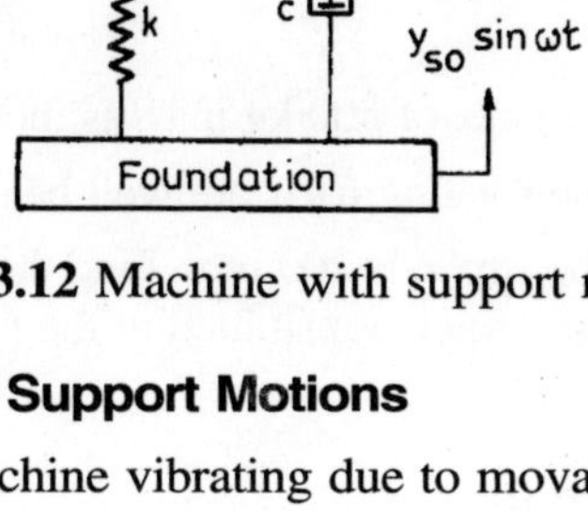

Fig. 3.12 Machine with support motions

3.6.1 Transmissibility Due to Support Motions

Let us consider the case of a machine vibrating due to movable support [Fig. 3.12] and the harmonic motion of the support is given by

$$y_s = y_{s0} \sin \omega t \tag{3.34}$$

The equation of motion of the machine of mass m is given by

$$m\ddot{y} + c\dot{y} + ky = c\dot{y}_s + ky_s \tag{3.35}$$

Assuming y_s, from Eq. (3.34) into Eq. (3.35), one obtains

$$m\ddot{y} + c\dot{y} + ky = cy_{s0}\omega \cos \omega t + ky_{s0} \sin \omega t$$

or $$m\ddot{y} + c\dot{y} + ky = \sqrt{(cy_{s0}\omega)^2 + (ky_{s0})^2} \sin(\omega t - \alpha) \tag{3.36}$$

There is a phase lag α between support motion and the existing force vector, given by the right hand side of Eq. (3.36).

If the solution is assumed as

$y = Y \sin(\omega t - \phi)$, then from Eq. (3.13)

$$Y = \frac{y_{s0}\sqrt{(k)^2 + (c\omega)^2}}{m\sqrt{(p^2 - \omega^2)^2 + (2\omega n)^2}}$$

or $$Y = \frac{y_{s0}k\sqrt{1 + \left(\frac{c\omega}{k}\right)^2}}{mp^2\sqrt{\left[1 - \left(\frac{\omega}{p}\right)^2\right]^2 + \left(2 \cdot \frac{n}{p} \cdot \frac{\omega}{p}\right)^2}}$$

or $$\frac{Y}{y_{s0}} = \frac{\sqrt{1 + (2\eta\zeta)^2}}{\sqrt{(1 - \eta^2)^2 + (2\eta\zeta)^2}} \tag{3.37}$$

In this case, Y / y_{s0} the ratio of the maximum mass displacement to maximum support displacement is known as the transmissibility. Comparing Eq. (3.37) with Eq. (3.33), the transmissibility for both force and motion is identical.

Example 3.6

A vertical single-cylinder diesel engine, of 500 kg mass is mounted on springs with k = 200 kN/m and dampers with $\zeta = 0.2$. The rotating parts are well balanced. The mass of the equivalent reciprocating parts is 10 kg and the stroke is 200 mm. Find the dynamic amplitude of the vertical motion, the transmissibility and the force transmitted to the foundation, if engine is operated at 200 rpm.

The magnitude of the unbalanced force is $F_0 = me\omega^2$

$$m = 10 \text{ kg}, e = 100 \text{ mm} = 0.1 \text{ m and } \omega = \frac{200\times 2\pi}{60} = \frac{20\pi}{3} \text{ rad/s}$$

Therefore, $F_0 = 10\times 0.1\times\left(\frac{20\pi}{3}\right)^2 = 438.65 \text{ N}$

$$k = 200{,}000 \text{ N/m}$$

$$c = 2\times 500\times 0.2\times\sqrt{\frac{200{,}0000}{500}} = 4000 \text{ Nm/s}$$

The dynamic amplitude is

$$x_{max} = \frac{438.65}{\sqrt{\left[200{,}000 - 10\times\left(\frac{20\pi}{3}\right)^2\right]^2 + \left(4000\times\frac{20\pi}{3}\right)^2}}$$

$$= 2.06\times 10^{-3} \text{ m}$$

Natural frequency of the system is

$$p = \sqrt{\frac{200{,}000}{500}} = 20 \text{ rad/s}$$

$$\eta = \frac{\omega}{p} = \frac{20\pi}{3\times 20} = 1.047$$

Transmissibility is given by Eq. (3.33) as

$$T_R = \frac{\sqrt{1+(2\eta\zeta)^2}}{\sqrt{(1-\eta^2)^2+(2\eta\zeta)^2}}$$

$$= \frac{\sqrt{1+(2\times 0.2\times 1.047)^2}}{\sqrt{(1-1.047^2)^2+(2\times 0.2\times 1.047)^2}} = 2.523$$

The force transmitted to the foundation is

$$F_T = 2.523\times 43\,8.65$$

$$= 1106.71 \text{ N.}$$

3.7 ENERGY DISSIPATION BY DAMPING

We have seen in the earlier section that the amplitude of motion goes on reducing with each successive cycle in damped free vibration. This suggests the occurrence of loss of energy.

For viscous damping, the damping force is

$$F_0 = c\dot{x} \tag{3.38}$$

Let us assume the displacement of the spring-mass system as

$$x = A \sin(pt - \in) \tag{3.39}$$

and the velocity is

$$\dot{x} = Ap \cos(pt - \in) \tag{3.40}$$

Further, $$dx = \dot{x}\, dt \tag{3.41}$$

The energy dissipated by viscous damping in one such cycle is given by

$$E_d = \int F_D\, dx = \int_0^{\frac{2\pi}{p}} (cx)\, dx = \int_0^{\frac{2\pi}{p}} (c\dot{x})\, \dot{x}\, dt$$

$$= \int_0^{\frac{2\pi}{p}} c \cdot (Ap)^2 \cos^2(\omega t - \in)\, dt$$

or, $$E_d = \pi c \omega A^2 \tag{3.42}$$

We know that $p = \sqrt{\dfrac{k}{m}}$ and $c = 2\zeta\sqrt{km}$. Substituting these values in Eq. (3.42) yields

$$E_d = 2\zeta \pi k A^2 \left(\frac{\omega}{p}\right) \tag{3.43}$$

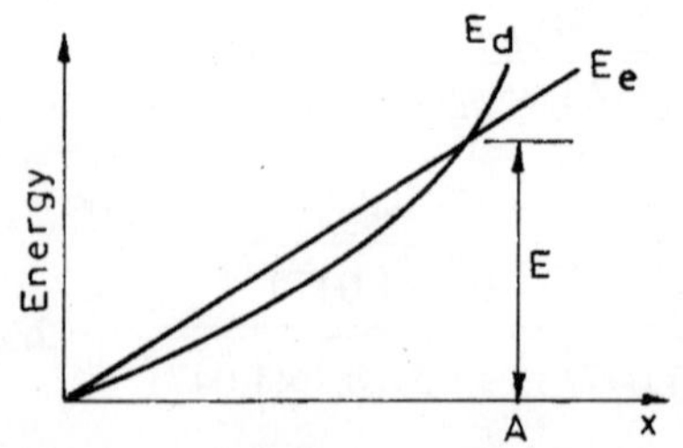

Fig. 3.13 Energies in viscous damping

The energy dissipated is proportional to the square of the amplitude of motion (Fig. 3.13).

If we consider the steady state of vibration with the external force $F(t) = F_0 \sin \omega t$, then the input energy of the external force per cycle is

$$E_e = \int F(t)\, dx = \int_0^{\frac{2\pi}{p}} F(t)\, \dot{x}\, dt$$

$$= \int_0^{\frac{2\pi}{p}} [F_0 \sin \omega t][pA \cos \omega (pt - \in)]\, dt$$

$$= \pi F_0 A \sin \in \tag{3.44}$$

Thus, Eq. (3.44) reveals that the energy due to the external force is proportional to the displacement amplitude.

Using Eq. (3.19), we can show that

$$\sin \in = 2\zeta \left(\frac{\omega}{p}\right) \frac{Ak}{F_0} \tag{3.45}$$

Combining Eqs. (3.44) and (3.45), we get

$$E_e = 2\pi\zeta \frac{\omega}{p} kA^2 \tag{3.46}$$

Eqs. (3.43) and (3.46) reveal that energy dissipation in steady-state vibration is due to various damping in steady-state vibration.

The graphical representation of energy dissipation in viscous damping is performed as follows.

The velocity of motion is expressed as

$$x = \omega A \cos(\omega t - \in) = \pm \omega A \sqrt{1 - \sin^2(\omega t - \varepsilon)} = \pm \sqrt{A^2 - x^2} \tag{3.47}$$

The damping force is related to displacement x as follows

$$F_d = c\dot{x} = \pm c\omega \sqrt{A^2 - x^2} \tag{3.48}$$

Eq. (3.48) is rearranged as

$$\left(\frac{F_d}{cpA}\right)^2 + \left(\frac{x}{A}\right)^2 = 1 \tag{3.49}$$

Eq. (3.49) is represented by an ellipse shown in Fig. 3.14. The loop of the curve is known as hysteresis loop, the area inside gives an estimate of the dissipated energy.

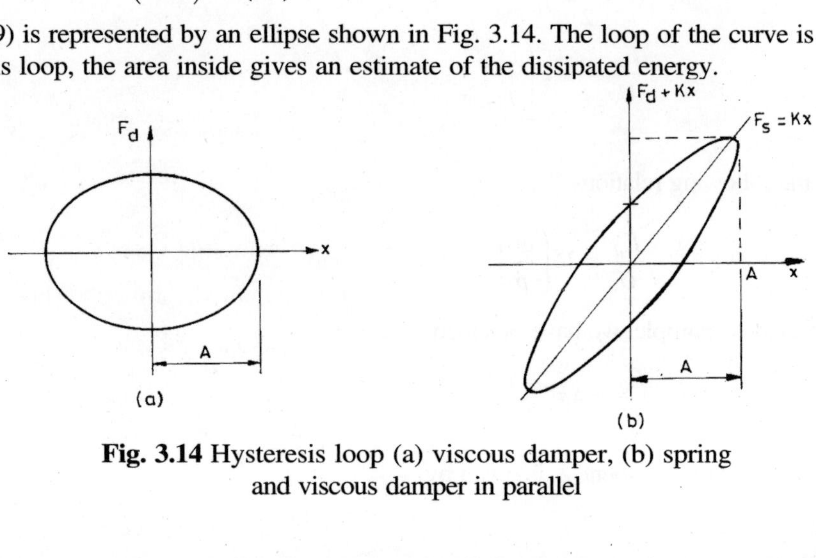

Fig. 3.14 Hysteresis loop (a) viscous damper, (b) spring and viscous damper in parallel

It may be of some interest to investigate the total resisting force

$$F_s + F_d = kx + c\omega\sqrt{A^2 - x^2} \tag{3.50}$$

where F_s is the strain energy of the spring.

The graphical representation of Eq. (3.50) is shown in Fig. 3.14(b).

In this case the hysterisis loop gets rotated. From Eq. (3.50) it is evident that loop area is proportional to ω, which suggests that the hysteresis loop will not be formed, instead a single-value curve will appear if the harmonic load is applied slowly enough (i.e., $\omega \cong 0$).

We now look into two measures of damping. Specific damping capacity and the specific damping factor. Specific damping capacity is defined as the energy loss per cycle, E_d divided by the peak potential energy $U \cdot (E_d / U)$ and $U = \frac{kA^2}{2}$.

The specific damping factor, also known as the loss factor or loss coefficient is defined as the damping energy loss per radian $E_d / 2\pi$ divided by the peak potential energy *U*,

$$\bar{\zeta} = \frac{kA^2}{2} \tag{3.51}$$

These two measures of damping are useful in comparing damping capacity of materials.

Example 3.7

Prove the following

$$\frac{E_d}{U} = 4\pi\zeta\left(\frac{\omega}{p}\right)$$

From Eq. (3.43)

$$E_d = 2\pi\zeta\frac{\omega}{p}kA^2$$

$$U = \frac{1}{2}kA^2$$

Therefore,

$$\frac{E_d}{U} = 4\pi\zeta\left(\frac{\omega}{p}\right)$$

Example 3.8:

Prove the following relations

$$\frac{E_d}{U} = 2\delta\left(\frac{\omega}{p}\right)$$

From the previous example, we have obtained

$$\frac{E_d}{U} = 4\pi\zeta\left(\frac{\omega}{p}\right) \tag{a}$$

Now, the logarithmic decrement δ is given by

$$\delta = 2\pi\zeta \tag{b}$$

Combining Eqs. (a) and (b), we get

$$\frac{E_d}{U} = 2\delta\left(\frac{\omega}{p}\right)$$

3.8 EQUIVALENT VISCOUS DAMPING

We have seen in Section 3.2 that the amplitude of vibration reduces due to the presence of damping and its major influence remains within the band $0.5 \le \eta \le 1.5$. At resonance, the damping considerably limits the amplitude of motion.

For the condition of resonance $(\omega / p = 1)$, the amplitude given by Eq.(3.14) reduces to

$$A = \frac{F_0}{2k\zeta} = \frac{F_0 p^2}{2knp} = \frac{F_0 k\, 2m}{2kmc\, p} = \frac{F_0}{cp} \tag{3.52}$$

In the above equation, concept of equivalent damping may be introduced to approximate the resonant amplitude. The equivalent damping C_{eq} can be obtained from Eq (3.42) by equating the energy dissipated by viscous damping to that of the actual damping force (will be of nonviscous type) undergoing harmonic motion

$$E_d = \pi\, C_{eq}\, pA^2$$

where E_d is the value of the energy associated with the particular value of damping force.

Example 3.9

A mass moving through a fluid experiences a damping force that is proportional to the square of the velocity. Determine C_{eq} for these forces acting on a system undergoing harmonic motion of amplitude A and ω. Find also its resonant amplitude.

Assuming harmonic motion, the time is measured from the position of largest negative displacement, the displacement is given by

$$x = -A\cos\omega t$$

The damping force is expressed as

$$F_d = \pm a \cdot \dot{x}^2$$

where negative sign is considered with positive and vice versa.

The energy dissipated per cycle is

$$E_d = \int F_D\, du = \int_0^{\frac{2\pi}{\omega}} F_d\, \dot{x}\, dt$$

$$= 2\int_0^{\frac{\pi}{\omega}} (a\dot{x}^2)\, \dot{x}\, dt$$

$$= 2a\omega^2 A^3 \int_0^{\pi/\omega} \sin^3 \omega t \, dt$$

$$= \frac{8}{3} a\omega^2 A^3$$

The equivalent viscous damping is given by

$$\pi C_{eq}\omega A^2 = \frac{8}{3} a\omega^2 A^3$$

or,

$$C_{eq} = \frac{8}{3\pi} a\omega A$$

From Eq. (3.52) substituting $c = C_{eq}$, we get

$$A = \frac{F_0}{cp} = \frac{3\pi F_0}{8a\,pAa}$$

or,

$$A = \left(\frac{3\pi F_0}{8a^2 p}\right)^{1/2}$$

3.9 SELF-EXCITED VIBRATIONS

We have seen that the external dynamic loading acting on the system are independent of the response of the system, such as the displacement, velocity and acceleration. But there are certain systems where the external dynamic load is a function of the motion itself. Such systems are called self-excited vibrating systems. Typical examples of this category are the flutter of aeroplanes, aerodynamic vibration of bridges, the shimming of automobile wheels and flow-induced vibrations.

Self-excited vibrations may be linear or nonlinear. If the vibration causes an increase in energy of the system, the amplitude will go on increasing and the system will become dynamically unstable. Only a convergence of the displacement of the motion or its steady behaviour with time will result in a dynamically stable system.

As an example, let us consider a viscously damped single d.o.f. linear system subjected to an internal force which is function of velocity. The equation of motion with usual notation is given by

$$m\ddot{x} + c\dot{x} + kx = F_0\dot{x} \tag{3.53}$$

Eq. (3.53) when rearranged is given by

$$\ddot{x} + \left(\frac{c - F_0}{m}\right)\dot{x} + \frac{k}{m}x = 0 \tag{3.54}$$

Substituting an assumed solution $x = Ae^{\lambda t}$ where A is a constant in Eq. (3.54), we get,

$$\lambda^2 + \frac{(c - F_0)}{m} + \frac{k}{m} = 0 \tag{3.55}$$

which leads to the following solution

$$\lambda_{1,2} = -\left(\frac{c - F_0}{2m}\right) \pm \sqrt{\left(\frac{c - F_0}{2m}\right)^2 - \frac{k}{m}} \tag{3.56}$$

For positive damping $c > F_0$. This will result in a vibrating system similar to damped free vibrations and the system will remain stable.

Let us investigate the possibility of negative damping. In this case, $c > F_0$. Three cases arise.

Case I:

$$\left(\frac{c - F_0}{2m}\right)^2 > \frac{k}{m} \tag{3.57}$$

In this case, both λ_1 and λ_2 are real and positive.

The solution of Eq. (3.54) is

$$x = C_1 e^{\lambda_1 t} + C_2 e^{\lambda_2 t} \tag{3.58}$$

The system is unstable, as the solution indicates a diverging nonoscillating motion.

Case II

$$\left(\frac{c - F_0}{2m}\right)^2 = \frac{k}{m} \tag{3.59}$$

In this case, both λ_1 and λ_2 are real, equal and positive. The system is unstable, as it indicates a diverging non-oscillating motion.

Case III

$$\left(\frac{c - F_0}{2m}\right)^2 < \frac{k}{m} \tag{3.60}$$

Here, λ_1 and λ_2 are complex conjugate.

The solution is expressed as

$$x = Ae^{\left(\frac{F_0 - c}{2m}\right)t} \sin\left\{\sqrt{\frac{k}{m} - \left(\frac{c - F_0}{2m}\right)^2}\, t + \in\right\} \tag{3.61}$$

Eq. (3.61) indicates a diverging oscillating motion, as the exponent is positive. Thus the system is unstable in this case as well.

Therefore, the condition for the dynamic stability is given by

$$F_0 \le c \tag{3.62}$$

Example 3.10

A cantilever pipe of length L, cross-section area A_0, flexural rigidity EI_0 and mass is shown in the Fig (3.15). If the fluid of mass density ρ flows through the pipe, determine the velocity of the fluid v at which the system will be unstable.

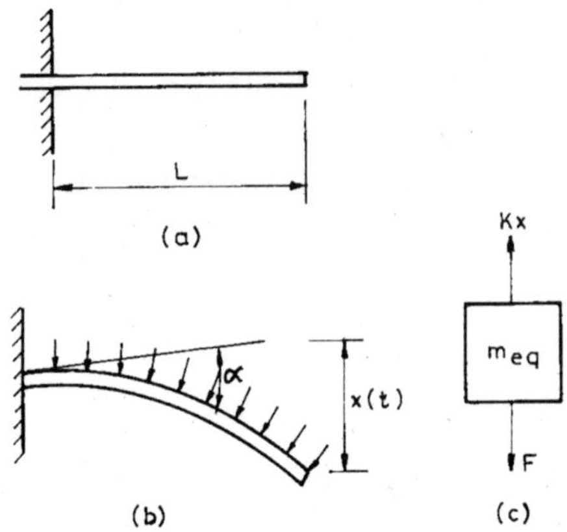

Fig. 3.15 Example 3.10

The cantilever pipe is treated as a single degree of freedom system. The freebody diagram is shown in Fig. 3.15(c). The spring constant k is

$$k = \frac{8EI_0}{L^3} \tag{a}$$

The force in the pipe due to fluid velocity is $\rho A_0 v^2$. As the pipe deflects, the component of the force at an angle α is

$$F = 2\rho A_0 v^2 \sin\alpha \tag{b}$$

The deflection being small, Eq. (b) becomes

$$F = 2\rho A_0 v^2 \alpha \tag{c}$$

The slope at the free end of the cantilever due to a uniformly load is

$$\alpha = \frac{wl^3}{6EI_0} = \frac{4}{3L}\left(\frac{wL^4}{8EI_0}\right) = \frac{4x}{3L} \tag{d}$$

The equation of motion of the equivalent mass is

$$m_{eq}\ddot{x} + kx = F \tag{e}$$

Substituting proper values, Eq. (e) becomes

$$\frac{m}{2}\ddot{x} + \left(\frac{8EI_0}{L^3} - 2\rho A_0 v^2 \frac{4}{3L}\right)x = 0$$

The system is unstable when x is negative, that is, when

$$v > \sqrt{\frac{3EI_0}{\rho A_0 L^2}} \tag{f}$$

3.10 VIBRATION MEASURING SEISMIC INSTRUMENTS

An engineer may be interested to measure any one or all the three quantities of interest in the motion of a vibrating body. They are the displacement, the velocity and the acceleration. Suitable instruments have been developed to measure each of the above quantities. Principles associated with the instrument is essentially based on the spring-mass system.

An idealised system for the measurement is shown in Fig. 3.16. The base is undergoing a motion $y = y_0 \sin \omega t$. As a result of which spring-mass system vibrates. If the displacement of the mass m is $x(t)$, then the equation of motion is

$$m\ddot{x} + c(\dot{x} - \dot{y}) + k(x - y) = 0 \tag{3.63}$$

The relative displacement between the mass and the vibrating body is $z = x - y$, then Eq. (3.63) becomes

$$m(\ddot{x} - \ddot{y}) + c(\dot{x} - \dot{y}) + k(x - y) = -m\ddot{y}$$

or

$$m\ddot{z} + c\dot{z} + kz = -my_0\omega^2 \sin \omega t \tag{3.64}$$

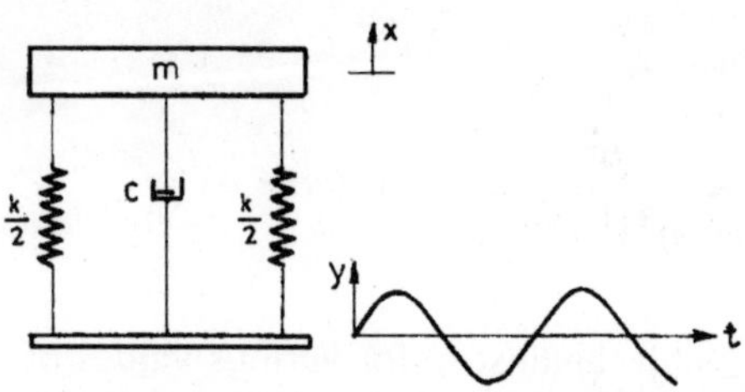

Fig. 3.16 An idealized system

Equation (3.64) is similar to Eq. (3.1) and its steady-state solution is

$$z = \frac{my_0\omega^2}{k} \cdot \frac{1}{\sqrt{(1-\eta^2)^2 + (2\eta\zeta)^2}} \sin(\omega t - \phi) \tag{3.65}$$

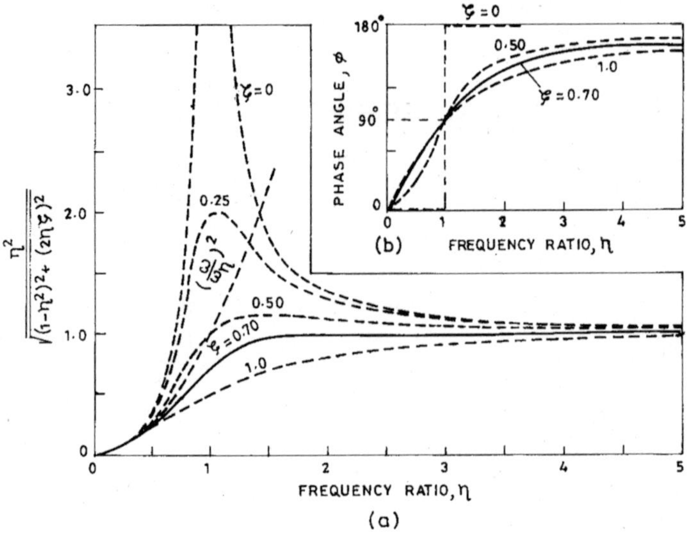

Fig. 3.17 Variation of x and ϕ with η

and ϕ is given by Eq. (3.1lb).

Equation (3.65) can be written as

$$z = y_0 \frac{\eta^2}{\sqrt{(1-\eta^2)^2 + (2\eta\zeta)^2}} \sin(\omega t - \phi) \tag{3.66}$$

Plot of $\eta^2 / \sqrt{(1-\eta^2)^2 + (2\eta\zeta)^2}$ against η for various values of ζ has been shown in Fig. 3.17. Based on Eq. (3.66), two instruments have been developed. They are given below.

3.10.1 Vibrometer

Vibrometer is an instrument for measuring displacements of the vibrating body. From Fig. 3.17, it is clear that for $\eta >> 1, \eta^2 / \sqrt{(1-\eta^2)^2 + (2\eta\zeta)^2}$ tends to be unity and the phase angle ϕ tends to 180°. As such for $\eta >> 1$, Eq. (3.66) becomes

$$z = - y_0 \sin \omega t \tag{3.67}$$

Therefore, the relative displacement directly measures the base displacement, which in fact is the objective.

To obtain high values of η, the natural frequency of the spring-mass system has to be of a very low order in comparison to the exciting frequency, preferably one-third to one-half of the lowest frequency to be recorded. In order to obtain a low natural frequency, the system should have a large mass and soft spring. This will entail in a large static displacement. Damping is induced in the system to reduce transient vibrations. The instrument is sensitive and delicate. Vibrometers are rarely used, other than for seismological measurements.

3.10.2 Accelerometers

Basically vibrometers and accelerometers are similar in appearance. The main difference lies in the stiffness of the spring of both instruments. Accelerometers record the acceleration of the body.

If $\eta << 1, 1/\sqrt{(1-\eta^2)^2 + (2\eta\zeta)^2}$ or μ, the magnification factor tends to unity (See Fig. 3.3) and ϕ tends to zero (Fig. 3.4). Therefore, for $\eta << 1$,

Eq. (3.40) becomes

$$z = \frac{m}{k}\omega^2 y_0 \sin\omega t \tag{3.68}$$

Now $$y = y_0 \sin\omega t \tag{3.69}$$

Therefore $$\ddot{y} = -\omega^2 y_0 \sin\omega t \tag{3.70}$$

Equation (3.68) can be written as

$$z = -\frac{1}{p^2}\ddot{y} \tag{3.71}$$

Equation (3.71) reveals that the relative displacement to some scale represents the acceleration of the vibrating body.

For $\eta << 1$, the system will have a very high natural frequency, which means that the spring will be very stiff. As a result of which, the instrument is very rugged and is very much suited for measuring high accelerations, such as in strong earthquakes. But p being high, Eq. (3.71) indicates that the output will be of a low order, and as such it is to be sufficiently amplified. Figure 3.3 indicates that the useful frequency range for an undamped accelerometer is very small. From Fig. 3.18, it is seen that at $\zeta = 0.7$, $\mu = 1$ for $0 \le \eta \le 0.2$; the maximum error being less than 0.01 per cent. For an accelerometer with natural frequency 200 Hz, the useful range of frequency is 0 to 40 Hz.

Introduction of damping in accelerometer is due to another reason as well. The motion to be recorded is impure, in the sense that it contains harmonics higher than the fundamental frequency. As the fundamental frequency is on the lower side, it may be probable that one of these harmonics may be close to the natural frequency of the instrument. The only way to take care of it is to introduce damping in the instrument.

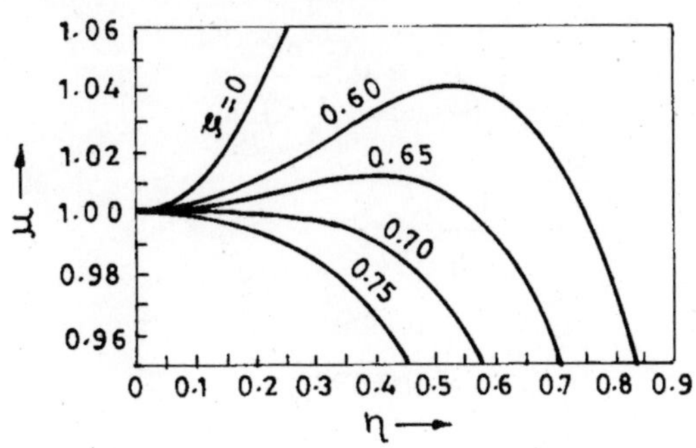

Fig. 3.18 Variation of μ with η

3.10.3 Discussion about the Instruments

Let $\eta = 1, \phi = 90^\circ$. Substituting these values in Eq. (3.65), it becomes

$$z = \frac{m y_0 \omega^2}{k 2\eta\zeta} \cos\omega t$$

or

$$z = \frac{\omega\, y_0}{2p\zeta} \cos\omega t \tag{3.72}$$

Differentiating Eq. (3.69) with respect to *t*, we get

$$\dot{y} = y_0 \omega \cos\omega t \tag{3.73}$$

Combining Eq. (3.72) and Eq. (3.73), we get

$$z = \frac{\dot{y}}{2p\zeta} \tag{3.74}$$

The relative displacement in this case is proportional to the relative velocity. However, as the range of frequencies of this system is very small, this principle is never used for the measurement of velocities of the vibrating body.

The relative displacement which is to be measured, is converted into electrical pulse. The motion to be measured is detected by a transducer. The base of the transducers is firmly attached to the body, whose motion is to be measured. The input acceleration or displacement is usually converted into voltage. Output from the transducer passes on to an amplifier, where the value gets modified. It is then passed on to an oscilloscope for necessary display.

Example 3.11

An undamped vibration pickup having a natural frequency 1 cps is used to measure a harmonic vibration of 3 cps. If the amplitude indicated by the pickup is 0.05 cm, what is the correct amplitude?

Referring to Eq. (3.66)

$$\zeta = 0, \quad \eta = \frac{3}{1} = 3, \quad z_0 = 5.05$$

and if y_0 be the correct amplitude, then

$$z_0 = y_0 \frac{\eta^2}{1-\eta^2}$$

or

$$y_0 = \frac{1-\eta^2}{\eta^2} z_0$$

$$= \frac{1-9}{9} \times 0.05$$

$$= 0.044 \text{ cm.}$$

3.11 RESPONSE OF STRUCTURES DUE TO TRANSIENT VIBRATION

In this section, simple forms of non-harmonic excitation have been considered. The forces considered are transient in nature, such as impact loading, blast loading and earthquakes. This will give an insight into the system parameters and excitation parameters on which response is dependent.

The force can have any type of variation with time. For many of the problems, maximum motions are attained within a short time span after the force is applied. Analysis of these systems with this type of force is often complicated. Presence of damping makes the analysis even more involved. Numerical techniques are better suited for the solution of these problems and they are taken up later. In the present section, some typical cases of transient vibration is discussed.

3.11.1 Response of SDF System to an Ideal Step Input

A SDF system is subjected to an ideal step input of Fig. 3.19. This is equivalent to a force F_0 suddenly applied and remains constant at all times. The equation of motion is given by

$$m\ddot{x} + c\dot{x} + kx = F_0 \quad \text{for} \quad t \geq 0 \tag{3.75}$$

The solution of Eq. (3.75) is

$$x = \frac{F_0}{k} + e^{-nt}\left(A\cos\sqrt{p^2 - n^2}\,t + B\sin\sqrt{p^2 - n^2}\,t\right) \tag{3.76}$$

The system starts at rest. Therefore at $t = 0$, $x = 0$ and $\dot{x} = 0$

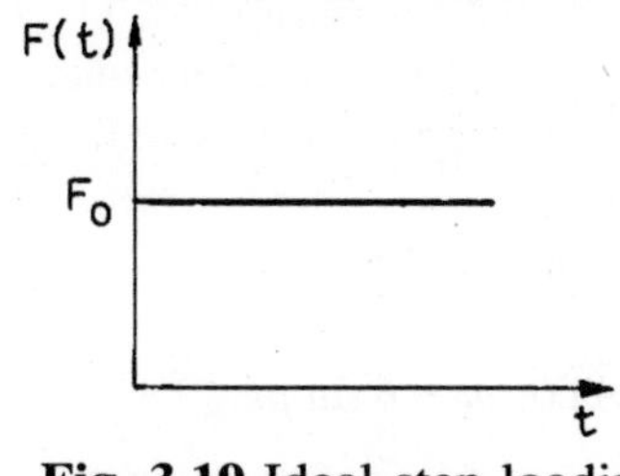

Fig. 3.19 Ideal step loading

Substituting these conditions in Eq. (3.76), from the solution of the equation we get

$$A = -\frac{F_0}{k} \quad \text{and} \quad B = \frac{-nF_0}{k\sqrt{p^2 - n^2}}$$

Substituting the above values of *A* and *B* in Eq. (3.76) we get

$$x = \frac{F_0}{k}\left[1 - e^{-nt}\left\{\cos\sqrt{p^2 - n^2}\,t + \frac{n}{\sqrt{p^2 - n^2}}\sin\sqrt{p^2 - n^2}\,t\right\}\right] \tag{3.77}$$

The dynamic load factor μ can be defined as

$$DLF = \frac{x}{F_0/k} = 1 - e^{-nt}\left[\cos\sqrt{p^2 - n^2}\,t + \frac{n}{\sqrt{p^2 - n^2}}\sin\sqrt{p^2 - n^2}\,t\right] \tag{3.78}$$

Time variation of *DLF*, has been shown in Fig. 3.20. As the load has been suddenly applied, the dynamic displacement shows an initial high value and after the oscillations are damped out, it becomes equal to unity, i.e. equal to the static displacement value.

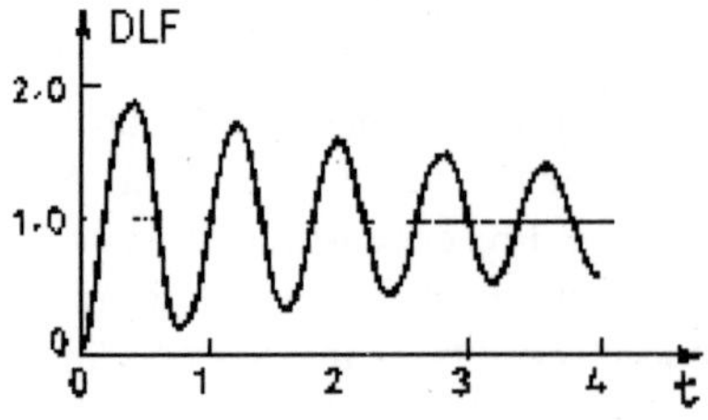

Fig. 3.20 Time variation of DLF

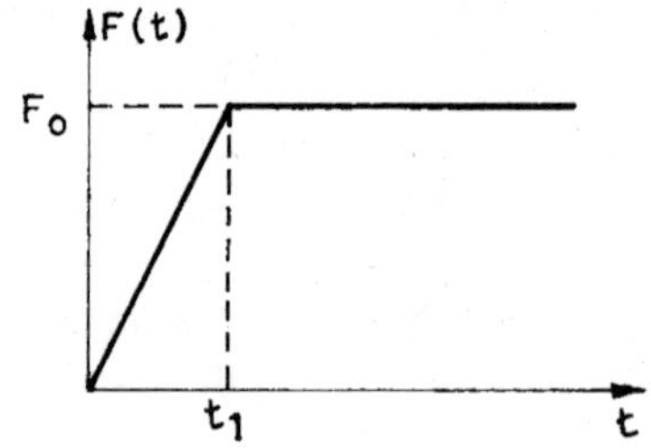

Fig. 3.21 Gradually applied load

3.11.2 Response of SDF System to Gradually Applied Load

The forcing function is shown in Fig. 3.21. For $0 \le t \le t_1$, the force is linearly varying with time and for $t > t_1$, the force remains a constant value. The solution of this problem is to be obtained separately for these two time ranges.

(i) For $t \le t_1$

$$F(t) = \frac{F_0 t}{t_1} \tag{3.79}$$

The equation of motion for the undamped system is

$$m\ddot{x} + kx = \frac{F_0 t}{t_1} \tag{3.80}$$

The solution of the above equation is

$$x = \frac{F_0}{kt_1} t + A \cos pt + B \sin pt \tag{3.81}$$

Substituting the initial conditions at $t = 0$, $x = 0$ and = 0 in Eq. (3.81), yields

$$A = 0 \text{ and } B = -\frac{F_0}{pkt_1}$$

Substituting the values of A and B in Eq. (3.81) yields

$$x(t) = x_t = \frac{F_0}{kpt_1}[pt - \sin pt] \tag{3.82}$$

At $t = t_1$, Eq. (3.82) gives

$$x_{t_1} = \frac{F_0}{kpt_1}[pt_1 - \sin pt_1] \tag{3.83}$$

Differentiation of Eq. (3.82) with respect to t and substitution of $t = t_1$, yields

$$\dot{x}_{t_1} = \frac{F_0}{kpt_1}[p - p\cos pt_1] \tag{3.84}$$

(ii) For $t > t_1$, $F(t) = F_0$

Equation of motion is

$$m\ddot{x} + kx = F_0 \tag{3.85}$$

Solution of Eq. (3.85) is

$$x = \frac{F_0}{k} + A\cos pt' + B\sin pt' \tag{3.86}$$

t' is the parameter measured from the instant $t = t_1$. At $t' = 0$, $x = x_{t_1}$ and $\dot{x} = \dot{x}_{t_1}$.

Substituting initial conditions and determining the values of constants A and B of Eq. (3.86), yields

$$x = \frac{F_0}{k} - \frac{F_0}{kpt_1}(\sin pt_1 - pt_1)\cos pt' + \frac{F_0}{kt_1}[1 - \cos pt_1]\sin pt'$$

or

$$x = \frac{F_0}{k}\left[1 - \frac{2}{pt_1}\sin\frac{pt_1}{2}\cos p\left(t' + \frac{t_1}{2}\right)\right] \tag{3.87}$$

The plot of response against time is shown in Fig. 3.22.

3.12 RESPONSE TO SDF SYSTEMS TO A GENERAL TYPE OF FORCING FUNCTION

A solution of SDF system to a general type of forcing function is sought in this section. To start with, let us consider an undamped system subjected to an arbitrary external force $F(t)$. The external force $F(t)$ can be divided into a large number of rectangular pulses, each having an area $F(\tau)\,d\tau$ as shown in Fig. 3.23. The solution of the problem is sought by the superimposition of the results obtained from individual pulses.

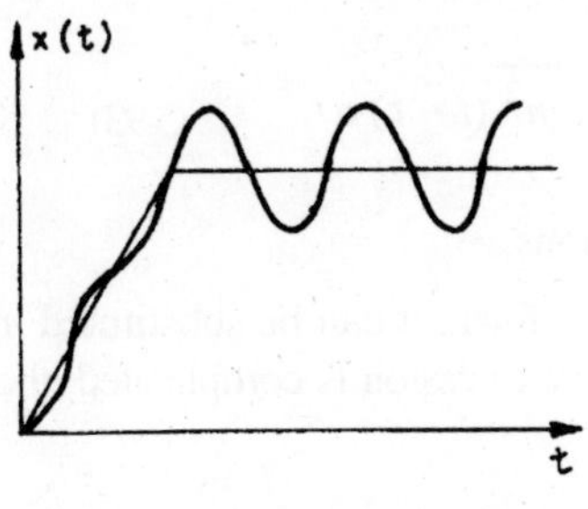

Fig. 3.22 Response variation

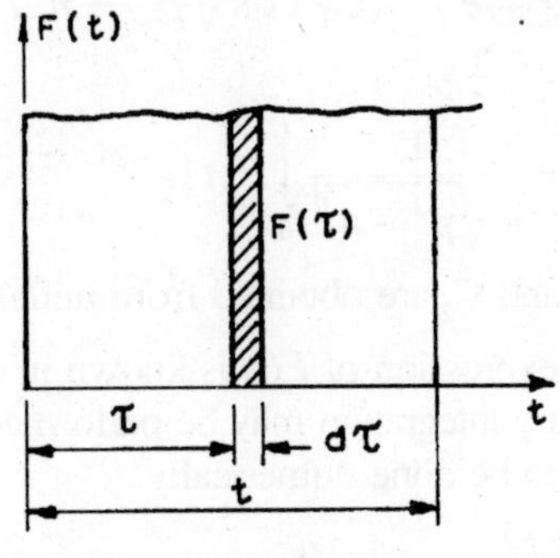

Fig. 3.23 General loading

In effect, it means finding out the solution of one pulse and integrating the results from time equal to zero to the instant when it is required to find the response.

We shall first consider the effect of a simple pulse having a force $F(\tau)$ and duration $d\tau$. From the impulse momentum equation, the velocity imparted to the system is given by

$$\dot{x}_\tau = \frac{F(\tau)\, d\tau}{m} \tag{3.88}$$

Considering the system to be at rest before the application of the pulse, the equation of motion is given by (from Eq. 2.11)

$$x = \frac{F(\tau)\, d\tau}{mp} \sin pt \tag{3.89}$$

Having obtained the displacement under the effect of one impulse, we may consider this impulse situated at a time instant τ. Then the displacement after the application of the impulse is given by Eq. (3.89), except that in place of t we should put $(t-\tau)$, which is the time after impulse application. With this

$$x = \frac{F(\tau)\, d\tau}{mp} \sin p(t-\tau) \tag{3.90}$$

Total displacement then is given by

$$x = \int_0^t \frac{F(\tau)}{mp} \sin p(t-\tau)\, d\tau \tag{3.91}$$

The above integral gives the complete solution, after the constants of integrations are evaluated from the initial conditions.

The expression is known as Duhamel integral or convolution integral.

If the initial conditions are not zero, then the solution becomes

$$x = x_0 \cos pt + \frac{\dot{x}_0}{p} \sin pt + \frac{1}{mp} \int_0^t F(\tau) \sin p(t-\tau)\, d\tau \tag{3.92}$$

If viscous damping is included in the system, then Eq. (3.92) becomes

$$x = e^{-nt} \left(C_1 \cos \sqrt{p^2 - n^2}\, t + C_2 \sin \sqrt{p^2 - n^2}\, t\right)$$

$$+ \frac{1}{m\sqrt{p^2 - n^2}} \int_0^t F(\tau) e^{-n(t-\tau)} \sin \sqrt{p^2 - n^2}\, (t-\tau)\, d\tau \tag{3.93}$$

where C_1 and C_2 are obtained from initial conditions.

If the expression of $F(t)$ is known in analytical form, it can be substituted in Eq. (3.93), and the necessary integration may be performed. If $F(t)$ expression is complicated, then the integration may have to be done numerically.

Example 3.12

A SDF system is subjected to a suddenly applied load with a limited duration t_d as shown in Fig. 3.24. Using Duhamel integral, determine the response of the undamped system. The system starts at rest.

The response is determined for two time ranges.

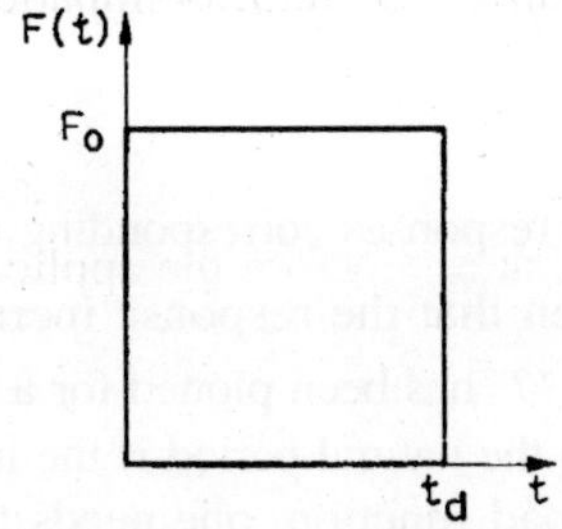

Fig. 3.24 Suddenly applied load

(i) $t < t_d$

$$F(\tau) = F_0$$

From Eq. (3.91)

$$x = \int_0^t \frac{F_0}{mp} \sin p(t - \tau)\, d\tau = \frac{F_0}{k}(1 - \cos pt)$$

At $t = t_d$

$$x_{t_d} = \frac{F_0}{k}(1 - \cos pt_d) \qquad \text{(a)}$$

and

$$\dot{x}_{t_d} = \frac{F_0}{k}(p \sin pt_d)$$

(ii) $t > t_d$, $F(\tau) = 0$

From Eq. (3.92)

$$x = x_{t_d} \cos p(t - t_d) + \frac{\dot{x}_{t_d}}{p} \sin p(t - t_d)$$

or

$$x = \frac{F_0}{k}(1 - \cos pt_d) \cos p(t - t_d) + \frac{F_0}{k} \sin pt_d \sin p(t - t_d)$$

or

$$x = \frac{F_0}{k}[\cos p(t - t_d) - \cos pt]$$

3.13 DYNAMIC LOAD FACTOR AND RESPONSE SPECTRUM

The dynamic load factor (DLF) is defined as the ratio of dynamic displacement at any instant of time to the static displacement. It is non-dimensional and independent of the magnitude of the load. In many structural problems, maximum value of dynamic load factor is of importance.

For the Example 3.12, the dynamic load factor is given by

$$\text{DLF} = 1 - \cos pt = 1 - \cos 2\pi \frac{t}{T} \quad \text{for} \quad t \le t_d$$

$$\text{DLF} = \cos p(t - t_d) - \cos pt$$

$$= \cos 2\pi \left(\frac{t}{T} - \frac{t_d}{T} \right) - \cos 2\pi \frac{t}{T} \quad \text{for} \quad t > t_d \tag{3.94}$$

where T is the time period.

For the rectangular pulse, two responses corresponding to $\frac{t_d}{T} = \frac{1}{10}$ and $\frac{t_d}{T} = \frac{5}{4}$ have been shown in Fig. 3.25. It can be seen that the response increases with the increase of t_d/T. Maximum response for varying t_d/T has been plotted for a rectangular pulse in Fig. 3.26. The ratio of the duration of the pulse to the natural period is the important parameter. So to read the maximum response for a given load function, one needs to know the natural period of the system. However, the chart does not include damping. Maximum dynamic load factor usually corresponds to the first peak response, where the damping existing in structures does not decrease it appreciably. As such, damping in the system will not have any significant effect.

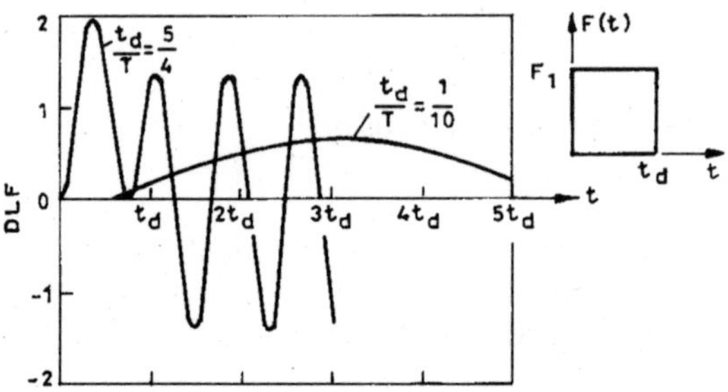

Fig. 3.25 Time variation of DLF for a rectangular pulse

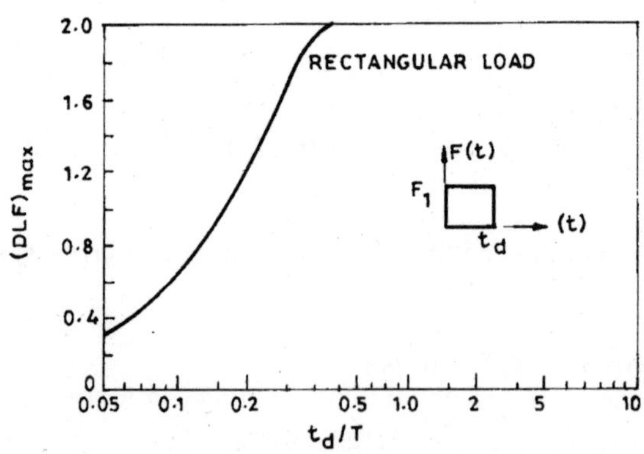

Fig. 3.26 Time variation of (DLF)max for a rectangular pulse

3.14 RESPONSE DUE TO PERIODIC FORCES

Forces acting on different systems are often periodic. Any periodic force can be resolved into a series of harmonic components. We start by introducing real Fourier series and studying the response of the system to periodic forces [Fig. 3.27].

3.14.1 Real Fourier Series

Consider a periodic function $F(t)$ of period T. This function can be separated into harmonic components by means of a Fourier series expansion.

The periodic function $F(t)$ is given by

$$F(t) = a_0 + \sum_{n=1}^{\infty} a_n \cos n\omega t + \sum_{n=1}^{\infty} b_n \sin n\omega t \quad (3.95)$$

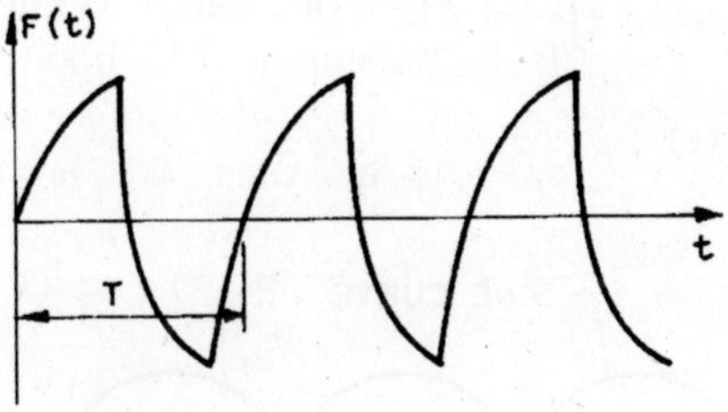

Fig. 3.27 Periodic force

where

$$\omega = \frac{2\pi}{T} = \text{fundamental frequency}$$

$$a_0 = 1/T \int_{\tau}^{T+\tau} F(t)\, dt$$

$$a_n = 2/T \int_{\tau}^{T+\tau} F(t) \cos(n\omega t)\, dt \quad (3.96)$$

$$b_n = 2/T \int_{\tau}^{T+\tau} F(t) \sin(n\omega t)\, dt$$

where τ is any arbitrary time.

Though the Fourier series of $F(t)$ contain an infinite number of terms, in practice the expression is truncated to contain a relatively small number of terms.

Example 3.13

Find the Fourier series representation of the periodic force of Fig. 3.28. Draw the frequency spectrum for the waveform of the figure.

The function $F(t)$ is given by the expression

$$F(t) = A\left|\sin \omega_0 t\right|$$

where

$$\omega_0 = \frac{2\pi}{\tau}$$

Different constants of the Fourier series can be evaluated as follows:

$$a_0 = 1/\tau \int_0^{\tau} F(t)\, dt$$

$$= 1/\tau \int_0^{\tau} A\,|\sin \omega_0 t|\, dt = 2/\tau \int_0^{\tau/2} A \sin \omega_0 t\, dt$$

$$= \frac{2A}{\omega_0 \tau}\left[-\cos \omega_0 t\right]_0^{\tau/2} = \frac{2A}{\pi}$$

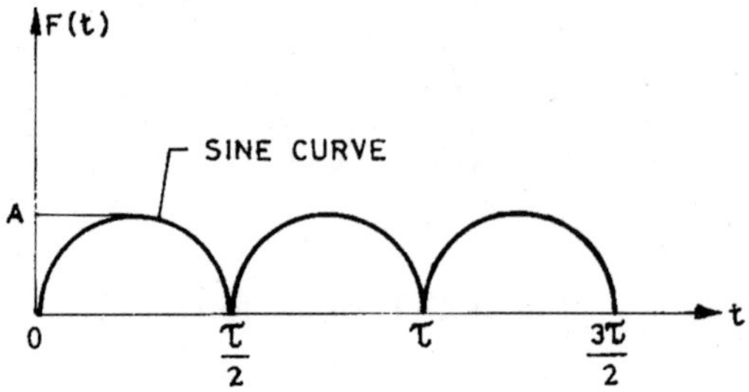

Fig. 3.28 Example 3.13

$$b_n = 2/\tau \int_0^{\tau} F(t) \sin n\omega_0 t\, dt$$

$$= 2/\tau \int_0^{\tau} A\,|\sin \omega_0 t|\,\sin n\omega_0 t\, dt$$

$$= 0$$

$$a_n = \frac{2}{\tau}\int_0^{\tau} A\,|\sin \omega_0 t|\,\cos n\omega_0 t\, dt$$

$$= \frac{4}{\tau}\int_0^{\tau/2} A \sin \omega_0 t \cos n\omega_0 t\, dt$$

$$= \frac{2}{\tau} A \int_0^{\tau/2} [\sin (n+1)\,\omega_0 t - \sin (n-1)\,\omega_0 t]\, dt$$

$$= \frac{-2A}{\tau}\left[\frac{\cos (n+1)\pi}{(n+1)\omega_0} - \frac{\cos (n-1)\pi}{(n-1)\omega_0} - \frac{1}{(n+1)\omega_0} + \frac{1}{(n-1)\omega_0}\right]$$

For odd values of n, $n + 1$ and $n - 1$, the above expressions are zero. For even values of n, the value of a_n is given by

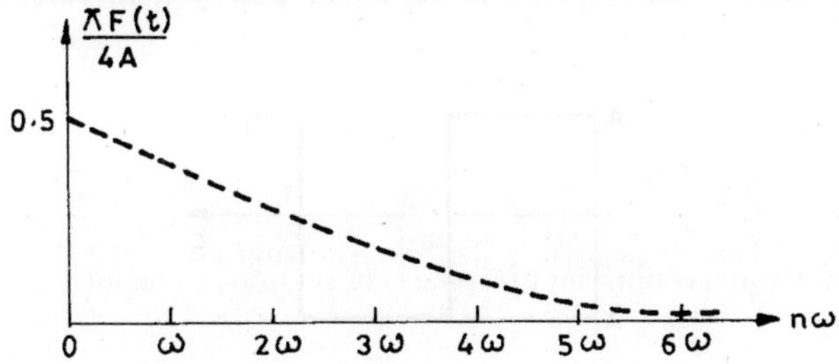

Fig. 3.29 Plot of $\pi F(t)/4A$ vs ω

$$a_n = -\frac{2A}{\tau\omega_0}\left[-\frac{1}{n+1}+\frac{1}{n-1}-\frac{1}{n+1}+\frac{1}{n-1}\right]$$

$$= -\frac{2A}{2\pi}\left(\frac{4}{n^2-1}\right) = \frac{-4A}{\pi}\left(\frac{1}{n^2-1}\right)$$

Therefore

$$F(t) = a_0 + \sum_{n=2,4\cdots}^{\infty} a_n \cos n\omega_0 t$$

or

$$F(t) = \frac{2A}{\pi} + \sum_{n=2,4}^{\infty} -\frac{4A}{\pi}\left(\frac{1}{n^2-1}\right)\cos n\omega_0 t \tag{3.97}$$

where even values of n should be used. It may be noted that is included in Eq. (3.97). The frequency spectrum of the plot given by Eq. (3.97) has been shown in Fig. 3.29.

3.14.2 Response of SDF System to Periodic Forces Represented by Real Fourier Series

The sine and cosine terms alternately present in the Fourier series representing the periodic force can be combined in the following form

$$F(t) = \sum F_n \sin(n\omega_0 t - \alpha_n) \tag{3.98}$$

This force when considered to be applied to the SDF system, each harmonic component is assumed to act separately, and the final response is obtained by the superposition of the harmonic components. Thus, the response of individual component is given by Eq. (3.13), as

$$x_n = \frac{F_n/m}{\sqrt{(p^2-\omega_n^2)^2+(2n'\omega_n)^2}}\sin(\omega_n t - \beta_n) \tag{3.99}$$

where,

$$\omega_n = n\omega_0 \quad \text{and} \quad n' = c/2m$$

The steady-state response is given by

$$x = \sum_n x_n(t) \tag{3.100}$$

Example 3.14

A SDF system having a natural frequency p is subjected to a square wave excitation of Fig.

3.30. Determine the steady-state response of the undamped system. Take $p = 8\omega$

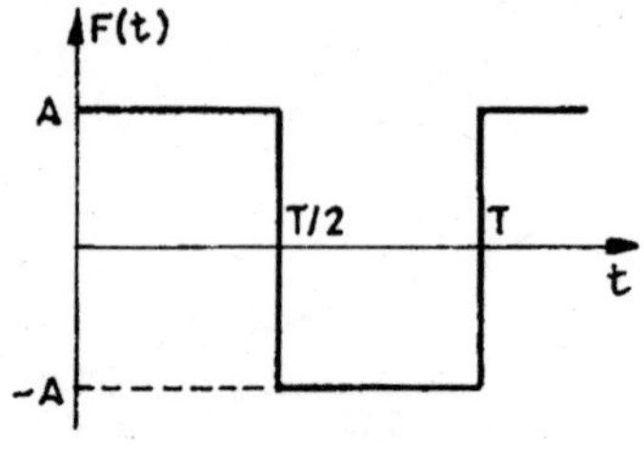

Fig. 3.30 Square wave

The Fourier series representation of the square wave excitation can be shown to be

$$F(t) = \frac{4A}{\pi} \sum_{n=1,3,\ldots}^{\infty} \left(\frac{1}{n}\right) \sin(n\omega t) \tag{3.101}$$

Taking the nth term of Eq. (3.101), the response to SDF system is given by Eq. (3.13), after neglecting damping

$$x_n = \frac{4A}{\pi n \left[1 - \left(\frac{n\omega}{p}\right)^2\right] k} \sin(n\omega t) \tag{3.102}$$

Therefore, the steady-state response given by Eq. (3.100) is

$$x = \frac{4A}{\pi k} \sum_{n=1,3,\ldots}^{\infty} \frac{\sin n\omega t}{n\left[1 - \left(\frac{n\omega}{p}\right)^2\right]} \tag{3.103}$$

For the present problem $p = 8\omega$. Therefore, Eq. (3.103) becomes

$$x = \frac{4A}{\pi k} \sum_{n=1,3,\ldots}^{\infty} \frac{\sin n\omega t}{n\left[1 - \left(\frac{n}{8}\right)^2\right]} \tag{3.104}$$

The excitation spectra and the response spectra have been plotted in Fig. 3.31

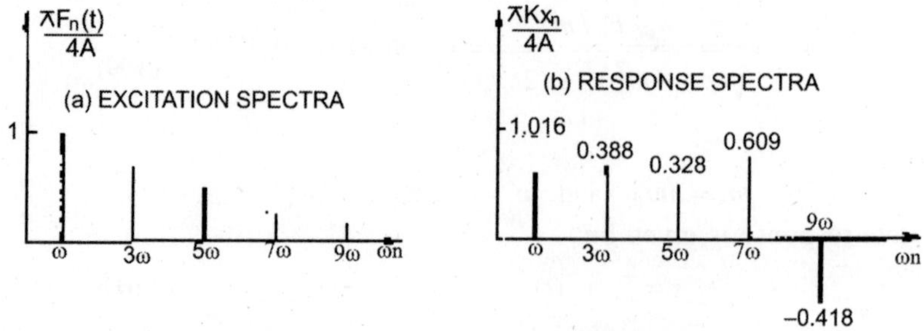

Fig. 3.31 Excitation and response spectra

3.14.3 Complex Fourier Series

A periodic function $F(t)$ can be represented in the following form in terms of complex quantity

$$F(t) = \sum_{-\infty}^{\infty} C_n e^{in\omega t} \tag{3.105}$$

where $\omega = \frac{2\pi}{T}$ is the fundamental frequency.

The value of C_n can be obtained as

$$C_n = \frac{1}{T} \int_{\tau}^{T+\tau} F(t) e^{-in\omega t}\, dt \tag{3.106}$$

Therefore $\qquad C_0 = \frac{1}{T} \int_{\tau}^{T+\tau} F(t)\, dt$ = average value of $F(t)$

In this form of Fourier series, the coefficients are generally complex and for real functions, C_{-n} is the complex conjugate of C_n.

Example 3.15

Determine the complex Fourier series expression for the square wave of Fig. 3.30.

The coefficient of the complex Fourier series expression is determined as follows:

$$C_n = \frac{1}{T} \int_0^{T/2} Ae^{-in\omega t}\, dt + \frac{1}{T} \int_{T/2}^{T} -Ae^{-in\omega t}\, dt$$

$$= \frac{A}{in\omega T} \left[-e^{-in\omega T/2} + 1 + e^{-in\omega T} - e^{-in\omega T/2} \right]$$

$$= \frac{A}{in2\pi} \left[1 - 2e^{-in\pi} + e^{-2\pi in} \right]$$

Further

$$e^{-in\pi} = \cos n\pi - i \sin n\pi = \left. \begin{matrix} 1 \text{ when } n \text{ is even} \\ -1 \text{ when } n \text{ is odd} \end{matrix} \right\} \tag{3.107}$$

Therefore, $\qquad C_n = \frac{iA}{2\pi n} \left[2e^{-in\pi} - 1 - e^{-2in\pi} \right] = -\frac{2iA}{n\pi}$ when n is odd.

Substitution of the value of C_n in Eq. (3.105) yields

$$F(t) = \sum_{n=1,3,\ldots}^{\infty} -\frac{2iA}{n\pi} e^{in\omega t} = -\frac{2iA}{\pi} \sum_{n=1,3,\ldots}^{\infty} \frac{1}{n} e^{in\omega t} \tag{3.108}$$

3.14.4 Response of SDF System to Periodic Forces Represented by Complex Fourier Series

For a SDF system subjected to an exciting force $F_0 e^{i\omega t}$, the equation of motion is

$$m\ddot{x} + c\dot{x} + kx = F_0 e^{i\omega t} \quad (3.109)$$

The steady-state solution of the SDF system is assumed in the complex form as follows

$$x = H(\omega) F_0 e^{i\omega t} \quad (3.110)$$

$H(w)$ is termed as complex-frequency-response function.

Substituting the value of x and the necessary derivatives from Eq. (3.110) into Eq. (3.109), we get

$$H(\omega) = \frac{1}{-m\omega^2 + ic\omega + k}$$

$$= \frac{1/k}{\left[1 - \left(\frac{\omega}{p}\right)^2\right] + i\left(2\zeta \frac{\omega}{p}\right)} \quad (3.111)$$

or

$$H(\omega) = \frac{1/k}{[1-\eta^2] + i\,(2\zeta\eta)} \quad (3.112)$$

Therefore, the response of individual component of $C_n e^{in\omega t}$ is given by

$$x_n = H(n\omega) C_n e^{in\omega t} \quad (3.113)$$

Applying the principle of superposition, the total steady-state response is given by

$$x = \sum_{n=-\infty}^{\infty} H(n\omega)\, C_n\, e^{in\omega t} \quad (3.114)$$

Example 3.16

Determine the steady state response of an undamped SDF system for an excitation in the nature of a square wave of Fig. 3.30, by using complex Fourier series. Given $p = 8\omega$. Also, draw the response spectra.

For a square wave excitation, the complex Fourier series is given by Eq. (3.108) as

$$F(t) = -\frac{2iA}{\pi} \sum_{n=1,3,\cdots}^{\infty} \frac{1}{n} e^{in\omega t}$$

The complex frequency response function is given by

$$H(n\omega) = \frac{1/k}{\left[1 - \left(\frac{n\omega}{p}\right)^2\right] + i\left(2\zeta \frac{n\omega}{p}\right)} \quad (3.115)$$

Now $\zeta = 0$ and $p = 8\omega$.

Therefore $H(n\omega) = \dfrac{1/k}{1 - \left(\dfrac{n}{8}\right)^2}$

Therefore, total response is given by [Eq. (3.114)]

$$x(t) = \sum_{n=1,3,\ldots}^{\infty} -\frac{2iA}{n\pi} \frac{1/k}{\left[1 - \left(\dfrac{n}{8}\right)^2\right]} e^{in\omega t} \tag{3.116}$$

or

$$x(t) = \sum_{n=1,3,\ldots}^{\infty} x_n e^{in\omega t} \tag{3.117}$$

where

$$|x_n| = |H(n\omega)\, F_n(\omega)|$$

$$= \frac{2A}{\pi kn\left[1 - \left(\dfrac{n}{8}\right)^2\right]}$$

The sketch of $|x_n|$ is given in Fig. 3.32.

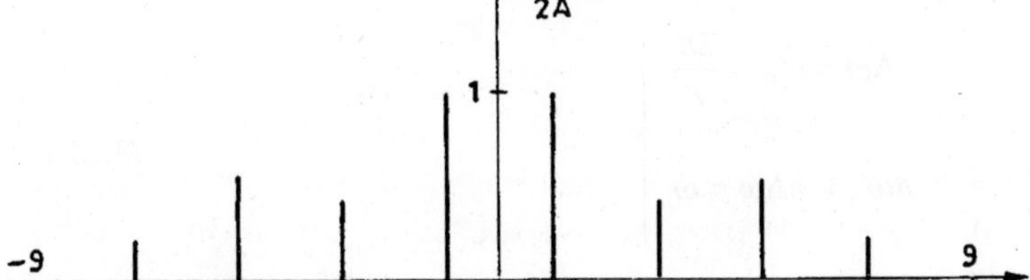

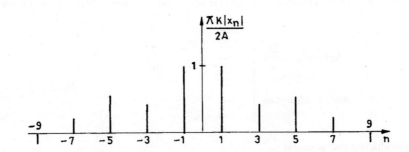

Fig. 3.32 Plot of $(\pi k|x_n|/2A)$ Vs n

3.15 RESPONSE DUE TO NONPERIODIC EXCITATION

In many physical problems such as earthquake and wave loading, the period of forcing function cannot be observed. As such they are non-periodic. For this class of problems, Fourier transform method is used. Further, for certain category of problems, it is more convenient to perform frequency domain analysis than time domain analysis.

The Fourier series concept can be extended to incorporate the nonperiodic loading. An arbitrary non-periodic loading is shown in Fig. 3.33. The coefficient C_n of this loading is given by Eq. (3.106) can be obtained in the interval $0 < t < T$, if it is made periodic as shown by the dotted line. However, this repetitive loading has been assumed to be imaginary and can be eliminated by extending the loading period to infinity. So the Fourier series expression is reformulated to take into account this infinite time range.

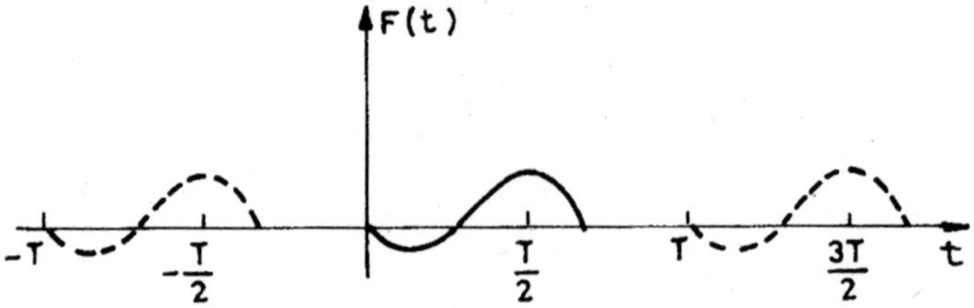

Fig. 3.33 Non-periodic loading

Use the notation defined below:

$$\frac{1}{T} = \frac{\omega_0}{2\pi} \tag{3.118}$$

As T approaches infinity, $\omega_0 = \dfrac{2\pi}{T}$ tends to zero. Hence each component of the series is nearly equal to the adjacent one. In other words, the smallness of the quantity

$$(n+1)\omega_0 - n\omega_0 = \frac{2\pi}{T} \tag{3.119}$$

is limited by the size of T.

The difficulties of this procedure could be removed by introducing minor adjustments, that is

$$\left.\begin{aligned} \Delta\omega = \omega_0 = \frac{2\pi}{T} \\ \text{and} \qquad n\omega_0 = n\Delta\omega = \omega_n \end{aligned}\right\} \tag{3.120}$$

Further, introducing the notation

$$F(n\omega_0) = TC_n$$

or

$$\Delta\omega \cdot F(n\omega_0) = 2\pi C_n \tag{3.121}$$

Equations (3.106) and (3.105) can be written as

$$F(\omega_n) = TC_n = \int_{-T/2}^{T/2} F(t) \exp(-i\omega_n t)\, dt \tag{3.122}$$

and
$$F(t)=\frac{\Delta\omega}{2\pi}\sum_{n=-\infty}^{\infty}F(\omega_n)\exp(i\omega_n t) \quad (3.123)$$

It may be noted that limits of the integral are arbitrary, so long its one complete period is considered.

Now, we consider the loading period to be tending to infinity, which suggests that the frequency increment tends ω to $d\omega$. The discrete frequencies becomes a continuous function of w.

Thus, in the limit, the Fourier series expression of Eq. (3.123) becomes the following Fourier integral

$$F(t)=\frac{1}{2\pi}\int_{-\infty}^{\infty}F(\omega)e^{i\omega t}\,d\omega \quad (3.124)$$

Equation (3.122) can be written as

$$F(\omega)=\int_{-\infty}^{\infty}F(t)e^{-i\omega t}\,dt \quad (3.125)$$

$F(w)$ is called the Fourier transform of $F(t)$ and may be directly evaluated from Eq. (3.125). In general, $F(\omega)$ is complex, but $F(t)$ is real. $F(-\omega)$ is the complex conjugate of $F(\omega)$. Two Fourier integrals given by Eqs. (3.124) and (3.125) are known as Fourier transform pairs.

Example 3.17

For the rectangular forcing function of Fig. 3.34, determine the Fourier transform.

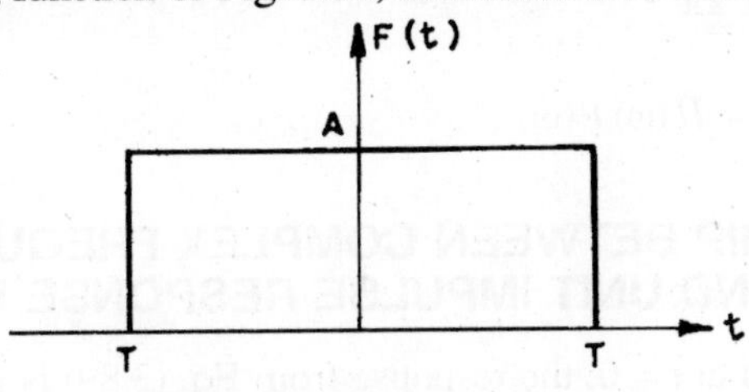

Fig. 3.34 Example 3.17

Fourier transform of the forcing function is given by

$$\begin{aligned}
F(\omega) &= \int_{-\infty}^{\infty}F(t)e^{-i\omega t}\,dt \\
&= \int_{-T}^{T}Ae^{-i\omega t}\,dt \\
&= \frac{-A}{i\omega}\left[e^{-i\omega T}-e^{i\omega T}\right] \\
&= \frac{2A}{\omega}\sin\omega T
\end{aligned}$$

$$= \quad 2AT\left(\frac{\sin\omega T}{\omega T}\right)$$

The response of a SDF system as from Eq. (3.110) is given by

$$x = H(\omega)F(t) \tag{3.126}$$

Substitution of $F(t)$ from Eq. (3.124) into the above equation, yields

$$x(t) = \frac{1}{2\pi}\int_{-\infty}^{\infty} H(\omega)\,F(\omega)\,e^{i\omega t}\,d\omega \tag{3.127}$$

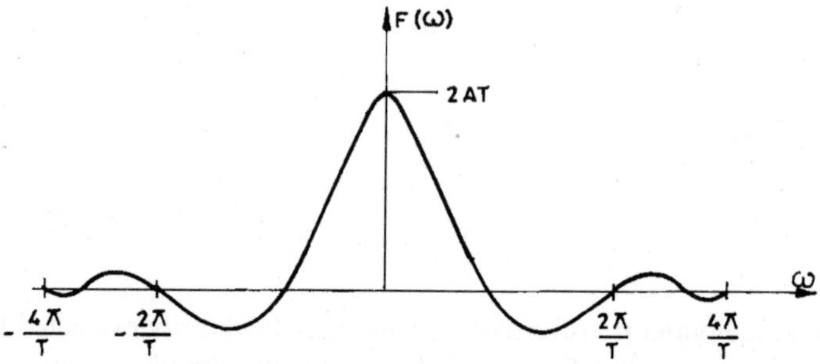

Fig. 3.35 Variation of $F(w)$ with w

or
$$x(t) = \frac{1}{2\pi}\int_{-\infty}^{\infty} X(\omega)\,e^{i\omega t}\,d\omega \tag{3.128}$$

where
$$X(\omega) = H(\omega)\,F(\omega) \tag{3.129}$$

3.16 RELATIONSHIP BETWEEN COMPLEX FREQUENCY RESPONSE FUNCTION AND UNIT IMPULSE RESPONSE FUNCTION

For a unit impulse applied at $t = 0$, the response from Eq. (3.89) is given by

$$x(t) = \frac{\sin pt}{mp} \tag{3.130}$$

In this case
$$\int_{-\infty}^{\infty} F(t)\,dt = 1$$

For a damped system the response due to unit impulse is given by

$$h(t) = \frac{1}{m\sqrt{p^2 - n^2}}\exp(-nt)\sin\sqrt{p^2 - n^2}\,t \tag{3.131}$$

This response function is known as unit impulse response function. This response function describes the response of the system in time domain, whereas the complex frequency response function describes its response in the frequency domain.

By the definition of unit impulse, the Fourier transform of the unit impulse is given by

$$F(\omega) = \int_{-\infty}^{\infty} F(t)\, e^{-i\omega t}\, dt = 1$$

Therefore

$$x(t) = h(t) = \frac{1}{2\pi} \int_{-\infty}^{\infty} H(\omega)\, F(\omega)\, e^{i\omega t}\, d\omega$$

or

$$h(t) = \frac{1}{2\pi} \int_{-\infty}^{\infty} H(\omega)\, e^{i\omega t}\, d\omega \tag{3.132}$$

$h(t)$ is thus the inverse Fourier transform of $H(\omega)$ and therefore

$$H(\omega) = \int_{-\infty}^{\infty} h(t)\, e^{-i\omega t}\, dt \tag{3.133}$$

Thus, $h(t)$ and $H(\omega)$ are Fourier transform pairs.

3.17 SUPPORT MOTION

Determination of the response of the system due to movement of the support rather than the application of external force constitutes an important class of problems. Due to earthquake, the supports of the structure move. The analysis due to these effects can be performed by the displacement approach or the acceleration approach. Let us consider both the approaches one after another.

3.17.1 Displacement Approach

Consider the frame of Fig. 3.36 subjected to support motion x_s. The beam of the frame is considered to be rigid and having total mass m. The mass of the column is negligible. The system is assumed to be undamped. The equation of motion is

$$m\ddot{x} + k\,(x - x_s) = 0 \tag{3.134}$$

or

$$m\ddot{x} + kx = kx_s \tag{3.135}$$

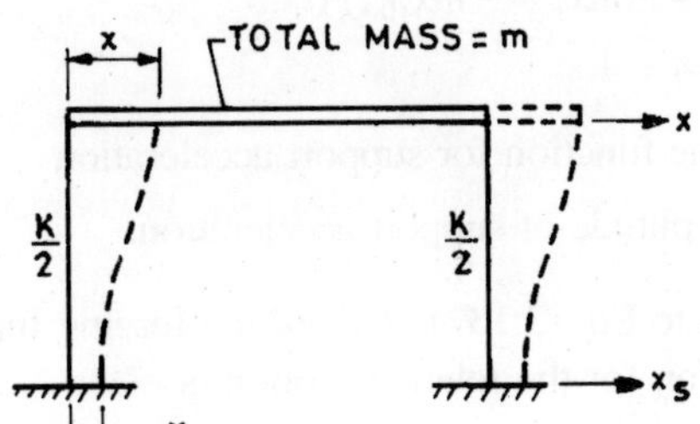

Fig. 3.36 A portal frame

Before proceeding further, Eq. (3.91) is modified. Let $F(\tau) = F_1 f(\tau)$, where F_1 is the amplitude of the force and $f(\tau)$ time function. Noting that $x_{st} = F_1/k$, Eq. (3.91) is rewritten as

$$x = x_{st}\, p \int_0^t f(\tau) \sin p\,(t-\tau)\, d\tau \tag{3.136}$$

Equation (3.135) is of the following form

$$m\ddot{x} + kx = F(t) \tag{3.137}$$

If we assume that the support displacement is sinusoidally varying with time, then $x_s = x_{s0} \sin \omega t$, and solution of Eq. (3.135) can be obtained from Eq.(3.136)

$$x = x_{s0}\, p \int_0^t \sin \omega\tau \sin p\,(t-\tau)\, d\tau \tag{3.138}$$

or

$$x = x_{s0}\, p^2 \left[\frac{\sin \omega t}{p^2 - \omega^2} - \frac{\omega}{p} \frac{\sin pt}{p^2 - \omega^2} \right] \tag{3.139}$$

The displacement of the mass, which is of interest to us is the relative displacement and is given by

$$z = x - x_s = x_{s0}\,\omega^2 \left[\frac{\sin \omega t}{p^2 - \omega^2} - \frac{p}{\omega} \frac{\sin pt}{p^2 - \omega^2} \right] \tag{3.140}$$

3.17.2 Acceleration Approach

The response of the structure due to support motion can also be obtained, if the input to the problem is the support acceleration rather than the displacement. Here the support acceleration is given by

$$\ddot{x}_s = \ddot{x}_{s0}\, f_a(t) \tag{3.141}$$

where $\ddot{x}_{s0}$ is the amplitude of support acceleration and $f_a(t)$ is the time function for support acceleration. Equation (3.134) is rewritten as

$$m\ddot{x} + k(x - x_s) = 0$$

or

$$m(\ddot{x} - \ddot{x}_s) + k(x - x_s) = -m\ddot{x}_s$$

or

$$m\ddot{z} + kz = -m\ddot{x}_s = -m\ddot{x}_{s0} f_a(t) \tag{3.142}$$

where $z = x - y$

$f_a(t)$ = time function for support acceleration

$\ddot{x}_{s0}$ = amplitude of support acceleration.

Equation (3.142) is identical to Eq. (3.137), if F_1 of the forcing function is replaced by $-m\ddot{x}_{s0}$. Therefore, the general solution for the relative motion is

$$z = -\frac{\ddot{x}_{s0}}{p} \int_0^t f_a(\tau) \sin p\,(t-\tau)\, d\tau \tag{3.143}$$

if the system starts at rest. The effect of damping can similarly be incorporated.

3.18 RESPONSE OF SDF SYSTEMS RELATED TO EARTHQUAKES

A SDF system is represented by three constants-the mass m, the stiffness k and damping c. They are combined for the case of viscous damping into two constants p and n. The quantity of interest due to earthquake for this system is its displacement, velocity and acceleration, specially their maximum values. If the earthquake treated on a deterministic basis is the input, then curves can be prepared indicating maximum values of these quantities for the given disturbance, having prescribed values of m, k and c. These diagrams are called response spectrum. Thus, the response spectrum is a graph indicating the maximum responses of all SDF systems, due to the given disturbance. The abscissa of the spectrum is the natural frequency of the system and the ordinate is the maximum response. Thus, in order to determine the response for the particular input, we need to know only the natural frequency of the SDF system, of whose response we are interested. This can be illustrated with an example.

Let us consider the ground acceleration as shown in Fig. 3.37(a). The corresponding ground displacement obtained by integrating twice the ground acceleration has been shown in Fig. 3.37(b). The spectra for maximum relative displacement and maximum absolute acceleration are to be obtained.

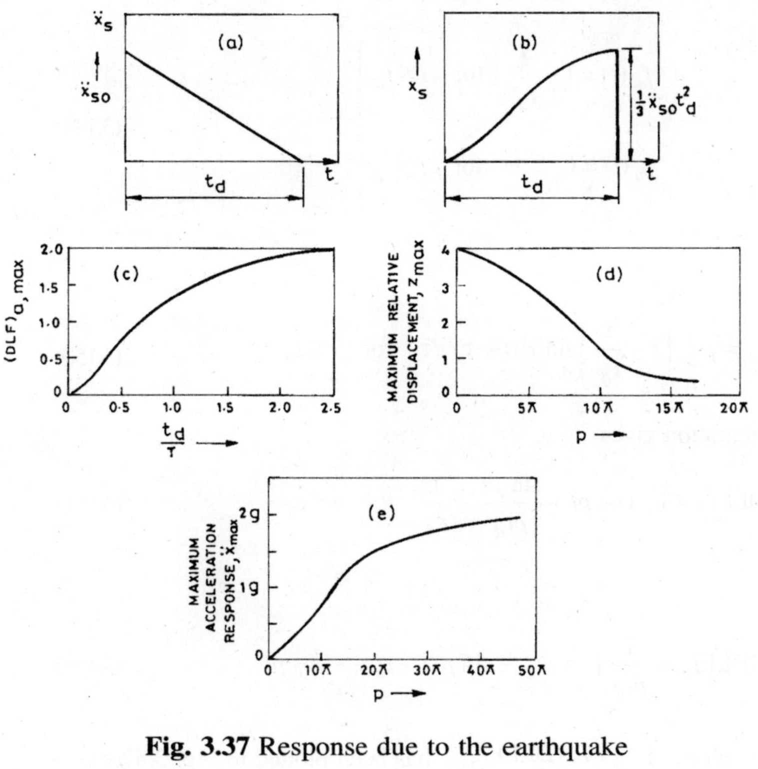

Fig. 3.37 Response due to the earthquake

Equation (3.143) is rewritten as

$$z = -\frac{\ddot{x}_{s0}}{p^2}\, p \int_0^t f_a(\tau) \sin p(t-\tau)\, d\tau$$

or $$z = -\frac{\ddot{x}_{s0}}{p^2}\,(\text{DLF})_a \tag{3.144}$$

Therefore $$\left| z_{\max} \right| = -\frac{\ddot{x}_{s0}}{p^2}\,(\text{DLF})_{a,\,\max} \tag{3.145}$$

It can easily be shown that maximum relative displacement $z_{\max}$, and maximum absolute acceleration $\ddot{x}_{\max}$ are directly related. At the instant of time when the relative displacement is maximum, absolute acceleration is also maximum.

Therefore, from Eq. (3.134)

$$m\ddot{x}_{\max} + kz_{\max} = 0 \tag{3.146}$$

or $$\ddot{x}_{\max} = -p^2 z_{\max} \tag{3.147}$$

Substituting the value of $z_{\max}$ from Eq. (3.145) into Eq. (3.147) we get

$$\left| \ddot{x}_{\max} \right| = \ddot{x}_{s0}\,(\text{DLF})_{a,\,\max} \tag{3.148}$$

Based on the ground acceleration of Fig. 3.37(a)

$$\left.\begin{aligned} f_a(t) &= 1 - \frac{t}{t_d} \quad \text{for} \quad t \le t_d \\ f_a(t) &= 0 \qquad\quad \text{for} \quad t > t_d \end{aligned}\right\} \tag{3.149}$$

Therefore

$$(\text{DLF})_a = p\int_0^t \left(1 - \frac{\tau}{t_d}\right)\sin p(t-\tau)\,d\tau \quad \text{for} \quad t \le t_d \tag{3.150}$$

which on integration gives

$$(\text{DLF})_a = 1 - \cos pt + \frac{\sin pt}{pt_d} - \frac{t}{t_d} \quad \text{for} \quad t \le t_d \tag{3.151}$$

For $t > t_d$

$$(\text{DLF})_a = \frac{1}{pt_d}\,[\sin pt_d - \sin p(t - t_d)] - \cos pt \tag{3.152}$$

For different values of t_d/T, $(\text{DLF})_{a,\,\max}$ has been plotted in Fig. 3.37(c).

If $\ddot{x}_{s0} = g$ and $t_d = 0.1$, then the variation of the maximum relative displacement with the natural frequency has been plotted in Fig. 3.37(d) and the curve of maximum absolute acceleration versus natural frequency has been plotted in Fig. 3.37(e).

Figures 3.37(d) and (e) give some interesting reading. From Eq. (3.145), it can be seen that as p approaches zero, $(\text{DLF})_{a,\,\max}$ also approaches zero. Therefore, $z_{\max}$ becomes indeterminate. However, it can be shown that as p approaches, zero, $z_{\max}$ approaches maximum ground motion.

The results of the response spectra at two extreme values of natural frequencies can be physically explained. A very high value of natural frequency indicates a very stiff structure. The structure with a high value of p will move more or less as a rigid body, and the mass will move more or less wholly with the ground, so that the maximum relative displacement is nearly zero and the absolute maximum acceleration of the mass approaches the maximum ground motion. Similarly, a very low value of natural frequency indicates a very flexible structure. Therefore, due to ground motion, the mass remains more or less stationary, while the ground moves below it. The relative displacement, of the mass, therefore, becomes nearly equal to the ground displacement, and the maximum acceleration of the mass becomes negligible.

3.19 TECHNIQUES FOR ANALYSING EARTHQUAKE RESPONSE

Unfortunately, the records obtained from the earthquake are not as simple as the one indicated in Fig. 3.37(a). The accelerograph of the earthquake having amplitudes and periods varying with time indicate a random movement. The acceleration associated with ground movement has all three components. However, the horizontal component of it is most important, as it has the most damaging effect for the majority of the structures.

So far very few records of earthquakes have been obtained. Only during the last four decades, earthquake records are taken on a more systematic basis in some countries. Still, their number is not sufficient to be used for statistical purposes. So, prediction of future earthquake at a place becomes problematic. However, past earthquake recordings have indicated that they all follow a particular pattern. As such, the same trend can be assumed for future earthquakes.

Due to the limited extent of data available on earthquake motions, a simplified approach is adopted in design. A standard earthquake may be assumed for this purpose. If one is interested in determining the response on the basis of this earthquake, one has to numerically integrate the record to compute DLF.

However, numerical integration of the earthquake records is time consuming. For design purposes, some standard spectra are adopted. In IS 1893, the acceleration and velocity spectra based on studies of four strongest earthquakes have been indicated. To take into account the seismicity of various zones, the ordinates of the spectra are to be multiplied by a coefficient specified in the table of the code. The magnitude of this coefficient depends on a number of factors. They are the magnitudes, duration, form of the expected earthquake, distance of the place from the expected earthquake, soil conditions and the deformation characteristics of the structure. The average velocity and acceleration curves of ground motions adopted by IS 1893 have been given in Fig. 3.38.

If S_v is the value of the maximum relative velocity as given by the velocity spectra, then the following relationship is valid

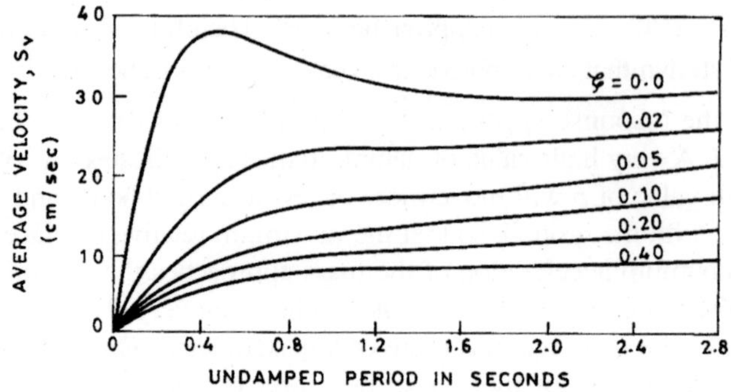

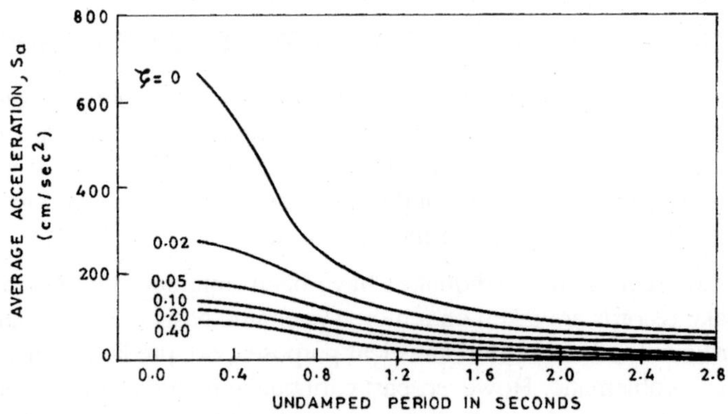

Sl. No.	Location and Date	Magnitude	Multiplying Factor
1	El Centro, California May 18, 1940	6.7	2.7
2	El Centro, California Dec. 30, 1934	6.5	1.9
3	Olympia, Washington April 13, 1949	7.1	1.9
4	Taft, California July 21, 1952	7.7	1.6

Fig. 3.38 Velocity and acceleration spectra

$$\left.\begin{aligned} \dot{z}_{\max} &\cong S_v \\ z_{\max} &\cong \frac{S_v}{p} \\ \ddot{x}_{\max} &\cong pS_v \end{aligned}\right\} \tag{3.153}$$

The above relations are only approximate, but are fairly accurate for design purposes. Eqs. (3.153) are derived as follows for the undamped cases. The relative velocity is given by

$$\dot{z} = \dot{x} - \dot{y} = \frac{d}{dt}(x - y) = \frac{d}{dt}z \tag{3.154}$$

Substituting the value of z from Eq. (3.153) into Eq. (3.154), one obtains

$$\dot{z} = \frac{d}{dt}\left[-\frac{\ddot{x}_{s0}}{p}\int_0^t f_a(\tau)\sin p(t-\tau)\,d\tau\right] \tag{3.155}$$

$$= -\frac{1}{p}\int_0^t \ddot{x}_{s0}\,[pf_a(\tau)\cos p(t-\tau)\,d\tau \tag{3.156}$$

As S_v indicates the maximum value of relative velocity

$$S_v = \dot{z}_{\max} = \left\|\left[\int_0^t \ddot{x}_{s0} f_a(\tau)\cos p(t-\tau)\,d\tau\right]\right\|_{\max} \tag{3.157}$$

The maximum value of the integrand of Eq. (3.157) remains unchanged if sine term is replaced by the cosine term.

Therefore,

$$z_{\max} = \frac{S_v}{p} \tag{3.158}$$

Further, noting the relations given by Eqs. (3.147) and (3.158), the following relation can be derived

$$S_a = \ddot{x}_{\max} = pS_v \tag{3.159}$$

Example 3.18

Determine the base shear in all columns and base moment in fixed end column of the structure shown in Fig. 3.39 for a horizontal earthquake motion for 2 percent viscous damping. The columns may be assumed to be massless and the horizontal girder may be assumed to be rigid. Assume uniform ground motion at the foot of the columns. Use average acceleration curves of Fig. 3.38. The structure is assumed to be located in a zone for which the coefficient of seismicity is 0.4.

The stiffness of two extreme columns will be same and they are given by

$$k_1 = k_4 = \frac{12EI}{L^3} = \frac{12 \times E \times I_0}{3^3} = \frac{12EI_0}{27}$$

The stiffness of the two inside columns will be same and they are given by

$$k_2 = k_3 = \frac{3EI}{L^3} = \frac{3EI_0}{64}$$

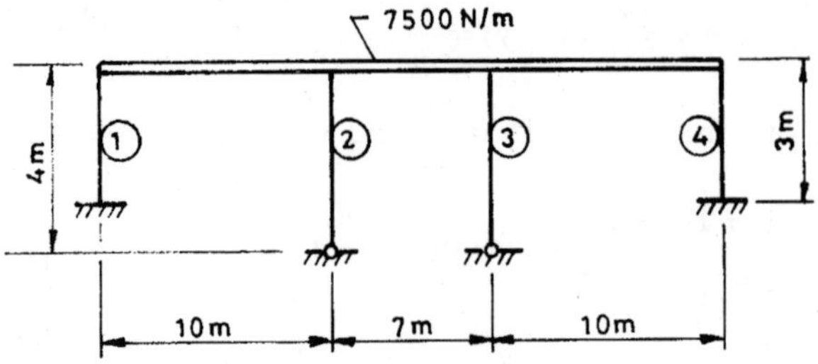

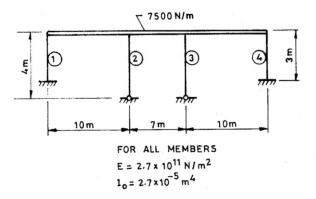

Fig. 3.39 Example 3.18

Therefore, total stiffness of the system is given by

$$k = \left[\frac{2\times 12}{27} + \frac{2\times 3}{64}\right]\times 2\times 10^{11}\times 2.7\times 10^{-5} \text{ N/m}$$

$$= 5.30625\times 10^{6} \quad \text{N/m}$$

Total mass is given by

$$m = \frac{7500\times 27}{9.8} = 20{,}663.27 \text{ kg}$$

Therefore, the undamped angular frequency of the system is given by

$$p = \sqrt{\frac{5.30625\times 10^{6}}{20663.27}} = 16.02 \text{ rad/s}$$

Time period is given by

$$T = \frac{2\pi}{p} = \frac{2\pi}{16.02} = 0.392 \text{ s}$$

Based on $T = 0.392$, $\zeta = 0.02$ and coefficient of seismicity given as 0.4, the ground acceleration is given by [from Fig. 3.38]

$$\ddot{x}_{max} = S_a = 270\times 0.4 = 108 \text{ cm/s}^2$$

From Eq. (3.153), the maximum relative displacement is given by

$$z_{max} = \frac{S_a}{p^2} = \frac{108}{16.02\times 16.02} = 0.421 \text{ cm}$$

Base shear in column (1) = Base shear in column (4)

$$= \frac{12EI_0}{L^3}\times z_{max}$$

$$= \frac{12 \times 2 \times 10^{11} \times 2.7 \times 10^{-5}}{3^3} \times 0.421 \times 10^{-2}$$

$$= 10{,}104 \text{ N}$$

Base shear in column (2) = Base shear in column (3)

$$= \frac{3EI_0}{L^3} \times z_{\max}$$

$$= \frac{3 \times 2 \times 10^{11} \times 2.7 \times 10^{-5}}{4^3} \times 0.421 \times 10^{-2}$$

$$= 1065.66 \text{ N}$$

Base moment for column (1) or (4) is

$$M = \frac{6 \times 2 \times 10^{11} \times 2.7 \times 10^{-5}}{9} \times 0.421 \times 10^{-2}$$

$$= 15156 \text{ N}$$

References

3.1 R.A. Anderson, Fundamentals of Vibration, McMillan Co., New York, 1967.
3.2 J.M. Biggs, Introduction to Structural Dynamics, McGraw-Hill Book Inc., New York, 1964.
3.3 R.W. Clough and J. Penzien, Dynamics of Structures, 2nd Edition, McGraw-Hill Book Inc., New York, 1993.
3.4 Roy R. Craig, Jr., Structural Dynamics, John Wiley & Sons, 1981.
3.5 J.P. Den Hartog, Mechanical Vibrations, Fourth Edition, McGraw-Hill Book Inc., 1956.
3.6 D.G. Fertis, Dynamics and Vibrations of Structures, 2nd edition, John Wiley & Sons, New York, 2000.
3.7 L.S. Jacobsen and R.S. Ayer, Engineering Vibrations, McGraw-Hill Book Inc, 1958.
3.8 L. Meirovitch, Elements of Vibration Analysis, McGraw-Hill Book Inc., New York, 1975.
3.9 W.T. Thomson, Theory of Vibration with Applications, Prentice Hall, New Jersey, 1972.
3.10 M. Paz, Structural Dynamics Theory and Computations, Second Edition, CBS Publication and Distribution, New Delhi, 1983.

EXERCISE 3

3.1 A machine weighing 600 N is supported by springs of stiffness and has a damper attached, whose coefficient is c = 0.010 N-s/mm. A harmonic force of amplitude 20 N is applied by the machine. Determine the maximum and resonant amplitudes of steady-state vibration.

3.2 With the help of a harmonic-loading machine, the mass, damping and stiffness of a SDF system can be determined. But for so doing, it has to be operated twice at two different frequencies. If for a single bay portal, the machine was operated at frequencies $\omega_1 = 20$ rad/s and $\omega_2 = 30$ rad/s with a force amplitude of 2.5 kN, the response amplitude and phase relationships measured for the two cases were

$$x_1 = 3.34 \times 10^{-3} \text{ mm} \quad \text{and} \quad \phi_1 = 14°$$

$$x_2 = 6.94 \times 10^{-3} \text{ mm} \quad \text{and} \quad \phi_2 = 52°$$

Determine the mass, stiffness and damping of the SDF system.

3.3 For the basic SDF system of Fig. 2.4, m = 500 kg, k = 5000 N and 100 s/mm. If the system is started from rest, determine the static displacement after 4 cycles.

3.4 A viscously damped mass of 2 kg undergoes a resonant amplitude of 1.4 m with a period of 0.25 s when subjected to a harmonically excited force of 25 kg. Determine the damping coefficient.

3.5 If for problem 1, the harmonic force has a frequency of 5 Hz, what will be the percentage increase in the amplitude of forced vibration when damping is removed?

3.6 A SDF system having viscous damping has a spring of stiffness 500 N/m. When the weight is displaced and released, the period of vibration is 2 s and ratio of successive amplitudes is 4 to 1. Determine the amplitude of the motion and the phase-angle when a force is applied to the system.

3.7 The mass of a SDF system is subjected to a force $C\omega^2 \sin \omega t$. Allowing viscous damping, obtain an expression for steady-state forced vibration. Also, determine the value of ω, at which the amplitude is maximum, in terms of damping ratio.

3.8 A harmonic motion is applied to the system shown in the figure. Derive the equations of motion.

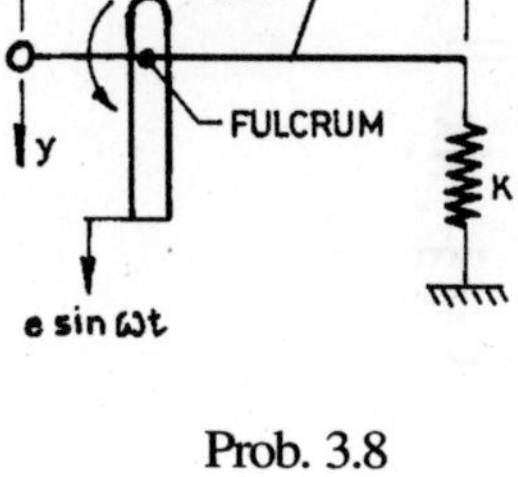

Prob. 3.8

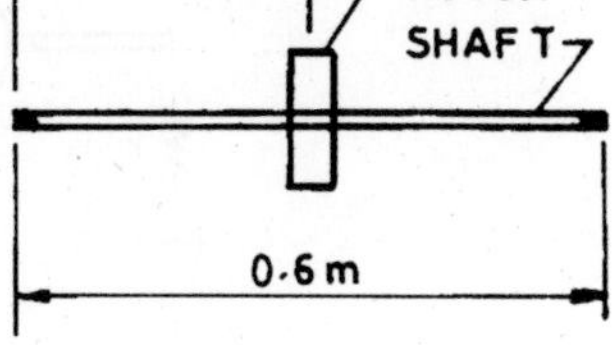

Prob. 3.9

3.9 A 15 kg rotor has an unbalance of 3×10^{-3} kgm. Neglecting the mass of the shaft, find the amplitude of vibration and force at each bearing for a 20 mm diameter shaft.

3.10 Assuming a shaft to be simply supported between bearings, determine the lowest critical speed if a solid disk of mass 5 kg is keyed to the centre of the shaft. The diameter of the shaft 3 mm and the distances between bearings 200 mm.

3.11 An aircraft radio weighing 110 N is to be isolated from engine vibrations in the frequency range between 20 Hz and 40 Hz. What statical deflection must the isolators have for 85% isolation?

3.12 Turbines run through dangerous speed at resonance each time they started or stopped when they are made to operate at critical speed. Assuming the critical speed ω_n to be reached with amplitude p_0, determine the equation for the amplitude building with time for an undamped system.

3.13 A water rotating eccentric weight exciter is used to produce forced vibration of a spring supported mass, as shown in the Fig. When the speed of rotation is varied, a resonant amplitude of 0.50 cm is recorded. When the speed of rotation is increased considerably beyond the resonant frequency, the amplitude tended to be constant at 0.75 cm. Determine the damping factor of the system.

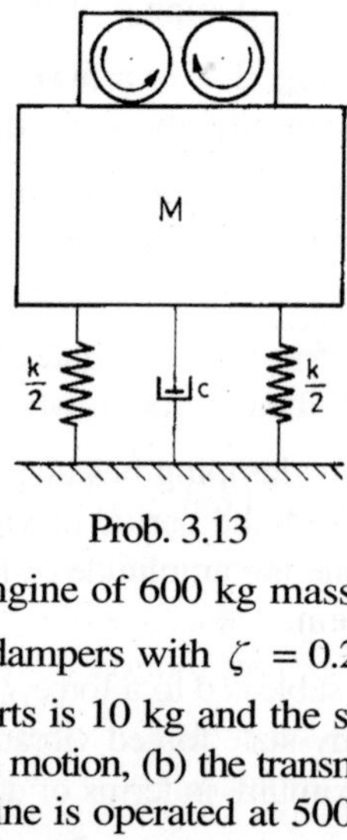

Prob. 3.13

3.14 A vertical single cylinder engine of 600 kg mass is mounted on springs having total stiffness $k = 150$ kN/m and dampers with $\zeta = 0.2$. The engine is balanced. The mass of the equivalent rotating parts is 10 kg and the stroke is 150 mm. Determine, (a) the dynamic amplitude of vertical motion, (b) the transmissibility and (c) the force transmitted to the foundation, if the engine is operated at 500 rpm.

3.15 A 50 kg motor is supported on four equal springs, each having stiffness of 4.5 kN/m. The motor has a radius of gyration equal to 200 mm. The motor operates at a speed of 1200 rpm. Find the transmissibility for vertical and torsional vibrations.

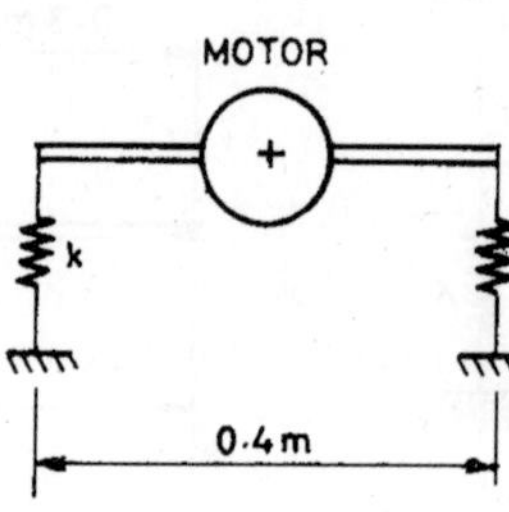

Prob. 3.15

3.16 Sources of vibration in the range of 200-300 Hz are to be isolated from an equipment of mass 15 kg and having insignificant damping. What will be the specification of the spring, if 50 per cent isolation is to be achieved?

3.17 A machine of mass 40 kg is supported by springs having static deflection 1.5 mm. A portion of the machine rotating at the rate of 600 rpm has a static unbalance of 280 kN. Assuming the damping ratio of 0.2, determine the transmissibility of forces.

3.18 Determine equivalent viscous damping for Coulomb damping.

3.19 Determine the expression for power developed by a force $P = P_0 \sin(\omega t + \phi)$ acting on a displacement $x = A \sin \omega t$.

3.20 An accelerometer with damping $\zeta = 0.70$ is used to measure a periodic motion with the displacement given by the equation

$$x = A \sin p_1 t + B \sin p_2 t$$

3.21 A vehicle idealised as shown, is moving over a surface having a sine curve having an amplitude Y_s, and wave length l. Determine the transmissibility of the motion.

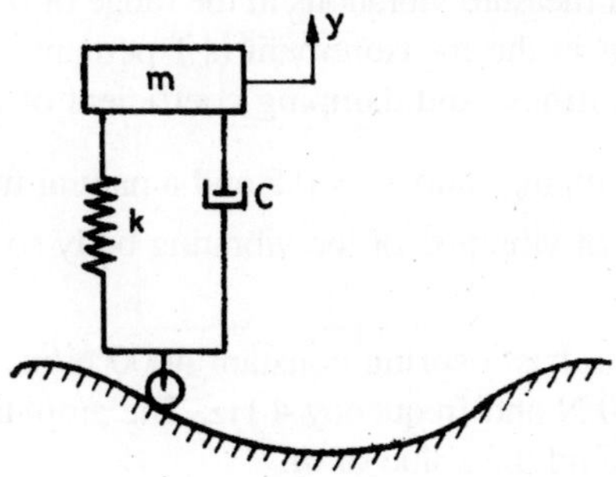

Prob. 3.21

3.22 An instrument is attached to a rubber mounting having a static deflection of 3.6 mm. The supporting structure vibrates at a frequency of 30 Hz. Determine the percentage reduction in the transmitted support motion.

3.23 A machine of mass 400 kg is supported on springs with a statical deflection of 5 mm. If the machine has a rotating unbalance of 0.2 kgm, determine the force transmitted to the floor at 1000 rpm.

3.24 Write the equation for the force vibration of a single – DOF system in terms of the loss factor at resonance.

3.25 A viscously damped spring-mass system is initially at rest. If the system is actuated by a harmonic force of frequency $\omega = p = \sqrt{\frac{k}{m}}$, determine the equation of its motion.

3.26 Two gears at two ends are being driven by an electric motor placed centrally on the shaft. If the torque experienced by the rotor is $T = T_0 \cos \omega t$, determine the resulting torsional vibration. The ends may be assumed to be changed for the purpose of the analysis. The following values are assumed:

$T_0 = 200$ Nm, $\omega = 100$ Hz,

J_0 = mass moment of whole of the rotor = 0.08 kgm2,
diameter of the shaft = 0.032 m,
length of the shaft on the either side of the motor = 0.45 m,
G = shear modulus of the shaft = 0.8×10^{11} N/m^2

3.27 A vibrating system with a moving base has a mass of 400 kg, a spring of stiffness 50,000 N/m and a damper whose damping constant is not known. It has been found that when the amplitude of support vibration is 3 mm at the natural frequency of the supported system, the amplitude of the vibration of the mass is 12 mm. Determine (1) the damping constant of the system, (2) the dynamic force amplitude at the base and (3) the amplitude of the mass relative to the base.

3.28 The radial clearance between the stator and the rotor of a variable-speed electric motor is 1 mm. The mass of the rotor is 30 kg and its unbalance (*me*) is 7.5 kg-mm. The rotor is centrally placed between the two bearings, that is, the bearings are 1 m on either side of the rotor. The operating speed of the machine varies between 600 and 6000 rpm. Assuming negligible damping, determine the shaft diameter so that the rotor is always clear of the stator at operating condition.

3.29 An accelerometer can measure vibrations in the range of 0 to 50 Hz. The maximum error that is likely to occur in the measurement is 1 percent. The suspended mass m is 0.1 kg. Find the spring stiffness and damping coefficient of the accelerometer.

3.30 A velometer has a damping ratio ζ = 0.5 and a natural frequency of 25 Hz. Determine the lowest frequency of vibration of the vibrating body so that the error of measurement of velocity is 1%.

3.31 A spring-mass system has a spring constant 4000 N/m. It is subjected to a harmonic force of amplitude 50 N and frequency 4 Hz. The amplitude of the forced vibration of the mass is 20 mm. Find the value of m.

3.32 Determine the ratio of the maximum amplitude to resonant amplitude for a viscously damped system when it is subjected to steady state vibration in harmonic force.

3.33 A machine weighing 30,000 N exerts a harmonic force of 3000 N amplitude, at 10 Hz at its supports. After installing the machine on a spring-type isolator, the force exerted on the support is reduced to 300 N. Determine the spring stiffness k.

3.34 Determine the response of a viscously damped system to the step function shown in the figure.

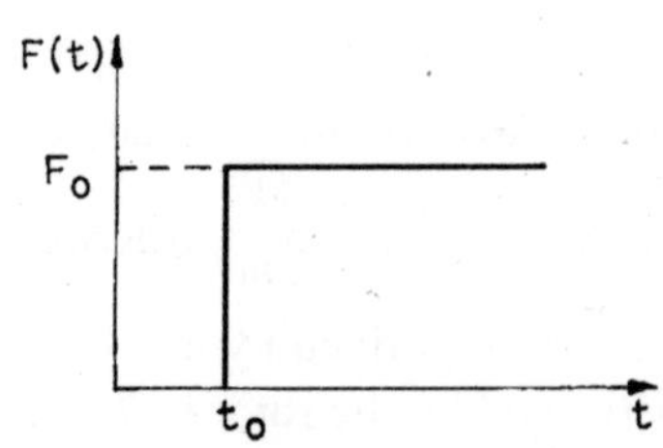

Prob. 3.34

3.35 Determine the response of the SDF system having m, k and c as the mass, stiffness and damping due to a periodic force shown in Fig. 3.28.

3.36 Determine the response of an undamped SDF system, subjected to a periodic loading shown in the figure.

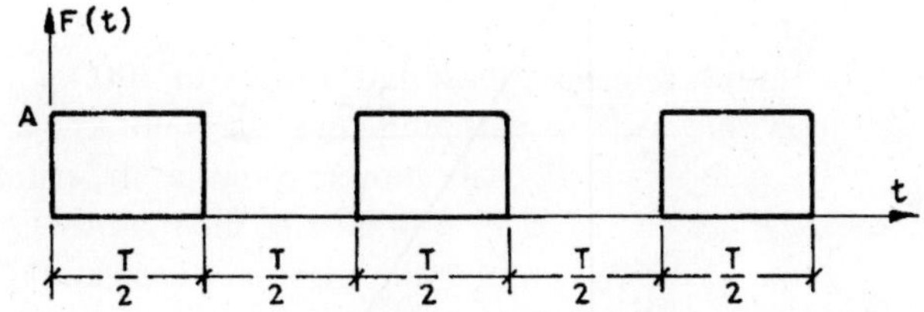

Prob. 3.36

3.37 If the displacement for the support point for a spring-mass system is given in Fig. 3.30 [change $F(t)$ for $x_s(t)$], determine the relative motion and absolute motion of the mass.

3.38 Obtain the Fourier series representation of the function given in the figure.

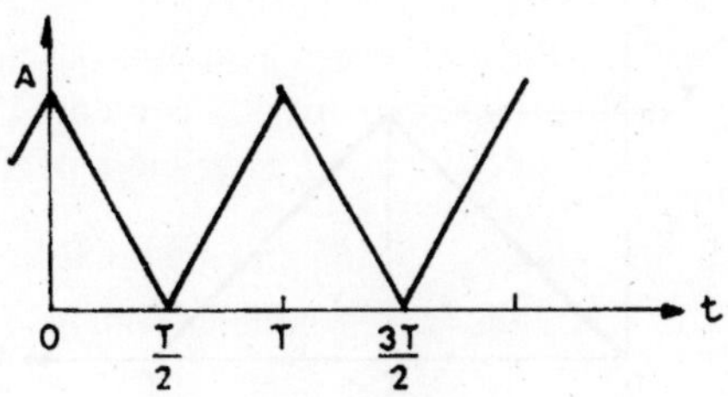

Prob. 3.38

3.39 Determine the response of a spring-mass system having viscous damping due to the linearly increasing force as shown in the figure.

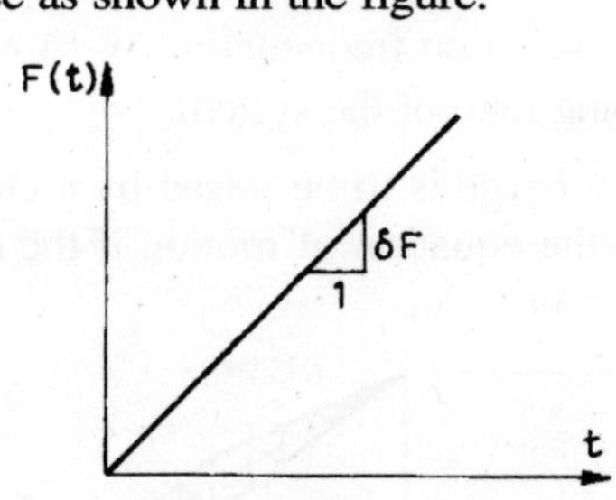

Prob. 3.39

3.40 An undamped SDF system is subjected to a base acceleration as shown in the figure. Determine the response spectrum of z.

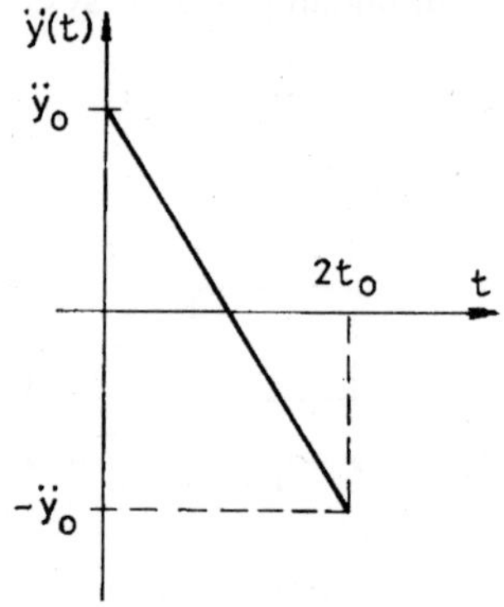

Prob. 3.40

3.41 If the frame of Fig. 3.36 is subjected to a support motion as shown in the figure, determine the relative motion.

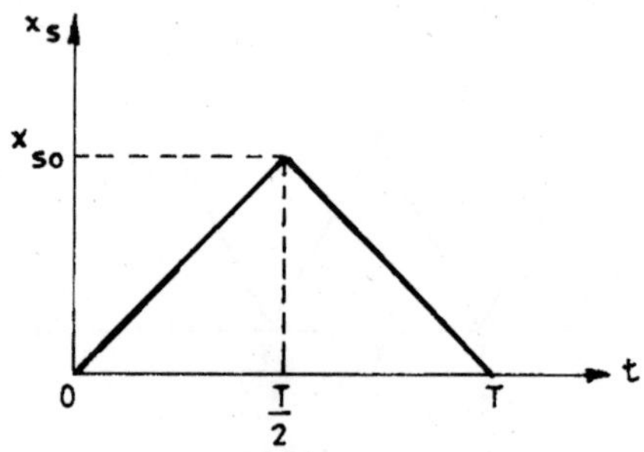

Prob. 3.41

3.42 The displacement amplitude $A\ (x) = A\cos(\omega t - \phi)$ of a SDF system due to harmonic force is known for two excitation frequencies. At , A = 125 mm; at $\omega = 5\omega_0$, A = 0.5 mm. Estimate the damping ratio of the system.

3.43 A weight W placed on a barge is to be raised by a crane. The stiffness of the crane boom is K_c. Determine the equation of motion if the extended point on the boom is given a displacement $x = Vt$.

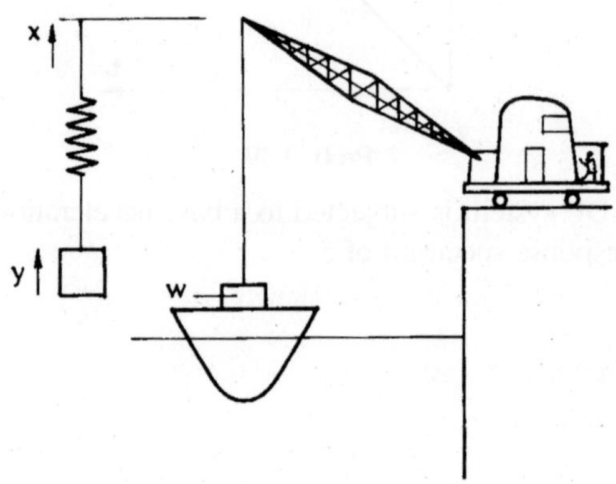

Prob. 3.43

3.44 A machine component vibrates with a frequency of 60 Hz. If an accelerometer having negligible damping and having weight and spring constant of 6 N and 150 N/mm indicates a total travel of 8 mm, what are the maximum amplitude and maximum acceleration of the machine component?

3.45 The natural frequency of a vibrometer is 30 rad/s. When it is used to record the vibration of a machine component, the reading obtained was 2 mm, at operating frequency of 150 rad/s, $z = 0.02$. What is the maximum displacement?

3.46 Based on response spectra of Fig. 3.34, determine the maximum displacement and base shear for a SDF system having $m = 50$ kg, $k = 800$ N/mm and $c = 20$ Ns/mm.

3.47 A damped system having m, k and c experiences a base acceleration as shown in the figure. The initial conditions are all zero. Derive expressions for maximum relative displacement and maximum absolute acceleration.

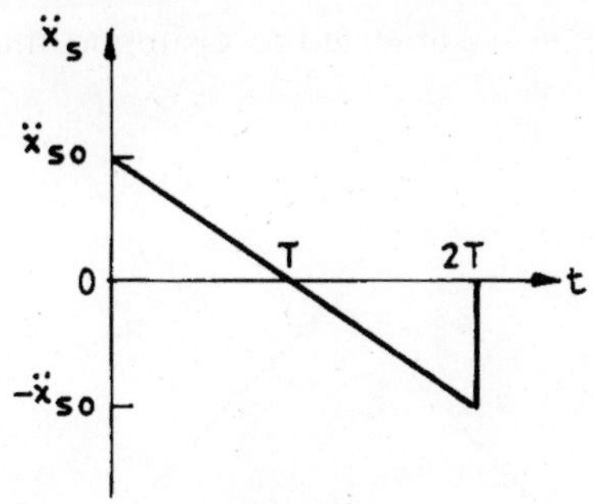

Prob. 3.47

3.48 An automobile is moving over a multispan elevated roadway supported at every 30 metres. Due to long-term creep the roadway has become a wavy surface with 150 mm deflection of the middle of each span. The profile can be assumed as sinusoidal with an amplitude of 75 mm and length of 30 m. When fully loaded, the weight of the automobile is 18 kN. The effective stiffness of the automobile suspension is 14000 N/m and the damping ratio of the system is 0.4. Determine (a) the amplitude of the vertical motion, (b) when the vehicle is travelling at 60 km/hr and the speed of the vehicle that would produce a resonant condition.

3.49 For a viscously damped SDF system, show that the energy dissipated per cycle is given by

$$\Delta W = \frac{\pi F_0^2}{k} = \frac{2\eta\zeta}{(1-\eta^2)^2 + (2\eta\zeta)^2}$$

3.50 A machine component is vibrating with the motion

$$y = 0.5 \sin 30\pi t + 1.0 \sin \pi t$$

Determine the vibration record that would be obtained with an accelerometer having $\zeta = 0.15$ and $\omega = 1000$ Hz.

Chapter 4

Numerical Methods in Structural Dynamics Applied to SDF Systems

4.1 INTRODUCTION

The analysis of SDF systems requires the solution of the differential equation. For simple cases of forcing functions, the solution of the equation can be obtained in closed bound form. However, the dynamic loading obtained from the records for many practical cases is not easily mathematically amenable and as such numerical procedures are to be applied for obtaining the solution. In time domain analysis, the numerical techniques can be classified as follows:

(1) Direct integration techniques
(2) Numerical evaluation of the Duhamel's integral.

In direct integration techniques, the basic differential equation of a SDF system is integrated using a step-by-step procedure [4.1 - 4.6]. The equations in this case are not transformed to any other form. In the case of Duhamel's integral, numerical integration is performed over the integrand. The numerical procedures are much more general than the rigorous approach, which are of limited usefulness due to severe restrictions imposed on it by practical problems.

4.2 DIRECT INTEGRATION TECHNIQUES

In direct integration techniques, the time interval is divided into a discrete number of steps Δt apart. The solution is attempted at each time step, i.e. the differential equation is solved progressively from given or known initial conditions. Shorter the time step, the more accurate is the solution. A few of the direct integration techniques are given below.

4.2.1. Finite Difference Method

The central difference scheme is adopted here. Time span under consideration is divided into n equal time intervals $\Delta t(= T/n)$. The solution is obtained in a progressive manner from time $t = 0$ to $t = T$, at the time instants $0, \Delta t, 2\Delta t, 3\Delta t, \ldots, i\Delta t, \ldots, T$.

Let $x_i, \dot{x}_i$ and $\ddot{x}_i$ be the displacement, velocity and acceleration respectively, at the time instant $i\Delta t$.

These terms, expressed in finite difference form are

$$\ddot{x}_i = \frac{1}{\Delta t^2}\left(x_{i+1} - 2x_i + x_{i-1}\right) \tag{4.1}$$

Also $$\ddot{x}_i = \frac{1}{2\Delta t}\left(\dot{x}_{i+1} - \dot{x}_{i-1}\right) \tag{4.2}$$

and $$\dot{x}_i = \frac{1}{2\Delta t}\left(x_{i+1} - x_{i-1}\right) \tag{4.3}$$

From Eqs. (4. 1) and (4.3), it can be shown that

$$x_{i+1} = x_i + \dot{x}_i \Delta t + \frac{1}{2}\ddot{x}_i \Delta t^2 \tag{4.4}$$

$$\dot{x}_{i+1} = 2\ddot{x}_i \Delta t + \dot{x}_{i-1}. \tag{4.5}$$

Equations (4.1) to (4.5), together with the differential equation of motion are used for obtaining the necessary numerical solution. Initiation of the computation is dependent on the initial acceleration. Two conditions are encountered.

4.2.1.1 When the Initial Acceleration is not zero

We start with subscript 1 to indicate initial conditions. It is assumed that the initial acceleration remains constant during the first interval. Based on this assumption, the following relations can be written:

and $$\left.\begin{aligned} \ddot{x}_2 &= \ddot{x}_1 \\ \dot{x}_2 &= \ddot{x}_1 \Delta t \\ x_2 &= \frac{1}{2}\ddot{x}_1 \Delta t^2 \end{aligned}\right\} \tag{4.6}$$

The last two equations of Eq. (4.6) are derived from the integration of the acceleration terms. For example

$$x_2 = \int_0^{\Delta t}\int_0^{t} \ddot{x}_1 \, dt\, dt = \frac{1}{2}\ddot{x}_1 \Delta t^2 \tag{4.7}$$

The acceleration $\ddot{x}_1$ to be used in the above equation is obtained from the equation of motion. However, this is dependent on the initial velocity and the initial displacement. The equation of motion can be rewritten in the following form:

$$\ddot{x}_i = \frac{1}{m}F_i(t) - \frac{c}{m}\dot{x}_i - \frac{k}{m}x_i \tag{4.8}$$

The quantities x_2 and $\dot{x}_2$ are determined from Eq. (4.6), which are substituted into Eq. (4.8) to obtain $\ddot{x}_2$. Based on the known values of $x_2, \dot{x}_2$ and $\ddot{x}_2$, the displacement x_3 is obtained for the next time step from Eq. (4.4) and $\dot{x}_3$ from Eq. (4.5). The process is repeated for other time steps. In order to obtain satisfactory solution, the time interval $\Delta t < T/10$.

4.2.1.2 When the Initial Acceleration is zero

When the initial acceleration is zero, Eq. (4.6) cannot be used, as the iterative process cannot be started. As such, recourse has to be taken to a new assumption. In this case, the acceleration in the first interval is assumed to vary linearly.

Referring to Fig. 4.1

$$x_2 = \int_0^{\Delta t}\int_0^{t}\left(\ddot{x}_1 + \frac{\ddot{x}_2 - \ddot{x}_1}{\Delta t}\, t\right) dt\, dt$$

or

$$x_2 = \frac{1}{6}\left(2\ddot{x}_1 + \ddot{x}_2\right)\left(\Delta t^2\right) \tag{4.9}$$

Similarly

$$\dot{x}_2 = \frac{1}{2}\left(\ddot{x}_1 + \ddot{x}_2\right)\Delta t \tag{4.10}$$

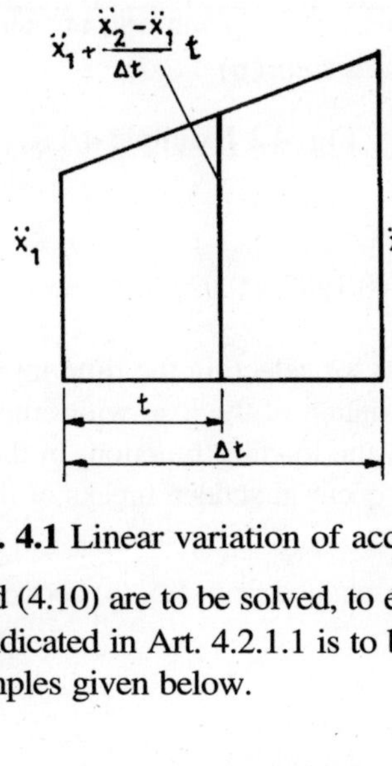

Fig. 4.1 Linear variation of acceleration

With $\ddot{x}_1 = 0$, Eqs. (4.8) (4.9) and (4.10) are to be solved, to evaluate the necessary quantities to start with. Then the procedure indicated in Art. 4.2.1.1 is to be followed. Both the above cases have been dealt with in the examples given below.

Example 4.1

Solve numerically the differential equation for a SDF system starting from rest with the force applied on it, as shown in Fig. 4.2. The equation of motion for the SDF system of Fig. 4.2(a) is given by

$$2\ddot{x} + 2000x = F(t)$$

or

$$\ddot{x} = \frac{1}{2}F(t) - 1000x \tag{a}$$

Time period of the system is

$$T = 2\pi\sqrt{\frac{m}{k}} = 2\pi\sqrt{\frac{2}{2000}} = 0.1987\,s$$

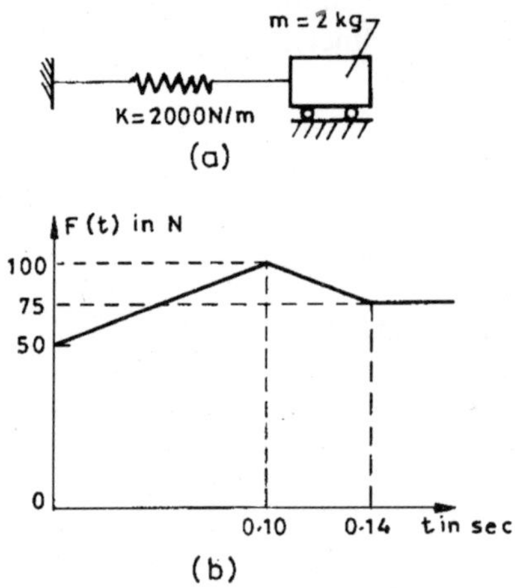

Fig. 4.2 Example 4.1

Time interval Δt is chosen as

$$\Delta t = \frac{1}{10} \times 0.1987 \cong 0.02\, s$$

Another criterion is very important for selecting the time interval, i.e. time interval should be such that it should represent the variation of the load with time and should be capable of taking into account the sudden change in the loading function. In the present problem with the time interval of 0.02 s, the time stations occur at sudden breaks of the load function at $t = 0.10$ s and 0.14 s.

The initial acceleration can be obtained from Eq. (a) above, given in Fig. 4.2(b). Noting that $x_1 = \dot{x}_1 = 0$

$$\ddot{x}_1 = \frac{1}{2} \times 50 - 1000 \times 0 = 25$$

From Eq. (4.4), $x_2 = \frac{1}{2} \times 25 \times (0.02)^2 = 0.005$ and

from Eq. (4.6), $\dot{x}_2 = 25 \times 0.02 = 0.5.$

Therefore at $t = 0.02$

$$\ddot{x}_2 = \frac{1}{2} \times 60 - 1000 \times 0.005 = 25$$

At $t = 2 \times 0.02$, x_3 and $\dot{x}_3$ be obtained from Eqs. (4.4) and (4.5)

$$x_3 = 0.005 + 0.5 \times 0.02 + \frac{1}{2} \times 25 \times (0.02)^2 = 0.02$$

$$\dot{x}_3 = 2 \times 25 \times 0.02 + 0 = 1$$

Similarly, from Eq. (a), at $t = 2 \times 0.02$

$$\ddot{x}_3 = \frac{1}{2} \times 70 - 1000 \times 0.02 = 15$$

At $t = 3 \times 0.02$

$$x_4 = 0.02 + 1 \times 0.02 + \frac{1}{2} \times 15(0.02)^2 = 0.043$$

$$\dot{x}_4 = 2 \times 15 \times 0.02 + 0.5 = 1.1$$

and the process is repeated and results up to $t = 0.14$ s is presented in the Table 4.1

Table 4.1 Response Evaluation by Finite Difference Method

t_s	$F(t)$	x	$\dot{x}$	$\ddot{x}$
0	50	0	0	25
0.02	60	0.005	0.5	25
0.04	70	0.02	1.0	15
0.06	80	0.043	1.1	– 3.00
0.08	90	0.0644	0.8	–19.40
0.10	100	0.0765	0.024	–26.52
0.12	87.5	0.0717	–1.037	–27.95
0.14	75	0.0454	–2.155	–7.87

Example 4.2

The equation of motion for a spring-mass system is $0.4\ddot{x} + 64x = F(t)$, where the excitation force is shown in Fig. 4.3. The system starts at rest. Using finite difference method, determine the response.

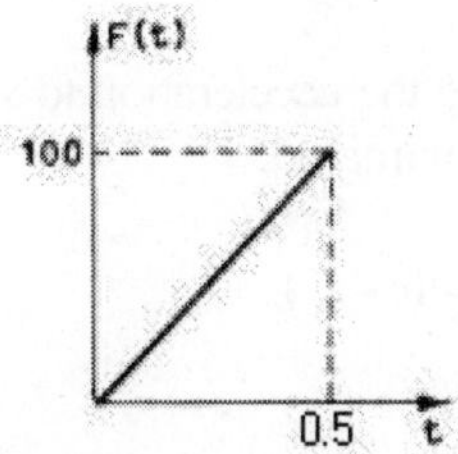

Fig. 4.3 Example 4.2

Time period of the system is given by

$$T = 2\pi\sqrt{\frac{m}{k}} = 2\pi\sqrt{\frac{0.4}{64}} = 0.497 \cong 0.5\,\text{s}$$

Time interval selected is $\Delta t = \frac{1}{10}T = 0.05\,\text{s}$

Given, $x_1 = \dot{x}_1 = 0$ and the equation of motion is rewritten as

$$\ddot{x} = 2.5F(t) - 160x \qquad \text{(a)}$$

From Eq. (4.9), at $t = 0.05$

$$x_2 = \frac{1}{6}\ddot{x}_2(0.05)^2 = 0.000417\ddot{x}_2 \quad \text{(b)}$$

and from Eq.(a), $t = 0.05$

$$\ddot{x}_2 = 2.5 \times 10 - 160x_2$$

or

$$\ddot{x}_2 = 25 - 160x_2 \quad \text{(c)}$$

Solving Eqs. (b) and (c), we get

$$x_2 = 0.009765$$

$$\ddot{x}_2 = 23.4375$$

From Eq. (4.10), $\dot{x} = \frac{1}{2}\times 23.4375 \times 0.05 = 0.5859$

At $t = 2 \times 0.05$ s, from Eq. (4.4)

$$x_3 = 0.009765 + 0.5859 \times 0.05 + \frac{1}{2}(26.79)(0.05)^2 = 0.06835$$

$$= 0.0780$$

From Eq. (4.5), $\dot{x}_3 = 2 \times 23.4375 \times 0.05 + 0 = 2.3438$

From Eq. (a), $\ddot{x}_3 = 2.5 \times 20 - 160 \times 0.06835 = 39.06$

and the process is repeated as given in Example 4.1.

4.2.2 Linear Acceleration Method

One of the basic techniques for computing numerically the response of a SDF system is the linear acceleration method. As the name indicates, the acceleration here is assumed to vary linearly. Some of the numerical methods developed later derives its base from the linear acceleration method.

Referring to Fig. 4.4 and assuming the acceleration to vary linearly, the acceleration between time stations i and $i + 1$ can be approximated.

$$\ddot{x} = \ddot{x}_i + \frac{\ddot{x}_{i+1} - \ddot{x}_i}{\Delta t}(t - t_i) \quad (4.11)$$

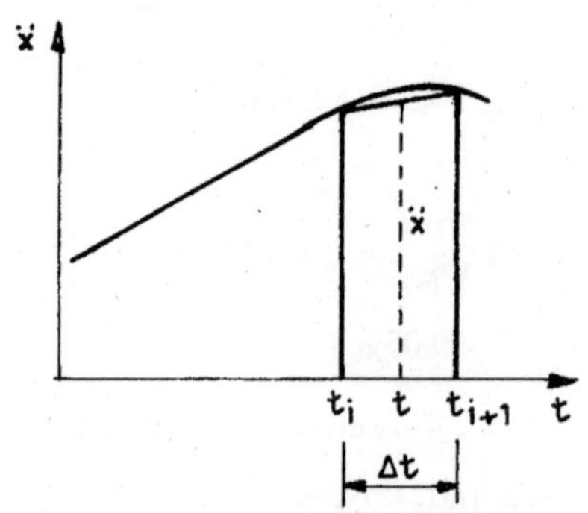

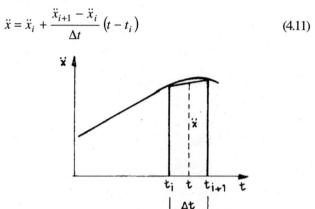

Fig. 4.4 Variation of acceleration

The velocity at any instant of time is

$$\dot{x}_{i+1} = \dot{x}_i + \int_0^{\Delta t} \ddot{x}\, dt$$

or $$\dot{x}_{i+1} = \dot{x}_i + \int_0^{\Delta t} \left[\ddot{x}_i + \frac{\ddot{x}_{i+1} - \ddot{x}_i}{\Delta t}(t - t_i) \right] dt$$

$$\dot{x}_{i+1} = \dot{x}_i + \frac{1}{2}\left[\ddot{x}_{i+1} + \ddot{x}_i\right]\Delta t \tag{4.12}$$

Similarly, the displacement at time t_{i+1} can be derived [an expression similar to Eq. (4.9)]

$$x_{i+1} = x_i + \dot{x}_i \Delta t + \frac{\Delta t^2}{6}\left[\ddot{x}_{i+1} + 2\ddot{x}_i\right] \tag{4.13}$$

Eqs. (4.12) and (4.13) indicate that the values of x_{i+1} and $\dot{x}_{i+1}$ are dependent on which $\ddot{x}_{i+1}$, in turn is related to x_{i+1} and $\dot{x}_{i+1}$ through the equation of motion. Thus, respective values at any time instant is to be evaluated in an iterative manner.

Various schemes may be adopted for the iterative procedure. Some are given below:

(1) By assuming $\ddot{x}_{i+1} = \ddot{x}_i$

(2) A linear variation of acceleration is assumed in two successive time intervals.

The assumed value of $\ddot{x}_{i+1}$ is substituted into Eqs. (4.12) and (4.13), along with the already computed values of x_i, $\dot{x}_i$ and $\ddot{x}_i$, to obtain $\dot{x}_{i+1}$ and x_{i+1}. These values of x_{i+1} and $\dot{x}_{i+1}$ are, substituted into the equation of motion, to obtain $\ddot{x}_{i+1}$. If this value of $\ddot{x}_{i+1}$ does not tally with the previous value of $\ddot{x}_{i+1}$, then the process is to be repeated with the new value of $\ddot{x}_{i+1}$. This is to be continued till two successive values of $\ddot{x}_{i+1}$ agree within reasonable limits. This method is demonstrated in the following example.

Example 4.3

Solve the problem of Example 4.1 (given in Fig. 4.2) by the linear acceleration method.

The equation of motion for this problem is

$$\ddot{x} = \frac{1}{2}F(t) - 1000x \tag{a}$$

$$x_i = \dot{x}_i = 0 \qquad \text{and} \qquad \Delta t = 0.02\, s$$

Assume $\ddot{x}_1 = \ddot{x}_2 = 25$.

Then substituting these values into Eqs. (4.12) and (4.13), we get

$$\dot{x}_2 = 0 + \frac{1}{2} \times 0.02 \times 2 \times 25 = 0.5$$

and $$x_2 = 0 + 0 + \frac{1}{6} \times (0.02)^2\, [2 \times 25 + 25] = 0.005$$

Substituting the above values x_2 and $\dot{x}_2$ into Eq. (a), gives

$$\ddot{x}_2 = 30 - 1000 \times 0.005 = 25$$

The value $\ddot{x}_2$ is identical to the assumed value. Therefore, calculation may be started for the next time step, i.e $t = 2 \times 0.02$s.

Let $\ddot{x}_2 = \ddot{x}_3 = 25.$

Then from Eqs. (4.12) and (4.13)

$$\dot{x}_3 = \dot{x}_2 + \frac{1}{2}(\ddot{x}_2 + \ddot{x}_3)\Delta t$$

$$= 0.5 + \frac{1}{2}(2 \times 25) \times 0.02 = 1.0$$

and $$x_3 = x_2 + \dot{x}_2 \Delta t + \frac{1}{6}(2\ddot{x}_2 + \ddot{x}_3)(\Delta t)^2$$

$$= 0.005 + 0.5 \times 0.02 + \frac{1}{6}(2 \times 25 + 25)(0.02)^2$$

$$= 0.02$$

From Eq. (a), $\ddot{x}_3 = 35 - 1000 \times 0.02 = 15$

This value of $\ddot{x}_3$ does not tally with the assumed value of 25.

Now, assuming $\ddot{x}_3 = 15$, the above steps are repeated.

$$\dot{x}_3 = 0.5 + \frac{1}{2}(15 + 25) \times 0.02 = 0.9$$

$$x_3 = 0.005 + 0.5 \times 0.02 + \frac{1}{6}(2 \times 25 + 15) \times 0.02^2$$

$$= 0.0193$$

$$\ddot{x}_3 = 35 - 1000 \times 0.0193 = 15.7$$

Two successive values of $\ddot{x}_3$ differ somewhat.

Assume $\ddot{x}_3 = 15.7.$

$$\dot{x}_3 = 0.5 + \frac{1}{2}(25 + 15.7) \times 0.02 = 0.907$$

$$x_3 = 0.005 + 0.5 \times 0.02 + \frac{1}{6}(2 \times 25 + 15.7) \times 0.02^2 = 0.0194$$

$$\ddot{x}_3 = 35 - 1000 \times 0.0194 = 15.6$$

Iteration may be continued if higher accuracy is desired, and once the satisfaction of the value is obtained, one proceeds on to the next time step.

4.2.3 Runge—Kutta Method

There are various methods available for the numerical solution of the differential equation. One of the popular methods in this direction is the Runge-Kutta method.

The equation of motion of a SDF system is a second-order differential equation. In Runge-Kutta method, a second-order differential equation is to be reduced to two first-order equations. The equation of motion of the SDF system is given by

$$\ddot{x} = \frac{1}{m}\left[F(t) - kx - c\dot{x}\right] = f(x, \dot{x}, t) \tag{4.14}$$

Let us put $\dot{x} = y$. Then the above equation can be expressed in the form of following two first order equations.

$$\begin{aligned} \dot{x} &= y \\ \dot{y} &= f(x, y, t) \end{aligned} \tag{4.15}$$

Assuming $h = \Delta t$, both x and y can be expressed in the terms of Taylor's series as

$$\begin{aligned} x &= x_i + \left(\frac{dx}{dt}\right)_i h + \left(\frac{d^2x}{dt^2}\right)_i \frac{h^2}{2} + \cdots\cdots \\ y &= y_i + \left(\frac{dy}{dt}\right)_i h + \left(\frac{d^2y}{dt^2}\right)_i \frac{h^2}{2} + \cdots\cdots \end{aligned} \tag{4.16}$$

The expressions of Eq. (4.16) can be replaced in terms of the average slope by neglecting higher order derivatives

$$\begin{aligned} x &= x_i + \left(\frac{dx}{dt}\right)_{i_{av}} h \\ y &= y_i + \left(\frac{dy}{dt}\right)_{i_{av}} h \end{aligned} \tag{4.17}$$

By using Simpson's rule, the average slope in the interval h becomes,

$$\left(\frac{dy}{dt}\right)_{i_{av}} = \frac{1}{6}\left[\left(\frac{dy}{dt}\right)_{t_i} + 4\left(\frac{dy}{dt}\right)_{t_i+\frac{h}{2}} + \left(\frac{dy}{dt}\right)_{t_i+h}\right] \tag{4.18}$$

Runge-Kutta method is very much similar to the preceding computations except the middle term of Eq. (4.18) is split into two terms and the computation of t, x, y, and f at point i is as follows :

t	x	$y = \dot{x}$	$f = \dot{y} = \ddot{x}$
$T_1 = t_i$	$X_1 = x_i$	$Y = y_i$	$F_1 = f(T_1, X_1, Y_1)$
$T_2 = t_i + \frac{h}{2}$	$X_2 = x_i + Y_1 \frac{h}{2}$	$Y_2 = y_i + F_1 \frac{h}{2}$	$F_2 = f(T_2, X_2, Y_2)$
$T_3 = t_i + \frac{h}{2}$	$X_3 = x_i + Y_2 \frac{h}{2}$	$Y_3 = y_i + F_2 \frac{h}{2}$	$F_3 = f(T_3, X_3, Y_3)$
$T_4 = t_i + h$	$X_4 = x_i + Y_3 h$	$Y_4 = y_i + F_3 h$	$F_4 = f(T_4, X_4, Y_4)$

The above quantities are used in the recurrence formula given below :

$$x_{i+1} = x_i + \frac{h}{6}[Y_1 + 2Y_2 + 2Y_3 + Y_4]$$
$$y_{i+1} = y_i + \frac{h}{6}[F_1 + 2F_2 + 2F_3 + F_4] \qquad (4.19)$$

The second term on the right hand side of Eq. (4.19) of the two equations are average values of $\frac{dx}{dt}$ and $\frac{dy}{dt}$ respectively.

Example 4.4

Solve Example 4.1 by Runge-Kutta method.
The equation of motion is

$$\ddot{x} = \frac{1}{2}F(t) - 1000x$$

Let $y = \dot{x}$, then

$$\dot{y} = f(x,t) = \frac{1}{2}F(t) - 1000x$$

With $h = 0.02$, the following table is calculated.

	t	x	$y = \dot{x}$	f
t_1	0	0	0	25
	0.1	0	0.25	30
	0.1	0.0025	0.30	32.5
t_2	0.2	0.0060	0.75	34

x_2 and y_2 can be calculated as follows :

$$x_2 = 0 + \frac{0.02}{6}(0 + 0.5 + 0.6 + 0.75) = 0.00617$$

$$y_2 = 0 + \frac{0.02}{6}(25 + 60 + 65 + 34) = 0.547$$

The process is continued.

4.2.4 Newmark's β-Method

Newmark's scheme of numerical integration is an extension of acceleration methods and the derivation of the basic equations is essentially those given in Sec. 4.2.2. However, the method is somewhat more general and can be applied to a nonlinear system as well [4.4]. The step-by-step solution procedure is as follows:

(1) The quantities at t_{i+1} are to be evaluated on the basis of known values of $x_i, \dot{x}_i$ and $\ddot{x}_i$. Assume the value of $\ddot{x}_{i+1}$.

(2) Compute

$$\dot{x}_{i+1} = \dot{x}_i + \left[(1-\delta)\ddot{x}_i + \delta \ddot{x}_{i+1}\right]\Delta t \qquad (4.20)$$

$$x_{i+1} = x_i + \dot{x}_i \Delta t + \left(\frac{1}{2} - \beta\right)\ddot{x}_i \Delta t^2 + \beta \ddot{x}_{i+1} (\Delta t)^2 \qquad (4.21)$$

(3) Based on the values of $\dot{x}_{i+1}$ and x_{i+1} in step (2), a new $\ddot{x}_{i+1}$ is computed

$$\ddot{x}_{i+1} = \frac{F_{i+1}(t)}{m} - \frac{c}{m}\dot{x}_{i+1} - \frac{k}{m}x_{i+1} \qquad (4.22)$$

(4) Steps (2) and (3) are repeated, beginning with a newly computed value of $\ddot{x}_{i+1}$, or with an extrapolated value, until satisfactory convergence is attained.

With $\delta = \frac{1}{2}$ and $\beta = \frac{1}{6}$, Eqs. (4.20) and (4.21) reduce to Eqs. (4.12) and (4.13) respectively, corresponding to the linear acceleration method.

Newmark has suggested that a choice of β between $\frac{1}{4}$ and $\frac{1}{6}$ will give sufficiently satisfactory and accurate value (with $\delta = \frac{1}{2}$). The value of $\beta = \frac{1}{6}$ corresponds to a parabolic variation of the acceleration.

In Newmark's β-method, equilibrium conditions at time station (i+1) is considered, unlike the finite difference method, where the equilibrium is considered at time *i*. A great elegance of the method lies, in that no starting procedures are needed, since displacements, velocities and accelerations at time t_{i+1} are expressed in terms of same quantities at time t_i only. Greater accuracy and faster convergence in this method are attained by shortening the period $\Delta t/T$.

Example 4.5

Compute the response of the SDF system, which is initially at rest by Newmark's β-method. The data in consistent units are as follows:

$k = 72$, $m = 8$, $\zeta = 0.1$ and $F(t) = 480\,t$.

The natural angular frequency of the system is

$$p = \sqrt{\frac{72}{8}} = 3$$

$$n = 0.1 \text{ X } 3 = 0.3$$

$$T = \frac{2\pi}{p} = \frac{2\pi}{3} = 2.0944 \text{ s}$$

Assume $\Delta T = 0.1T \cong 0.2$ s, $\beta = \frac{1}{6}$ and $\delta = \frac{1}{2}$

At $t = 0$, $x_1 = 0$, $\dot{x}_1 = 0$ and $\ddot{x}_1 = 0$, further, $F(t) = 0$ at $t = 0$

$$\dot{x}_2 = \dot{x}_1 + \frac{1}{2}(\ddot{x}_1 + \ddot{x}_2)\Delta t = 0.1\,\ddot{x}_2 \quad \text{(a)}$$

$$x_2 = x_1 + \dot{x}_1\Delta t + \frac{1}{3}\ddot{x}_1\Delta t^2 + \frac{1}{6}\ddot{x}_2\Delta t^2$$

$$x_2 = 0.0067\,\ddot{x}_2 \quad \text{(b)}$$

From equation of motion [Eq. (4.16)]

$$\ddot{x}_2 = \frac{F_2(t)}{m} - 2\zeta p\dot{x}_2 - \frac{k}{m}x_2$$

$$\ddot{x}_2 = \frac{96}{8} - 0.6\dot{x}_2 - 9x_2 \quad \text{(c)}$$

Substituting x_2 and $\dot{x}_2$ from Eqs. (a) and (b) into Eq. (c) gives

$$\ddot{x}_2 = 10.71,\ \dot{x}_2 = 1.071 \text{ and } x_2 = 0.072.$$

Solution of equations as presented above is not general, which would have been the case had the solution been obtained by successive approximation.

At $t = 2 \times 0.2$.

Let $\ddot{x}_3 = \ddot{x}_2 = 10.71$

Then, $\dot{x}_3 = 1.071 + \frac{1}{2}[10.71 + 10.71] \times 0.2 = 3.213$

$$x_3 = 0.072 + 1.071 \times 0.2 + \frac{1}{3} \times 10.71 \times 0.2^2 + \frac{1}{6} \times 10.71 \times 0.2^2 = 0.5$$

and $\ddot{x}_3 = 24 - 0.6 \times 3.213 - 9 \times 0.5 = 17.572$

Let $\ddot{x}_3 = 17.572$

$$\dot{x}_3 = 1.071 + \frac{1}{2}[10.71 + 17.572] \times 0.2 = 3.899$$

$$x_3 = 0.072 + 1.071 \times 0.2 + \frac{1}{3} \times 10.71 \times 0.2^2 + \frac{1}{6} \times 17.572 \times 0.2^2 = 0.546$$

and $$\ddot{x}_3 = 24 - 0.6 \times 3.899 - 9 \times 0.546 = 16.746$$

A new iteration is to be performed with $\ddot{x}_3 = 16.746$ and steps are to be continued.

4.3 NUMERICAL EVALUATION OF DUHAMEL'S INTEGRAL

The response of a SDF system subjected to a general type of forcing function, as given by Duhamel's integral is as follows [Eq. (3.91)].

$$x(t) = \int_0^t \frac{F(\tau)}{mp} \sin p(t - \tau)\, d\tau \tag{3.91}$$

For practical cases, this integral is to be evaluated numerically, as $F(t)$ is such that analytical solutions become difficult to obtain. The form of Eq. (3.91) is not suited well for numerical evaluation. If we use this expression as it stands, then the values of x and $\dot{x}$ obtained from the previous instants are not required for computing x and $\dot{x}$ at any time instant, and fresh computations for the previous instants are required at any time instant. The procedure, thus demands an impractically large number of computations [4.5]. This is explained with the help of an example of a SDF system, in consistent units $m = 2$, $p = 31.6$, $x_0 = 0$, $\dot{x}_0 = 0$, $\Delta t = 0.02$ and the forcing function is shown in Fig. 4.2(b).

At $$t = t_1 = \Delta\tau = 0.02$$

$$x_1 = \frac{F_0}{mp} \sin pt_1\, \Delta\tau$$

$$= \frac{50}{2 \times 31.6} \sin(31.6 \times 0.02) \times 0.02 = 0.00932$$

At $$t = t_2 = 2\Delta\tau = 0.04$$

$$x_2 = \int_0^{t_2} \frac{F(\tau)}{mp} \sin p(t - \tau)\, d\tau$$

$$= \frac{1}{mp}\left[F_0 \sin pt_2\, \Delta\tau + F_1 \sin pt_1\, \Delta\tau\right]$$

$$= \frac{1}{63.2}\left[\, 50 \sin(31.6 \times 0.04) \times 0.02 + 60 \times \sin(31.6 \times 0.02) \times 0.02\,\right] = 0.026$$

It is clear from the above calculation that the value of x_i is not utilised in evaluating x_2. So the previous value is not utilised and this involves a lengthy calculation.

For efficient numerical integration, Eq. (3.91) is recast in a different form.

Equation (3.91) is rewritten as

$$x(t) = \int_0^t \frac{F(\tau)}{mp} [\sin pt \cos p\tau - \cos pt \sin p\tau] \, d\tau$$

$$= \frac{\sin pt}{mp} \int_0^t F(\tau) \cos p\tau \, d\tau - \frac{\cos pt}{mp} \int_0^t F(\tau) \sin p\tau \, d\tau$$

or

$$x(t) = A(t) \sin pt - B(t) \cos pt \tag{4.23}$$

where

$$A(t) = \frac{1}{mp} \int_0^t F(\tau) \cos p\tau \, d\tau \tag{4.24}$$

$$B(t) = \frac{1}{mp} \int_0^t F(\tau) \sin p\tau \, d\tau \tag{4.25}$$

Thus the evaluation of $x(t)$ is primarily numerical integration of $A(t)$ and $B(t)$. This can be done by various means. Here we adopt Simpson's rule.

The variation of the displacement with time is of interest. Time is divided into a number of equal intervals, each of duration $\triangle\tau$, and the response at these sequences of time is to be evaluated. It must be noted that though the time interval is $\triangle\tau$, the response is obtained at $2\triangle\tau$, time intervals, if Simpson's rule is adopted.

Applying Simpson's rule, numerical integration of Eq. (4.24) is

$$A(t) = A(t - 2\Delta\tau) + \frac{\Delta\tau}{3mp} [F(t - 2\Delta\tau) \cos p(t - 2\Delta\tau)$$

$$+ 4F(t - \Delta\tau) \cos p(t - \Delta\tau) + F(t) \cos pt] \tag{4.26}$$

$A(t - 2\triangle\tau)$ is the value of the integral at time instant $(t - 2\triangle\tau)$ obtained by the summation of the proceeding values.

$B(t)$ of Eq. (4.25) can be numerically integrated in a similar manner.

The procedure is explained with the help of an example.

Example 4.6

A tower of Fig. 4.5 is subjected to a dynamic load. Evaluate numerically by Duhamel's integral its response.

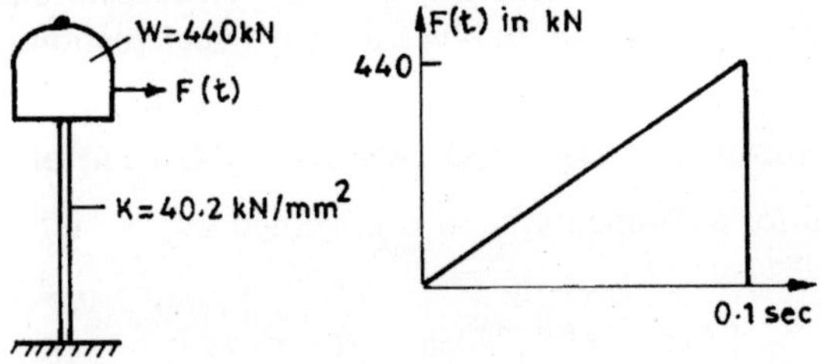

Fig. 4.5 Example 4.6

The frequency of vibration is given by

$$p = \sqrt{\frac{kg}{W}} = \sqrt{\frac{40.2 \times 9810}{440}} = 29.94 \text{ rad/s}$$

$$T = \frac{2\pi}{p} = \frac{2\pi}{29.94} = 0.021\text{s}$$

The time increment chosen for the numerical integration is $\triangle\tau = 0.01$s.

The calculation has been presented in a tabular form and is given in Table 4.2. Calculation as presented can be continued up to $\tau = 0.1$ s. After that, when internal force is not acting, A and B become constant values. But the motion still remains harmonic, as is given by Eq. (4.23).

Table 4.2 Numerical Integration of Duhamel's Integral

τ s	$F(\tau)$	sin $p\tau$	cos $p\tau$	Evaluation of A				
				(2)×(4)	Multiplier	(5)×(6)	ΔA*	A
(1)	(2)	(3)	(4)	(5)	(6)	(7)	(8)	(9)
0.00	0	0	1.0	0	1	0		0
0.01	44	0.295	0.956	42.06	4	168.24	0.0006	
0.02	88	0.564	0.826	72.69	1	72.69		0.0006
0.03	132	0.782	0.623	82.23	4	328.92	0.0012	
0.04	176	0.931	0.365	64.24	1	64.24		0.0018

Evaluation of B							
(2)×(3)	Multiplier	(10)×(11)	ΔB*	B	(9)×(3)	(14)×(4)	(15) – (16)
(10)	(11)	(12)	(13)	(14)	(15)	(16)	(17)
0	1	0		0	0	0	0
12.98	4	51.92	0.0002				
49.63	1	49.63		0.0002	0.00034	0.00016	0.00018
103.22	4	412.88	0.0016				
163.86	1	163.86		0.0018	0.00168	0.00066	0.00102

*$\triangle A$ and $\triangle B$ have been calculated by summing up the three consecutive values of the previous column as given in bracket and multiplying it by $\frac{\Delta\tau}{3mp}$.

4.3.1 Numerical Evaluation of Damped System by Duhamel's Integral

The Duhamel's integral for a damped system is governed by

$$x(t) = \frac{1}{mp_d}\int_0^t F(\tau)\, e^{-\zeta p(t-\tau)} \sin p_d (t-\tau)\, d\tau \tag{4.27}$$

For numerical evaluation we proceed as in the undamped case and obtain $x(t)$ from Eq. (4.27)

$$x(t) = \{A_d(t) \sin p_d t - B_d(t) \cos p_d t\}\frac{e^{-\zeta pt}}{mp_d} \tag{4.28}$$

where
$$A_d(t_i) = A_d(t_{i-1}) + \int_{t_{i-1}}^{t_i} F(\tau)\, e^{-\zeta p\tau} \cos p_d\tau\, d\tau \tag{4.29}$$

$$B_d(t_i) = B_d(t_{i-1}) + \int_{t_{i-1}}^{t_i} F(\tau)\, e^{\zeta p\tau} \sin p_d\tau\, d\tau \tag{4.30}$$

Assuming $F(\tau)$ to be piecewise linear function as shown in Fig. 4.6 we may write

$$F(\tau) = F(t_{i-1}) + \frac{\Delta F_i}{\Delta t_i}\left(\tau - t_{i-1}\right) \qquad t_{i-1} \le \tau \le t_i \tag{4.31}$$

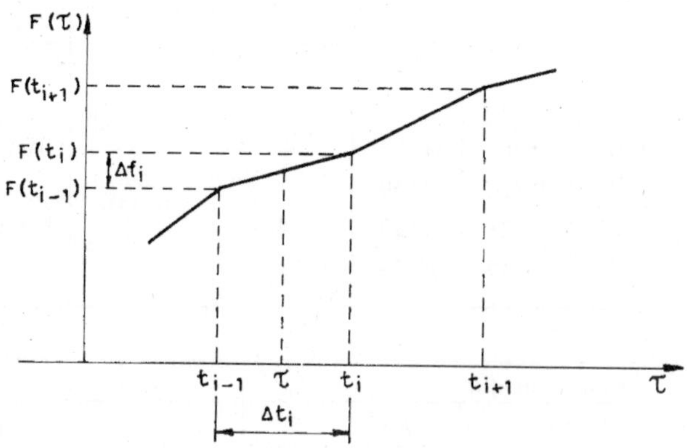

Fig. 4.6 Variation of $F(\tau)$

Eq. (4.31) when substituted in Eqs. (4.29) and (4.30) will require the evaluation of the following integrals

$$I_1 = \int_{i-1}^{t_i} e^{\zeta p\tau} \cos p_d\tau\, d\tau = \frac{e^{\zeta p\tau}}{(\zeta p)^2 + (p_d)^2}\left(\zeta p \cos p_d\tau + p_d \sin p_d\tau\right)\Big|_{t_{i-1}}^{t_i} \tag{4.32}$$

$$I_2 = \int_{t_{i-1}}^{t_i} e^{\zeta p\tau} \sin p_d\tau \, d\tau = \frac{e^{\zeta p\tau}}{(\zeta p)^2 + (p_d)^2}(\zeta p \sin p_d\tau - p_d \cos p_d\tau)\Big|_{t_{i-1}}^{t_i} \quad (4.33)$$

$$I_3 = \int_{t_{i-1}}^{t_i} \tau e^{\zeta p\tau} \sin p_d\tau \, d\tau = \tau - \frac{\zeta p}{(\zeta p)^2 + p_d^2} I_2' + \frac{p_d}{(\zeta p)^2 + p_d^2} I_1' \Big|_{t_{i-1}}^{t_i} \quad (4.34)$$

$$I_4 = \int_{t_{i-1}}^{t_i} \tau e^{\zeta p\tau} \cos p_d\tau \, d\tau = \frac{\zeta p}{(\zeta p)^2 + p_d^2} I_1' + \frac{p_d}{(\zeta p)^2 + p_d^2} I_2' \Big|_{t_{i-1}}^{t_i} \quad (4.35)$$

where I_1' and I_2' are the integrals of Eqs. (4.32) and (4.33) before the evaluation of limits. $A_d(t_i)$ and $B_d(t_i)$ are evaluated as follows.

$$A_d(t_i) = A_d(t_{i-1}) + \left(F_{(t_{i-1})} - t_{i-1}\frac{\Delta F_i}{\Delta t_i} \right) I_1 + \frac{\Delta F_i}{\Delta t_i} I_4 \quad (4.36)$$

$$B_d(t_i) = B_d(t_{i-1}) + \left(F_{(t_{i-1})} - t_{i-1}\frac{\Delta F_i}{\Delta t_i} \right) I_2 + \frac{\Delta F_i}{\Delta t_i} I_3 \quad (4.37)$$

On substitution of Eqs. (4.36) and (3.37) in Eq. (4.27), we obtain the displacement at time t_i as

$$x(t_i) = \frac{e^{-\zeta p t_i}}{m p_d} \{A_d(t_i) \sin p_d t_i - B_d(t_i) \cos p_d t_i\} \quad (4.38)$$

4.4 NUMERICAL COMPUTATION IN FREQUENCY DOMAIN

The evaluation of integrals of Fourier transform in closed form is tedious and poses considerable difficulty. Numerical integration of them is the only practical solution. The numerical treatment is divided into two steps: (a) Discrete Fourier Transform (DFT), which corresponds to Fourier transform pairs given by Eq. (3.124) and Eq. (3.125) are derived and then, (b) efficient numerical algorithm (Fast Fourier Transform or FFT) are evaluted for the DFTs developed.

4.4.1 Discrete Fourier Transform

Though the function is non-periodic, a period T is to be assumed to start with. This is dictated by the lowest frequency that is to be considered in the analysis. From Eq. (3.120)

$$\omega_0 = \omega\Delta = \frac{2\pi}{T} \quad (4.39)$$

The period is divided into N equal intervals of $\triangle T$ and the function is sampled at time $t_m = m\triangle t$.

Equation (3.123) is written as follows

$$F(t_m) = \frac{\Delta\omega}{2\pi} \sum_{n=0}^{N-1} F(\omega_n) e^{(in\Delta\omega m\Delta T)} \quad (4.40)$$

Using the relation of Eq. (4.39) into Eq. (4.40), we get

$$F(t_m) = \frac{\Delta\omega}{2\pi} \sum_{n=0}^{n-1} F(\omega_n) e^{(2\pi inm/N)} \tag{4.41}$$

Equation (3.122) can be replaced by the following finite sum

$$F(\omega_n) = \Delta t \sum_{n=0}^{n-1} F(t_m) e^{(-2\pi inm/N)} \tag{4.42}$$

Equations (4.41) and (4.42) are defined as discrete Fourier transform (DFT) pairs. They correspond to the transforms given by Eqs. (3.124) and (3.125).

4.4.2 **Fast Fourier Transform (FFT)**

FFT is a computer algorithm for calculating DFTs. The DFT of a finite sequence $\{F(t_m)\}$, $m = 0, 1, 2, \cdots, (N-1)$ is a new finite sequence $\{F(\omega_n)\}$ defined by (from Eq. [4.42]).

$$F(\omega_n) = \frac{T}{N} \sum_{m=0}^{N-1} F(t_m) e^{i(-2\pi nm/N)} \tag{4.43}$$

If we want to calculate values of $F(\omega_n)$ by a direct approach, N multiplications of the form $F(t_m)\, e^{-i(2\pi nm/N)}$ for each of the N values of $F(w_n)$ and the total work of calculating full sequence will require N^2 multiplications. As we shall see shortly, FFT reduces this work to a number of operations equal to $N \log_2 N$. If $N = 2^{20}$, then $N^2 = 1.1 \times 10^{12}$, whereas $N \log_2 N = 2.1 \times 10^7$, which is only about 1/52,000th of the number of operations. The FFT therefore offers an enormous reduction of computer time. Further, the number of operations being performed by the computer being less, round off errors due to truncation will be less and accuracy will be increased.

The FFT works by partitioning full sequence $\{F(\omega_n)\}$ into a number of shorter sequences. Instead of computing the DFT of the original sequence, only the DFTs of the shorter sequences are worked out. These are combined in an ingenious way to yield full DFT of $F(\omega_n)$.

In FFT, full sequence of $F(t_m)$ will be partitioned into a number of shorter sequences. Instead of calculating DFTs for the original sequence, only the DFTs of the shorter sequences are to be calculated. As to how the DFTs of the shorter sequences are combined to yield the full DFT is explained below, and this is the technique of FFT.

$F(t_m)$ is written singly as Fm in short. $\{F_m\}$, $m = 0, 1, 2, (N-1)$ is a sequence shown in Fig. 4.7(a), where N is even. This is partitioned into two shorter sequences $\{y_m\}$ and $\{z_m\}$ where

$$\left.\begin{aligned} y_m &= F_{2m} \\ z_m &= F_{2m+1} \end{aligned}\right\} m = 0, 1, 2, \cdots, \left(\frac{N}{2} - 1\right) \tag{4.44}$$

The DFTs of these two short sequences are

$$\left.\begin{aligned} Y_n &= \frac{T}{N/2}\sum_{m=0}^{N/2-1} y_m \exp\{-i\,(2\pi mn)/(N/2)\} \\ Z_n &= \frac{T}{N/2}\sum_{m=0}^{N/2-1} z_m \exp\{-i\,(2\pi mn)/(N/2)\} \end{aligned}\right\} n = 0, 1, 2, \cdots, \left(\frac{N}{2}-1\right) \quad (4.45)$$

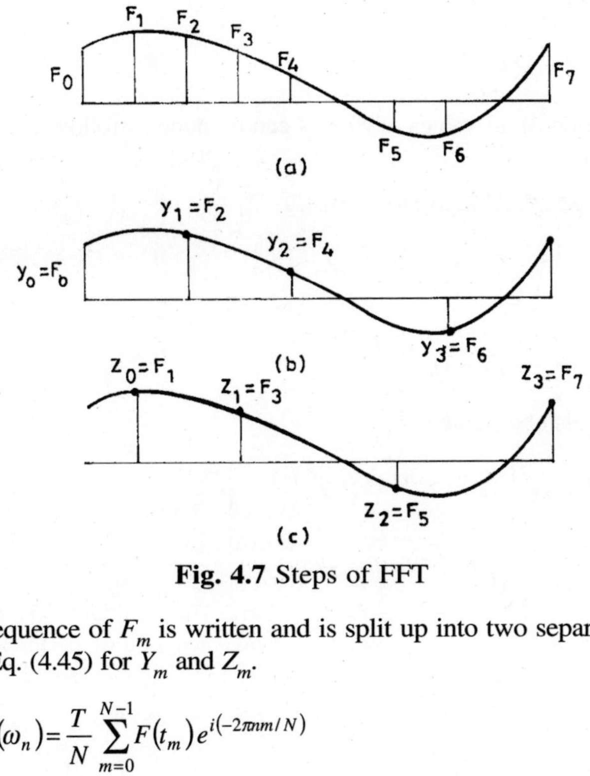

Fig. 4.7 Steps of FFT

Now the original sequence of F_m is written and is split up into two separate sums, similar to those occurring in Eq. (4.45) for Y_m and Z_m.

$$F(\omega_n) = \frac{T}{N}\sum_{m=0}^{N-1} F(t_m)\, e^{i(-2\pi nm/N)}$$

$$= \frac{T}{N}\left\{\sum_{m=0}^{N/2-1} F_{2m}\, e^{-i[2\pi(2m)n/N]} + \sum_{m=0}^{N/2-1} F_{2m+1}\, e^{-i[2\pi(2m+1)n/N]}\right\}$$

$$= \frac{T}{N}\left\{\sum_{m=0}^{N/2-1} y_m\, e^{-i(2\pi mn)/(N/2)} + \sum_{m=0}^{N/2-1} z_m\, e^{-i[2\pi n/N]}\, e^{-i(2\pi mn)/(N/2)}\right\}$$

or

$$F(\omega_n) = \frac{1}{2}\left\{Y_n + e^{-i(2\pi n/N)}\, Z_n\right\} \quad (4.46)$$

for $n = 0, 1, 2, \cdots, \left(\frac{N}{2}-1\right)$

It is thus seen that the DFTs of the original sequence is obtained from the two half-sequence of Y_n and Z_n directly. Thus Eq. (4.46) forms the very basis of the FFT method. The half-sequences of $\{Y_m\}$ and $\{Z_m\}$ may further be partitioned into quarter-sequences and so on, till the last sequence may contain only one term.

Equation (4.46) indicates that $F(\omega_n)$ can be calculated for $0 < n < \frac{N}{2} - 1$, i.e. only half of the coefficients of $F(\omega_n)$ can thus be obtained, But $F(\omega_n)$ need to be calculated for the values of n from $\frac{N}{2}$ to $(N - 1)$. In order to do that, advantage may be taken of the fact that Y_n and Z_n are periodic in n and repeat themselves with period $N/2$, so that

and
$$\left.\begin{aligned} Y_{n-N/2} &= Y_n \\ Z_{n-N/2} &= Z_n \end{aligned}\right\} \tag{4.47}$$

Therefore, computation of all values of $F(\omega_n)$ can be done as follows:

$$\left.\begin{aligned} F(\omega_n) &= \frac{1}{2}\left\{Y_n + e^{-i(2\pi n/N)} Z_n\right\} \quad \text{for} \quad n = 0, 1, 2, \ldots, \left(\frac{N}{2} - 1\right) \\ F(\omega_n) &= \frac{1}{2}\left\{Y_{n-N/2} + e^{-i(2\pi n/N)} Z_{n-N/2}\right\} \end{aligned}\right\} \tag{4.48}$$

for $\quad n = \frac{N}{2}, \frac{N}{2} + 1, \cdots, (N - 1)$

Equation (4.48) can also be written as

$$\left.\begin{aligned} F(\omega_n) &= \frac{1}{2}\left\{Y_n + e^{-i(2\pi n/N)} Z_n\right\} \\ F(\omega_{n+N/2}) &= \frac{1}{2}\left\{Y_n + e^{-i(2\pi/N)(n+N/2)} Z_n\right\} \end{aligned}\right\} \tag{4.49}$$

where $n = 0, 1, 2, \cdots, \left(\frac{N}{2} - 1\right)$

Noting that $e^{-i\pi} = 1$

$$\left.\begin{aligned} F(\omega_n) &= \frac{1}{2}\left\{Y_n + e^{-i(2\pi n/N)} Z_n\right\} \\ F\left(\omega_{n+\frac{N}{2}}\right) &= \frac{1}{2}\left\{Y_n - e^{-i(2\pi n/N)} Z_n\right\} \end{aligned}\right\} \tag{4.50}$$

for $\quad n = 0, 1, 2, \cdots, \left(\frac{N}{2} - 1\right)$

Further, we define a new complex variable as follows

$$W_N^n = e^{-i(2\pi n/N)} \tag{4.51}$$

Therefore, the computational recipe can be expressed as

$$\left.\begin{aligned} F(\omega_n) &= \frac{1}{2}\left\{Y_n + W_N^n Z_n\right\} \\ F\left(\omega_{n+N/2}\right) &= \frac{1}{2}\left\{Y_n - W_N^n Z_n\right\} \end{aligned}\right\} \tag{4.52}$$

for $n = 0, 1, 2, \cdots, \left(\frac{N}{2} - 1\right)$ The form exists in most of the FFT programs.

References

4.1 J.M. Biggs, Introduction to Structural Dynamics, McGraw-Hill Book Inc., New York, 1961.

4.2 K.J. Bathe, Finite Element Procedures in Engineering Analysis, Prentice-Hall, New Jersey, 1996.

4.3 J.L. Humar, Dynamics of Structures, Prentice-Hall, New Jersey, 1990.

4.4 N.H. Newmark, A method of computation of structural dynamics, Journal of Engineering Mechanics Division, ASCE, V. 88, 1959, pp. 67-94.

4.5 R.W. Clough and J. Penzien, Dynamics of Structures, 2nd Edition, McGraw-Hill Books Inc., New York, 1993.

4.6 A.K. Chopra, Dynamics of structures, Prentice-Hall, New Delhi, 1996.

EXERCISE 4

4.1 A SDF system has $k = 7150$ N/m and the mass is 20 kg. Determine the response of the mass by numerical integration, if an external force is applied to the system. The system is undamped and is initially at rest.

4.2 Solve Prob. 4.1 by applying Newmark's β-method using $\beta = 1/5$.

4.3 The spring-mass system has $m = 0.5$ kg and $p = 8.88$ rad/s. It is excited by a force shown in the figure. Determine the response using the finite difference technique.

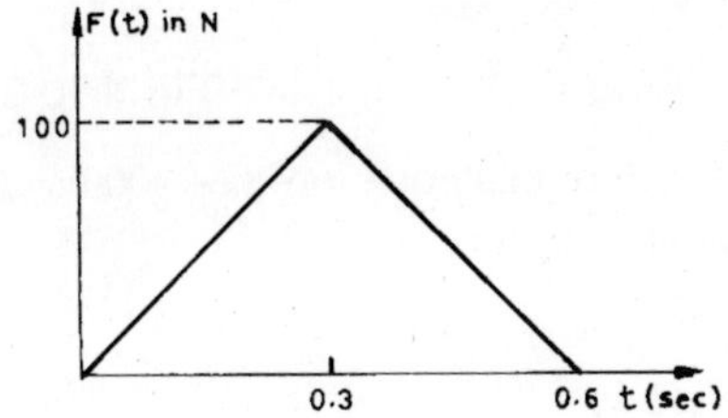

Prob. 4.3.

4.4 Solve Prob. 4.3 by numerically evaluating Duhamel's integral.

4.5 If the SDF system of Prob. 4.3 is subjected to a half sine pulse of amplitude 100 N and duration 0.6s, determine numerically the response.

4.6 Repeat Prob. 4.3 using linear acceleration method.

4.7 An undamped spring-mass system has a base excitation of $10(1 - 2t)$. If for the system, $p = 12$ rad/s, determine numerically the variation of relative displacement with time,

4.8 A spring-mass system having $m = 5$ kg, $p = 0.5$ rad/s and $c = 0.5$ Ns/m is subjected to an impulse 10 N*s*, which has a triangular shape with time duration of 0.5s. Determine the response of the system by using numerical integration. What is the maximum displacement of the mass?

4.9 Solve numerically the differential equation

$$10\ddot{x} + 5000x = F(t)$$

with initial conditions $x_1 = \dot{x}_1 = 0$ and forcing function as shown in the figure.

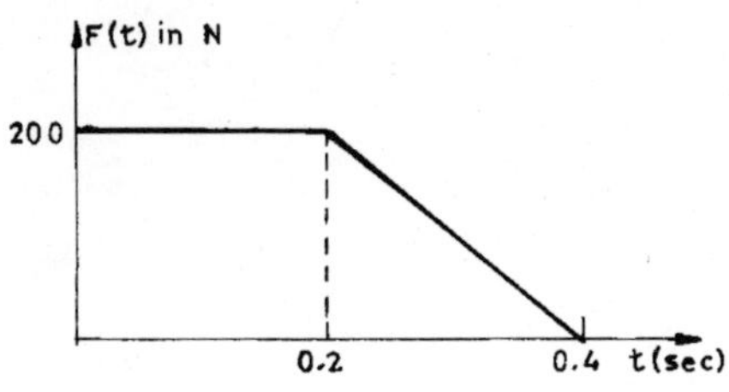

Prob. 4.9

4.10 Determine dynamic response of the tower by a numerical evaluation of the Duhamel's integral due to blast loading.

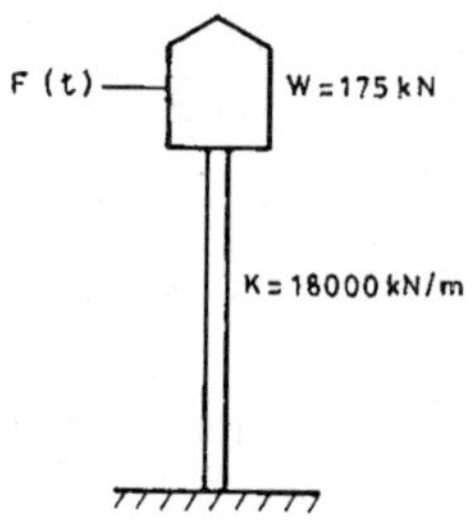

Fig. 4.10 (a)

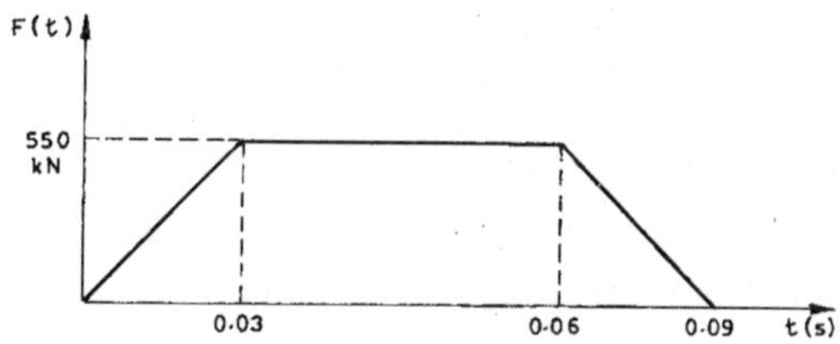

Fig. 4.10 (b)

4.11 Solve numerically the differential equation $8\ddot{x} + 4000x = F(t)$, when the system starts at rest and $F(t)$ is indicated in figure.

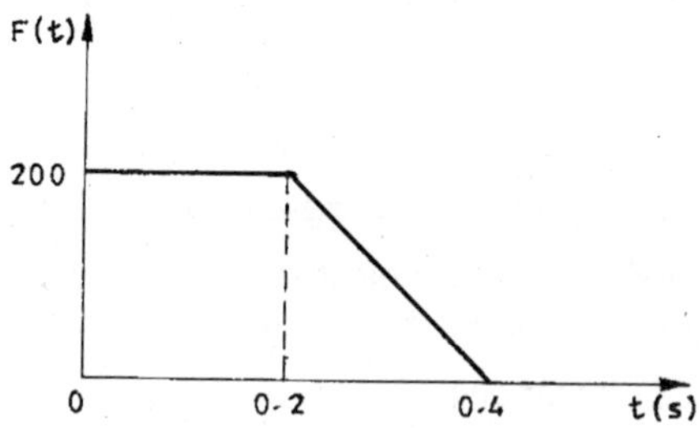

Fig. 4.11

4.12 A SDF system has the following properties : $m = 4.5$ kg – s^2/mm, $k = 1800$ N/mm, $T = 1$ sec and $\zeta = 0.05$. Determine the response $x(t)$ of the system due to $F(t)$ defined by half-cycle sine pulse of Fig. No. 4.12.

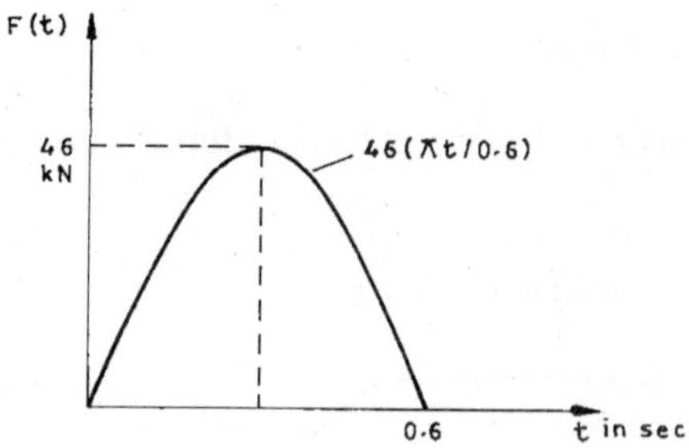

Chapter 5

Vibration of Two Degrees of Freedom System

5.1 INTRODUCTION

Up to the last chapter, we have dealt with systems having only single degree of freedom. We gradually pass on to the more advanced topics. We embark on this chapter on systems, which are referred to as two degrees of freedom system. As has already been explained, if a system requires two independent coordinates to describe the motion, it is said to have two degrees of freedom.

Two degrees of freedom form the simplest class of systems, referred to multiple degrees of freedom. The later category is presented in the next chapter and treatment there has been meted out with the help of matrix notations. But for a thorough understanding of these systems, two degrees of freedom systems have been taken up separately and treated in an explicit manner. Free vibration of two degrees of freedom systems has been taken up first, and then their forced vibration characteristics are dealt with.

5.2 FREE VIBRATION OF UNDAMPED TWO DEGREES OF FREEDOM SYSTEMS

The system shown in Fig. 5.1 has two masses m_1 and m_2, connected by two springs having stiffness k_1 and k_2. It is two degrees of freedom system, as its configuration is fully described by two displacements, x_1 and x_2 as shown in the Fig. 5.1.

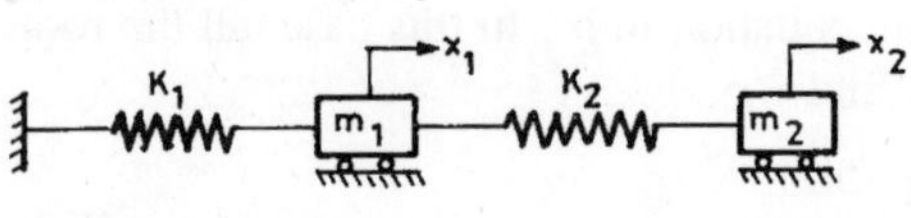

Fig. 5.1 Two degrees of freedom system

Freebody diagrams of the masses m_1 and m_2 are shown in Fig. 5.2. The equations of motion of the two masses are

$$\left.\begin{aligned} m_1\ddot{x}_1 + k_1 x_1 + k_2(x_1 - x_2) = 0 \\ m_2\ddot{x}_2 + k_2(x_2 - x_1) = 0 \end{aligned}\right\} \tag{5.1}$$

Equation (5.1) involves coupled ordinary linear homogeneous differential equations, with constant coefficients. The following solutions are assumed

$$\left.\begin{aligned} x_1 &= Ae^{ipt} \\ x_2 &= Be^{ipt} \end{aligned}\right\} \tag{5.2}$$

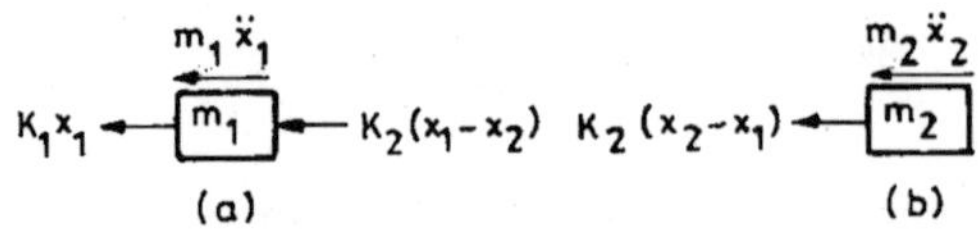

Fig. 5.2 Free body diagrams of (a) mass m_1, (b) mass m_2

Substitution of x_1 and x_2 from Eq. (5.2) into Eq. (5.1) yields

$$\left.\begin{aligned} A\left[-m_1 p^2 + (k_1 + k_2)\right] - Bk_2 = 0 \\ -Ak_2 + B\left(-m_2 p^2 + k_2\right) = 0 \end{aligned}\right\} \tag{5.3}$$

Equation (5.3) involves two homogeneous equations. One obvious solution of Eqs. (5.3) is $A = B = 0$. This only defines the equilibrium condition of the system and tells nothing about vibrations. From the theory of homogeneous equations, we know that a non-trivial solution of A and B is possible only if the determinant of the coefficients of A and B vanishes. This can be achieved in another way. From Eqs. (5.3), the following relationship can be obtained

$$\frac{A}{B} = \frac{k_2}{-m_1 p^2 + k_1 + k_2} = \frac{-m_2 p^2 + k_2}{k_2} \tag{5.4}$$

Cross-multiplication of the terms of Eq. (5.4), yields

$$\left(-m_2 p^2 + k_2\right)\left(-m_1 p^2 + k_1 + k_2\right) = k_2^2 \tag{5.5}$$

which on simplification, yields

$$p^4 - \left(\frac{k_2}{m_2} + \frac{k_1 + k_2}{m_1}\right)p^2 + \frac{k_1 k_2}{m_1 m_2} = 0 \tag{5.6}$$

Equation (5.6) is a quadratic equation in p^2. In this case, all the roots are real. Let the four roots of Eq. (5.6) be $p_1, -p_1, p_2$ and $-p_2$.

Therefore, according to Eq. (5.2)

$$\left.\begin{aligned} x_1 = A_1 e^{ip_1 t} + A_2 e^{-ip_1 t} + A_3 e^{ip_2 t} + A_4 e^{-ip_2 t} \\ x_2 = B_1 e^{ip_1 t} + B_2 e^{-ip_1 t} + B_3 e^{ip_2 t} + B_4 e^{-ip_2 t} \end{aligned}\right\} \tag{5.7}$$

Once the values of p_1^2 and p_2^2 are determined, B and A have known ratios, as given by Eq. (5.4). Equation (5.4) can be written as

$$B = \frac{k_2}{-m_2 p^2 + k_2} A = CA \tag{5.8}$$

Equation (5.8) yields for each of the two values of p^2, a corresponding value of C, both of which are real numbers. Therefore, the second of Eq. (5.7) can be written as

$$x_2 = C_1\left(A_1 e^{ip_1 t} + A_2 e^{-ip_1 t}\right) + C_2\left(A_3 e^{ip_2 t} + A_4 e^{-ip_2 t}\right) \tag{5.9}$$

Noting that $e^{\pm i\theta} = \cos\theta \pm i\sin\theta$, Eq. (5.7) reduces, after modifying the second equation as per Eq. (5.9) to

$$\left.\begin{aligned} x_1 &= D_1 \cos p_1 t + D_2 \sin p_1 t + D_3 \cos p_2 t + D_4 \sin p_2 t \\ x_2 &= C_1\left(D_1 \cos p_1 t + D_2 \sin p_1 t\right) + C_2\left(D_3 \cos p_2 t + D_4 \sin p_2 t\right) \end{aligned}\right\} \tag{5.10}$$

Equations (5.10) can also be written as

$$\left.\begin{aligned} x_1 &= E_1 \sin\left(p_1 t + \phi_1\right) + E_2 \sin\left(p_2 t + \phi_2\right) \\ x_2 &= C_1 E_1 \sin\left(p_1 t + \phi_1\right) + C_2 E_2 \sin\left(p_2 t + \phi_2\right) \end{aligned}\right\} \tag{5.11}$$

The system has two natural frequencies corresponding to p_1 and p_2. Equation (5.10) or (5.11) indicates that the motion of the masses m_1 and m_2 consists of superposition of two harmonic motions. If the system is vibrating with the circular frequency p_1 alone, then x_1 and x_2 are related by the constant C_1. In other words, during the vibrating stage, if the position of one of the masses is defined, the position of the other is automatically known. The same inference can be made, when the system is vibrating with angular frequency p_2 alone. Each of these configurations corresponding to each natural frequency is called the normal mode or the principal mode. Four constants related to the equation of motion can be evaluated from the initial conditions.

Example 5.1

From the two degrees of freedom system of Fig. 5.1, $k_1 = k$, $k_2 = 2k$, $m_1 = m$ and $m_2 = 2m$. Find the angular frequencies, the corresponding mode shapes and the equations of motion.

Substituting the values given in the problem, Eq. (5.6) becomes

$$p^4 - \left(\frac{k}{m} + \frac{3k}{m}\right)p^2 + \frac{k^2}{m^2} = 0$$

or

$$p^4 - 4\frac{k}{m}p^2 + \frac{k^2}{m^2} = 0 \tag{a}$$

Solution of Eq. (a) is

$$\begin{aligned} p_{1,2}^2 &= \frac{4 \pm \sqrt{16-4}}{2}\frac{k}{m} \\ &= 0.268\frac{k}{m},\ 3.73\frac{k}{m} \end{aligned}$$

Therefore,

$$p_1 = 0.518\sqrt{\frac{k}{m}},\ p_2 = 1.93\sqrt{\frac{k}{m}}$$

Values of C_1 and C_2 can be obtained from Eq. (5.8) as

$$C_1 = \frac{2}{-2\times 0.268+2} = 1.366$$

$$C_2 = \frac{2}{-2\times 3.73+2} = -\ 0.366$$

The resulting equations of the masses are

$$x_1 = D_1 \cos 0.518 \sqrt{\frac{k}{m}}\, t + D_2 \sin 0.518 \sqrt{\frac{k}{m}}\, t$$
$$+ D_3 \cos 1.93 \sqrt{\frac{k}{m}}\, t + D_4 \sin 1.93 \sqrt{\frac{k}{m}}\, t$$

$$x_2 = 1.366 D_1 \cos 0.518 \sqrt{\frac{k}{m}}\, t + 1.366 D_2 \sin 0.518 \sqrt{\frac{k}{m}}\, t$$
$$-\ 0.366 D_3 \cos 1.93 \sqrt{\frac{k}{m}}\, t -\ 0.366 D_4 \sin 1.93 \sqrt{\frac{k}{m}}\, t$$

The normal mode shape has been given in Fig. 5.3.

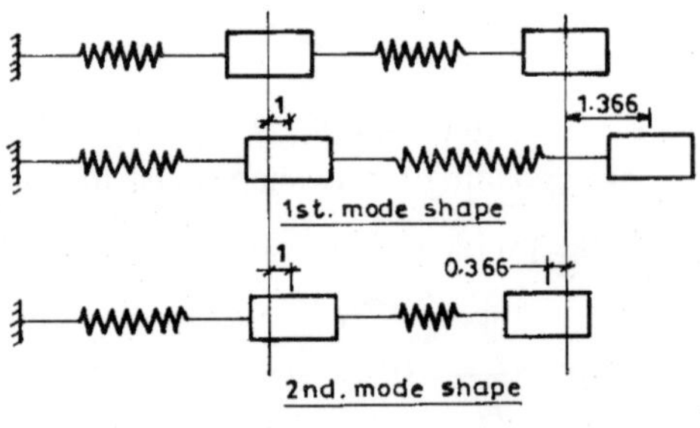

Fig. 5.3 Mode shapes of Example 5.1.

5.3 TORSIONAL VIBRATION OF TWO DEGREES OF FREEDOM SYSTEM

Consider two rotors having mass moment of inertia I_1 and I_2 respectively. They are separated by a distance *L*. Torsional displacements of the two rotors be θ_1 and θ_2 (Fig. 5.4).

Two equations of motion for the system are

$$\left.\begin{aligned} I_1\ddot{\theta}_1 + k(\theta_1 - \theta_2) = 0 \\ I_2\ddot{\theta}_2 + k(\theta_2 - \theta_1) = 0 \end{aligned}\right\} \quad (5.12)$$

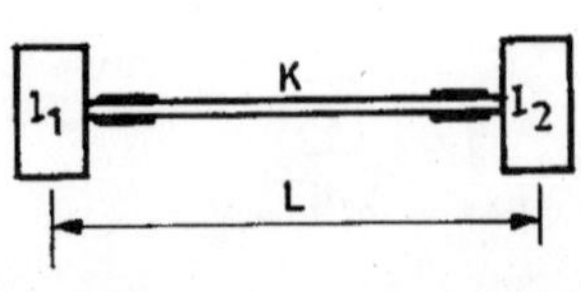

Fig. 5.4 A two rotor system

Assume the solution in the following form:

$$\left.\begin{aligned}\theta_1 &= Ae^{ipt}\\ \theta_2 &= Be^{ipt}\end{aligned}\right\} \tag{5.13}$$

Substituting the solutions of Eq. (5.13) into Eq. (5.12), we get

$$\left.\begin{aligned}A\left[-I_1 p^2 + k\right] - Bk &= 0\\ -Ak + B\left[-I_1 p^2 + k\right] &= 0\end{aligned}\right\} \tag{5.14}$$

From Eq. (5.14) the following relation can be written

$$\frac{A}{B} = \frac{k}{-I_1 p^2 + k} = \frac{-I_2 p^2 + k}{k} \tag{5.15}$$

On cross-multiplication of the terms of Eq. (5.15), we get

$$\left(-I_1 p^2 + k\right)\left(-I_2 p^2 + k\right) = k^2$$

or

$$p^2\left(p^2 - k\frac{I_1 + I_2}{I_1 I_2}\right) = 0 \tag{5.16}$$

Solution of Eq. (5.16) is

$$p_1^2 = 0$$

and

$$p_2^2 = k\frac{I_1 + I_2}{I_1 I_2} \tag{5.17}$$

Substituting $p_1^2 = 0$ into Eq. (5.15), we get

$$\frac{A}{B} = 1 \tag{5.18}$$

which indicates that two masses rotate by the same amount. This represents what is termed as a rigid body mode. Further, as $p_1^2 = 0$, there is no vibration corresponding to this angular frequency.

Substituting the value p_2^2 from Eq. (5.17) into Eq. (5.15), we get

$$\frac{A}{B} = -\frac{I_2}{I_1}$$

or

$$C = \frac{B}{A} = -\frac{I_1}{I_2} \tag{5.19}$$

Second mode shape has been plotted in Fig. 5.5. The above system is referred to as a semi-definite system, which by definition is one having one of the natural frequencies as zero.

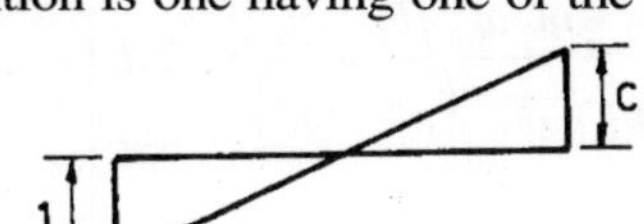

Fig. 5.5 Second mode shape

5.4 FORCED VIBRATION OF TWO DEGREES OF FREEDOM UNDAMPED SYSTEM

Let the mass m_1 of Fig. 5.1 be subjected to an external harmonic force $F_1 \sin \omega t$. We consider only the steady state forced vibration of this system. The equations of motion of the masses are as follows:

$$\left.\begin{aligned} m_1 \ddot{x}_1 + k_1 x_1 + k_2 (x_1 - x_2) &= F_1 \sin \omega t \\ m_2 \ddot{x}_2 + k_2 (x_2 - x_1) &= 0 \end{aligned}\right\} \tag{5.20}$$

The particular solution can be assumed as

$$\left.\begin{aligned} x_1 &= A \sin \omega t \\ x_2 &= B \sin \omega t \end{aligned}\right\} \tag{5.21}$$

Substituting the values of x_1, x_2, $\ddot{x}_1$ and $\ddot{x}_2$ from Eq. (5.21) into Eq. (5.20), we get

$$\left.\begin{aligned} (k_1 + k_2 - m_1 \omega^2) A - k_2 B &= F_1 \\ -k_2 A + (k_2 - m_2 \omega^2) B &= 0 \end{aligned}\right\} \tag{5.22}$$

Solving the values of A and B from Eq. (5.22), we have

$$\left.\begin{aligned} A &= \frac{(k_2 - m_2 \omega^2) F_1}{\left[(k_1 + k_2 - m_1 \omega^2)(k_2 - m_2 \omega^2) - k_2^2\right]} \\ B &= \frac{k_2 F_1}{\left[(k_1 + k_2 - m_1 \omega^2)(k_2 - m_2 \omega^2) - k_2^2\right]} \end{aligned}\right\} \tag{5.23}$$

The denominator of the expressions in Eq. (5.23) is in the same form as that of the frequency equation. If p_1^2 and p_2^2 are the roots of the frequency equation given by Eq. (5.6), then Eq. (5.23) can be written as

$$\left.\begin{aligned} A &= \frac{(k_2 - m_2 \omega^2) F_1}{m_1 m_2 (p_1^2 - \omega^2)(p_2^2 - \omega^2)} \\ B &= \frac{k_2 F_1}{m_1 m_2 (p_1^2 - \omega^2)(p_2^2 - \omega^2)} \end{aligned}\right\} \tag{5.24}$$

The steady-state response of the system is given by

$$\left.\begin{aligned} x_1 &= \frac{(k_2 - m_2 \omega^2) F_1}{m_1 m_2 (p_1^2 - \omega^2)(p_2^2 - \omega^2)} \sin \omega t \\ x_2 &= \frac{k_2 F_1}{m_1 m_2 (p_1^2 - \omega^2)(p_2^2 - \omega^2)} \sin \omega t \end{aligned}\right\} \tag{5.25}$$

Equation (5.25) reveals that if $\omega = p_1$ or $\omega = p_2$, i.e. the frequency of the external force is equal to one of the natural frequencies of the system, then the denominator is equal to zero and amplitudes would be infinite. There are, thus, two conditions of resonance for a two degree of freedom system, each corresponding to one of the two natural frequencies of free vibration.

5.5 VIBRATION ABSORBER

In order to make x_1 of Eq. (5.25) equal to zero, the numerator has to be made zero, which suggests

$$\left(k_2 - m_2\omega^2\right)F_1 = 0$$

or

$$\omega^2 = \frac{k_2}{m_2} \tag{5.26}$$

This is the principle of vibration absorber. This suggests that the mass, which is subjected to a force, $F_1 \sin\omega t$ will not vibrate at all, if there is a second spring and a second mass, which is so designed that k_2/m_2 is equal to square of the forcing frequency.

Introducing

$$\omega_1^2 = \frac{k_1}{m_1} \text{ and } \omega_2^2 = \frac{k_2}{m_2}$$

and assuming the motion to be harmonic, the equation of the amplitude given by Eq. (5.23) can be shown as

$$\left.\begin{aligned} \frac{Ak_1}{F_1} &= \frac{\left[1-\left(\dfrac{\omega}{\omega_2}\right)^2\right]}{\left[1+\dfrac{k_2}{k_1}-\left(\dfrac{\omega}{\omega_1}\right)^2\right]\left[1-\left(\dfrac{\omega}{\omega_2}\right)^2\right]-\dfrac{k_2}{k_1}} \\ \frac{Bk_1}{F_1} &= \frac{1}{\left[1+\dfrac{k_2}{k_1}-\left(\dfrac{\omega}{\omega_1}\right)^2\right]\left[1-\left(\dfrac{\omega}{\omega_2}\right)^2\right]-\dfrac{k_2}{k_1}} \end{aligned}\right\} \tag{5.27}$$

In Fig. 5.6 is presented the plot of $\frac{Ak_1}{F_1}$ against $\frac{\omega}{\omega_2}$ for a mass ratio $\mu = m_2/m_1$ and a ratio of $\frac{\omega_2}{\omega_1}$. Two natural frequencies of the system has been plotted in Fig. 5.7 against mass ratio μ.

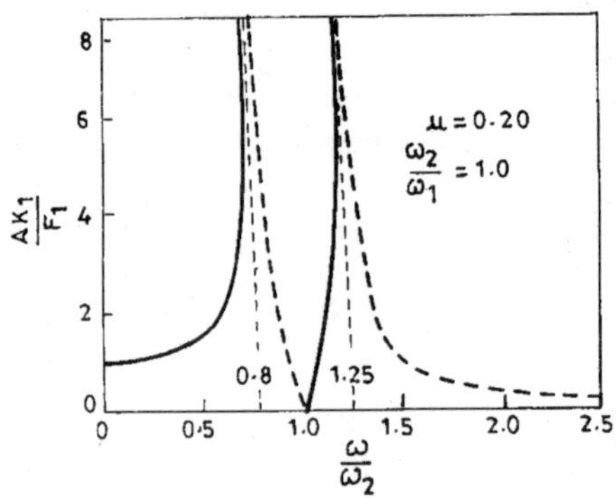

Fig. 5.6 Vibration of $\frac{Ak_1}{F_1}$ with $\frac{\omega}{\omega_2}$

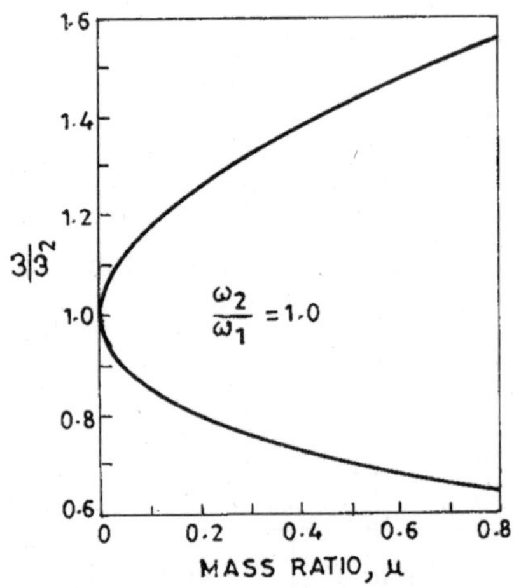

Fig. 5.7 Vibration of $\frac{\omega}{\omega_2}$ with mass ratio.

It is clearly evident in Fig. 5.6 that when $\frac{\omega}{\omega_2} = 1$, A = 0.

However, at $\omega = \omega_2$, the mass of the absorber does undergo displacement and is given by [from second of Eq. (5.27)]

$$\frac{Bk_1}{F_1} = -\frac{k_1}{k_2}$$

or

$$B = -\frac{F_1}{k_2} \tag{5.28}$$

The negative sign of Eq. (5.28) implies that the force exerted to the second spring is opposite to the external or impressed force.

The maximum force acting on the mass m_2 is

$$k_2 B = -F_1 \tag{5.29}$$

The force on the absorber mass is equal and opposite to the external force. The sizes of k_2 and m_2 are dictated by the permissible value of B.

Example 5.2

For the system of Fig. 5.8, $W_1 = 900$ N and the absorber weight 225 N. If W_1 is excited by a 40 N-mm unbalance rotating at 1800 rpm, determine the proper value of absorber spring k_2. What will be the amplitude of W_2?

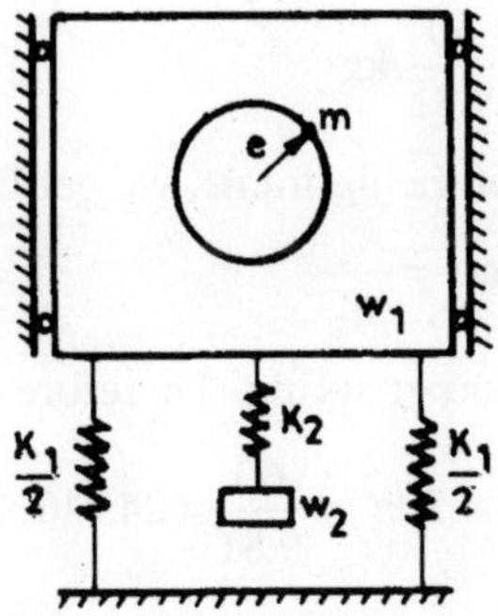

Fig. 5.8 Example 5.2.

Based on Art. 3.3, it can be shown that equation of motion of the two masses are

$$\left.\begin{aligned} \frac{W_1}{g}\ddot{x}_1 + k_1 x_1 + k_2(x_1 - x_2) &= me\omega^2 \sin\omega t \\ \frac{W_2}{g}\ddot{x}_2 + k_2(x_2 - x_1) &= 0 \end{aligned}\right\} \tag{a}$$

Equation (a) is similar to Eq. (5.20), except that F_1 is replaced by $me\omega^2$. Therefore, solutions of Eq. (a) are identical to Eq. (5.25), except that F_1 is substituted as $(me\omega^2)$

$$\omega = \frac{2\pi \times 1800}{60} = 188.5\, rad/s$$

For the system to act as vibration absorber, from Eq. (5.26)

$$\omega^2 = \frac{k_2}{m_2}$$

Therefore

$$k_2 = \omega^2 m_2 = 188.5^2 \times \frac{225}{981} = 814{,}960\, N/m$$

The amplitude of W_2 is given by Eq. (5.28) as

$$B = -\frac{me\omega^2}{k_2} = -\frac{40}{1000} \times \frac{188.5^2}{814960} = -0.0017\, m$$

Example 5.3

A jig contains a screen that reciprocates with a frequency of 600 cpm. The jig weighs 230 N and has a fundamental frequency of 400 cpm. If an absorber weighing 60 N is to be installed to eliminate the vibration of the jig frame, determine the absorber spring stiffness. What will be the resulting two natural frequencies of the system?

The external frequency $\omega = \dfrac{2\times\pi\times600}{60} = 20\pi$ rad/s

The mass of the jig frame $m_1 = \dfrac{230}{9.81}$ kg

The mass of the absorber $m_2 = \dfrac{60}{9.81}$ kg

In order to eliminate the vibration of the jig frame, we get from Eq. (5.26)

$$\omega^2 = \frac{k_2}{m_2}$$

where k_2 is the stiffness of the absorber spring. Therefore

$$k_2 = \omega^2 m_2 = 20\pi \times 20\pi \times \frac{60}{9.81} = 24{,}146 \text{ N/m}$$

$$= 24.146 \text{ N/mm}$$

If k_1 is the stiffness of the jig frame

$$\omega_1^2 = \frac{k_1}{m_1}$$

where

$$\omega_1 = \frac{400\times2\pi}{60} = \frac{40\pi}{3} \text{ rad/s}$$

Therefore

$$k_1 = \frac{40\pi}{3}\times\frac{40\pi}{3}\times\frac{230}{9.81} = 41{,}137 \text{ N/m}$$

Natural frequencies of the two degrees of freedom system can be obtained by Eq. (5.6) as

$$p^4 - \left(\frac{k_2}{m_2} + \frac{k_1+k_2}{m_1}\right)p^2 + \frac{k_1 k_2}{m_1 m_2} = 0$$

or
$$p^4 - \left(\frac{24146\times9.81}{60} + \frac{(24146+41137)\times9.81}{230}\right)p^2$$

$$+\frac{24146\times41137\times9.81^2}{60\times230} = 0$$

or
$$p^4 - 6732.33p^2 + 6926865 = 0$$

or
$$p_{1,2}^2 = 1267.55, \quad 5464.78$$

or $p_{1,2} = 35.60, \quad 73.92 \text{ rad/s}$

5.6 FREE VIBRATION OF TWO DEGREES OF FREEDOM SYSTEM WITH VISCOUS DAMPING

Two degrees of freedom damped system with viscous damping has been shown in Fig. 5.9. The equations of motion of the two masses for free vibration are

$$\left.\begin{aligned} m_1\ddot{x}_1 + c_1\dot{x}_1 + c_2(\dot{x}_1 - \dot{x}_2) + k_1x_1 + k_2(x_1 - x_2) = 0 \\ m_2\ddot{x}_2 + c_2(\dot{x}_2 - \dot{x}_1) + k_2(x_2 - x_1) = 0 \end{aligned}\right\} \quad (5.30)$$

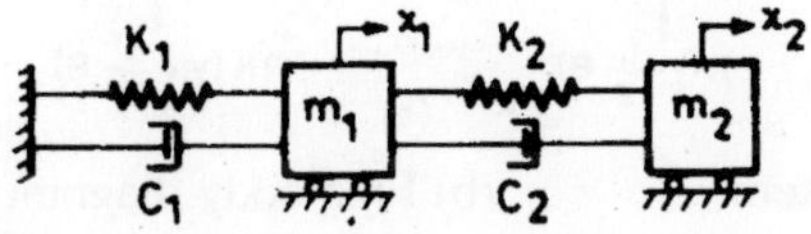

Fig. 5.9 Two degrees of freedom system

Assuming solutions in the following form

$$\left.\begin{aligned} x_1 = Ae^{ipt} \\ x_2 = Be^{ipt} \end{aligned}\right\} \quad (5.31)$$

and substituting them in Eq. (5.30), results in

$$\left.\begin{aligned} A\left[-m_1p^2 + i(c_1 + c_2)p + (k_1 + k_2)\right] + B\left[-ipc_2 - k_2\right] = 0 \\ -A\left[-ipc_2 - k_2\right] + B\left[-m_2p^2 + ipc_2 + k_2\right] = 0 \end{aligned}\right\} \quad (5.32)$$

A non-trivial solution of Eq. (5.32) is obtained, only when determinant formed by the coefficients is equal to zero. Expanding the determinant, one obtains

$$\left[-m_1p^2 + i(c_1 + c_2)p + (k_1 + k_2)\right]\left[-m_2p^2 + ipc_2 + k_2\right] - \left[ipc_2 + k_2\right]^2 = 0 \quad (5.33)$$

Equation (5.33) is an equation of fourth degree of p and has four roots, all of which are complex. The general solution is written as

$$\left.\begin{aligned} x_1 = A_1e^{ip_1t} + A_2e^{ip_2t} + A_3e^{ip_3t} + A_4e^{ip_4t} \\ x_2 = B_1e^{ip_1t} + B_2e^{ip_2t} + B_3e^{ip_3t} + B_4e^{ip_4t} \end{aligned}\right\} \quad (5.34)$$

We further know from Art. 5.2 and Eq. (5.32) that *As* and *Bs* are related.
Further, p_1 and p_2 as also p_3 and p_4 are complex conjugates.
This has been discussed in more details in section 5.8.

5.7 COORDINATE COUPLING

An idealised system having two masses m and $2m$ are connected to the ends of a rigid massless bar (Fig. 5.10). Two springs having stiffnesses k and $2k$ respectively, are connected to two ends of the bar. To describe the motion of the system, it has two degrees of freedom—the vertical translation y and rotation θ as shown. The origin is considered at the point of static equilibrium. For free vibration of the system, the forces acting on it are shown in Fig. 5.10(b).

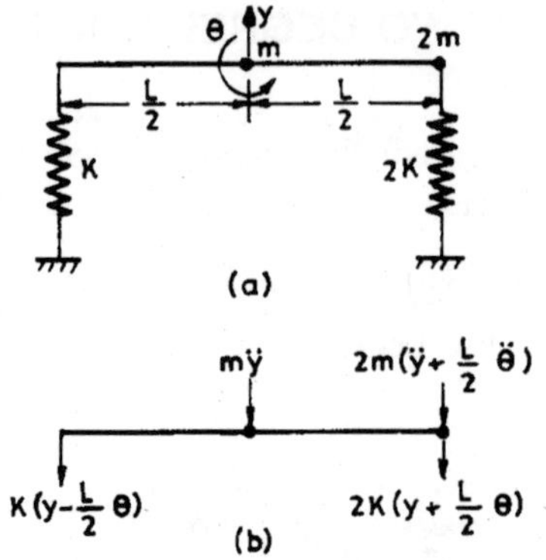

(a) The system (b) Free body diagram of the system

Fig. 5.10 A two mass system

Equation of motion for the system for vertical motion is

$$m\ddot{y}+2m\left(\ddot{y}+\frac{L}{2}\ddot{\theta}\right)+k\left(y-\frac{L}{2}\theta\right)+2k\left(y+\frac{L}{2}\theta\right)=0$$

or, $$3m\ddot{y}+mL\ddot{\theta}+3ky+\frac{kL}{2}\theta=0 \tag{5.35}$$

Taking moment of the forces about the origin as shown in Fig. 10(b)

$$2m\left(\ddot{y}+\frac{L}{2}\ddot{\theta}\right)\frac{L}{2}+2k\left(y+\frac{L}{2}\theta\right)\frac{L}{2}-k\left(y-\frac{L}{2}\theta\right)\frac{L}{2}=0$$

or, $$mL\ddot{y}+\frac{1}{2}mL^2\ddot{\theta}+\frac{kL}{2}y+\frac{3}{4}kL^2\theta=0 \tag{5.36}$$

Writing Eqs. (5.35) and (5.36) in matrix form, we get

$$\begin{bmatrix} 3m & mL \\ mL & \dfrac{mL^2}{2} \end{bmatrix}\begin{Bmatrix} \ddot{y} \\ \ddot{\theta} \end{Bmatrix}+\begin{bmatrix} 3k & \dfrac{kL}{2} \\ \dfrac{kL}{2} & \dfrac{3}{4}kL^2 \end{bmatrix}\begin{Bmatrix} y \\ \theta \end{Bmatrix}=\begin{Bmatrix} 0 \\ 0 \end{Bmatrix} \tag{5.37}$$

Equation (5.37) indicates the existence of coupling between the independent coordinates. If the mass matrix is non-diagonal, then it is dynamic coupling. If the stiffness matrix is non-diagonal, it is called static coupling. Equation (5.37) reveals both dynamic and static coupling. Equation (5.1) reveals the static coupling only. Two or multiple degrees of freedom systems always have coupled equations. The techniques of uncoupling the equations have been shown in the next chapter.

Example 5.4

A car body shown in Fig. 5.11 has the following values for the different quantities

$W = 150$ N, $L_1 = 1.35$ m, $L_2 = 1.65$ m, $k_1 = 360$ N/m , $k_2 = 370$ N/m, $J_c = 27$ m^4

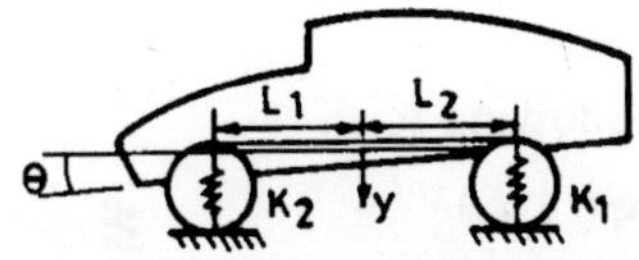

Fig. 5.11 Example 5.4

In its simplest form, car is treated as a rigid body. It is connected by springs at the front and the rear. The system will have two degrees of freedom in y and θ. The equations of motion of the system will be

$$\left.\begin{aligned} m\ddot{y}+k_1(y-L_1\theta)L_1+k_2(y+L_2\theta)=0 \\ J_c\ddot{\theta}-k_1L_1(x-L_1\theta)+k_2(y+L_2\theta)L_2=0 \end{aligned}\right\} \quad \text{(a)}$$

Assuming harmonic motion, frequency determinant can be shown to be as follows:

$$\begin{vmatrix} (k_1+k_2-p^2m) & -(k_1L_1-k_2L_2) \\ -(k_1L_1-k_2L_2) & (k_1L_1^2+k_2L_2^2-p^2J_c) \end{vmatrix}=0 \quad \text{(b)}$$

Substituting the values given in the problem, Eq. (b) becomes

$$\begin{vmatrix} 730-15.29p^2 & 124.5 \\ 124.5 & (1663-27p^2) \end{vmatrix}=0 \quad \text{or}$$

$$412.8p^4 - 45677p^2 + 1231750 = 0$$

or $$p^2 = 46.57, 64.07$$

or $$p = 6.82, 8.0 \text{ rad /s}$$

5.8 FREE VIBRATION OF DAMPED TWO DEGREES OF FREEDOM SYSTEM

We move on to deal with free vibration of two degrees of freedom system when damping is present. A two-degrees of freedom damped system has been shown in Fig. 5.12

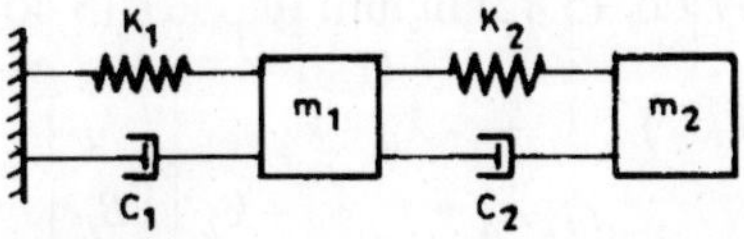

Fig. 5.12 A two-degree of freedom damped system

The equations of motions of two masses are

$$\left.\begin{aligned} m_1\ddot{x}_1+(c_1+c_2)\dot{x}_1+(k_1+k_2)x_1-c_2\dot{x}_2-k_2x_2=0 \\ m_2\ddot{x}_2+c_2\dot{x}_2+k_2x_2-c_2\dot{x}_1-k_2x_1=0 \end{aligned}\right\} \quad (5.38)$$

We can proceed as we have done in Section 5.2 by assuming a harmonic form of response. But due to the presence of damping, complex quantities will appear in the roots of the equations [5.1, 5.2].

Let us assume a solution of the following form

$$x_1(t) = Ae^{\lambda t}$$
$$x_2(t) = Be^{\lambda t} \tag{5.39}$$

Substitution of x_1 and x_2 from Eq. (5.39) into Eq. (5.38), yields

$$\left[m_1\lambda^2 + (c_1 + c_2)\lambda + (k_1 + k_2)\right]A - (c_2\lambda + k_2)B = 0$$
$$-(c_2\lambda + k_2)A + \left[m_2\lambda^2 + c_2\lambda + k_2\right]B = 0 \tag{5.40}$$

In order to obtain a nontrivial solution of Eq. (5.40), the characteristic determinant of the coefficients is equal to zero, that is

$$\begin{vmatrix} m_1\lambda^2 + (c_1 + c_2)\lambda + (k_1 + k_2) & -(c_2\lambda + k_2) \\ -(c_2\lambda + k_2) & m_2\lambda^2 + c_2\lambda + k_2) \end{vmatrix} = 0$$

or,

$$m_1m_2\lambda^4 + \{m_1c_1 + (c_1 + c_2)m_2\}\lambda^3 + \{m_1k_2 + c_1c_2 + (k_1 + k_2)m_2\}\lambda^2$$
$$+ (c_1k_2 + k_1c_2)\lambda + k_1k_2 = 0 \tag{5.41}$$

Eq. (5.41) will have four roots, which according to the theory of algebraic equations must be either real or negative or complex with negative real parts. Complex roots of algebraic equations always occur in conjugate pairs.

Two complex conjugate pairs are assumed of the following form:

$$\lambda_{11} = -a_1 + ip_{d_1}$$
$$\lambda_{12} = -a_1 - ip_{d_1}$$
$$\lambda_{21} = -a_2 + ip_{d_2} \tag{5.42}$$
$$\lambda_{22} = -a_2 - ip_{d_2}$$

where damped natural frequencies are $p_{d_i} = p_i\sqrt{1-\zeta^2}$ and a_i are positive numbers.

Substituting each root given by Eq. (5.42) in turn into Eq. (5.40), yields

$$\begin{bmatrix} m_1\lambda_{ij}^2 + (c_1 + c_2)\lambda_{ij} + (k_1 + k_2) & -c_2\lambda_{ij} - k_2 \\ -c_2\lambda_{ij} - k_2 & m_2\lambda_{ij}^2 + c_2\lambda_{ij} + k_2 \end{bmatrix} \begin{Bmatrix} A_{ij} \\ B_{ij} \end{Bmatrix} = \begin{Bmatrix} 0 \\ 0 \end{Bmatrix} \tag{5.43}$$

From Eq. (5.43), we get four possible ratios

$$\frac{A_{ij}}{B_{ij}} = \gamma_{ij} = \frac{c_2\lambda_{ij} + k_2}{m_1\lambda_{ij}^2 + (c_1 + c_2)\lambda_{ij} + (k_1 + k_2)} = \frac{m_2\lambda_{ij}^2 + c_2\lambda_{ij} + k_2}{c_2\lambda_{ij} + k_2} \tag{5.44}$$

$i = 1, 2$ and $j = 1, 2$

The ratios γ_{11}, γ_{12}, γ_{21} and γ_{22} are modal conjugate pairs that relate the modal amplitudes which are generally of complex modes.

$$A_{ij} = \gamma_{ij} \, B_{ij} \tag{5.45}$$

The complete solution of the equation is

$$\begin{aligned} x_1(t) &= \gamma_{11} B_{11} e^{\lambda_{11} t} + \gamma_{12} B_{12} e^{\lambda_{12} t} + \gamma_{21} B_{21} e^{\lambda_{21} t} + \gamma_{22} B_{22} e^{\lambda_{22} t} \\ x_2(t) &= B_{11} e^{\lambda_{11} t} + B_{12} e^{\lambda_{12} t} + B_{21} e^{\lambda_{21} t} + B_{22} e^{\lambda_{22} t} \end{aligned} \tag{5.46}$$

(B_{11}, B_{12}) and (B_{21}, B_{22}) are complex conjugate pairs that are to be determined from initial conditions.

Now,

$$B_{11} e^{\lambda_{11} t} + B_{12} e^{\lambda_{12} t} = e^{-a_1 t}\left(C_1 \cos p_{d_1} t + C_2 \sin p_{d_1} t\right) \tag{5.47}$$

Note that C_1 and C2 are real numbers

where $$C_1 = B_{11} + B_{12}, \quad C_2 = i\left(B_{11} - B_{12}\right) \tag{5.48}$$

Similarly,

$$\gamma_{11} = \alpha_1 + i\beta_1, \quad \gamma_{12} = \alpha_1 - i\beta_1 \tag{5.49}$$

Then

$$\gamma_{11} B_{11} e^{\lambda_{11} t} + \gamma_{12} B_{12} e^{\lambda_{12} t} = e^{-a_1 t}\left[(C_1\alpha_1 - C_2\beta_2) cos\, p_{d_1} t + (C_1\beta_1 - C_2\alpha_1) sin\, p_{d_1} t\right] \tag{5.50}$$

Similar treatment is meted out to second pair of complex conjugate.

$$C_3 = B_{21} + B_{22}, \qquad C_4 = i\left(B_{21} - B_{22}\right) \tag{5.51}$$

$$\gamma_{21} = \alpha_2 + i\beta_2, \qquad \gamma_{21} = \alpha_2 - i\beta_2 \tag{5.52}$$

Therefore, Eq. (5.46) can be written as

$$\begin{aligned} x_1(t) &= e^{-a_1 t}\left(\gamma_1 C_1 \cos p_{d_1} t + \gamma_1' C_2 \sin p_{d_1} t\right) \\ &\quad + e^{-a_2 t}\left(\gamma_2 C_3 \cos p_{d_2} t + \gamma_2' C_4 \sin p_{d_2} t\right) \end{aligned}$$

$$\begin{aligned} x_2(t) &= e^{-a_1 t}\left(C_1 \cos p_{d_1} t + C_2 \sin p_{d_1} t\right) \\ &\quad + e^{-a_2 t}\left(C_3 \cos p_{d_2} t + C_4 \sin p_{d_2} t\right) \end{aligned} \tag{5.53}$$

where the amplitude ratios are given by

$$\begin{aligned} \gamma_1 &= \frac{C_1\alpha_1 - C_2\beta_1}{C_1}, \qquad \gamma_1' = \frac{C_1\beta_1 + C_2\alpha_1}{C_2} \\ \gamma_2 &= \frac{C_3\alpha_2 - C_4\beta_2}{C_3}, \qquad \gamma_2' = \frac{C_2\beta_2 + C_4\alpha_2}{C_4} \end{aligned} \tag{5.54}$$

Eq. (5.53) denotes the damped response of two masses, which can be compared with Eq. (5.10) for the undamped case. Eq. (5.53) can be written as

$$x_1(t)= B_1' e^{-a_1 t} \cos\left(p_{d_1} t - \phi'_{d_1}\right) + B_2' e^{-a_2 t} \cos\left(p_{d_2} t - \phi'_{d_2}\right)$$
$$x_2(t)= B_1 e^{-a_1 t} \cos\left(p_{d_1} t - \phi'_{d_1}\right) + B_2 e^{-a_2 t} \cos\left(p_{d_2} t - \phi'_{d_2}\right) \tag{5.55}$$

where

$$B_1 = \sqrt{C_1^2 + C_2^2}\,, \qquad B_2 = \sqrt{C_3^2 + C_4^2}$$
$$B_1' = B_1\sqrt{\alpha_1^2 + \beta_1^2}\,, \qquad B_2' = B_2\sqrt{\alpha_2^2 + \beta_2^2} \tag{5.56}$$

$$\phi_{d_1} = \tan^{-1}\left(\frac{C_2}{C_1}\right), \qquad \phi_{d_2} = \tan^{-1}\left(\frac{C_4}{C_3}\right)$$
$$\phi'_{d_1} = \tan^{-1}\left(\frac{\gamma_1' C_2}{\gamma_1 C_1}\right), \quad \phi'_{d_2} = \tan^{-1}\left(\frac{\gamma_2' C_4}{\gamma_2 C_3}\right) \tag{5.57}$$

Natural modes that exist have a phase relationship, that is, they are out of phase.

Some simplifications are possible for the case of small viscous damping where it is approximated as follows:

$$p_{d_1} = p_1, \quad \gamma_1' = \gamma_1$$
$$p_{d_2} = p_2, \quad \gamma_2' = \gamma_2 \tag{5.58}$$

Based on the assumption of Eq. (5.58), the responses of the two masses are

$$x_1(t) = \gamma_1 e^{-a_1 t}\left(C_1 \cos p_1 t + C_2 \sin p_1 t\right) + \gamma_2 e^{-a_2 t}\left(C_3 \cos p_2 t + C_4 \sin p_2 t\right)$$
$$x_2(t) = e^{-a_1 t}\left(C_1 \cos p_1 t + C_2 \sin p_1 t\right) + e^{-a_2 t}\left(C_3 \cos p_2 t + C_4 \sin p_2 t\right) \tag{5.59}$$

where C_1, C_2, C_3 and C_4 are to be determined from initial conditions.

Example 5.5

For the two-degree of freedom system shown in Fig. 5.12, the system parameters are given by $m_1 = m_2 = 2$ kg, $c_1 = c_2 = 2$ Ns/m and $k_1 = k_2 = 2$ N/m. Determine the eigenvalues, mode shapes and the equation of motion of the two masses.

The equation of motion is given by

$$\begin{bmatrix} 2 & 0 \\ 0 & 2 \end{bmatrix}\begin{Bmatrix} \ddot{x}_1 \\ \ddot{x}_2 \end{Bmatrix} + \begin{bmatrix} 4 & -2 \\ -2 & 2 \end{bmatrix}\begin{Bmatrix} \dot{x}_1 \\ \dot{x}_2 \end{Bmatrix} + \begin{bmatrix} 4 & -2 \\ -2 & 2 \end{bmatrix}\begin{Bmatrix} x_1 \\ x_2 \end{Bmatrix} = \begin{Bmatrix} 0 \\ 0 \end{Bmatrix} \tag{a}$$

The characteristic matrix is given by [Eq. (5.40)].

$$\begin{bmatrix} 2\lambda^2 + 4\lambda + 4 & -2(\lambda + 1) \\ -2(\lambda + 1) & 2\lambda^3 + 2\lambda + 2 \end{bmatrix}\begin{Bmatrix} \text{A} \\ \text{B} \end{Bmatrix} = \begin{Bmatrix} 0 \\ 0 \end{Bmatrix} \tag{b}$$

The characteristic equation then is

$$\lambda^4 + 3\lambda^3 + 4\lambda^2 + 2\lambda + 1 = 0 \tag{c}$$

The roots of Eq. (a) are

$$\begin{aligned} \lambda_{11} &= -1.309 + 0.951\,i \\ \lambda_{12} &= -1.309 - 0.951\,i \end{aligned} \tag{d}$$

$$\begin{aligned} \lambda_{21} &= -0.191 + 0.588\,i \\ \lambda_{22} &= -0.191 - 0.588\,i \end{aligned}$$

Therefore,

$$\frac{A_{11}}{B_{11}} = \gamma_{11} = \frac{\lambda_{11} + 1}{\lambda_{11}^2 + 2\lambda_{11} + 2} = \frac{-0.309 + 0.951i}{0.191 - 0.588i} \tag{e}$$

Multiplying both the numerator and dominator by the complex conjugate of the dominator yields

$$\gamma_{11} = -1.618 \tag{f}$$

where a small imaginary part remains as a round-off error. This negligible imaginary value suggests that the phase difference between the modes is negligible, that is, the masses vibrate as if no damping is present in the system.

Similarly, we have $\gamma_{12} = -1.618, \gamma_{21} = 0.618, \gamma_{22} = 0.618$.

We have

$$a_1 = 1.309, \quad p_{d_1} = 0.951, \quad a_2 = 0.191, \quad p_{d_2} = 0.588$$

and $$\alpha_1 = -1.618, \quad \alpha_2 = 0.618, \quad \beta_1 = 0, \quad \beta_2 = 0$$

These parameters when substituted into Eq. (5.53), yields

$$x_1(t) = e^{-1.309t}\left[\frac{C_1(1.618) - C_2\,\beta_1}{C_1} C_1 \cos 0.951\,t \right.$$

$$\left. + \frac{C_1(\beta_1) + C_2(-1.618)}{C_2} C_2 \sin 0.951\,t\right]$$

$$+ e^{-0.191t}\left[\frac{C_3(-0.618) - C_4(\beta_2)}{C_3} C_3 \cos 0.588\,t \right.$$

$$\left. + \frac{C_3(\beta_2) + C_4(0.618\,t)}{C_4} C_4 \sin 0.588\,t\right]$$

$$x_2(t) = e^{-1.309t}\left[C_1 \cos 0.951t + C_2 \sin 0.951t\right]$$

$$+ e^{-0.191t}\left[C_3 \cos 0.588t + C_4 \sin 0.588t\right]$$

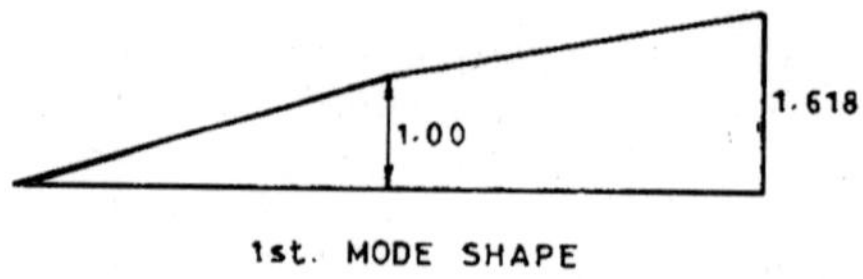

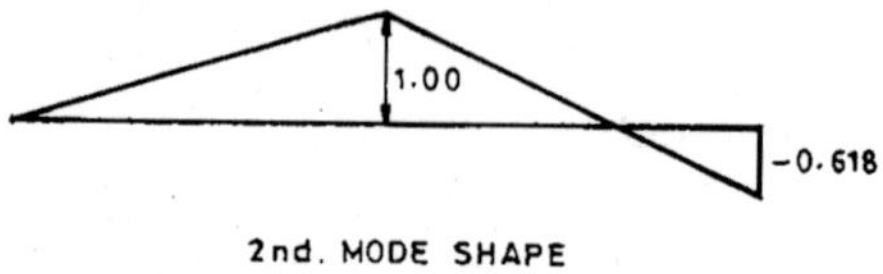

Fig. 5.13 Mode Shapes of Example 5.5

References

5.1 W. Weaver, S.P. Timoshenko and D.H. Young, Vibration Problems in Engineering, Fifth Edition, Wiley Interscience, 1990.

5.2 H. Benaroya, Mechanical Vibration, Practice Hall, New Jersey, 1998.

EXERCISE 5

5.1 Determine the natural frequencies of the system shown in the figure.

Prob. 5.1.

5.2 Determine the natural frequencies of the system shown in the figure.

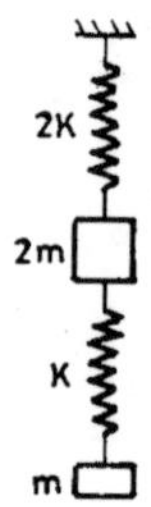

Prob. 5.2.

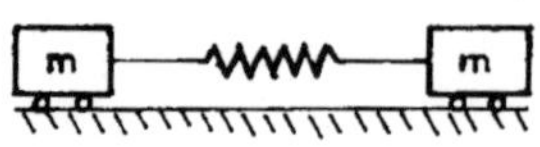

Prob. 5.3.

5.3 Determine the natural frequencies of the semi-definite system

5.4 Determine the natural frequencies and mode shapes of the system shown.

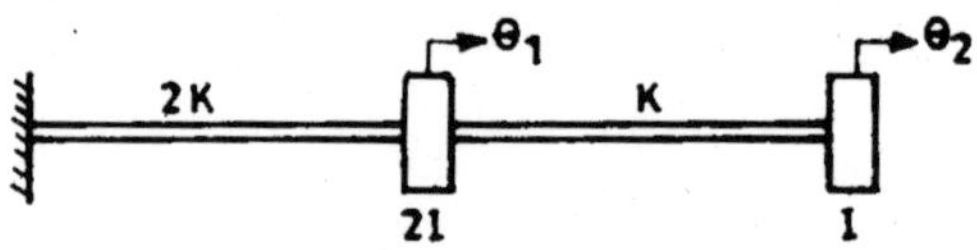

Prob. 5.4.

5.5 Determine the natural frequencies and mode shapes of the system.

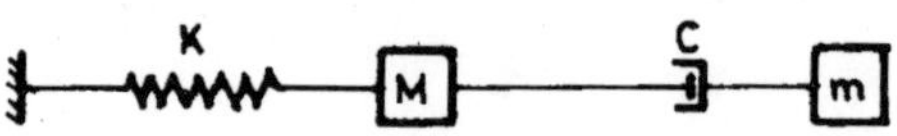

Prob. 5.5.

5.6 Find the natural frequencies and mode shapes for the torsional system shown in figure, for $J_1 = J_o,\ J_2 = 2J_o, k_1 = k_2 = k_o$

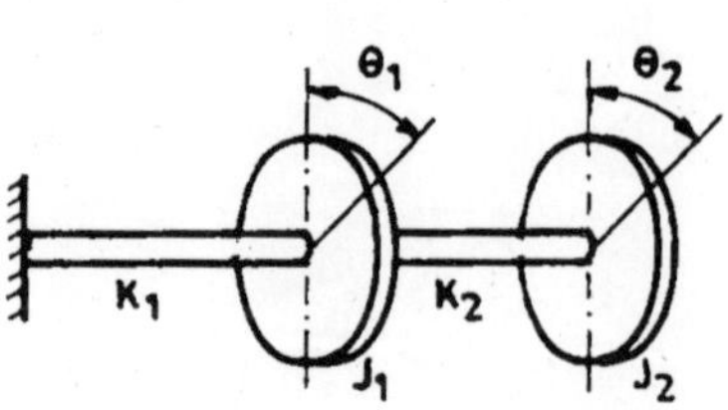

Prob. 5.6

5.7 Determine the natural frequency of the torsional system and draw the mode shape. $G = 81\ N/mm^2$.

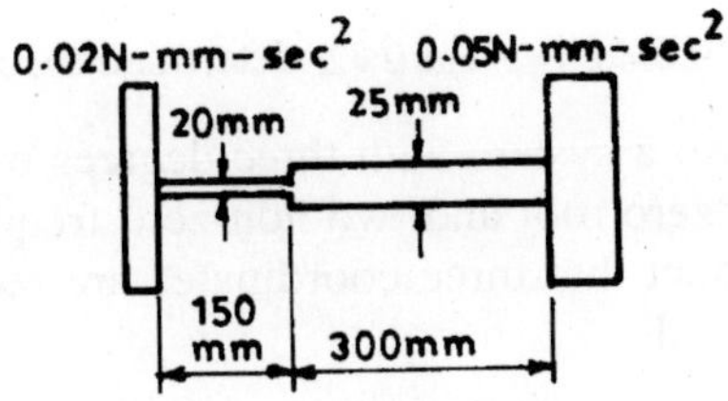

Prob. 5.7

5.8 Write the equations of motion for the two degrees of freedom system of Fig. 5.1, when left support moves by $x_s = x_{s0} \sin \omega t$.

5.9 Determine the principal coordinates of the system shown in figure.

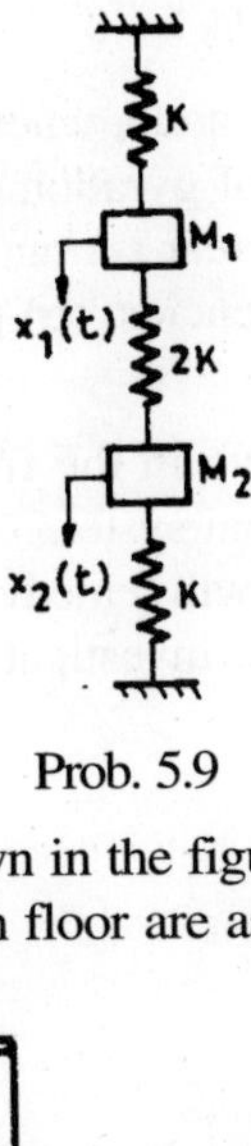

Prob. 5.9

5.10 For the two storeyed frame shown in the figure, determine the natural frequencies and mode shapes. The beams at each floor are assumed to be infinitely rigid.

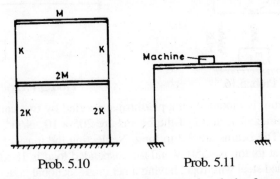

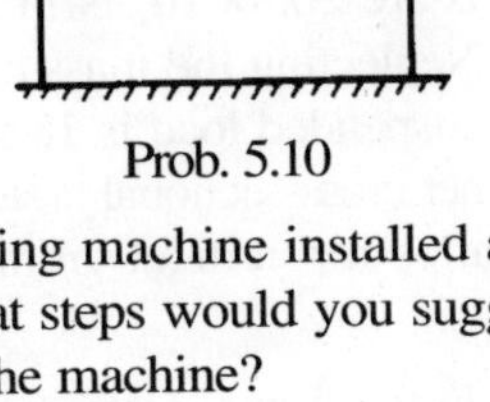

Prob. 5.10

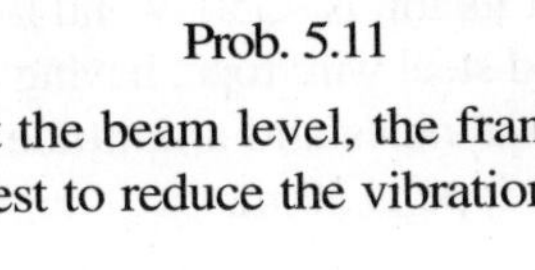

Prob. 5.11

5.11 Due to a rotating machine installed at the beam level, the frame of the figure sways violently. What steps would you suggest to reduce the vibration without changing the frequency of the machine?

5.12 A coupled system has the following equations of motion

$$\ddot{x}+1000x-100\theta = 0$$

$$\ddot{\theta}+1000x-100\theta = 0$$

Given that at $t=0, x=10, \theta=0, \dot{x}=5$ and $\dot{\theta}=2$, determine the resultant motion of the system.

5.13 In the figure is shown a system with three degrees of freedom. The characteristic equation yields one zero root and two non-zero frequencies. Discuss the physical significance of the fact that three coordinates are required, but only two natural frequencies are obtained.

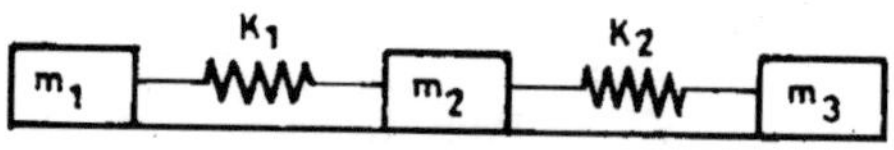

Prob. 5.13

5.14 Find the response of the system of Fig. 5.1, if both the masses are subjected to an external force $F_o \sin \omega t$.

Assume $k_1 = k_2 = k$ and $m_1 = m_2 = m$

5.15 A vehicle has a mass of 2000 kg and a wheelbase of 3.5 m. The mass centre is 1.5 m from the front axle. The radius of gyration of the wheel about the c.g. is 1.5 m. The spring constants of the front and rear springs are 40 kN/m and 50 kN/m, respectively. Determine (a) the natural frequencies, (b) principal modes of vibration and (c) the motion $x(t)$ and $q(t)$.

5.16 A constant speed machine shown in the figure is having excessive vibration. The values of m_1 and k_1 of the original system cannot be changed. Show that a dynamic absorber, consisting of m_2 and k_2 will remedy the problem. Assuming $m_2/m_1 = 0.5$, plot the response curve of the system. Investigate the effect of the mass ratio m_2/m_1.

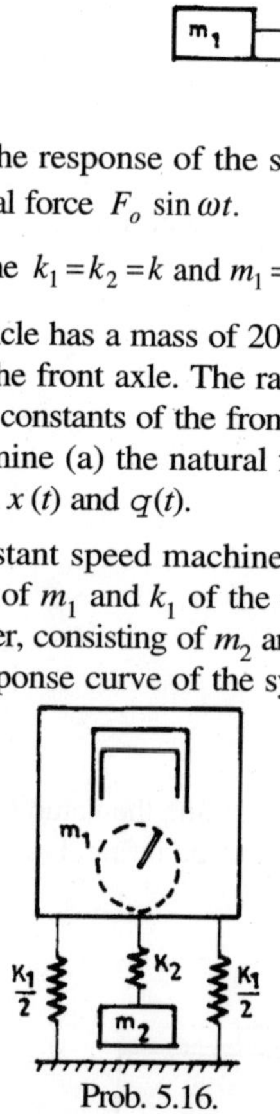

Prob. 5.16.

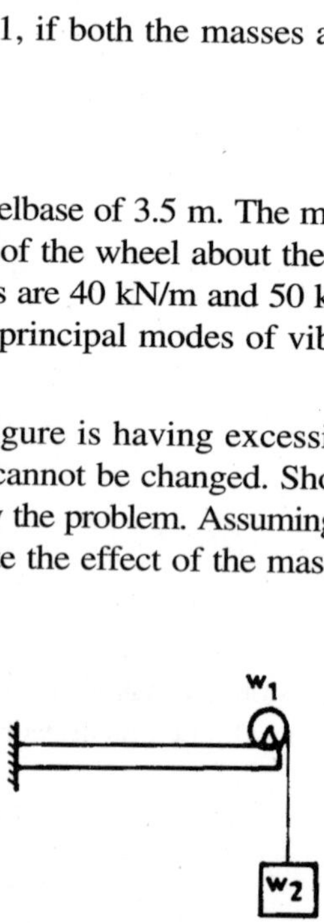

Prob. 5.17.

5.17 A hoisting drum is mounted on a platform supported by two cantilever beams at their two ends. Values of E and I of the beams are 205×10^2 N/m^2, and 2672×10^4 mm^4 respectively. The beams are 3.2 m long. Neglecting the mass of the beam, the weight concentrated at its top is 4550 N and the suspended load is 18 kN. The hoist rope is a 12 mm stranded steel wire rope, having a net cross-sectional area of 0.65 cm^2. Calculate the natural frequencies and amplitude ratios for vertical vibration, if the suspended length L of the rope is 30 m.

5.18 Consider a two storeyed structure shown in the figure. Determine the displacement of the masses as a function of time for free vibration, if the top storey is given an initial displacement of 50 mm.

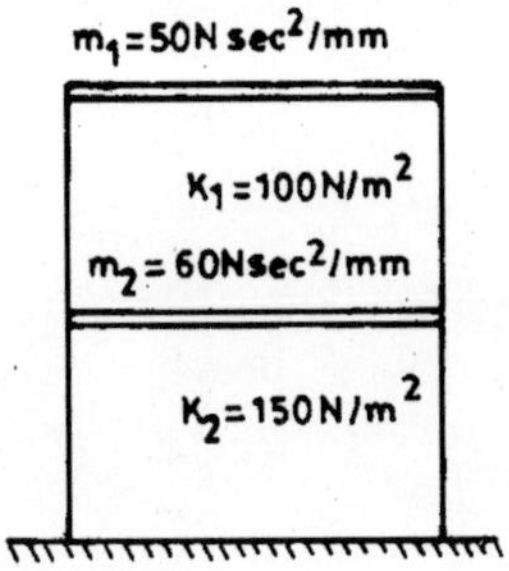

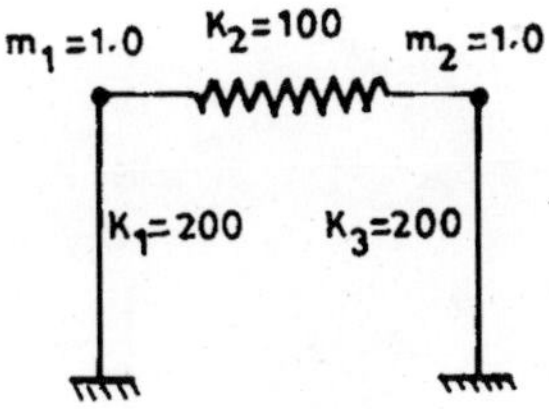

Prob. 5.18 Prob. 5.19

5.19 A structure linked together by spring element is shown. Determine the displacement of the masses for free vibration, if an initial displacement of 50 mm, 180° out of phase is given for both the masses.

5.20 A cantilever beam is modelled by lumping the mass as shown. The density of the material of the beam is ρ and E is the modulus of elasticity. Determine the natural frequencies and mode shapes of this model.

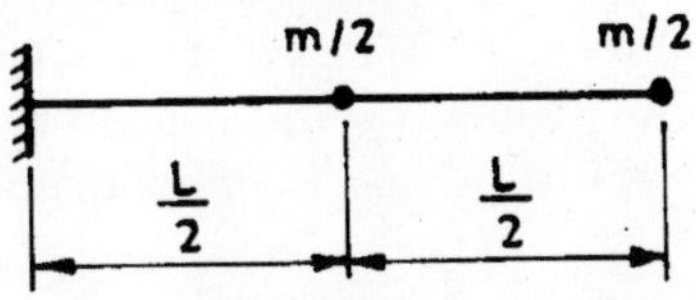

Prob. 5.20

5.21 Determine the steady-state vibration of the system shown in figure, assuming that $F_1(t) = \bar{F}_1 \cos \omega t$ and $F_2(t) = \bar{F}_2 \cos \omega t$.

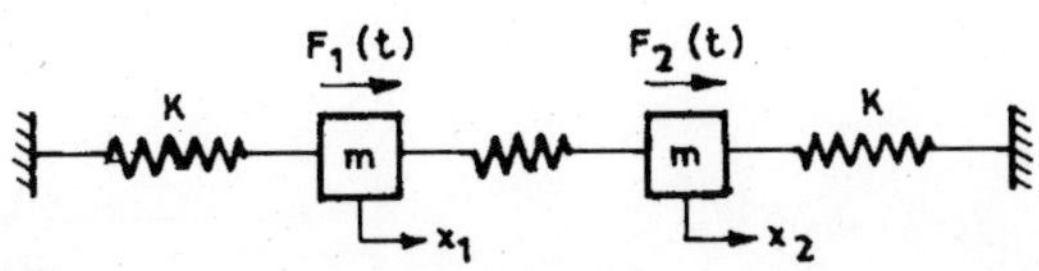

Prob. 5.21

5.22 In a refrigeration plant a section of the pipe carrying the refrigerant vibrated violently at a compressor speed of 200 rpm. To eliminate this problem, spring-mass system is clamped to the pipe to act as an absorber. In the trial test the 4 kg absorber turned to 250 cpm resulted in two natural frequencies of 190 and 250 cpm. If the absorber system is to be designed so that the natural frequencies lie outside the region 170 and 300 cycles/m, what must be the weight and spring stiffness?

Chapter 6

Free Vibration of Multiple Degrees of Freedom System

6.1 INTRODUCTION

After getting a taste of two degrees of freedom system, we now proceed to a more general treatment of the problem. In the previous chapter, it is revealed that the two degrees of freedom system is more involved in computation than the single degree of freedom system. When more number of masses are considered, the mathematical formulation becomes much more complicated.

The theory of multiple degrees of freedom system (MDF system) is important, because they are applicable not only to systems which may be classified as discrete, but also to the numerical solution of the continuous systems converted into discrete systems, by lumping the mass at significant points. In many cases, a structure represented by its model of a single degree of freedom does not describe it adequately. A truer representation of such cases will be the consideration of multiple degrees of freedom system of the structure. Many structures such as framed structures have essentially lumped masses, since the mass of the columns is often negligible compared to floors and the system can be treated as multiple degrees of freedom system. The free vibration analysis of the multiple degrees of freedom system is presented in this chapter.

6.2 EQUATIONS OF MOTION OF MDF SYSTEMS

A frame structure with rigid floors and multiple discs mounted on a shaft are shown in Fig. 6.1. Both these systems can be reduced to equivalent spring-mass-damper system shown in Fig. 6.2(a).

The freebody diagrams of the three masses are shown in Fig. 6.2(b). Applying D'Alembert's principle, the equations of motion are written as follows:

$$\left.\begin{aligned} m_1\ddot{x}_1 + c_1\dot{x}_1 + c_2(\dot{x}_1 - \dot{x}_2) + k_1x_1 + k_2(x_1 - x_2) &= F_1(t) \\ m_2\ddot{x}_2 + c_2(\dot{x}_2 - \dot{x}_1) + c_3(\dot{x}_2 - \dot{x}_3) + k_2(x_2 - x_1)& \\ + k_3(x_2 - x_3) &= F_2(t) \\ m_3\ddot{x}_3 + c_3(\dot{x}_3 - \dot{x}_2) + k_3(x_3 - x_2) &= F_3(t) \end{aligned}\right\} \quad (6.1)$$

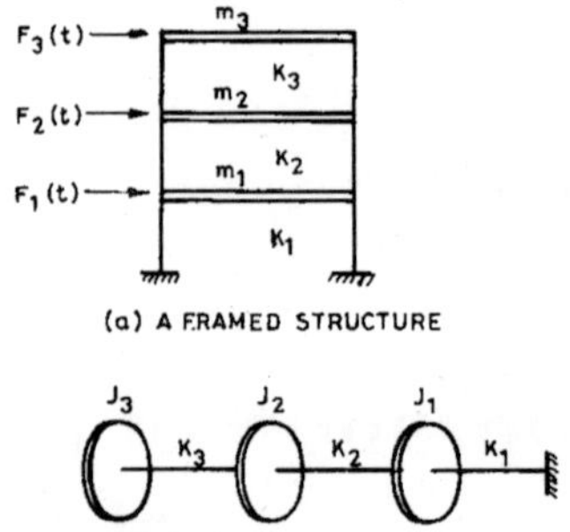

(a) A FRAMED STRUCTURE

(b) MULTIPLE DISCS ON A SHAFT

Fig 6.1

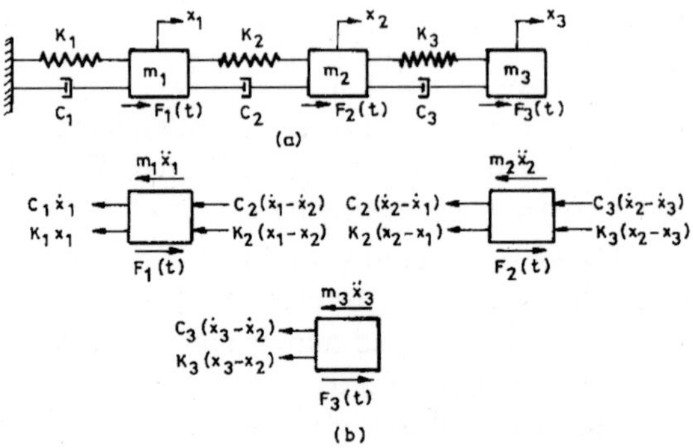

Fig. 6.2

$$
\left.\begin{array}{c}
m_1\ddot{x}_1 + (c_1 + c_2)\dot{x}_1 - c_2\dot{x}_2 + (k_1 + k_2)x_1 - k_2x_2 = F_1(t) \\
m_2\ddot{x}_2 - c_2\dot{x}_1 + (c_2 + c_3)\dot{x}_2 - c_3\dot{x}_3 - k_1x_1 \\
+ (k_2 + k_3)x_2 - k_3x_3 = F_2(t) \\
m_3\ddot{x}_3 - c_3\dot{x}_2 + c_3\dot{x}_3 - k_3x_2 + k_3x_3 = F_3(t)
\end{array}\right\} \quad (6.2)
$$

Equation (6.2) can be written in matrix form as follows:

$$
\begin{bmatrix} m_1 & 0 & 0 \\ 0 & m_2 & 0 \\ 0 & 0 & m_3 \end{bmatrix}\begin{Bmatrix} \ddot{x}_1 \\ \ddot{x}_2 \\ \ddot{x}_3 \end{Bmatrix} + \begin{bmatrix} c_1 + c_2 & -c_2 & 0 \\ -c_2 & c_2 + c_3 & -c_3 \\ 0 & -c_3 & c_3 \end{bmatrix}\begin{Bmatrix} \dot{x}_1 \\ \dot{x}_2 \\ \dot{x}_3 \end{Bmatrix}
$$

$$
+ \begin{bmatrix} k_1 + k_2 & -k_2 & 0 \\ -k_2 & k_2 + k_3 & -k_3 \\ 0 & -k_3 & k_3 \end{bmatrix}\begin{Bmatrix} x_1 \\ x_2 \\ x_3 \end{Bmatrix} = \begin{Bmatrix} F_1(t) \\ F_2(t) \\ F_3(t) \end{Bmatrix} \quad (6.3)
$$

Equation (6.3) can be written in compact form as

$$
[M\}\{\ddot{x}\} + [C]\{\dot{x}\} + [K]\{x\} = \{F(t)\} \quad (6.4)
$$

where [*M*] is known as the mass matrix,

$[C]$ is the damping matrix,
$[K]$ is the stiffness matrix, and
$\{F(t)\}$ is the loading matrix.

If the system has n degrees of freedom, sizes of $[M]$, $[C]$ and $[K]$ will be of the order $n \times n$. Equation (6.4) represents the general form of the equation of a system of n degrees of freedom.

6.2.1 Mass Matrix

There are two ways of forming the mass matrix of the structure. They are: (1) lumped mass matrix and (2) consistent mass matrix. The mass matrix of Eq. (6.3) is a lumped mass matrix. In this system, masses are lumped at nodal points. They are assumed to be independently placed without any interaction between them. In the lumped mass system, the mass matrix is diagonal, which gives a considerable advantage in the computation of different quantities related to the vibration analysis.

If a nodal point has more than one translational degree of freedom, the same mass will be associated with each degree of freedom. If there is a rotational degree of freedom, then the corresponding diagonal element in the mass matrix will be considered as zero, provided rotational inertia of the mass is not lumped.

The consistent mass matrix will be discussed later.

6.2.2 The Stiffness Matrix

For linear elastic systems, the stiffness matrix $[K]$ is a symmetric matrix. The structure consists of a number of elements. Total stiffness matrix is formed by assembling the stiffness matrix of the individual elements. The assembly can be done in a systematic manner in a computer by writing a suitable computer program. The stiffness matrix of a structure for dynamic analysis is calculated by utilising the methods used in standard structural analysis procedures.

6.2.3 The Damping Matrix

The damping matrix $[C]$ given by Eq. (6.3) has been formed by assuming the system to have viscous damping. The damping in a system depends on various factors. These values are to be obtained experimentally. Unfortunately, not much data are available on full-scaled structures. Further, prescribing the damping matrix in the form given by Eq. (6.3), poses considerable difficulties in the analysis. As such, $[C]$ matrix is reduced to simpler forms for facilitating the analysis.

6.2.4 Loading Matrix

The dynamic loads are assumed to act at nodal points corresponding to the displacement degrees of freedom. Loads when acting in between nodal points, or when they are distributed, are converted to equivalent values acting at respective nodal points.

6.3 FREE UNDAMPED VIBRATION ANALYSIS OF MDF SYSTEMS

The equation of motion for MDF system for free vibration can be written from Eq. (6.4) as

$$[M]\{\ddot{x}\} + [K]\{x\} = \{O\} \tag{6.5}$$

Assuming the solution in the following form

$$\{x\} = ae^{ipt}\{\phi\} \tag{6.6}$$

where a is a scalar of dimension L,

p is a scalar of dimension T^{-1},

$\{\phi\}$ is a non-dimensional vector, such that

$$\{\phi\}^T = \{\phi_1, \phi_2, \phi_3, \cdots, \phi_n\}$$

Therefore,

$$\{\ddot{x}\} = -ap^2 e^{ipt}\{\phi\} \tag{6.7}$$

Substitution of the values of $\{\ddot{x}\}$ and $\{x\}$ from Eqs. (6.6) and (6.7) into Eq. (6.5) yields

$$-p^2[M]\{\phi\} + [K]\{\phi\} = \{O\} \tag{6.8}$$

or

$$[M]\{\phi\} - \lambda[K]\{\phi\} = \{O\} \tag{6.9}$$

where $\lambda = \dfrac{1}{p^2}$

or

$$([M] - \lambda[K])\{\phi\} = \{O\} \tag{6.10}$$

Equation (6.9) will have a non-trivial solution only if the determinant corresponding to $([M] - \lambda[K])$ vanishes, that is

$$\left|[M] - \lambda[K]\right| = 0 \tag{6.11}$$

The determinant is known as the frequency determinant. The solution of Eq. (6.11) will give n values of λ or $1/p^2$, where n is the number of degrees of freedom. $p_1, p_2, ..., p_r, ..., p_n$ are called natural frequencies of the system. Corresponding to each value of p_r, a vector $\{\phi^{(r)}\}$ can be evaluated. These are variously termed as natural mode, normal mode or principal mode of vibration.

Equation (6.9) is rewritten in terms of a typical eigenvalue problem. Premultiplying both sides of Eq. (6.9) by $[K]^{-1}$, yields

$$[K]^{-1}[M]\{\phi\} = \lambda[I]\{\phi\} \tag{6.12}$$

or

$$[D]\{\phi\} = \lambda[I]\{\phi\} \tag{6.13}$$

where $[D] = [K]^{-1}[M]$

Equation (6.13) is in the form of a typical eigenvalue problem. Equation (6.13) contains a set of n homogeneous equations in $\{\phi\}$ with λ as unknown. The quantities λ or $1/p^2$ are called eigenvalues or characteristic values. The vector $\{\phi\}$ is called the characteristic vector or the eigenvector or the modal vector.

One of the various approaches of obtaining the solution is to make the following determinant zero.

$$\left|[D] - \lambda[I]\right| = 0 \tag{6.14}$$

The determinant when expanded gives a polynomial of degree n. Thus

$$\lambda^n + C_1\lambda^{n-1} + C_2\lambda^{n-2} + \cdots + C_{n-1}\lambda + C_n = 0 \tag{6.15}$$

Solution of the polynomial equation known as characteristic equation or frequency equation given by Eq. (6.15) will yield n values of λ. Once λs are determined, next ϕs are to be obtained. It may be noted that unique solutions of fs do not exist. We obtain only ratios among ϕs. In other words, there exists for each mode, a unique solution for $(n - 1)$ of the ϕs, if we assign an arbitrary value to one of them.

Before we embark upon the solution techniques of eigenvalue problem, let us look into the orthogonality relationship.

6.4 ORTHOGONALITY RELATIONSHIP

For a MDF system, let p_r and p_s, be two natural frequencies and $\{\phi^{(r)}\}$ and $\{\phi^{(s)}\}$ be the corresponding modal vectors.

Applying Eq. (6.8) for the above two frequencies, the following equations are obtained

$$\left.\begin{aligned} -p_r^2[M]\{\phi^{(r)}\}+[K]\{\phi^{(r)}\}=\{0\} \\ -p_s^2[M]\{\phi^{(s)}\}+[K]\{\phi^{(s)}\}=\{0\} \end{aligned}\right\} \quad (6.16)$$

Premultiplying the first of Eq. (6.16) by $\{\phi^{(s)}\}^T$ and the second by $\{\phi^{(r)}\}^T$, we get

$$\left.\begin{aligned} -p_r^2\{\phi^{(s)}\}^T[M]\{\phi^{(r)}\}+\{\phi^{(s)}\}^T[K]\{\phi^{(r)}\}=\{0\} \\ -p_s^2\{\phi^{(r)}\}^T[M]\{\phi^{(s)}\}+\{\phi^{(r)}\}^T[K]\{\phi^{(s)}\}=\{0\} \end{aligned}\right\} \quad (6.17)$$

Taking transpose of the second of Eq. (6.17), we get

$$-p_s^2\{\phi^{(s)}\}^T[M]\{\phi^{(r)}\}+\{\phi^{(s)}\}^T[K]\{\phi^{(r)}\}=\{0\} \quad (6.18)$$

Comparing Eq. (6.18) with the first of Eq. (6.17), we get for $r \neq s$,

$$\{\phi^{(s)}\}^T[M]\{\phi^{(r)}\}=\{0\} \quad (6.19)$$

and

$$\{\phi^{(s)}\}^T[K]\{\phi^{(r)}\}=\{0\} \quad (6.20)$$

If $[M]$ is a diagonal matrix, Eq. (6.20) can be written as

$$\sum_{i=1}^{n} m_i\phi_i^{(r)}\phi_i^{(s)}=0 \quad (6.21)$$

Equations (6.19) to (6.21) are known as orthogonality relationship. As we shall see later in this chapter, this relationship is used for solving eigenvalue problems, as also for the treatment of forced vibration. The orthogonality relationship is a fundamental property of MDF systems.

We now try to give a physical meaning to the orthogonality relationship [6.1]. Let there be two masses m_1 and m_2. The mode shapes corresponding to two masses be $\phi_1^{(1)}$ and $\phi_2^{(1)}$ for the first natural frequency and $\phi_1^{(2)}$ and $\phi_2^{(2)}$ for the second natural frequency. According to the orthogonality relationship given by Eq. (6.21), we get

$$m_1\phi_1^{(1)}\phi_1^{(2)}+m_2\phi_2^{(1)}\phi_2^{(2)}=0 \quad (6.22)$$

Multiplying both sides of Eq. (6.22) by $-a^2 e^{ip_1 t} e^{ip_2 t} p_1^2$, we get

$$-(m_1 a p_1^2 e^{ip_1 t}\cdot\phi_1^{(1)})(a e^{ip_2 t}\cdot\phi_1^{(2)}) - (m_2 a p_1^2 e^{ip_1 t}\cdot\phi_2^{(1)})(a e^{ip_2 t}\cdot\phi_2^{(2)}) = 0 \quad (6.23)$$

We know from Eq. (6.6), that

$$\left.\begin{aligned} x_1 &= a e^{ipt}\phi_1 \\ x_2 &= a e^{ipt}\phi_2 \end{aligned}\right\} \quad (6.24)$$

Substituting these values of x_1, x_2 and their differentiations $\ddot{x}_1$ and $\ddot{x}_2$, Eq. (6.23) can be written as

$$m_1\ddot{x}_1^{(1)}\cdot x_1^{(2)} + m_2\ddot{x}_2^{(1)} x_2^{(2)} = 0 \quad (6.25)$$

where the following interpretations are made

$m_1\ddot{x}_1^{(1)}$ = inertia force of mass m_1 for the first mode,

$x_1^{(2)}$ = displacement of mass m_1 for the second mode,

$m_2\ddot{x}_2^{(1)}$ = inertia force of mass m_2 for the first mode and

$x_2^{(2)}$ = displacement of mass m_2 for the second mode.

Equation (6.25) can thus be interpreted as the work done by the inertia forces occurring in the first mode, in going through the displacements of the second mode, is equal to zero. We know from elementary mechanics that if the line of action of the force and displacement is orthogonal, the work done by the force is equal to zero. Equation (6.25) reveals that normal modes are orthogonal. Thus, this is known as orthogonality relationship.

Another physical interpretation of the orthogonality is described in the following:

The equivalent static force acting at the rth mode

$$\{F^{(r)}\} = [K]\{\phi^{(r)}\} a e^{ip_r t} \quad (6.26)$$

and the corresponding displacement in the s-th mode is

$$\{x^{(s)}\} = \{\phi^{(s)}\} a e^{ip_s t} \quad (6.27)$$

The work done by the equivalent static force in *r*-th mode in undergoing displacement in the *s*-th mode is

$$W = \left[(\{\phi^{(r)}\})^T [K]\{\phi^{(s)}\}\right] a^2 e^{ip_r t} e^{ip_s t} \quad (6.28)$$

which is zero due to the orthogonality relation given by Eq. (6.20).

Therefore, another interpretation of orthogonality properties is that the work done by the equivalent static forces in the r-th mode in undergoing displacement in the s-th mode is zero [6.2]

6.5 EIGENVALUE PROBLEM

For the solution of the eigenvalue problem, the most direct approach is to compute the zeroes of the characteristic polynomial and then obtain eigenvectors for each root of λ. The value of the determinant $\left|[D]-\lambda[I]\right|$ in Eq. (6.14) involves a computation, which if the size of the determinant is large, may lead to round-off errors. Hence, when the size of [D] is large, the determination of the roots of the determinant is not attempted.

For the general case, no explicit formulas are available for computation of roots of $f(\lambda)$ given by Eq. (6.15), if the degree of the polynomial is larger than four. As such, various methods have been developed for the solution of the problem. As the problem demands the determination of the roots of the polynomial, the proposed methods are basically iterative in nature. However, for the economic evaluation of the eigensystem, [M] and [K] matrices may be suitably modified. The various methods which have been proposed can be broadly divided into four groups. They are: (1) vector iteration method, (2) transformation method, (3) determinant search method and (4) other methods.

6.6 DETERMINATION OF ABSOLUTE DISPLACEMENT OF FREE VIBRATION OF MDF SYSTEMS

The equations of motion for the MDF systems given by Eq. (6.4) are coupled equations. To obtain the absolute displacement of the masses, these equations are to be solved. In its present form, the solution of the equations is difficult. As such, these sets of equations are uncoupled by the use of principal or normal coordinates.

From Eq. (6.4), it is revealed that a MDF system having n degrees of freedom is represented by n equations based on n independent coordinates of the masses. In Eq. (6.4), coordinates have been chosen as the displacements of the masses from their static equilibrium position. This choice has been made, because it is convenient. However, any other set of independent coordinate system can be chosen as well. By a proper selection, it is always possible to select a set of coordinates, such that the coupling of the equation can be removed, as a result of which each equation will contain one independent variable. Thus n sets of coupled equations are reduced to n sets of independent equations. Coordinates which enable the equations to be solved independent of one another, each of which has its own amplitude, frequency and phase angle are called principal coordinates or normal coordinates. The equation of transformation is given by

$$\{x\}=[\Phi]\{\xi\} \tag{6.29}$$

where $$[\Phi]=[\{\Phi^{(1)}\}\{\Phi^{(2)}\}.........\{\Phi^{(n)}\}]$$

i.e. each column of $[\Phi]$ represents one normal mode and $[\xi]$ is the principal coordinate.

Equation (6.8) is rewritten as

$$-p^2[M]\{\phi\}+[K]\{\phi\}=\{0\}$$

or $$p^2[M]\{\phi\}=[K]\{\phi\} \tag{6.30}$$

For different frequencies, we can write a set of equations of the type of Eq. (6.30).

$$\left.\begin{array}{l} p_1^2 [M]\{\phi^{(1)}\} = [K]\{\phi^{(1)}\} \\ p_2^2 [M]\{\phi^{(2)}\} = [K]\{\phi^{(2)}\} \\ \vdots \qquad\qquad \vdots \\ p_n^2 [M]\{\phi^{(n)}\} = [K]\{\phi^{(n)}\} \end{array}\right\} \tag{6.31}$$

Combining all the equations of Eq. (6.31), the following equation may be written

$$[M][\Phi][p^2] = [K][\Phi] \tag{6.32}$$

where $[p^2]$ is a diagonal matrix with the angular frequency squared term in each diagonal. Substituting $\{x\}$ from Eq. (6.29) into Eq. (6.5), we get

$$[M][\Phi]\{\ddot{\xi}\} + [K][\Phi]\{\xi\} = \{0\} \tag{6.33}$$

Premultiplying both sides of Eq. (6.33) by $[\Phi]^T$, we get

$$[\Phi]^T[M][\Phi]\{\ddot{\xi}\} + [\Phi]^T[K][\Phi]\{\xi\} = \{0\} \tag{6.34}$$

Combining Eqs. (6.32) and (6.34), we get

$$[\Phi]^T[M][\Phi](\{\ddot{\xi}\} + [p^2]\{\xi\}) = \{0\} \tag{6.35}$$

Now, $[M]$ is diagonal and using orthogonality relationship

$$[\Phi]^T[M][\Phi] = \begin{bmatrix} \overline{m}_1 & & & & & \\ & \overline{m}_2 & & & 0 & \\ & & \cdot & & & \\ & & & \overline{m}_r & & \\ & & & & \cdot & \\ & 0 & & & & \overline{m}_n \end{bmatrix} \tag{6.36}$$

where $$\overline{m}_r = \{\phi^{(r)}\}^T[M]\{\phi^{(r)}\} = \sum_{i=1}^{n} m_i \{(\phi_i^{(r)})^2\} \tag{6.37}$$

All the off-diagonal elements of $[\Phi]^T[M][\Phi]$ are zeroes, because of the orthogonality relationship.

Equation (6.35), therefore, consists of a set of n uncoupled equations. The rth equation can be written as

$$\overline{m}_r(\ddot{\xi}_r + p_r^2\xi_r) = 0$$

or $$\ddot{\xi}_r + p_r^2\xi_r = 0 \tag{6.38}$$

The solution of Eq. (6.38) is

$$\xi_r = A_r \sin p_r t + B_r \cos p_r t \tag{6.39}$$

Thus, a n-degree of freedom system represented by Eq. (6.5) is reduced to n single degree of freedom systems, represented by Eq.(6.38), with the full use transformed coordinates instead of $\{x\}$. n uncoupled equations of the type of Eq. (6.38) can be solved easily. The coordinate of $\{x\}$ is obtained by

$$x_i(t) = \sum_{r=1}^{n} \phi_i^{(r)} \xi_r \tag{6.40}$$

Since the values of ϕs are arbitrary, it is possible to choose ϕs, such that $\overline{m}_r$, is unity for all values of r = 1, 2, 3, . . ., n. This is shown with the help of the following example.

Example 6.1

The $[\phi]$ and $[M]$ matrices for a three degrees of freedom system is as follows:

$$[\Phi] = \begin{bmatrix} 1.0000 & 1.0000 & 1.0000 \\ 1.2289 & -4.1286 & 0.1479 \\ 1.2054 & 4.2788 & -0.4847 \end{bmatrix}$$

$$[M] = \begin{bmatrix} 2m & -m & 0 \\ -m & 3m & 0 \\ 0 & 0 & 2m \end{bmatrix}$$

Modify the $[\Phi]$ matrix such that

$$[\Phi]^T [M][\Phi] = \overline{m}[I]$$

Let us multiply each column of the matrix by c_1, c_2 and c_3 respectively.

Therefore

$$\{\phi^{(1)}\}^T [M] \{\phi^{(1)}\} = c_1^2 \{1.0000 \quad 1.2289 \quad 1.2054\} \begin{bmatrix} 2m & -m & 0 \\ -m & 3m & 0 \\ 0 & 0 & 2m \end{bmatrix} \begin{Bmatrix} 1.0000 \\ 1.2289 \\ 1.2054 \end{Bmatrix}$$

$$= 6.9788\, m c_1^2 = \overline{m}$$

Hence $$c_1 = \sqrt{\frac{1}{6.9788} \frac{\overline{m}}{m}}$$

Therefore $$c_1 \{\phi^{(1)}\} = \begin{Bmatrix} 0.3785 \\ 0.4652 \\ 0.4563 \end{Bmatrix} \sqrt{\frac{\overline{m}}{m}}$$

Following the above steps, we can show that

$$c_2 \{\phi^{(2)}\} = \begin{Bmatrix} 0.1010 \\ -0.4169 \\ 0.4321 \end{Bmatrix} \sqrt{\frac{\overline{m}}{m}}$$

and $$c_3 \{\phi^{(3)}\} = \begin{Bmatrix} 0.6680 \\ 0.0987 \\ -0.3237 \end{Bmatrix} \sqrt{\frac{\overline{m}}{m}}$$

The modal matrix then is

$$[\Phi] = \sqrt{\frac{\overline{m}}{m}} \begin{bmatrix} 0.3785 & 0.1010 & 0.6680 \\ 0.4652 & -0.4169 & 0.0987 \\ 0.4563 & 0.4321 & -0.3237 \end{bmatrix} \tag{a}$$

The results can now be checked by substituting $[\Phi]$ of Eq. (a)

$$[\Phi]^T [M][\Phi] = \overline{m}[I]$$

Example 6.2

For a two-degree of freedom system, the modal vectors are given by

$$[\Phi] = \begin{bmatrix} 1.000 & 1.000 \\ 1.618 & -0.618 \end{bmatrix}$$

The initial conditions at $t = 0$ are

$$(x_1)_0 = -(x_2)_0 = x_0$$

and $$(\dot{x}_1)_0 = v, \qquad (\dot{x}_2)_0 = 0$$

The natural frequencies of the system are p_1 and p_2. Determine the equations of displacements in free vibration.

On the basis of principal coordinates chosen as per Eq. (6.29), the two uncoupled equations in terms of the principal coordinates are as follows [Eq. (6.38)]

$$\left.\begin{aligned} \ddot{\xi}_1 + p_1^2 \xi_1 = 0 \\ \ddot{\xi}_2 + p_2^2 \xi_2 = 0 \end{aligned}\right\} \tag{a}$$

Solutions of Eq. (a) are

$$\left.\begin{aligned} \xi_1 = A_1 \cos p_1 t + B_1 \sin p_1 t \\ \xi_2 = A_2 \cos p_2 t + B_2 \sin p_2 t \end{aligned}\right\} \tag{b}$$

Now

$$\begin{Bmatrix} x_1 \\ x_2 \end{Bmatrix} = \begin{bmatrix} 1.000 & 1.000 \\ 1.618 & -0.618 \end{bmatrix} \begin{Bmatrix} A_1 \cos p_1 t + B_1 \sin p_1 t \\ A_2 \cos p_2 t + B_2 \sin p_2 t \end{Bmatrix}$$

or

$$\left.\begin{aligned} x_1 &= (A_1 \cos p_1 t + B_1 \sin p_1 t) + (A_2 \cos p_2 t + B_2 \sin p_2 t) \\ x_2 &= 1.618(A_1 \cos p_1 t + B_1 \sin p_1 t) \\ &\quad - 0.618(A_2 \cos p_2 t + B_2 \sin p_2 t) \end{aligned}\right\} \tag{c}$$

Substituting the given initial values into Eq. (c) and their derivatives

$$\left.\begin{aligned} A_1 + A_2 = x_0 \\ 1.618 A_1 - 0.618 A_2 = -x_0 \\ B_1 p_1 + B_2 p_2 = v \\ 1.618 B_1 p_1 - 0.618 B_2 p_2 = 0 \end{aligned}\right\} \tag{d}$$

Solving Eq. (d)

$$A_1 = -0.1771x_0,\ A_2 = 1.171x_0,\ B_1 = \frac{0.276v}{p_1} \quad \text{and} \quad B_2 = \frac{0.724v}{p_2}$$

Hence, the equations of motion can be written as

$$x_1 = \left(-0.1771x_0 \cos p_1 t + \frac{0.276v}{p_1} \sin p_1 t \right)$$

$$+ \left(1.171x_0 \cos p_2 t + \frac{0.724v}{p_2} \sin p_2 t \right)$$

$$x_2 = 1.618 \left(-0.171x_0 \cos p_1 t + \frac{0.276v}{p_1} \sin p_1 t \right)$$

$$-0.618 \left(1.171x_0 \cos p_2 t + \frac{0.724v}{p_2} \sin p_2 t \right)$$

6.6.1 Normalisation of Modes

It has been seen that normal modes indicate a ratio between displacements. As such, the different elements may be varied in such a way that the constant ratios are maintained. There are infinite such possibilities. Scaling of the normal modes is sometimes done to standardize their elements associated with amplitudes in various DOFs which is known as normalization. In many cases the largest element in each mode is assigned a value of unity and other elements are accordingly adjusted in the normalization process. In some other cases it is convenient to normalize each mode so that the element corresponding to a particular d.o.f is unity. The modal matrix $[\Phi]$ may be so adjusted as to obtain the following relationship:

$$[\Phi]^T [M][\Phi] = [I] \tag{6.41}$$

$$[\Phi]^T [K][\Phi] = [\Delta] = \left[1/p_r^2\right] \tag{6.42}$$

6.7 EIGENVALUE SOLUTION TECHNIQUES

The vector iteration methods are very effectively used for the computation of the eigenvalues, as well as the corresponding eigenvectors at the same time. The aim of the method is to directly operate upon Eq. (6.5) or Eq. (6.13). Various methods come under this category. Stodola's method was the earliest one. Investigations of Jennings [6.31], Rutihauser [6.41] and Stewart [6.5] have led to the development of modern vector iteration methods. The advanced version applied to the finite element analysis is the Subspace Iteration Technique [6.10] and Simultaneous Iteration Technique [6.12].

One of the major advantages of the vector iteration methods is that, it may yield a few of the lower eigenvalues and eigenvectors. In most practical problems, only a few of the lower eigenvalues and eigenvectors are of importance. As such, the method avoids the determination of all n unknowns in the system.

Transformation methods are preferable when all the eigenvalues and eigenvectors are required.

Transformation methods operate on the following matrices

$$\left.\begin{array}{l}[\Phi]^T[K][\Phi]=[\Delta]\\ [\Phi]^T[M][\Phi]=[I]\end{array}\right\} \quad \text{and} \tag{6.43}$$

where as already stated [Φ] is called the modal matrix and formed by modal vectors.

$$[\Phi]=\left[\{\phi^{(1)}\},\{\phi^{(2)}\},\cdots,\{\phi^{(n)}\}\right] \tag{6.44}$$

where { $\phi^{(1)}$} is the mode shape for the first mode and so is the definition of { $\phi^{(2)}$}, { $\phi^{(3)}$} etc. It may be noted that [Φ] is a square matrix of order $n \times n$.

Further, from the orthogonality properties given by Eqs. (6.19) and (6.20), $[\Delta]$ is a square diagonal matrix containing p^2 values in the diagonal elements.

Due to transformations of the types given by Eqs. (6.43) and (6.44), this method of solution requires storage of large matrices.

Since [Φ], the mode shape matrix is unique, it can be constructed by iteration. Basically, the method involves the reduction of $[K]$ and $[M]$ into diagonal form, using successive pre-multiplication and post-multiplication of suitable matrices. A sequence of matrix transformations is thus obtained for the final evaluation. In this method, unless all the eigenvalues and eigenvectors are evaluated, the final output cannot be obtained.

Earlier transformation methods are due to Jacobi, Givens and Householder. It is very difficult to obtain stable methods, which are reliable as well as convergent for unsymmetric eigenvalue problems. The Eberline extension of Jacobi's transformation method is noted for its lack of universal stability. As such, attempts have been made to develop stable forms of the Eberline method. QR-transformation method is one of the methods falling in this group, in which the stiffness matrix $[K]$ is transformed into tri-diagonal form and then rotation matrices are employed.

The methods which come under frequency search method, start by assuming a trial frequency and then finding out a value of the determinant $\left|[D]-\lambda[I]\right|$. Different methods differ in the process of evaluation of the determinant. Eigenvalues within the prescribed intervals can be obtained by plotting the above determinant against assumed frequency. The zero-crossing points give the natural frequencies of the system [Fig. 6.3]. The earlier methods which come under the category of frequency search methods are Holzer method [6.6] and Myklestad method [6.7]. Transfer matrix method also belongs to this group.

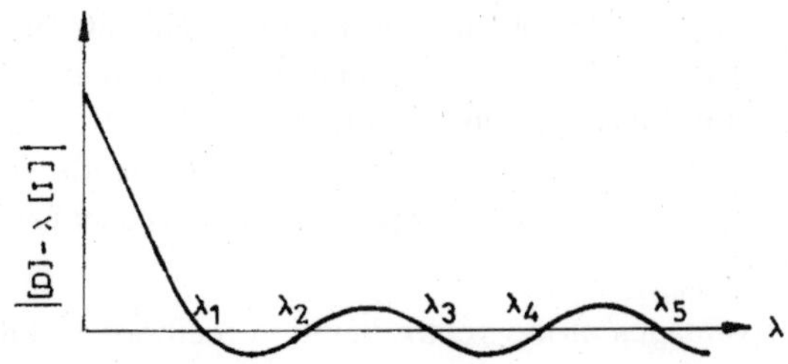

Fig. 6.3 Variation of the determinant

The methods which do not fall under the above categories are referred to as other methods. Sturm sequence is one such method,

A number of solution algorithms have been developed under each category. Some of them are discussed in the following.

6.8 DUNKERLEY'S EQUATION

A somewhat simpler approach has been suggested by Dunkerley, to obtain an approximation to the fundamental frequency. Good results are expected for the system, having very small amount of damping and the fundamental frequency differs to a significant extent from the higher harmonics.

When the determinant given by Eq. (6.14) is expanded for a 2-mass system, it becomes

$$\begin{vmatrix} \left(d_{11} - \dfrac{1}{p^2}\right) & d_{12} \\ d_{21} & \left(d_{22} - \dfrac{1}{p^2}\right) \end{vmatrix} = 0 \tag{6.45}$$

Expanding this determinant, we obtain

$$\left(\frac{1}{p^2}\right)^2 - (d_{11} + d_{22})\left(\frac{1}{p^2}\right) + (d_{11}\,d_{22} - d_{12}\,d_{21}) = 0 \tag{6.46}$$

If p_1 and p_2 are two natural frequencies of the system, then Eq. (6.46) can be written in terms of its factors as follows:

$$\left(\frac{1}{p^2} - \frac{1}{p_1^2}\right)\left(\frac{1}{p^2} - \frac{1}{p_2^2}\right) = 0 \tag{6.47}$$

or

$$\left(\frac{1}{p^2}\right)^2 - \left(\frac{1}{p_1^2} + \frac{1}{p_2^2}\right)\left(\frac{1}{p^2}\right) + \frac{1}{p_1^2\,p_2^2} = 0 \tag{6.48}$$

Equating the coefficients of $1/p^2$ terms in Eqs. (6.46) and (6.48), we get

$$\frac{1}{p_1^2} + \frac{1}{p_2^2} = d_{11} + d_{22} \tag{6.49}$$

If for the system, the values of p_1 and p_2 are such that $1/p_1^2 >> 1/p_2^2$, then

Eq. (6.49) becomes

$$\frac{1}{p_1^2} = d_{11} + d_{22} \tag{6.50}$$

Equation (6.50) is known as Dunkerley's equation. Equation (6.49) can be extended to systems with n masses and it will assume the following form

$$\frac{1}{p_1^2} \cong d_{11} + d_{22} + \ldots + d_{nn} = \sum_{i=1}^{n} d_{ii} \tag{6.51}$$

Dunkerley's equation gives a lower bound to the fundamental frequency.

From Eq. (6.13), it is seen that

$$d_{ii} = f_{ii} m_i \tag{6.52}$$

for a lumped mass system, where f_{ii} is the influence coefficient, that is, it denotes the deflection at i due to a unit force applied at i. The quantity d_{ii} can be given a physical meaning. If we set all the masses of the system to zero except m_i, the system becomes a single degree of freedom system, having a natural frequency given by

$$\frac{1}{p_{ii}^2} = f_{ii} m_i \tag{6.53}$$

Substituting Eqs. (6.52) and (6.53) into Eq. (6.51), Dunkerley's equation is expressed as follows:

$$\frac{1}{p_1^2} \cong \frac{1}{p_{11}^2} + \frac{1}{p_{22}^2} + \cdots + \frac{1}{p_{nn}^2} \tag{6.54}$$

Example 6.3

Estimate the fundamental frequency of torsional vibration by Dunkerley's equation, for the system having two discs fixed to the shaft as shown in Fig. 6.4.

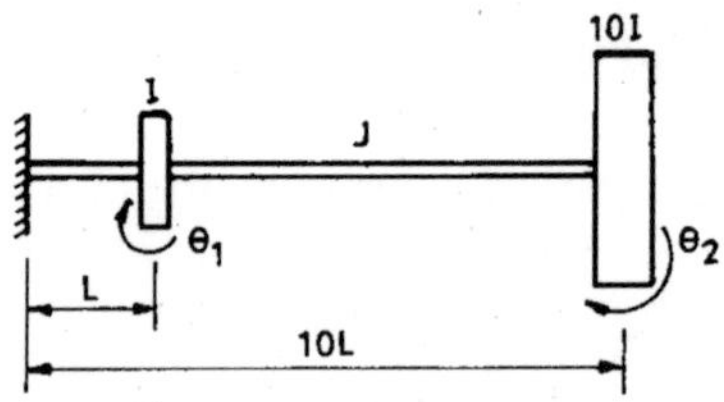

Fig. 6.4 Example 6.3

The influence coefficients corresponding to θ_1 and θ_2 are

$$f_{11} = \frac{L}{GJ}$$

$$f_{22} = \frac{10L}{GJ}$$

Then the natural frequencies of the shaft having only the first disc or a second disc is given by

$$\frac{1}{p_{11}^2} = f_{11} I = \frac{IL}{GJ}$$

$$\frac{1}{p_{22}^2} = f_{22}\, 10I = \frac{100IL}{GJ}$$

From Dunkerley's equation, we get

$$\frac{1}{p_1^2} \cong \frac{1}{p_{11}^2} + \frac{1}{p_{22}^2} = \frac{IL}{GJ} + \frac{100IL}{GJ} = 101\frac{IL}{GJ}$$

or, fundamental frequency is given by

$$p_1^2 = \frac{1}{101}\frac{GJ}{IL} = 0.0099\frac{GJ}{IL}$$

6.9 HOLZER METHOD

Holzer method falls under the determinant search technique. The method can be applied to rectilinear or angular motions for damped or undamped systems. The method is best suited for systems where the components are arranged along a basic axis.

The method starts by assuming a trial frequency. Displacement of unit amplitude is assumed at one end, and the forces and the displacements of different masses are then calculated in a progressive manner. If the force or the displacement at the other end is compatible with the conditions prevailing there, the assumed frequency is one of the natural frequencies of the system. The calculation steps can be presented in a systematic tabular form, and the method can be easily programmed in a computer. The method is explained for the spring-mass system of Fig. 6.5.

Assume a frequency p and a displacement $x_1 = 1.0$ at the top. The inertia force of mass m_1 is $m_1 p^2 .1\,(m\ddot{x} = -\,p^2 m_1 A_1 e^{ipt} = -\,p^2 m_1 x_1,$ if $x_1 = A_1 e^{ipt}$ is assumed). Deformation of the spring, k_1 due to this force is

$$\frac{m_1 p^2}{k_1} = 1 - x_2$$

or

$$x_2 = 1 - \frac{m_1 p^2}{k_1} \tag{6.55}$$

Similarly, the inertia force of mass m_2 is $m_2 p^2 x_2$. Considering the equilibrium of two masses m_1 and m_2 as shown in Fig. 6.5(b), the following relation can be written

$$\frac{m_1 p^2 + m_2 p^2 x_2}{k_2} = x_2 - x_3$$

or

$$x_3 = x_2 - \frac{p^2}{k_2}(m_1 + m_2 x_2) \tag{6.56}$$

x_3 can be obtained from Eq. (6.56). The procedure can be repeated. Thus, for the ith mass

$$x_i = x_{i-1} - \frac{p^2}{k_{i-1}} \sum_{j=1}^{i-1} m_j x_j \tag{6.57}$$

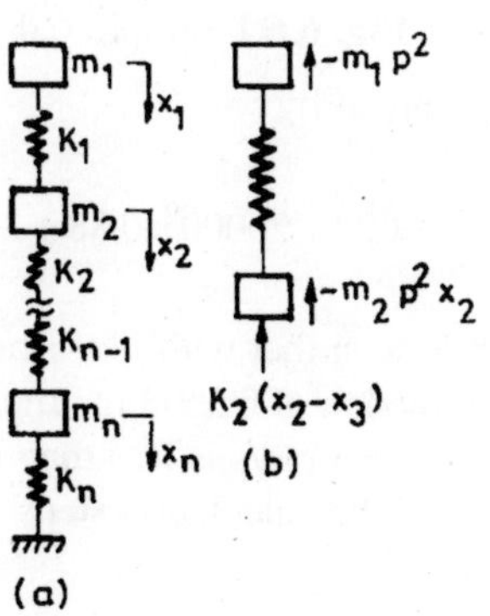

Fig. 6.5 Spring-mass system of n masses

The displacements of all the masses can thus be obtained. If the frequency assumed is correct, then $x_{n+1}=0$ for the system of Fig. 6.5. The procedure is to calculate x_{n+1} corresponding to the assumed frequency and a plot similar to Fig. 6.3, except that instead of $|[D]-\lambda[I]|$, it would be x_{n+1}. The frequencies which correspond to equal to zero, will be the natural frequencies of the system. The displacements at different levels will give the corresponding mode shapes.

Example 6.4

A three-storey frame with rigid beams is shown in Fig. 6.6. Determine the natural frequencies and mode shapes.

Assuming $x_1=1$ and a trial frequency of p = 3 rad/s, top beam inertia force is

$$F_1 = p^2 m_1 x_1 = 3^2 \times 20000 \times 1 = 18\times10^4 \text{ N}$$

The displacement at the second beam level [Eq. (6.44)] is

$$x_2 = 1 - \frac{18\times10^4}{1500\times10^3} = 0.88$$

The displacement at the third beam level is [from Eq. (6.45)]

$$x_3 = 0.6544 - \frac{3^2}{1500\times10^3}(20000 + 20000\times0.88)$$

$$= 0.6544 \text{ m}$$

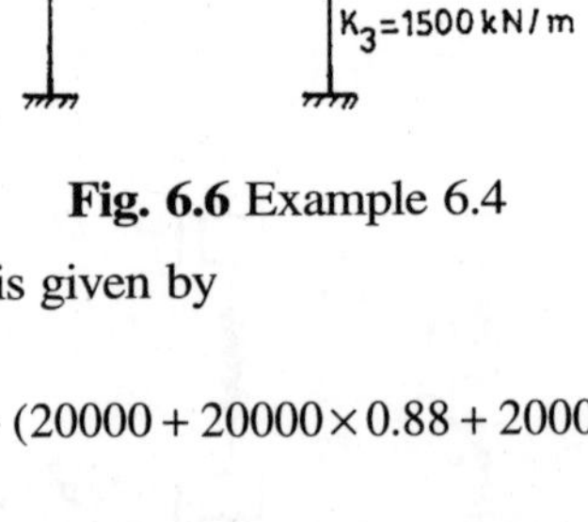

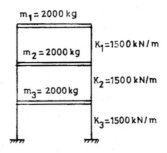

Fig. 6.6 Example 6.4

Similarly, the base displacement is given by

$$x_4 = 0.6544 - \frac{3^2}{1500\times10^3}(20000 + 20000\times0.88 + 20000\times0.6544) = 0.3503 \text{ m}$$

From the calculation made, it is seen that with p = 3 rad/s, displacements at different beam levels are all positive and for a base motion of 0.3503 m amplitude, produces a vibration amplitude of 1.0 m at the top. It suggests that p = 3 rad/s is less than the first mode natural frequency. As next trial, we assume p = 4 rad/s, and the calculation steps are repeated. Both these calculations are presented in tabular form below.

p p^2	$x_1 = 1.0$ $F_1 = m_1 p^2 x_1$	$x_2 = 1 - F_1/k_1$ $F_2 = F_1 + p^2 x_2 m_2$	$x_3 = x_2 - F_2/k_2$ $F_3 = F_2 + m_3 x_3 p^2$	$x_4 = x_3 - F_3/k_3$
3	1.0	0.88	0.6544	0.3503
9	18×10^4	33.84×10^4	45.62×10^4	
4	1.0	0.7867	0.4055	- 0.062
16	32×10^4	57.17×10^4	70.14×10^4	

The calculation can thus be continued for varying values of p and for each assumed value of p, x_{n+1} is calculated. The variation of x_{n+1} with p is plotted in Fig. 6.7. The values of p, where x_{n+1} is zero correspond to the natural frequencies of the system. For this example

$$p_1 = 3.85 \text{ rad/s}, \; p_2 = 10.8 \text{ rad/s and } p_3 = 15.61 \text{ rad/s}$$

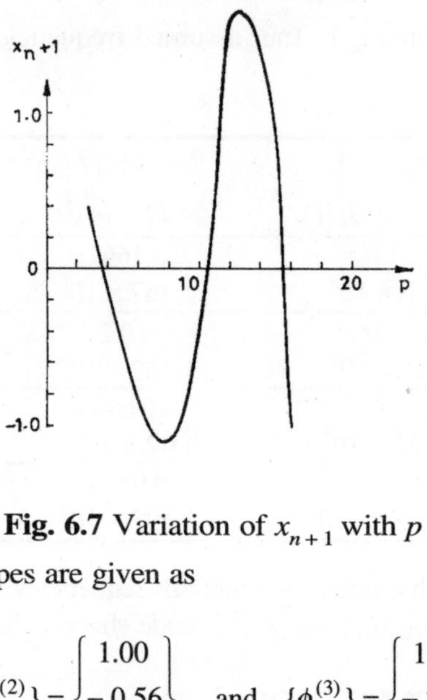

Fig. 6.7 Variation of x_{n+1} with p

The corresponding mode shapes are given as

$$\{\phi^{(1)}\} = \begin{Bmatrix} 1.00 \\ 0.80 \\ 0.44 \end{Bmatrix}, \quad \{\phi^{(2)}\} = \begin{Bmatrix} 1.00 \\ -0.56 \\ -1.25 \end{Bmatrix}, \quad \text{and} \quad \{\phi^{(3)}\} = \begin{Bmatrix} 1.00 \\ -2.25 \\ 1.80 \end{Bmatrix}$$

Example 6.5

Using Holzer's method, determine the natural frequencies and mode shapes for the system shown in Fig. 6.8.

The method discussed for a linear spring can be applied without any difficulty, to a system having a series of discs connected to a shaft. The mass m is to be replaced by the mass moment of inertia and the axial displacements are to be changed to rotation of the shaft. One proceeds in an identical manner as the previous example, by assuming a suitable frequency and $\theta_1 = 1$ at one end (say, the left hand end in this case).

In the previous example, we considered the frequency for which $x_{n+1} = 0$ as the correct frequency. In this case, the disc at the right hand end is free to rotate. Therefore, at the far end, we find out the necessary condition from the equilibrium of the system. The resulting torque at the far end is

$$T_n = \sum I_i p^2 \theta_i \tag{6.58}$$

For the natural frequency p^2 of the system, $T_n = 0$.

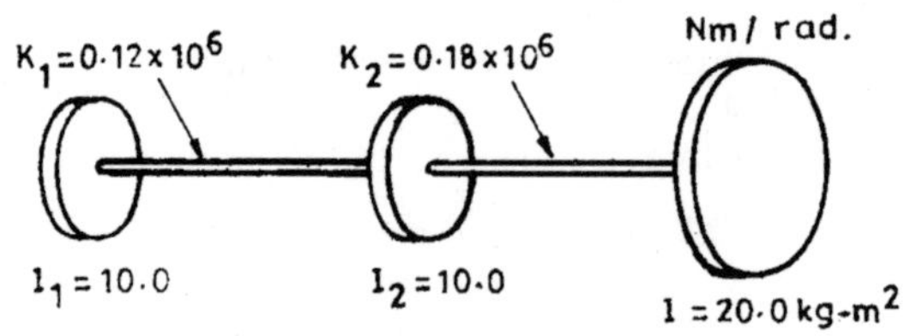

Fig. 6.8 Example 6.5

Calculation steps corresponding to four assumed frequencies have been presented in the tabular form below.

p p_2	$\theta_1 = 1.0$ $T_1 = p^2\theta_1 I_1$	$\theta_2 = 1 - T_1/k_1$ $T_2 = T_1 + p^2\theta_2 I_2$	$\theta_3 = \theta_2 - T_2/k_2$ $T_3 = T_2 + p_2\theta_3 I_2$
100 10^4	1.0 1.0×10^5	0.1667 1.167×10^5	− 0.4814 2.0410×10^4
120 1.44×10^4	1.0 1.0×10^5	− 0.2 1.152×10^5	− 0.84 -1.267×10^5
110 1.21×10^4	1.0 1.21×10^5	− 0.0083 1.2×10^5	− 0.675 -4.33×10^4
105 1.1025×10^4	1.0 1.1025×10^5	0.0813 1.192×10^5	− 0.581 -8.90×10^3

The calculations indicate that one of the natural frequencies will lie between 100 and 105. The calculation of the remaining frequency and mode shape is left as an exercise.

6.10 TRANSFER MATRIX METHOD

We start by introducing the concepts of the state vector. A state vector at a station *i* of an elastic system is a column matrix, the elements of which are the displacements and internal forces at point *i*. For the case of a spring mass system, the state vector is defined as

$$\{Z\}_i = \begin{Bmatrix} x_i \\ N_i \end{Bmatrix} = \begin{Bmatrix} x \\ N \end{Bmatrix}_i \tag{6.59}$$

where x_i is the linear displacement of the spring and N_i is the corresponding spring force.

Consider the spring-mass system of Fig. 6.9. The system is vibrating with an angular frequency *p*. Massless spring of stiffness k_i connects the masses m_i and m_{i-1}. $\{Z\}_i^R$ indicates the state vector to the right mass m_i and $\{Z\}_i^L$ denotes the state vector to the left of mass m_i.

The freebody diagram of this spring is shown in Fig.6.10.

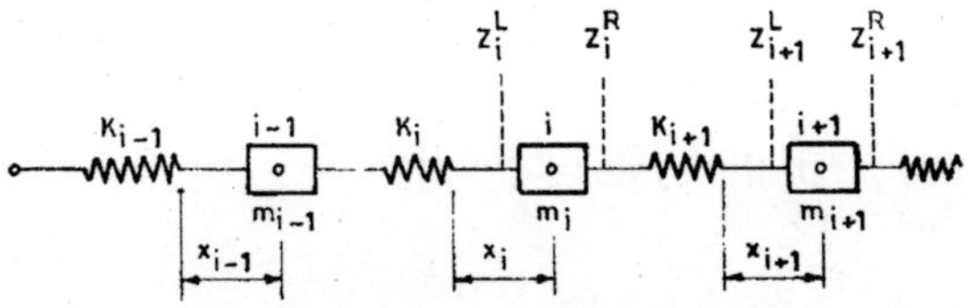

Fig. 6.9 State vector definition

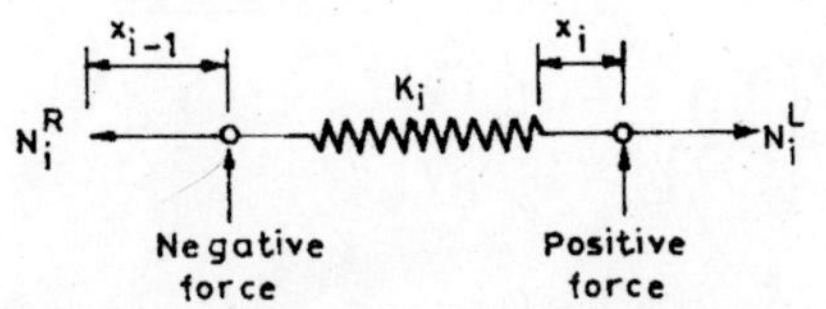

Fig. 6.10 Freebody diagram of the spring

Equilibrium of the spring yields

$$N_{i-1}^R = N_i^L \tag{6.60}$$

Further, the following relation follows from the stiffness property of the spring.

$$N_i^L = N_{i-1}^R = k_i\,(x_i - x_{i-1}) \tag{6.61}$$

Equations (6.60) and (6.61) can be rearranged in the following form

$$\left.\begin{aligned} x_i &= x_{i-1} + \frac{N_{i-1}^R}{k_i} \\ N_i^L &= (0)\,x_{i-1} + N_{i-1}^R \end{aligned}\right\} \tag{6.62}$$

Writing Eq. (6.62) in matrix form, we get

$$\begin{Bmatrix} x_1 \\ N_i^L \end{Bmatrix} = \begin{bmatrix} 1 & \dfrac{1}{k_1} \\ 0 & 1 \end{bmatrix} \begin{Bmatrix} x_{i-1} \\ N_{i-1}^R \end{Bmatrix} \tag{6.63}$$

or

$$\begin{Bmatrix} x \\ N \end{Bmatrix}_i^L = \begin{bmatrix} 1 & \dfrac{1}{k_1} \\ 0 & 1 \end{bmatrix} \begin{Bmatrix} x \\ N \end{Bmatrix}_{i-1}^R$$

or

$$\{Z\}_i^L = [T]_i\,\{Z\}_{i-1}^R \tag{6.64}$$

$[T]_i$ matrix relates the state vector $\{Z\}_i^L$ with $\{Z\}_{i-1}^R$ which is known as field transfer matrix or simply as field matrix.

The mass is considered to be rigid. Therefore, the displacement of the mass to the left is equal to that to the right.

$$x_i^R = x_i^L \tag{6.65}$$

The spring forces to the left and right of the mass are N_i^L and N_i^R, and there is an inertia force $m_i p^2 x_i$ acting. Therefore

$$N_i^R = N_i^L - m_1 p^2 x_i \tag{6.66}$$

It may be noted that while deriving Eq. (6.66), advantage has been taken of the fact that $\ddot{x} = -p^2 x$. Equations (6.65) and (6.66) can be written in matrix notations as follows:

$$\begin{Bmatrix} x \\ N \end{Bmatrix}_i^R = \begin{bmatrix} 1 & 0 \\ -m_i p^2 & 1 \end{bmatrix}_i^L \begin{Bmatrix} x \\ N \end{Bmatrix}_i^L \tag{6.67}$$

or

$$\{Z\}_i^R = [U]_i \{Z\}_i^L \tag{6.68}$$

$[U]_i$ is referred to as point transfer matrix, as it relates the two adjacent state vectors over a point.

6.10.1 Transfer Matrices as a Means of Elimination

A multiple degrees of freedom system is shown in Fig. 6.11. 0 and n are the boundaries of the system. The relation between the adjacent state vectors is as follows.

$$\left.\begin{matrix} \{Z\}_1^L = [T]_1 \{Z\}_0^R & \{Z\}_1^R = [U]_1 \{Z\}_1^L \\ \{Z\}_2^L = [T]_2 \{Z\}_1^R & \{Z\}_2^R = [U]_2 \{Z\}_2^L \\ \cdots & \cdots \\ \cdots & \cdots \\ \cdots & \cdots \\ \{Z\}_n^L = [T]_n \{Z\}_{n-1}^R & \{Z\}_n^R = [U]_n \{Z\}_n^L \end{matrix}\right\} \tag{6.69}$$

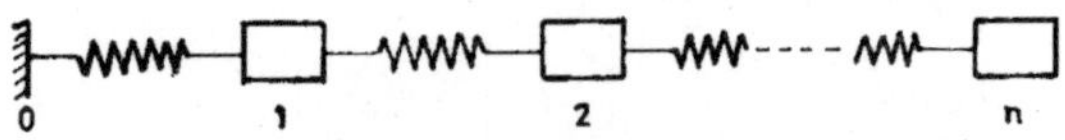

Fig. 6.11 A MDF system

Making proper substitutions in the state vectors from the relations given by Eq. (6.68), it can be shown that

$$\{Z\}_n = [U]_n [T]_n \; [U]_{n-1} [T]_{n-1}, \ldots, [U]_1 [T]_1 \{Z\}_0 \tag{6.70}$$

or

$$\{Z\}_n = [G]\{Z\}_0 \tag{6.71}$$

Equation (6.71) indicates that all intermediate state vectors are eliminated and a relationship is obtained between boundary state vectors. After putting the boundary conditions, a polynomial in p^2 will be obtained from Eq.(6.71). Solution of polynomial will yield the natural frequencies in the system.

Example 6.6

Find by transfer matrices the natural frequency of the system shown in Fig. 6.12.

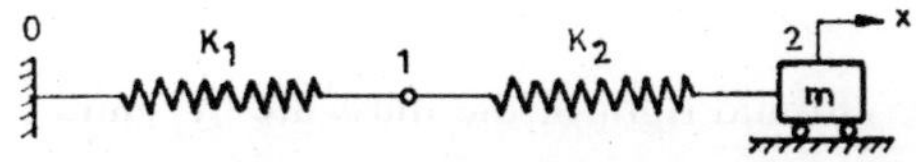

Fig. 6.12 Example 6.6

Indicating the joints as 0, 1 and 2, the following transfer matrices can be written

$$\begin{Bmatrix} x \\ N \end{Bmatrix}_1^L = \begin{bmatrix} 1 & 1/k_1 \\ 0 & 1 \end{bmatrix} \begin{Bmatrix} x \\ N \end{Bmatrix}_0^R$$

$$\begin{Bmatrix} x \\ N \end{Bmatrix}_1^R = \begin{bmatrix} 1 & 0 \\ 0 & 1 \end{bmatrix} \begin{Bmatrix} x \\ N \end{Bmatrix}_1^L$$

$$\begin{Bmatrix} x \\ N \end{Bmatrix}_2^L = \begin{bmatrix} 1 & 1/k_2 \\ 0 & 1 \end{bmatrix} \begin{Bmatrix} x \\ N \end{Bmatrix}_1^R$$

and $$\begin{Bmatrix} x \\ N \end{Bmatrix}_2^R = \begin{bmatrix} 1 & 0 \\ -mp^2 & 1 \end{bmatrix} \begin{Bmatrix} x \\ N \end{Bmatrix}_2^L$$

Therefore, from the above relations, the following equation can be written as

$$\begin{Bmatrix} x \\ N \end{Bmatrix}_2^R = \begin{bmatrix} 1 & 0 \\ -mp^2 & 1 \end{bmatrix} \begin{bmatrix} 1 & 1/k_2 \\ 1 & 1 \end{bmatrix} \begin{bmatrix} 1 & 0 \\ 0 & 1 \end{bmatrix}$$

$$\begin{bmatrix} 1 & 1/k_1 \\ 0 & 1 \end{bmatrix} \begin{Bmatrix} x \\ N \end{Bmatrix}_0^R$$

or $$\begin{Bmatrix} x \\ N \end{Bmatrix}_2^R = \begin{bmatrix} 1 & \left(\dfrac{1}{k_1} + \dfrac{1}{k_2}\right) \\ -mp^2 & -mp^2(1/k_1 + 1/k_2) + 1 \end{bmatrix} \begin{Bmatrix} x \\ N \end{Bmatrix}_0^R \quad \text{(a)}$$

The boundary conditions of the problem indicate

$$x_0 = 0 \text{ and } N_2^R = 0 \quad \text{(b)}$$

Therefore, the above equation yields

$$x_2^R = \left(\frac{1}{k_1} + \frac{1}{k_2}\right) N_0^R \quad \text{(c)}$$

and $$-mp^2\left(\frac{1}{k_1}+\frac{1}{k_2}\right)+1=0 \tag{d}$$

or $$p^2=\frac{k_1k_2}{(k_1+k_2)m}$$

Example 6.7

Find by the use of transfer matrices the natural frequencies of the system of Fig. 6.13. Also determine the mode shapes.

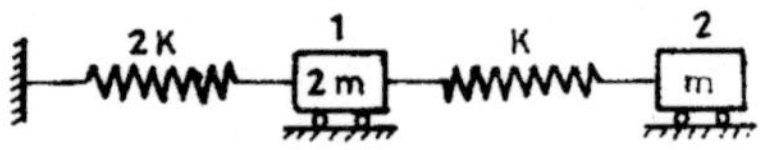

Fig. 6.13 Example 6.7

For the given problem

$$[T]_1=\begin{bmatrix}1 & 1/2k\\ 0 & 1\end{bmatrix},\ [U]_1=\begin{bmatrix}1 & 0\\ -2mp^2 & 1\end{bmatrix}$$

$$[T]_2=\begin{bmatrix}1 & 1/k\\ 0 & 1\end{bmatrix} \text{ and } [U]_2=\begin{bmatrix}1 & 0\\ -mp^2 & 1\end{bmatrix}$$

Therefore

$$\begin{Bmatrix}x\\ N\end{Bmatrix}_2^R=\begin{bmatrix}1 & 0\\ -mp^2 & 1\end{bmatrix}\begin{bmatrix}1 & 1/k\\ 0 & 1\end{bmatrix}\begin{bmatrix}1 & 0\\ -2mp^2 & 1\end{bmatrix}\begin{bmatrix}1 & 1/2k\\ 0 & 1\end{bmatrix}\begin{Bmatrix}x\\ N\end{Bmatrix}_0^R$$

or $$\begin{Bmatrix}x\\ N\end{Bmatrix}_2^R=\begin{bmatrix}\left(1-\frac{2mp^2}{k}\right) & \frac{1}{2k}+\frac{1}{k}\left(-\frac{mp^2}{k}+1\right)\\ -mp^2-2mp^2+\left(-\frac{mp^2}{k}+1\right) & -\frac{mp^2}{2k}+\left(-\frac{mp^2}{k}+1\right)^2\end{bmatrix}\begin{Bmatrix}x\\ N\end{Bmatrix}_0^R$$

The boundary conditions of the problem are

$$x_0=0 \text{ and } N_2^R=0$$

The frequency equation becomes

$$-\frac{mp^2}{2k}+\left(-\frac{mp^2}{k}+1\right)^2=0$$

or $$p^4-\frac{5k}{m}p^2+\frac{k^2}{m^2}=0$$

Solution of Eq. (a) yields

$$p_{1,2}^2 = \frac{k}{2m}, \frac{2k}{m}$$

The normal modes are calculated as follows:

$$\{Z\}_1^L = \begin{bmatrix} 1 & 1/2k \\ 0 & -1 \end{bmatrix} \begin{Bmatrix} 0 \\ N \end{Bmatrix}_0^R$$

Let $N_0 = 1$, then $\{Z\}_1^L = \begin{Bmatrix} 1/2k \\ 1 \end{Bmatrix}$

$$\{Z\}_1^R = \begin{bmatrix} 1 & 0 \\ -2mp^2 & 1 \end{bmatrix} \begin{Bmatrix} 1/2k \\ 1 \end{Bmatrix} = \begin{Bmatrix} 1/2k \\ -\dfrac{mp^2}{k} + 1 \end{Bmatrix}$$

$$\{Z\}_2^L = \begin{bmatrix} 1 & 1/k \\ 0 & 1 \end{bmatrix} \begin{Bmatrix} 1/2k \\ -\dfrac{mp^2}{k} + 1 \end{Bmatrix} = \begin{Bmatrix} \dfrac{1}{2k} + \dfrac{1}{k}\left(-\dfrac{mp^2}{k} + 1\right) \\ -\dfrac{mp^2}{k} + 1 \end{Bmatrix}$$

$$\{Z\}_2^R = \begin{bmatrix} 1 & 0 \\ -mp^2 & 1 \end{bmatrix} \begin{Bmatrix} \dfrac{1}{2k} + \dfrac{1}{k}\left(-\dfrac{mp^2}{k} + 1\right) \\ -\dfrac{mp^2}{k} + 1 \end{Bmatrix}$$

$$= \begin{Bmatrix} \dfrac{1}{2k} + \dfrac{1}{k}\left(-\dfrac{mp^2}{k} + 1\right) \\ -mp^2\left[\dfrac{1}{2k} + \dfrac{1}{k}\left(-\dfrac{mp^2}{k} + 1\right)\right] + \left(-\dfrac{mp^2}{k} + 1\right) \end{Bmatrix}$$

Substituting the value of the first frequency, $p_1^2 = \dfrac{k}{2m}$ in the above relations, we get

$$\{Z\}_1^L = \begin{Bmatrix} \dfrac{1}{2k} \\ 1 \end{Bmatrix}, \quad \{Z\}_1^R = \begin{Bmatrix} \dfrac{1}{2k} \\ \dfrac{1}{2} \end{Bmatrix}, \quad \{Z\}_2^L = \begin{Bmatrix} \dfrac{1}{k} \\ \dfrac{1}{2} \end{Bmatrix}$$

and
$$\{Z\}_2^R = \begin{Bmatrix} 1 \\ \overline{k} \\ 0 \end{Bmatrix}$$

Substituting the value of the second frequency, $p_2^2 = \dfrac{2k}{m}$ in the above relations, we get

$$\{Z\}_1^L = \begin{Bmatrix} \dfrac{1}{2k} \\ 1 \end{Bmatrix}, \quad \{Z\}_1^R = \begin{Bmatrix} \dfrac{1}{2k} \\ -1 \end{Bmatrix}, \quad \{Z\}_2^L = \begin{Bmatrix} -\dfrac{1}{2k} \\ -1 \end{Bmatrix}$$

and
$$\{Z\}_2^R = \begin{Bmatrix} -\dfrac{1}{2k} \\ 0 \end{Bmatrix}$$

The mode shapes are plotted in Fig. 6.14.

The results of the normal modes can be checked by the orthogonality property.

$$\sum_{i=1}^{1} m_1 \phi_i^{(1)} \phi_1^{(2)} = 2m \times 1 \times 1 + m \times 1 \times (-2) = 0$$

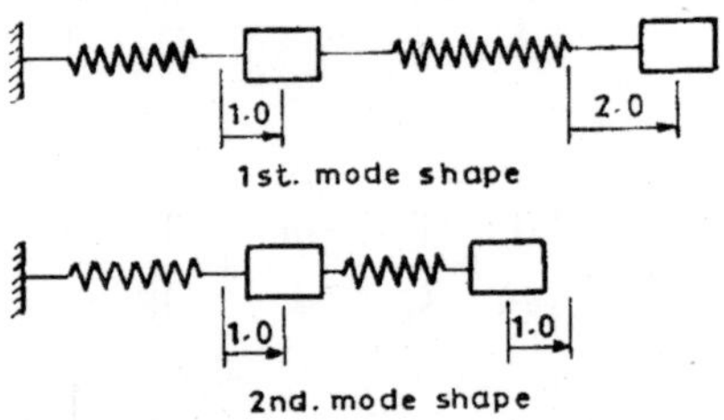

Fig. 6.14 Mode shapes of Example 6.7

6.11 MYKLESTAD METHOD

Like Holzer method, calculations are performed progressively from one station to another. In this section, Myklestad method has been presented for the vibrating beam. Let us consider that the beam is having lumped masses, and the sections in between the masses are massless. Figure 6.15 shows a typical section. Let us approach the problem in a general manner, by incorporating the shear deformation and rotary inertia effects. These effects have been discussed in details in Chapter 8, but here we adopt certain equations derived in that chapter. For free vibration of the beam, the equations are given as follows

$$V = -\eta AG \frac{\partial y_s}{\partial x} \tag{6.72}$$

$$M = EI \frac{\partial^2 y_b}{\partial x^2} \tag{6.73}$$

$$\frac{\partial M}{\partial x} = V + \rho I \frac{\partial^3 y_b}{\partial x \, \partial t^2} = V - \rho I \theta_b p^2 \tag{6.74}$$

$$\frac{\partial V}{\partial x} = myp^2 \tag{6.75}$$

$$y = y_b + y_s \tag{6.76}$$

where η = shear correction factor,
A = cross-sectional area of the member,
G = shear modulus of elasticity,
V = shear force at any section,
M = bending moment at any section,
y_b = bending deflection,
y_s = shearing deflection,
EI = flexural rigidity,
ρ = mass density of the material,
I = second moment of the area,
m = mass of the beam per unit length, and
p = angular frequency of the beam.

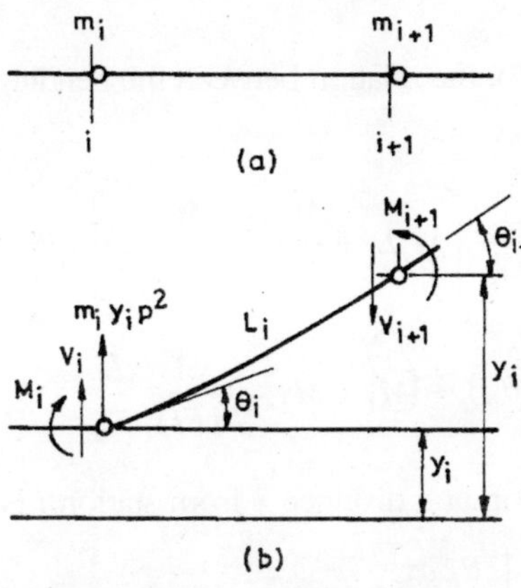

Fig. 6.15 A beam segment

The beam is divided into a number of segments. Expressions for the deflection, slope, bending moment and shear force at station $(i + 1)$ can be expressed in terms of those quantities at station i. From the equilibrium of the segment shown in Fig. 6.15,

$$V_{i+1} = V_i - m_i p^2 y_i \tag{6.77}$$

and

$$M_{i+1} = M_i + V_{i+1} L_i - (\rho I)_i (\theta_b)_i \, p^2 L_i \tag{6.78}$$

The last term of Eq. (6.78) is due to rotary inertia effect. From Eq. (6.72), the following relation can be expressed

$$y_{s_{i+1}} = y_{s_i} - \left(\frac{V}{\eta AG}\right)_i L_i \tag{6.79}$$

Bending moment M at any distance x from the left hand section i is

$$M = M_i + \frac{M_{i+1} - M_i}{L_i} x \tag{6.80}$$

The bending slope at a distance x from i is given by

$$\theta_b = \frac{1}{(EI)_i} \int M\, dx + C_1$$

or

$$\theta_b = \frac{1}{(EI)_i} \int \left[M_i + \frac{M_{i+1} - M_i}{L_i} x \right] dx + C_1$$

or

$$\theta_b = \frac{1}{(EI)_i} \left[M_i x + \frac{M_{i+1} - M_i}{L_i} \frac{x^2}{2} \right] + C_1 \tag{6.81}$$

when x = 0, $\theta_b = \theta_{b_i}$ which yields

$$C_1 = \theta_{b_i} \tag{6.82}$$

From Eqs. (6.81) and (6.82), the relation between the bending slopes at stations i and $(i + 1)$ can be obtained as

$$(\theta_b)_{i+1} = \frac{1}{(EI)_i} \left[M_i L_i + \frac{M_{i+1} - M_i}{2} L_i \right] + (\theta_b)_i$$

or

$$(\theta_b)_{i+1} = (\theta_b)_i + (M_i + M_{i+1}) \frac{1}{(EI)_i} \frac{L_i}{2} \tag{6.83}$$

Similarly, the bending deflection at a distance x from station i is

$$y_b = y_{b_i} + \int \theta_b\, dx \tag{6.84}$$

Substituting the value of θ_b from Eq.(6.81) into Eq.(6.83) and by performing the integration we get

$$y_b = y_{b_i} + \frac{1}{(EI)_i} \left(M_i \frac{x^2}{2} + \frac{M_{i+1} - M_i}{L_i} \frac{x^3}{6} \right) + \theta_{b_i} x \tag{6.85}$$

The relation of the bending deflections between stations i and $(i + 1)$ is as follows:

$$y_{b_{i+1}} = y_{b_i} + (\theta_b)_i L_i + \left(\frac{M_i}{3} + \frac{M_{i+1}}{6} \right) \frac{L_i^2}{(EI)_i} \tag{6.86}$$

Lastly, we can write

$$y_{i+1} = y_{b_{i+1}} + y_{s_{i+1}} \tag{6.87}$$

Sequence of calculations given by Eqs. (6.77), (6.78), (6.79), (6.83), (6.86) and (6.87) are performed for an assumed value of p^2.

A beam has two boundary conditions at each end. Therefore, out of four conditions required in a beam, two are known. The calculation is started by assuming a trial frequency p. The value of p that satisfies simultaneously both end conditions of the beam is its correct value.

We start the computation from one end. A value of the frequency p is to be assumed. Let it be p'. Let us take the case of a simply supported beam. At both ends, $y = M = 0$, but the slope and the shear force are not zero at that end. We assume a slope θ', at the left hand end (for convenience, it is taken as 1.0) and shear force at that end is considered as zero there. We calculate progressively from that end, till we reach the other end, which is a simple support. Let the deflection and the bending moment at the simply supported end be A_1 and A_2. Next, we assume a shear force V' (say equal to 1.0) and zero slope at the left hand end, and compute the deflection and the bending moment at the other end. Let A_3 and A_4 be the values of the deflection and the bending moment at the right hand end. If θ_0 and V_0 are the actual values of the slope and the shear force at the left hand end, then the deflection and the bending moment at the right hand end are

$$\left.\begin{aligned} y_e &= A_1 \frac{\theta_0}{\theta'} + A_3 \frac{V_0}{V'} \\ M_e &= A_2 \frac{\theta_0}{\theta'} + A_4 \frac{V_0}{V'} \end{aligned}\right\} \tag{6.88}$$

Equation (6.88) when written in matrix form becomes

$$\begin{Bmatrix} y_e \\ M_e \end{Bmatrix} = \begin{bmatrix} \dfrac{A_1}{\theta'} & \dfrac{A_3}{V'} \\ \dfrac{A_2}{\theta'} & \dfrac{A_4}{V'} \end{bmatrix} \begin{Bmatrix} \theta_0 \\ V_0 \end{Bmatrix} \tag{6.89a}$$

If the value coincides with one of the natural frequencies, then the determinant formed by the coefficients is equal to zero, that is

$$\Delta = \begin{vmatrix} \dfrac{A_1}{\theta'} & \dfrac{A_3}{V'} \\ \dfrac{A_2}{\theta'} & \dfrac{A_4}{V'} \end{vmatrix} = 0 \tag{6.89b}$$

The method comes under determinant search technique. A plot is made of Δ vs p^2. The values of p^2 for which $\triangle$ are zero, are the natural frequencies of the system.

Example 6.8

For the cantilever beam shown in Fig. 6.16, determine the natural frequencies. Assume, $EI = 50$, $L = 1$ and $m = 2$.

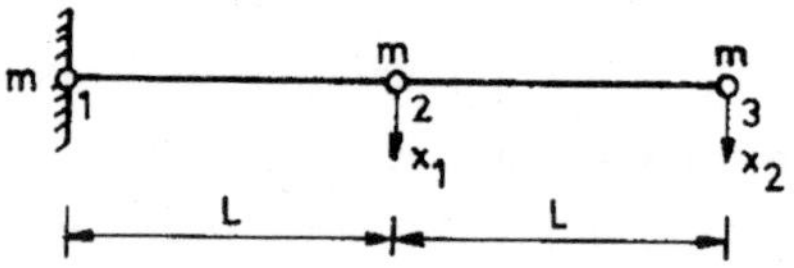

Fig. 6.16 Example 6.8

Assume $p^2 = 8.525$. The beam is divided into two segments. The clamped end is denoted as 1, the point at which mass m is placed inside the beam as station 2 and the free end as station 3. At station 1, $y_1 = \theta_1 = 0$. Let $M_1 = 1$ and $V_1 = 0$. Then from Eq. (6.77)

$$V_2 = V_1 - m_1 y_1 p^2 = 0 - 2\times 0\times 8.525 = 0$$

Calculation for other quantities can be carried out as follows:

$$M_2 = M_1 + V_2 L_1 = 1 + 0\times 1 = 1$$

$$\theta_2 = \theta_1 + (M_1 + M_2)\frac{L_1}{2EI} = 0 + \frac{1\times 2}{2\times 50} = 0.02$$

$$y_2 = y_1 + \theta_1 L_1 + \left(\frac{M_1}{3} + \frac{M_2}{6}\right)\frac{L_1^2}{EI} = 0 + 0 + \left(\frac{1}{3} + \frac{1}{6}\right)\frac{1}{50} = 0.01$$

Similarly, the different quantities at the free end will be

$$V_3 = V_2 - m_2 y_2 p^2 = 0 - 2\times 8.525\times 0.01 = -0.1705$$

$$M_3 = M_2 + V_3 L_2 = 1 - 0.1705\times 1 = 0.8295$$

$$\theta_3 = \theta_2 + (M_2 + M_3)\frac{L}{2EI} = 0.02 + (1 + 0.8295)\frac{1}{2\times 50} = 0.0383$$

$$y_3 = y_2 + \theta_2 L_2 + \left(\frac{M_2}{3} + \frac{M_3}{6}\right)\frac{L_2^2}{EI}$$

$$= 0.01 + 0.02\times 1 + \left(\frac{1}{3} + \frac{0.8295}{6}\right)\frac{1}{50} = 0.0394$$

Next, assuming $V_1 = 1$ and $M_1 = 0$ and following steps similar to those presented above

$V_3 \quad = \quad 0.943$

$M_3 \quad = \quad 1.943$

Therefore, the determinant given by Eq.(6.89b) will be

$$\Delta = \begin{vmatrix} -0.1705 & 0.943 \\ 0.8295 & 1.943 \end{vmatrix} = -1.1135$$

By assuming different frequencies, above steps may be repeated.

6.12 STODOLA'S METHOD

One of the most effective methods of computation of a few eigenvalues and corresponding eigenvectors is the Stodola's method. Stodola used it for the solution of the free vibration problem of a rotating shaft in 1904 [6.8]. Earlier in 1898 Vianello used it for determining the critical load in buckling for a rotating shaft. As such, the method is sometimes known as Stodola-Vianello's method. Equation (6.13) is rewritten as

$$[D]\{\phi\} = \lambda\{\phi\} \tag{6.90}$$

where $$[D] = [K]^{-1}[M]$$

and $$\lambda = \frac{1}{p^2}$$

A trial vector $\{\bar{\phi}^1\}$ is assumed and substituted into the left hand side of Eq. (6.13), which on matrix multiplication will give a new $\{\phi\}$, say $\{\bar{\phi}^2\}$.

$$[D]\{\bar{\phi}^1\} = \lambda\{\bar{\phi}^2\} \tag{6.91}$$

$\{\bar{\phi}^2\}$ will in all cases be different than $\{\bar{\phi}^1\}$, unless the correct mode shape $\{\bar{\phi}^2\}$ is assumed by chance. is assumed as the next trial and a different mode shape vector $\{\bar{\phi}^3\}$ will usually be obtained

$$[D]\{\bar{\phi}^2\} = \lambda\{\bar{\phi}^3\} \tag{6.92}$$

The process is repeated till the mode shapes converge, i.e. the assumed mode shape agrees with the derived mode shape, or their difference is within the tolerable limit. The corresponding λ will give the eigenvalue.

Matrix iteration gives the highest value of λ, i.e. the lowest value of p^2. For obtaining other modes, Eq.(6.13) is to be modified based on the orthogonality relationship. The orthogonality relationship given by Eq.(6.21), considering the first and second mode shapes is written as

$$m_1\phi_1^{(1)}\phi_1^{(2)} + m_2\phi_2^{(1)}\phi_2^{(2)} + \cdots + m_n\phi_n^{(1)}\phi_n^{(2)} = 0$$

or $$m_1\phi_1^{(2)} = -\frac{m_2\phi_2^{(1)}}{\phi_1^{(1)}}\phi_2^{(2)} - \frac{m_3\phi_3^{(1)}}{\phi_1^{(1)}}\phi_3^{(2)} - \cdots - \frac{m_n\phi_n^{(1)}}{\phi_1^{(1)}}\phi_n^{(2)} \tag{6.93}$$

The first mode shape being known, those values can be substituted in Eq. (6.88) and a relation between the displacement of the first mass to other masses can be obtained and this $\phi_1^{(2)}$ values in terms of other ϕ_s can be substituted in Eq. (6.13) and it can then be rearranged. Neglecting the first modified equation, this will give rise to $(n-1)$ independent equations for a n-degree of freedom system. This reduced matrix is known as the sweeping matrix, as this sweeps out ϕ_1 component. Matrix iteration as mentioned earlier can then be carried out with this matrix of size $(n-1) \times (n-1)$, to obtain the second natural frequency and the second mode shape.

Same procedure is to be repeated for other natural frequencies and mode shapes.

Example 6.9

Determine the natural frequencies and mode shapes for the framed structure shown in Fig. 6.17. The floor is considered to be absolutely rigid.

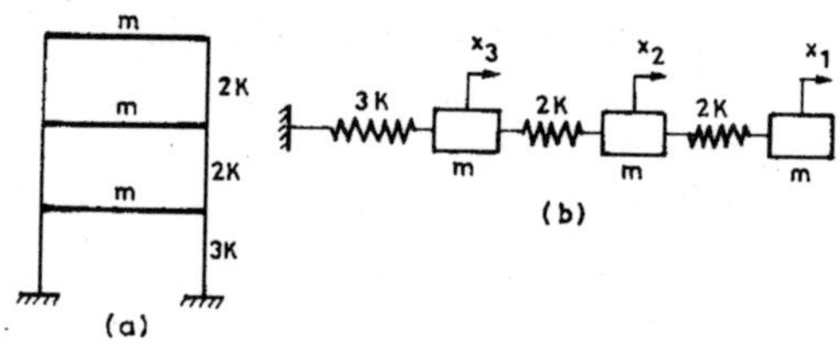

Fig. 6.17 Example 6.17

The equivalent spring-mass system of the framed structure has been shown in Fig. 6.17(b). The mass and the stiffness matrices of the problem based on the coordinate system are

$$[M]=\begin{bmatrix} m & 0 & 0 \\ 0 & m & 0 \\ 0 & 0 & m \end{bmatrix}=m\begin{bmatrix} 1 & 0 & 0 \\ 0 & 1 & 0 \\ 0 & 0 & 1 \end{bmatrix}$$

and

$$[K]=\begin{bmatrix} 2k & -2k & 0 \\ -2k & 4k & -2k \\ 0 & -2k & 5k \end{bmatrix}$$

By performing the matrix inversion of $[K]$, we obtain

$$[K]^{-1}=\frac{1}{6k}\begin{bmatrix} 8 & 5 & 2 \\ 5 & 5 & 2 \\ 2 & 2 & 2 \end{bmatrix}$$

Therefore

$$[D]=[K]^{-1}[M]=\frac{1}{6k}\begin{bmatrix} 8 & 5 & 2 \\ 5 & 5 & 2 \\ 2 & 2 & 2 \end{bmatrix} m\begin{bmatrix} 1 & 0 & 0 \\ 0 & 1 & 0 \\ 0 & 0 & 1 \end{bmatrix}=\frac{m}{6k}\begin{bmatrix} 8 & 5 & 2 \\ 5 & 5 & 2 \\ 2 & 2 & 2 \end{bmatrix}$$

The matrix iteration can now be started by assuming a trial vector

$$\{\phi\}^T = \{1 \quad 1 \quad 1\}$$

Therefore

$$\frac{m}{6k}\begin{bmatrix} 8 & 5 & 2 \\ 5 & 5 & 2 \\ 2 & 2 & 2 \end{bmatrix}\begin{Bmatrix} 1 \\ 1 \\ 1 \end{Bmatrix}=\frac{m}{6k}\begin{Bmatrix} 15 \\ 12 \\ 6 \end{Bmatrix}=\frac{m}{6k}\times 15\begin{Bmatrix} 1.0 \\ 0.8 \\ 0.4 \end{Bmatrix}$$

Next, a trial vector of $\{\phi\}^T = \{1.0 \quad 0.8 \quad 0.4\}$ is assumed and the next iteration is carried out.

$$\frac{m}{6k}\begin{bmatrix} 8 & 5 & 2 \\ 5 & 5 & 2 \\ 2 & 2 & 2 \end{bmatrix}\begin{Bmatrix} 1.0 \\ 0.8 \\ 0.4 \end{Bmatrix} = \frac{m}{6k}\begin{Bmatrix} 12.8 \\ 9.8 \\ 4.4 \end{Bmatrix} = \frac{12.8m}{6k}\begin{Bmatrix} 1.00 \\ 0.77 \\ 0.34 \end{Bmatrix}$$

It may be noted that for the second iteration, the assumed $\{\phi\}$ and the derived $\{\phi\}$ are much closer. The process is continued for a few more iterations. Thus, after a few iterations, it can be shown that

$$\frac{m}{6k}\begin{bmatrix} 8 & 5 & 2 \\ 5 & 5 & 2 \\ 2 & 2 & 2 \end{bmatrix}\begin{Bmatrix} 1.000 \\ 0.759 \\ 0.336 \end{Bmatrix} = \frac{m}{6k}\begin{Bmatrix} 12.47 \\ 9.47 \\ 4.19 \end{Bmatrix} = \frac{12.467m}{6k}\begin{Bmatrix} 1.000 \\ 0.759 \\ 0.336 \end{Bmatrix}$$

Therefore

$$\lambda_1 = \frac{1}{p_1^2} = \frac{12.467m}{6k}$$

or

$$p_1^2 = \frac{6k}{12.467m}$$

or

$$p_1 = 0.695\sqrt{\frac{k}{m}}$$

and the first mode shape is given by

$$\{\phi^{(1)}\} = \begin{Bmatrix} 1.00 \\ 0.759 \\ 0.336 \end{Bmatrix}$$

Next, the second natural frequency and the second mode shape is to be calculated. Considering the first and the second mode, the following is the orthogonality relationship.

$$m_1\phi_1^{(1)}\phi_1^{(2)} + m_2\phi_2^{(1)}\phi_2^{(2)} + m_3\phi_3^{(1)}\phi_3^{(2)} = 0 \qquad \text{(a)}$$

Substituting the values of $\phi_1^{(1)}$, $\phi_2^{(1)}$ and $\phi_3^{(1)}$ of the first mode in Eq. (a) and the values of the masses, we get

$$\phi_1^{(2)} = -0.759\phi_2^{(2)} - 0.336\phi_3^{(2)} \qquad \text{(b)}$$

Writing Eq. (6.13) in expanded form and omitting superscript 2 for convenience, we get

$$\left.\begin{aligned} 8\phi_1 + 5\phi_2 + 2\phi_3 &= \lambda\phi_1 \times \frac{6k}{m} \\ 5\phi_1 + 5\phi_2 + 2\phi_3 &= \lambda\phi_2 \times \frac{6k}{m} \\ 2\phi_1 + 2\phi_2 + 2\phi_3 &= \lambda\phi_3 \times \frac{6k}{m} \end{aligned}\right\} \qquad \text{(c)}$$

Substituting the values of from Eq.(b) into Eq.(c) and arranging the different terms

$$\left.\begin{aligned} 1.072\phi_2 + 0.688\phi_3 &= -\lambda(0.759\phi_2 + 0.336\phi_3)\times\frac{6k}{m} \\ 1.2\phi_2 + 0.32\phi_3 &= \lambda\phi_2 \times \frac{6k}{m} \\ 0.48\phi_2 + 1.33\phi_3 &= \lambda\phi_3 \times \frac{6k}{m} \end{aligned}\right\} \quad \text{(d)}$$

Considering the last two equations given by Eq. (d) and writing them in matrix form, we get

$$\frac{m}{6k}\begin{bmatrix} 1.2 & 0.32 \\ 0.48 & 1.33 \end{bmatrix}\begin{Bmatrix} \phi_2 \\ \phi_3 \end{Bmatrix} = \lambda\begin{Bmatrix} \phi_2 \\ \phi_3 \end{Bmatrix}$$

Performing matrix iterations given earlier, we obtain

$$\frac{m}{6k}\begin{bmatrix} 1.2 & 0.32 \\ 0.48 & 1.33 \end{bmatrix}\begin{Bmatrix} 1 \\ 1 \end{Bmatrix} = \frac{m}{6k}\begin{Bmatrix} 1.52 \\ 1.81 \end{Bmatrix} = \frac{m}{6k}\times 1.52\begin{Bmatrix} 1.00 \\ 1.19 \end{Bmatrix}$$

Performing a few more iterations, we obtain

$$\frac{m}{6k}\begin{bmatrix} 1.2 & 0.32 \\ 0.48 & 1.33 \end{bmatrix}\begin{Bmatrix} 1.00 \\ 1.44 \end{Bmatrix} = \frac{1.661m}{6k}\begin{Bmatrix} 1.00 \\ 1.44 \end{Bmatrix} \quad \text{(e)}$$

Substituting the values of ϕ_2 and ϕ_2 from Eq.(e) into Eq.(b), we get

$$\phi_1 = -0.759\times 1 - 0.336\times 1.44 = -1.243$$

The second natural frequency is given by

$$\lambda_2 = \frac{1}{p_2^2} = \frac{1.661m}{6k}$$

or

$$p_2 = 1.9\sqrt{\frac{k}{m}}$$

The second mode shape is given by

$$\{\phi^{(2)}\} = \begin{Bmatrix} -1.243 \\ 1.00 \\ 1.44 \end{Bmatrix}$$

As the mode shape is a ratio between the displacements, we can normalise it. On doing so

$$\{\phi^{(2)}\} = \begin{Bmatrix} 1.000 \\ -0.804 \\ -1.156 \end{Bmatrix}$$

Lastly, the third natural frequency and the third mode shape are to be determined. Writing the orthogonality relationship between the first mode and the third mode, we get

$$m_1\phi_1^{(1)}\phi_1^{(3)} + m_2\phi_2^{(1)}\phi_2^{(3)} + m_3\phi_3^{(1)}\phi_3^{(3)} = 0 \quad \text{(f)}$$

Substituting the values of the masses and those of the first mode shapes, we get

$$\phi_1^{(3)} = -0.759\phi_2^{(3)} - 0.336\phi_3^{(3)} \quad \text{(g)}$$

Similarly, writing the orthogonality relationship connecting the second mode and the third mode and substituting the values of the masses and those due to second mode, we obtain

$$\phi_1^{(3)} = 0.804\phi_2^{(3)} + 1.156\phi_3^{(3)} \quad \text{(h)}$$

Solving Eqs. (g) and (h), we get

$$\phi_2^{(3)} = -0.956\phi_3^{(3)} \quad \text{(i)}$$

Similarly from Eqs. (g) and (i), we get

$$\phi_1^{(3)} = 0.389\phi_3^{(3)} \quad \text{(j)}$$

Substituting $\phi_1^{(3)}$ and $\phi_2^{(3)}$ from Eqs.(j) and (i) into the last equation given by Eq. (d), we get

$$\frac{m}{6k} \times 2[0.389 - 0.956 + 1]\phi_3^{(3)} = \lambda_3\phi_3^{(3)}$$

or

$$0.867\frac{m}{6k}\phi_3^{(3)} = \lambda_3\phi_3^{(3)}$$

or

$$\lambda_3 = \frac{1}{p_3^2} = \frac{0.867m}{6k}$$

or

$$p_3 = 2.635\sqrt{\frac{k}{m}}$$

Third mode shape is given by

$$\{\phi^{(3)}\} = \begin{Bmatrix} 0.389 \\ -0.956 \\ 1.000 \end{Bmatrix}$$

which on normalisation gives

$$\{\phi^{(3)}\} = \begin{Bmatrix} 1.000 \\ -2.427 \\ 2.512 \end{Bmatrix}$$

The mode shapes for the framed structure have been shown in Fig. 6.18.

Example 6.10

For a simply supported beam having uniform flexural rigidity *EI* and span L as shown in Fig. 6.19, determine the first natural frequency by Stodola's method. The total mass of the beam is *M* and the mass per unit length is *m*. The beam is divided into four equal segments.

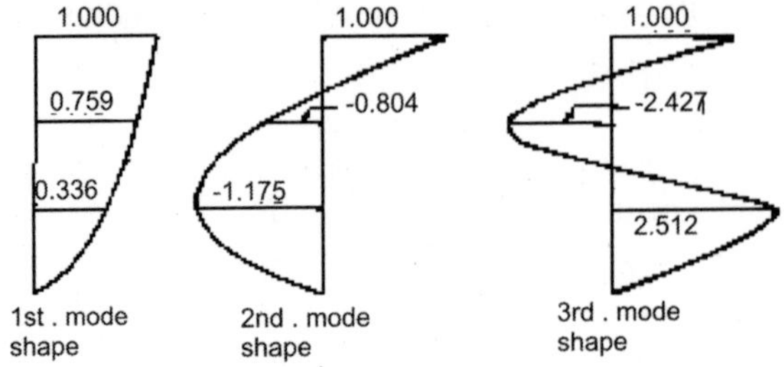

Fig. 6.18 Mode shapes of Example 6.9

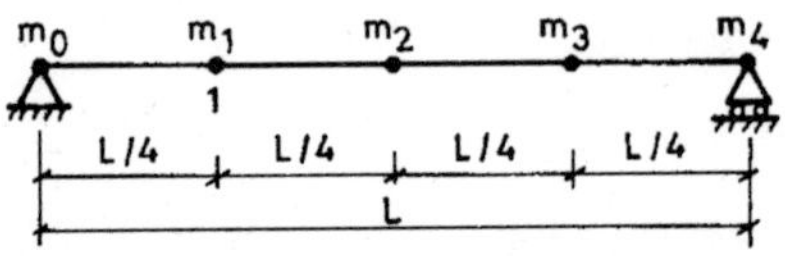

Fig. 6.19 A simply supported beam

The beam is divided into four equal segments. As such, the masses indicated in Fig. 6.19 will assume the following values

$$m_1 = m_2 = m_3 = \frac{mL}{4} = \frac{M}{4}$$

and

$$m_0 = m_4 = \frac{mL}{8} = \frac{M}{8}$$

The equivalent spring-mass model for this problem is complicated. So the most convenient approach to this problem is through the use of influence coefficients. The forces involved in a normal mode are the inertia forces. Total deflection at any point on the beam will be considered to be same as the deflection at that point due to individual inertial force. If the deflections at 1, 2 and 3 are y_1, y_2 and y_3, then the following equations can be written on this basis.

$$\left.\begin{aligned} y_1 &= -f_{11}m_1\ddot{y}_1 - f_{12}m_2\ddot{y}_2 - f_{13}m_3\ddot{y}_3 \\ y_2 &= -f_{21}m_1\ddot{y}_1 - f_{22}m_2\ddot{y}_2 - f_{23}m_3\ddot{y}_3 \\ y_3 &= -f_{31}m_1\ddot{y}_1 - f_{32}m_2\ddot{y}_2 - f_{33}m_3\ddot{y}_3 \end{aligned}\right\} \qquad \text{(a)}$$

where f_{11}, f_{12}, etc., are influence or flexibility coefficients.

Equation (a) can be written in matrix form as follows:

$$\begin{Bmatrix} y_1 \\ y_2 \\ y_3 \end{Bmatrix} = -\begin{bmatrix} f_{11} & f_{12} & f_{13} \\ f_{21} & f_{22} & f_{23} \\ f_{31} & f_{32} & f_{33} \end{bmatrix} \begin{bmatrix} m_1 & 0 & 0 \\ 0 & m_2 & 0 \\ 0 & 0 & m_3 \end{bmatrix} \begin{Bmatrix} \ddot{y}_1 \\ \ddot{y}_2 \\ \ddot{y}_3 \end{Bmatrix} \qquad \text{(b)}$$

or

$$\{y\} = -[F][M]\{\ddot{y}\}$$

or $$-[D]\{\ddot{y}\}=\{y\}$$

where $[D]=[F][M]$

The following solution is assumed

$$\{y\}=ae^{ipt}\{\phi\} \tag{c}$$

Substituting $\{y\}$ and $\{\ddot{y}\}$ from Eq.(c) into Eq.(b), we get

$$[D]\{\phi\}=\frac{1}{p^2}\{\phi\}=\lambda\{\phi\} \tag{d}$$

where $\lambda=\dfrac{1}{p^2}$

Now, the elements of [F] and [Ml matrices are to be generated. The mass matrix for this lumped case is a diagonal matrix, which is given below:

$$[M]=\begin{bmatrix}\frac{M}{4} & 0 & 0\\ 0 & \frac{M}{4} & 0\\ 0 & 0 & \frac{M}{4}\end{bmatrix}=\frac{M}{4}\begin{bmatrix}1 & 0 & 0\\ 0 & 1 & 0\\ 0 & 0 & 1\end{bmatrix}$$

The elements of [F] matrix are as follows:

$$f_{11}=f_{33}=\frac{3L^3}{256EI};\quad f_{13}=f_{31}=\frac{2.33L^3}{256EI};\quad f_{22}=\frac{5.33L^3}{256EI};$$

$$f_{21}=f_{12}=f_{32}=f_{23}=\frac{3.67L^3}{256EI}$$

Therefore

$$[F]=\frac{L^3}{256EI}\begin{bmatrix}3.00 & 3.67 & 2.33\\ 3.67 & 5.33 & 3.67\\ 2.33 & 3.67 & 3.00\end{bmatrix}$$

$$[D]=[F][M]=\frac{ML^3}{1024EI}\begin{bmatrix}3.00 & 3.67 & 2.33\\ 3.67 & 5.33 & 3.67\\ 2.33 & 3.67 & 3.00\end{bmatrix}$$

We can now start the matrix iteration

$$\frac{ML^3}{1024EI}\begin{bmatrix}3.00 & 3.67 & 2.33\\ 3.67 & 5.33 & 3.67\\ 2.33 & 3.67 & 3.00\end{bmatrix}\begin{Bmatrix}1.0\\ 1.0\\ 1.0\end{Bmatrix}=\frac{ML^3}{1024EI}\begin{Bmatrix}9.0\\ 12.67\\ 9.0\end{Bmatrix}$$

$$= \frac{9ML^3}{1024EI}\begin{Bmatrix}1.0\\1.41\\1.0\end{Bmatrix}$$

After a few more iterations, the following will be the value of p_1^2

$$p_1^2 = 97.32\frac{EI}{ML^3}$$

It is interesting to note that the exact solution for the fundamental frequency of the beam problem, by considering it as a continuous system is

$$p_1^2 = \frac{\pi^4 EI}{ML^3} = 97.41\frac{EI}{ML^3}$$

6.13 MATRIX DEFLATION PROCEDURE

The natural frequencies and higher modes for a multiple degrees of freedom system can be obtained by using Gram-Schmidt orthogonalisation and matrix deflation procedures. By inverse iteration, λ and $\{\phi\}$ will always converge to the first mode. In the sweeping matrix technique given in the previous article, the lower modes are eliminated by using orthogonality relationship. In Gram-Schmidt orthonormalisation, we use orthogonality relationship in a different way. We choose the second mode shape $\{\bar{\phi}\}$ given by

$$\{\bar{\phi}\} = \{\phi\} - \alpha\{\phi^{(1)}\} \tag{6.94}$$

and the value of α is such that $\{\bar{\phi}\}$ is orthogonal to $\{\phi^{(1)}\}$. Therefore

$$\{\phi^{(1)}\}^T[M]\{\bar{\phi}\} = \{0\}$$

or

$$\{\phi^{(1)}\}^T[M][\{\phi\} - \alpha\{\phi^{(1)}\}] = \{0\}$$

or

$$\alpha = \frac{\{\phi^{(1)}\}^T[M]\{\phi\}}{\{\phi^{(1)}\}^T[M]\{\phi^{(1)}\}}\{\phi\} \tag{6.95}$$

Substituting α from Eq. (6.95) into Eq. (6.94), we obtain

$$\{\bar{\phi}\} = \{\phi\} - \frac{\{\phi^{(1)}\}\{\phi^{(1)}\}^T[M]}{\{\phi^{(1)}\}^T[M]\{\phi^{(1)}\}}\{\phi\} \tag{6.96}$$

One can now use $\{\bar{\phi}\}$ as the starting vector instead of $\{\phi\}$ and the convergence to the second mode is thus achieved.

In order to determine $\{\phi^{(3)}\}$ by inverse iteration, we must remove both $\{\phi^{(1)}\}$ and $\{\phi^{(2)}\}$ as shown below

$$\{\bar{\phi}\} = \{\phi\} - \alpha_1\{\phi^{(1)}\} - \alpha_2\{\phi^{(2)}\} \tag{6.97}$$

In order to determine α_1 and α_2, the orthogonality relationships can be used as demonstrated above. The procedure may be repeated for other higher modes.

However, if the iteration is carried out with $\{\bar{\phi}\}$, then due to round-off errors, $\{\phi^{(1)}\}$ may be reintroduced in the trial vectors. Therefore, the modification as given by Eq. (6.83) or (6.96) is to be made for each cycle of iteration.

It is more convenient to work on a modified matrix

$$[D]\{\bar{\phi}\}=[D]\left([I]-\frac{\{\phi^{(1)}\}^T\{\phi^{(1)}\}[M]}{\{\phi^{(1)}\}^T[M]\{\phi^{(1)}\}}\right)\{\phi\} \qquad (6.98)$$

where $$[S]=[I]-\frac{\{\phi^{(1)}\}^T\{\phi^{(1)}\}[M]}{\{\phi^{(1)}\}^T[M]\{\phi^{(1)}\}}$$

This new dynamical matrix [*D*][*S*] does not contain any first-mode component, and this is expected to automatically converge to the second mode.

Example 6.11

Determine the second natural frequency and the second mode shape of the problem of Example 6.7 by matrix deflation method.

The first mode shape is given by

$$\{\phi^{(1)}\}=\begin{Bmatrix}1.000\\0.759\\0.336\end{Bmatrix}$$

To determine an appropriate starting iteration vector, we are required to deflate the unit full iteration vector $\{\phi^{(1)}\}$

$$\{\bar{\phi}\}=\begin{Bmatrix}1.000\\1.000\\1.000\end{Bmatrix}-\alpha_1\begin{Bmatrix}1.000\\0.759\\0.336\end{Bmatrix}$$

where given by Eq. (6.95) is

$$\alpha_1=\frac{\{1.000\ \ 0.759\ \ 0.336\}m\begin{bmatrix}1.0&0&0\\0&1.0&0\\0&0&1.0\end{bmatrix}\begin{Bmatrix}1.000\\1.000\\1.000\end{Bmatrix}}{\{1.000\ \ 0.759\ \ 0.336\}m\begin{bmatrix}1.0&0&0\\0&1.0&0\\0&0&1.0\end{bmatrix}\begin{Bmatrix}1.000\\0.759\\0.336\end{Bmatrix}}$$

or $\alpha_1 = 1.24$

Therefore

$$\{\bar{\phi}\}=\begin{Bmatrix}1.000\\1.000\\1.000\end{Bmatrix}-1.24\begin{Bmatrix}1.000\\0.759\\0.336\end{Bmatrix}=\begin{Bmatrix}-0.240\\0.059\\0.583\end{Bmatrix}$$

We can normalise $\{\bar{\phi}\}$ such that

$$\{\bar{\phi}\} = \begin{Bmatrix} -4.067 \\ 1.000 \\ 9.881 \end{Bmatrix}$$

The iteration process can now be started with this trial vector

$$\frac{m}{6k}\begin{bmatrix} 8 & 5 & 2 \\ 5 & 5 & 2 \\ 2 & 2 & 2 \end{bmatrix}\begin{Bmatrix} -4.067 \\ 1.000 \\ 9.881 \end{Bmatrix} = \frac{4.427m}{6k}\begin{Bmatrix} -1.756 \\ 1.000 \\ 3.078 \end{Bmatrix}$$

Without modifying the derived vectors, we proceed for two more iterations and let us see what happens.

$$\frac{m}{6k}\begin{bmatrix} 8 & 5 & 2 \\ 5 & 5 & 2 \\ 2 & 2 & 2 \end{bmatrix}\begin{Bmatrix} -1.756 \\ 1.000 \\ 3.078 \end{Bmatrix} = \frac{2.376m}{6k}\begin{Bmatrix} -1.217 \\ 1.000 \\ 1.955 \end{Bmatrix}$$

$$\frac{m}{6k}\begin{bmatrix} 8 & 5 & 2 \\ 5 & 5 & 2 \\ 2 & 2 & 2 \end{bmatrix}\begin{Bmatrix} -1.217 \\ 1.000 \\ 1.955 \end{Bmatrix} = \frac{2.825m}{6k}\begin{Bmatrix} -0.293 \\ 1.000 \\ 1.323 \end{Bmatrix} \qquad \text{(a)}$$

It is quite clear that convergence does not appear to be good in the third iteration step. It may be due to creeping of round-off errors. It has already been stated that the modification as given by Eq. (6.94) or Eq. (6.96) should be made at each iteration step. Let us modify the vector from the second to the third iteration step. Therefore

$$\alpha_1 = \frac{\{1.000 \quad 0.759 \quad 0.336\}m\begin{bmatrix} 1.0 & 0 & 0 \\ 0 & 1.0 & 0 \\ 0 & 0 & 1.0 \end{bmatrix}\begin{Bmatrix} -1.217 \\ 1.000 \\ 1.955 \end{Bmatrix}}{\{1.000 \quad 0.759 \quad 0.336\}m\begin{bmatrix} 1 & 0 & 0 \\ 0 & 1.0 & 0 \\ 0 & 0 & 1 \end{bmatrix}\begin{Bmatrix} 1.000 \\ 0.759 \\ 0.336 \end{Bmatrix}}$$

or $\alpha_1 = 0.117$

Therefore

$$\{\bar{\phi}\} = \begin{Bmatrix} -1.217 \\ 1.000 \\ 1.955 \end{Bmatrix} - 0.117\begin{Bmatrix} 1.000 \\ 0.759 \\ 0.336 \end{Bmatrix} = \begin{Bmatrix} -0.240 \\ 0.059 \\ 0.583 \end{Bmatrix}$$

We can normalise $\{\bar{\phi}\}$ such that

$$\{\bar{\phi}\} = \begin{Bmatrix} -1.464 \\ 1.000 \\ 2.103 \end{Bmatrix}$$

Therefore

$$\frac{m}{6k}\begin{bmatrix} 8 & 5 & 2 \\ 5 & 5 & 2 \\ 2 & 2 & 2 \end{bmatrix}\begin{Bmatrix} -1.464 \\ 1.000 \\ 2.103 \end{Bmatrix} = \frac{1.886m}{6k}\begin{Bmatrix} -1.319 \\ 1.000 \\ 1.982 \end{Bmatrix}$$

This is a much better convergence than in (a), as indicated above. Continuing in this way for a few more iterations, the quantities involved in second normal mode may be determined.

6.14 RAYLEIGH'S METHOD

Rayleigh's method is an approximate method to determine the fundamental frequency of a MDF system. The concept can be extended easily to continuous systems, as has been shown in a later chapter. For practical problems, the knowledge of the fundamental frequency of a system is always of great interest.

Let $[K]$ and $[M]$ be the stiffness and mass matrices respectively of a MDF system. From Eq. (6.6), we can write

$$\{x\} = Ae^{ipt}\{\phi\} \tag{6.99}$$

The maximum kinetic and potential energies can be written as

$$T_{\max} = \frac{1}{2}\{\dot{x}\}_{\max}^{T}[M]\{\dot{x}\}_{\max} \tag{6.100}$$

$$U_{\max} = \frac{1}{2}\{x\}_{\max}^{T}[K]\{x\}_{\max} \tag{6.101}$$

Substituting values of $\{x\}$ and $\{\dot{x}\}$ from Eq. (6.99) into Eqs. (6.100) and (6.101), we get

$$T_{\max} = \frac{1}{2}p^2A^2\{\phi\}^T[M]\{\phi\} \tag{6.102}$$

$$U_{\max} = \frac{1}{2}A^2\{\phi\}^T[K]\{\phi\} \tag{6.103}$$

Equating the maximum energies of the system, we get

$$p^2 = \frac{\{\phi\}^T[K]\{\phi\}}{\{\phi\}^T[M]\{\phi\}} \tag{6.104}$$

The above method is referred to as Rayleigh's method. As it is always possible to make a fairly reasonable estimate of the fundamental mode shape. Rayleigh's method gives a good idea about the fundamental frequency. The method is upper bound.

Example 6.12

Determine by Rayleigh's method, the fundamental frequency of the system shown in Fig. 6.17.

A reasonably good estimate for the fundamental mode of MDF system, is the shape obtained by applying a static load proportional to the mass, or the mass moment of inertia of the system. Based on this concept, the following is the mode shape for the given problem

$$\{\phi\} = \begin{Bmatrix} 1.00 \\ 0.80 \\ 0.40 \end{Bmatrix}$$

$\{\phi\}$ can be obtained with the value of Example 6.7. Therefore

$$\{\phi\}^T [K]\{\phi\} = \{1.0 \quad 0.8 \quad 0.4\} \begin{bmatrix} 2k & -2k & 0 \\ -2k & 4k & -2k \\ 0 & -2k & 5k \end{bmatrix} \begin{Bmatrix} 1.0 \\ 0.8 \\ 0.4 \end{Bmatrix}$$

$$= 0.88k$$

and $$\{\phi\}^T [M]\{\phi\} = \{1.0 \quad 0.8 \quad 0.4\} m \begin{bmatrix} 1 & & \\ & 1 & \\ & & 1 \end{bmatrix} \begin{Bmatrix} 1.0 \\ 0.8 \\ 0.4 \end{Bmatrix}$$

$$= 1.8m$$

Therefore, from Eq. (6.104)

$$p^2 = \frac{0.88k}{1.8m} = 0.49 \frac{k}{m}$$

or $$p = 0.699 \sqrt{\frac{k}{m}}$$

Comparing this value of p with those obtained in Example 6.9 where

$p = 0.695 \sqrt{\frac{k}{m}}$, the estimate is quite close.

6.15 RAYLEIGH-RITZ METHOD

A n-degree of freedom system can be related to the normal coordinates as follows:

$$\begin{Bmatrix} x_1 \\ x_2 \\ \vdots \\ \vdots \\ x_n \end{Bmatrix} = \begin{bmatrix} \phi_1^{(1)} & \cdots & \phi_1^{(k)} \\ \phi_2^{(1)} & \cdots & \phi_2^{(k)} \\ \cdots & \cdots & \cdots \\ \cdots & \cdots & \cdots \\ \phi_n^{(1)} & \cdots & \phi_n^{(k)} \end{bmatrix} \begin{Bmatrix} \xi_1 \\ \xi_2 \\ \vdots \\ \vdots \\ \xi_k \end{Bmatrix} \tag{6.105}$$

A reduced number of normal coordinates which is k has been chosen.

Equation (6.105) can be written as

$$\{x\} = [\overline{\Phi}]\{\xi\} \tag{6.106}$$

$$n\times1 \quad n\times k \quad k\times1$$

The kinetic and potential energies for the MDF system is given by

$$\left.\begin{aligned} T &= \frac{1}{2}\{\dot{\xi}\}^T[\overline{\Phi}]^T[M][\overline{\Phi}]\{\dot{\xi}\} \\ U &= \frac{1}{2}\{\xi\}^T[\overline{\Phi}]^T[K][\overline{\Phi}]\{\xi\} \end{aligned}\right\} \tag{6.107}$$

The generalised masses and stiffnesses associated with $\{\zeta\}$ are

$$\left.\begin{aligned} [\overline{M}] &= [\overline{\Phi}]^T[M][\overline{\Phi}] \\ [\overline{K}] &= [\overline{\Phi}]^T[K][\overline{\Phi}] \end{aligned}\right\} \tag{6.108}$$

The generalised inertia and elastic forces are given by

$$\left.\begin{aligned} \{F\}_{in} &= -\frac{d}{dt}\left\{\left(\frac{\partial T}{\partial \xi}\right)\right\} = -[\overline{M}]\{\ddot{\xi}\} \\ \{F\}_{el} &= -\left\{\frac{\partial u}{\partial \xi}\right\} = -[\overline{K}]\{\xi\} \end{aligned}\right\} \tag{6.109}$$

The work done by virtual displacement due to external forces is given by

$$\delta_W = \delta\{x\}^T\{F\} = \delta\{\xi\}^T[\overline{\Phi}]^T\{F\}$$

The generalised external force is given by

$$\{\overline{F}\} = [\overline{\Phi}]^T[F] \tag{6.110}$$

The equations of motion of the system is given by

$$\sum F = -[\overline{M}]\{\ddot{\xi}\} - [\overline{K}]\{\xi\} + \{F\} = \{0\} \tag{6.111}$$

$[\overline{M}]$ and $[\overline{K}]$ will be diagonal matrices only for the normal modes. As in this procedure, the calculations are based on the assumption of mode shapes, it is quite likely that Eq. (6.111) will be in coupled form. However, a good choice of assumed mode shapes will make off-diagonal terms relatively small. For free vibration of the system, the following solution is assumed

$$\{\xi\} = \{A\}\sin pt \tag{6.112}$$

Substituting the solution assumed in Eq. (6.112) into Eq. (6.111), the equation of motion becomes

$$([\overline{K}] - p^2[\overline{M}])\{A\} = \{0\} \tag{6.113}$$

Solution of eigenvalue problem given by Eq. (6.108) will yield the eigenvalues.

Example 6.13

For the problem given in Example 6.9, determine the first two natural frequencies by Rayleigh-

Ritz method, assuming first two mode shapes.

The first two mode shapes are assumed as

$$[\overline{\Phi}] = \begin{bmatrix} 1.00 & 1.00 \\ 0.80 & -0.80 \\ 0.40 & -1.20 \end{bmatrix}$$

$$T = \frac{1}{2}\begin{bmatrix}\dot{\xi}_1 & \dot{\xi}_2\end{bmatrix}\begin{bmatrix} 1.0 & 0.8 & 0.4 \\ 1.0 & -0.8 & -1.2 \end{bmatrix} m \begin{bmatrix} 1 & 0 & 0 \\ 0 & 1 & 0 \\ 0 & 0 & 1 \end{bmatrix}\begin{bmatrix} 1.0 & 1.0 \\ 0.8 & -0.8 \\ 0.4 & -1.2 \end{bmatrix}\begin{Bmatrix}\dot{\xi}_1 \\ \dot{\xi}_2\end{Bmatrix}$$

$$= \frac{1}{2}[\dot{\xi}_1 \quad \dot{\xi}_2]\, m \begin{bmatrix} 1.8 & -0.12 \\ -0.12 & 3.08 \end{bmatrix}\begin{Bmatrix}\dot{\xi}_1 \\ \dot{\xi}_2\end{Bmatrix}$$

Similarly, the potential energy is given by

$$U = \frac{1}{2}\{\xi_1 \quad \xi_2\}\begin{bmatrix} 1.0 & 0.8 & 0.4 \\ 1.0 & -0.8 & -1.2 \end{bmatrix}\begin{bmatrix} 2k & -2k & 0 \\ -2k & 4k & -2k \\ 0 & -2k & 5k \end{bmatrix}\begin{bmatrix} 1.0 & 1.0 \\ 0.8 & -0.8 \\ 0.4 & -1.2 \end{bmatrix}\begin{Bmatrix}\xi_1 \\ \xi_2\end{Bmatrix}$$

$$= \frac{1}{2}\{\xi_1 \quad \xi_2\}\, k \begin{bmatrix} 0.88 & -0.4 \\ -0.4 & 11.12 \end{bmatrix}\begin{Bmatrix}\xi_1 \\ \xi_2\end{Bmatrix}$$

For free vibration, we obtain from Eq. (6.118) the following eigenvalue problem

$$\left(k\begin{bmatrix} 0.88 & -0.4 \\ -0.4 & 11.12 \end{bmatrix} - p^2 m \begin{bmatrix} 1.80 & -0.12 \\ -0.12 & 3.08 \end{bmatrix}\right)\begin{Bmatrix}\xi_1 \\ \xi_2\end{Bmatrix} = \begin{Bmatrix} 0 \\ 0 \end{Bmatrix}$$

On solution of the characteristic equation, the following values of natural frequencies are obtained

$$p_1 = 0.695\sqrt{\frac{k}{m}},\ p_2 = 1.9\sqrt{\frac{k}{m}}$$

The above values are very close to those obtained in Example 6.9. One of the problems the analyst may face by applying Rayleigh-Ritz method to the MDF system is in approximating the mode shapes of the structure. The application of this method to continuous systems poses lesser difficulties and is thus more extensively used for those cases.

The point that is to be noted in Rayleigh-Ritz method is that, instead of one mode shape at a time as assumed in Stodola's method, a number of mode shapes is assumed simultaneously.

This fact is taken into consideration in the most practical method of solving large system structural vibration problems, that is subspace iteration method. For more details of subspace iteration method, see Bathe and Wilson [6.9].

6.16 SUBSPACE ITERATION METHOD

Subspace iteration method is one of the most powerful vector iteration method [6.9 - 6.11]. The steps to be followed in the method are described below.

1) Start with a trial vector $[\delta]_1$ having q columns where $q < p$

For $x = 1, 2, \ldots\ldots$, iterate the following

(a) $[K]\left[\bar{\delta}\right]_{r+1} = [M][\delta]_r$ (6.114)

(b) Perform the following calculations:

$$[K]_{r+1} = [\bar{\delta}]^T_{r+1}[K][\bar{\delta}]_{r+1} \tag{6.115a}$$

$$[M]_{r+1} = [\bar{\delta}]^T_{r+1}[M][\bar{\delta}]_{r+1} \tag{6.115b}$$

(c) Solve the following reduced eigenproblem

$$[K]_{r+1}[Q]_{r+1} = [M]_{r+1}[Q]_{r+1}[\Lambda]_{r+1} \tag{6.116}$$

for q eigenvalues $[\Lambda]_{r+1}$ (a diagonal matrix) and eigenvectors $[Q]_{r+1}$

(d) An improved approximation to the original system can be calculated as follows

$$\{\delta\}_{r+1} = [\bar{\delta}]_{r+1}[Q]_{r+1} \tag{6.117}$$

The eigenvalues $[\Lambda]_{r+1}$ and eigenvectors $[Q]_{r+1}$ converge to the lowest eigenvalues and eigenvectors as $r \to \infty$.

The first step in the subspace iteration method is the selection of starting vector $[\delta]_1$. For details, the reader may refer to References 6.13, 6.14 and 6.15.

6.17 SIMULTANEOUS ITERATION METHOD AND ALGORITHM

Simultaneous iteration method is similar in technique to subspace iteration method. [6.12]

In this method $[K]$ is decomposed as

$$[K] = [L][L]^T \tag{6.118}$$

where $[L]$ is a lower triangular matrix.

Using Eq. (6.115), the free vibration equation (6.118) can be written as

$$\left\{[L]^{-1}[M]\left([L]^{-1}\right)^T\right\}\left\{[L]^T\{\delta\}\right\} = \frac{1}{2}[L]^T\{\delta\} \tag{6.119}$$

or $$[A]\{x\} = \lambda\{x\} \quad (6.120)$$

where $$[A] = \left\{[L]^{-1}[M]\left([L]^{-1}\right)^T\right\}$$

$$\{x\} = [L]^T\{\delta\}$$

$$\lambda = \frac{1}{p^2}$$

The procedure for solving Eq. (6.120) by the simultaneous iteration method is summarised as follows:

1. Set a trial vector $[U]$ and orthonormalise it

2. Back substitute $[L][X]=[U]$
3. Multiply $[Y]=[M][X]$
4. Forward substitute $[L]^T[V]=[Y]$
5. Form $[B]=[U]^T[V]$
6. Construct [T] so that $t_{ii}=1$ and $t_{ij}=-2b_{ij}/[b_{ii}-b_{ij}+S(b_{ii}-b_{ij})^2]$
   where S = sign of $(b_{ii}-b_{ij})$
7. Multiply $[W]=[U][T]$
8. Perform Schmidt orthonormalisation to derive [U]
9. Check tolerance $[U]-[\overline{U}]$
10. If not satisfactory, go to step 2.

6.18 GEARED SYSTEMS

Two shafts are connected by a gear as shown in Fig. 6.20(a). The speed ratio of shaft 2 to shaft 1 is n. The system can be reduced to an equivalent shaft.

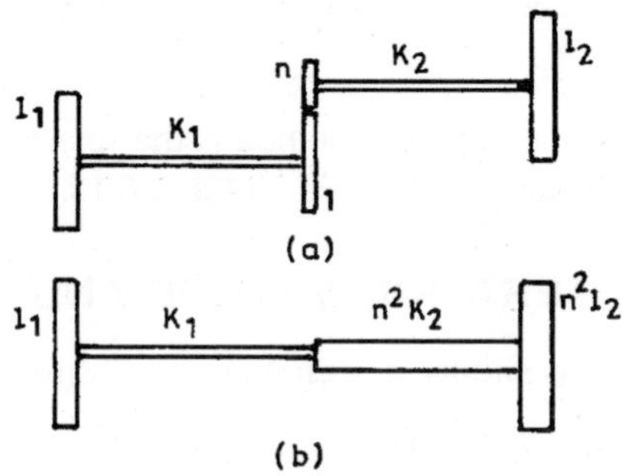

Fig. 6.20 Geared systems

If the rotations of two shafts at the gear are θ_1 and θ_2, then $\theta_2=n\theta_1$.

The kinetic energy of the system is given by

$$T=\frac{1}{2}I_1\dot{\theta}_1^2+\frac{1}{2}I_2\dot{\theta}_2^2$$

or

$$T=\frac{1}{2}I_1\dot{\theta}_1^2+\frac{1}{2}I_2n^2\dot{\theta}_1^2 \tag{6.121}$$

Thus the disc 2 with reference to disc 1 has an equivalent inertia, which is n^2I_2.

The potential energy of the system is

$$\left.\begin{aligned} U&=\frac{1}{2}K_1\theta_1^2+\frac{1}{2}K_2\theta_2^2 \\ \text{or}\quad U&=\frac{1}{2}K_1\theta_1^2+\frac{1}{2}K_2n^2\theta_1^2 \end{aligned}\right\} \tag{6.122}$$

The equivalent stiffness of shaft 2 with reference to shaft 1 is n^2k_2.

The equivalent single shaft system has been shown in Fig. 6.20(b). Thus the rule to be followed for the geared system is, multiply all stiffness and inertias of the geared shaft by n^2, where n is the speed ratio of the geared shaft to the reference shaft.

In dealing with multispeed systems, it is convenient to transform the system into one of constant speed using suitable equivalent masses and stiffness coefficients.

6.19 BRANCHED SYSTEMS

Branched systems are frequently encountered in engineering. Their examples include ship shafting installations, either as twin-screw or as twin-engine systems, drive shaft and the differential of an automobile.

By forming equivalent stiffnesses and inertia, branched systems can be reduced to the form of one-to-one gears, as shown in Fig. 6.21.

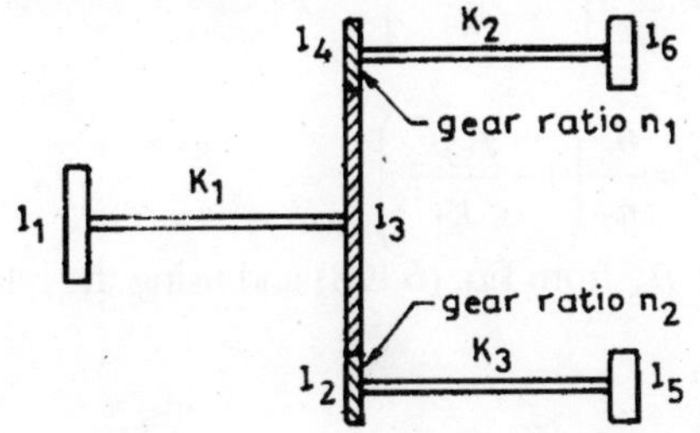

Fig. 6.21 Branched systems

The following steps are required for the calculation of natural frequency of the branched system of Fig. 6.21 by Holzer method:

(a) Start with an assumed frequency p.

(b) It is always preferable to start from the branched end of a system. Assume an amplitude of the upper branch, say $\theta_6 = 1$. Let us convert every stiffness and inertia with respect to the left hand shaft, which is the reference shaft.

$$I_{6e} = n_1^2 I_6, I_{4e} = n_1^2 I_4, I_{5e} = n_2^2 I_5, I_{2e} = n_2^2 I_2$$

$$K_{2e} = n_1^2 k_2, K_{3e} = n_2^2 k_3$$

(c) Applying Holzer's method

$$\theta_4 = 1 - \frac{n_1^2 I_6 p^2}{n_1^2 k_2} = 1 - \frac{I_6 p^2}{k_2} \tag{6.123}$$

θ_4 and θ_3 are related. Therefore,

$$\theta_3 = \frac{1}{n_1}\left(1 - \frac{I_6 p^2}{k^2}\right) = \bar{\theta}_3 \tag{6.124}$$

(d) Assume $\theta_5 = 1$.

Applying Holzer's method

$$\theta_2 = 1 - \frac{n_2^2 I_5 p^2}{n_2^2 k_3} = 1 - \frac{I_5 p^2}{k_3} \tag{6.125}$$

θ_2 is related to θ_3 of the centre gear. Therefore,

$$\bar{\theta}_3 = \frac{1}{n_2}\left(1 - \frac{I_5 p^2}{k_3}\right) = \bar{\theta}_3' \tag{6.126}$$

(e) In most cases, $\bar{\theta}_3$ will differ from $\bar{\theta}_3'$. Therefore, a new value of θ_5 is to be assumed. Let this value be

$$\theta_5 = \frac{\bar{\theta}_3}{\bar{\theta}_3'} = \theta_0 \tag{6.127}$$

Computation is again started from the lower branch.

$$\theta_2 = \theta_5\left(1 - \frac{I_5 p^2}{k_3}\right) \tag{6.128}$$

Therefore
$$\theta_3' = \frac{\theta_5}{n_2}\left(1 - \frac{I_5 p^2}{k_3}\right) \tag{6.129}$$

Substituting the value of θ_5 from Eq. (6.128) and using the relation given in Eq. (6.126), Eq. (6.129) becomes

$$\bar{\theta}_3' = \frac{\theta_3}{\bar{\theta}_3'} \times \bar{\theta}_3' = \bar{\theta}_3 \tag{6.130}$$

Now we obtain a unique value of rotation of the centre gear, whether we move from the lower branch or the upper branch.

(f) Then torques from the two branches are transferred to the single line part of the system. Total torque becomes

$$T = (n_1^2 I_6 \cdot 1 + n_2^2 I_5 \theta_0 + n_1^2 I_4 \theta_4 + n_2^2 I_2 \theta_2 + I_3 \theta_3)\, p^2 \tag{6.131}$$

Substituting $\theta_4 = n_1 \bar{\theta}_3$ and $\theta_2 = n_2 \theta_0$ into Eq. (6.131), yields

$$T = (n_1^2 I_6 \cdot 1 + n_2^2 I_5 \theta_0 + n_1^2 I_4 \bar{\theta}_3 + n_2^2 I_2 \theta_0 + I_3 \bar{\theta}_3)\, p^2 \tag{6.132}$$

(g) The rotation of disc I_p, is given by

$$\theta_1 = \bar{\theta}_3 - \frac{T}{k_1} \tag{6.133}$$

(h) Resulting torque at the far end is

$$T_{ext} = T + I_1 p^2 \theta_1 \tag{6.134}$$

If the correct frequency has been assumed, then T_{ext} will be zero. Steps (a) to (*h*) are to be followed based on one assumed frequency of p.

6.20 REDUCTION METHODS FOR DYNAMIC ANALYSIS

An analysis of the free vibration problem can be performed by solving the following eigenvalue problem

$$([K] - p^2[M])\{\phi\} = \{0\} \tag{6.135}$$

The direct method of generating [K] and [M] matrices is suited only to problems of smaller size. For bigger size problems advantage is taken of the fact that [K] and [M] are symmetric and banded for structural problems. These matrices may be stored in half-band and to save further storage space in skyline form. As most eigenvalue solvers are increasingly expensive (takes time of solution proportional to n^3 where n is the number of degrees of freedom), better methods have been sought which would take less time. It is desirable to apply methods in which the size of these matrices are reduced so that more economic solution of the eigenvalue is obtained. Two such methods will be discussed here. They are Guyan's reduction method of dynamic analysis [6.13, 6.14] and the component mode synthesis method [6.15, 6.16].

6.20.1 Static Condensation

Total degrees of freedom of the superelement is separated into two groups, namely the retained degrees of freedom and the internal degrees of freedom. Accordingly, the matrices may be split into submatrices

$$[K] = \begin{bmatrix} [K]_{rr} & [K]_{ri} \\ [K]_{ir} & [K]_{ii} \end{bmatrix}; \quad \{x\} = \begin{Bmatrix} \{x\}_r \\ \{x\}_i \end{Bmatrix} \quad \text{and} \quad \{F\} = \begin{Bmatrix} \{F\}_r \\ \{F\}_i \end{Bmatrix} \tag{6.136}$$

where subscripts *r* and *i* correspond to retained and internal degrees of freedom respectively.

$[K]_{rr}$ is the stiffness matrix associated with the retained joint displacements

$[K]_{ri}$ is the stiffness matrix associated with the retained displacements developed by unit values of internal displacements

Similarly, $[K]_{ir}$ and $[K]_{ii}$ can be defined.

[K] is symmetric, therefore $[K]_{ir} = ([K]_{ri})^T$

From Eq. (6.136), two matrix equations can be written

$$[K]_{rr}\{x\}_r + [K]_{ri}\{x\}_i = \{F\}_r \tag{6.137a}$$

$$[K]_{ir}\{x\}_r + [K]_{ii}\{x\}_i = \{F\}_i \tag{6.137b}$$

Adjusting the terms of Eq. (6.137b), we get

$$\{x\}_i = [K]_{ii}^{-1}\{F\}_i - [K]_{ii}^{-1}[K]_{ir}\{x\}_r \tag{6.138}$$

Eliminating $\{x\}_i$ from Eq. (6.137a) using Eq. (6.138), we get

$$[K]_{red}\{x\}_r = \{F\}_{red} \tag{6.139}$$

where

$$[K]_{red} = [K]_{rr} - [K]_{ri}\,[K]_{ii}^{-1}\,[K]_{ir} \tag{6.140}$$

and

$$\{F\}_{red} = \{F\}_r - [K]_{ri}\,[K]_{ii}^{-1}\,\{F\}_i \tag{6.141}$$

$[K]_{red}$ is the reduced stiffness matrix of the structure.

$\{F\}_{red}$ is the reduced load matrix.

6.20.2 **Guyan's Reduction Method of Dynamic Analysis**

In this case it is assumed that the same static relationship between retained and internal degrees of freedom remains valid in the dynamic problem. We rewrite Eq. (6.138) for an unloaded structure, that is, $\{F\}_i = \{0\}$

$$\{x\}_i = [W]\{x\}_r \tag{6.142}$$

where
$$[W] = -[K]_{ii}^{-1}[K]_{ir} \tag{6.143}$$

In dynamic analysis in order to ascertain the correct relation between internal and retained d.o.f., [*W*] matrix of Eq. (6.142) would have to be specified from the dynamic displacement pattern, which is not known a priori. [*W*] should also be time dependent. In order to use the reduction method as in static analysis, the normal approximation is to make [*W*] as the static displacements obtained in the internal d.o.f (also termed as slave d.o.f.) $\{x\}_i$, when an otherwise unloaded structure is given unit displacement in the retained d.o.f (also termed as master d.o.f.). Or in other words it means that an approximation is introduced which depicts that $\{x\}_i$ will get a static deformation pattern imposed from $\{x\}_i$ on an otherwise unloaded structure. The mass and damping matrices can be reduced by the same relationship. [*W*] is called as the influence matrix, since it relates the displacements at internal degrees of freedom to the retained d.o.f.

The dynamic equilibrium equation is given by

$$[M]\{\ddot{x}\} + [C]\{\dot{x}\} + [K]\{x\} = \{F(t)\} \tag{6.144}$$

Partitioning Eq. (6.144) yields

$$\begin{bmatrix} [M]_{rr} & [M]_{ri} \\ [M]_{ir} & [M]_{ii} \end{bmatrix} \begin{Bmatrix} \{\ddot{x}\}_r \\ \{\ddot{x}\}_i \end{Bmatrix} + \begin{bmatrix} [C]_{rr} & [C]_{ri} \\ [C]_{ir} & [C]_{ii} \end{bmatrix} \begin{Bmatrix} \{\dot{x}\}_r \\ \{\dot{x}\}_i \end{Bmatrix}$$

$$+ \begin{bmatrix} [K]_{rr} & [K]_{ri} \\ [K]_{ir} & [K]_{ii} \end{bmatrix} \begin{Bmatrix} \{x\}_r \\ \{x\}_i \end{Bmatrix} = \begin{Bmatrix} \{F\}_r \\ \{F\}_i \end{Bmatrix} \tag{6.145}$$

Eq. (6.142) may be written in another way;

$$\{x\} = \begin{Bmatrix} \{x\}_r \\ \{x\}_i \end{Bmatrix} = \begin{bmatrix} [I] \\ [W] \end{bmatrix} \{x\}_r = [H]\{x\}_r \tag{6.146}$$

Therefore,

$$[H] = \begin{bmatrix} [I] \\ [W] \end{bmatrix} \tag{6.147}$$

From Eq. (6.146), we get

$$\delta\{x\}^T = \delta\{x\}_r^T [H]^T \tag{6.148}$$

Premultiplying both sides of Eq. (6.144) by $\delta\{x\}^T$ and using Eq. (6.148) yields

$$\delta\{x\}_r^T ([H]^T[M][H]\{\ddot{x}\}_r + [H]^T[C][H]\{\dot{x}\}_r$$

$$[H]^T[K][H]\{x\}_r) = \delta\{x\}_r^T [H]^T \{F(t)\} \tag{6.149}$$

From Eq. (6.149) we get

$$[M]_{red}\{\ddot{x}\}_r + [C]_{red}\{\dot{x}\}_r + [K]_{red}\{x\}_r = \{F\}_{red} \qquad (6.150)$$

where

$$[M]_{red} = [H]^T[M][H], \quad [C]_{red} = [H]^T[C][H]$$

$$[K]_{red} = [H]^T[K][H] \quad \text{and} \quad \{F\}_{red} = [H]^T\{F\} \qquad (6.151)$$

Combining Eqs. (6.147) and (6.151), yield

$$[M]_{red} = [M]_{rr} + [W]^T[M]_{ir} + [M]_{ri}[W] + [W]^T[M]_{ii}[W] \qquad 6.152a)$$

$$[C]_{red} = [C]_{rr} + [W]^T[C]_{ir} + [C]_{ri}[W] + [W]^T[C]_{ii}[W] \qquad (6.152b)$$

$$[K]_{red} = [K]_{rr} + [W]^T[K]_{ir} \qquad (6.152c)$$

$$\{F\}_{red} = \{F\}_r + [W]^T\{R\}_i \qquad (6.152d)$$

The expression for the reduced stiffness matrix given by Eq. (6.152c) is identical to that given by Eq. (6.140), which is obtained by static condensation of the stiffness matrix.

If [*M*] and [*C*] matrices are diagonal, then $[M]_{red}$ and $[C]_{red}$ given by Eqs. (6.152a) and (6.152b) are considerably simplified, since all off-diagonal elements of the matrices will be zero.

This approach is known as master-slave reduction technique. The method is also termed as the static condensation of the dynamic problem.

Now, if the displacements corresponding to internal degrees of freedom are to be retracked, then

$$\{x\}_i = [K]_{ii}^{-1}\{F\}_i - [K]_{ii}^{-1}[K]_{ir}\{x\}_i = [K]_{ii}^{-1}\{F\}_i + [W]\{x\}_r \qquad (6.153)$$

Similarly, the velocities and accelerations if needed to be retracked is given by

$$\{\dot{x}\}_i = -[K]_{ii}^{-1}[K]_{ir}\{\dot{x}\}_r = [W]\{\dot{x}\}_r \qquad (6.154)$$

and

$$\{\ddot{x}\}_i = -[K]_{ii}^{-1}[K]_{ir}\{\ddot{x}\}_r = [W]\{\ddot{x}\}_r \qquad (6.155)$$

If there is no damping and no load, then Eq. (6.150) reduces to

$$[M]_{red}\{\ddot{x}\}_r + [K]_{red}\{x\}_r = \{0\} \qquad (6.156)$$

Eq. (6.156) indicates an eigenvalue problem.

Example 6.14

Use the reduction method to eliminate x_2 from the equations of motion of the system shown in Fig. 6.22

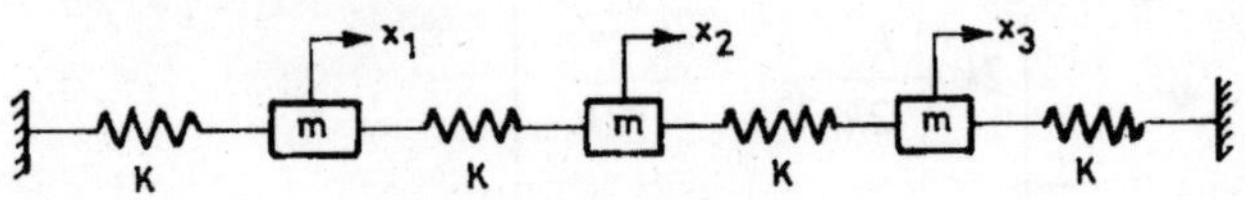

Fig. 6.22 Example 6.14

The equation of motion for free vibration is given by

$$\begin{bmatrix} m & 0 & 0 \\ 0 & m & 0 \\ 0 & 0 & m \end{bmatrix} \begin{Bmatrix} \ddot{x}_1 \\ \ddot{x}_2 \\ \ddot{x}_3 \end{Bmatrix} + \begin{bmatrix} 2k & -k & 0 \\ -k & 2k & -k \\ 0 & -k & 2k \end{bmatrix} \begin{Bmatrix} x_1 \\ x_2 \\ x_3 \end{Bmatrix} = \begin{Bmatrix} 0 \\ 0 \\ 0 \end{Bmatrix}$$

From the second equation of motion, we get after neglecting the inertia term

$$-kx_1 + 2kx_2 - kx_3 = 0$$

which yields

$$x_2 = \frac{1}{2k}(kx_1 + kx_3)$$

Therefore,

$$\begin{Bmatrix} x_1 \\ x_2 \\ x_3 \end{Bmatrix} = \begin{bmatrix} 1 & 0 \\ \frac{1}{2} & \frac{1}{2} \\ 0 & 1 \end{bmatrix} \begin{Bmatrix} x_1 \\ x_3 \end{Bmatrix}$$

Therefore,

$$[M]_{red} = [H]^T [M][H]$$

$$= \begin{bmatrix} m + \frac{mk^2}{4k^2} & \frac{k^2 m}{4k^2} \\ \frac{k^2 m}{4k^2} & m + \frac{mk^2}{4k^2} \end{bmatrix}$$

$$= \begin{bmatrix} \frac{5m}{4} & \frac{m}{4} \\ \frac{m}{4} & \frac{5m}{4} \end{bmatrix}$$

$$[K]_{red} = [H]^T [K][H]$$

$$= \begin{bmatrix} 2k - \frac{k^2}{2k} & -\frac{k^2}{2k} \\ -\frac{k^2}{2k} & 2k - \frac{k^2}{2k} \end{bmatrix}$$

$$= \begin{bmatrix} \frac{3k}{2} & -\frac{k}{2} \\ -\frac{k}{2} & \frac{3k}{2} \end{bmatrix}$$

It is suggested that the master degrees of freedom should be chosen in the region of high flexibility and slave degrees of freedom in regions of high stiffness [6.17, 6.18]. Reference [6.19] describes automatic procedure for this.

According to Ref. 6.17, retained (master) and internal (slave) d.o.f. can be selected on the basis of ratio of diagonal terms of [*K*] and [*M*] matrices of Eq. (6.144). Those d.o.f. which yield the largest values of the ratio K_{ii} / M_{ii} are selected as slave d.o.f.

Attempts have been made to reduce the error inherent in the static consideration method applied to dynamic problems. A dynamic consideration method has been proposed which is as follows [6.20]

To start the process, an approximate value is assigned to the first eigenvalue p_1^2, then the dynamic condensation is applied to the dynamic matrix $[D_1]=[K]-p_1^2[M]$, and then solving the reduced eigenvalue problem to determine the first and second eigenvalues p_2^2 and p_3^2. Next the dynamic condensation is applied to the dynamic matrix $[D_2]=[K]-p_2^2[M]$ to reduce the problem and the second and third eigenvalues p_1^2 and p_2^2 are calculated. The procedure is repeated with one exact eigenvalue and an approximation of next eigenvalue calculated at each step.

6.21 COMPONENT MODE SYNTHESIS METHOD

In this section another technique for reducing a number of degrees of freedom is introduced. The method is known as component mode synthesis method, substructure analysis or building block approach. These methods are very suitable for solving very large structural dynamics problems. This will prove to be useful for structures having natural components, for example, building with a surrounding soil and the space shuttle orbitor with its payloads.

Like the substructure method of static analysis, instead of catering large number of equations at one time, a set of smaller size problems are to be encountered in this method. A complete structure is divided into a number of substructures (component), each of which is idealised as a finite element model. Total number of degrees of freedom in each substructure is reduced by modal substitution. All the components are then assembled together and the complete structure is analysed. In the following, two types of component mode synthesis methods will be discussed. They are the fixed interface method and the free interface method.

6.21.1 Fixed Interface Method

Fig. 6.23 shows a clamped-clamped beam which has been divided into two substructures, I and II. The substructure has been idealised into finite elements.

Let the total displacements of each substructure is given by $\{x\}^s$. Total displacements are partitioned into two parts, that is

$$\{x\}^s = \begin{Bmatrix} \{x\}_I \\ \{x\}_B \end{Bmatrix} \tag{6.157}$$

where subscript I refers to internal nodes and subscript B refers to boundary nodes which are common to two or more substructures at the boundary.

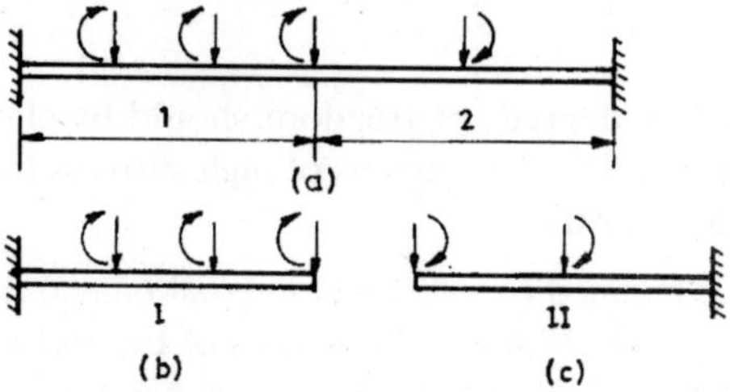

Fig. 6.23 Clamped-damped beam divided into two substructures

Using Eq. (6.28), Eq. (6.157) is rewritten

$$\{x\}^s = \begin{Bmatrix} \{x\}_I \\ \{x\}_B \end{Bmatrix} = \begin{bmatrix} [\Phi]_N & [\Phi]_C \\ [0] & [I] \end{bmatrix} \begin{Bmatrix} \{\xi\}_N \\ \{x\}_B \end{Bmatrix} \tag{6.158}$$

or,
$$\{x\}^s = [T]_s \{\xi\}_s \tag{6.159}$$

Considering interface boundaries fixed, an eigensolution of the following equation will give $[\Phi]_N$ matrix

$$([K]_{II} - p^2 [M]_{II}) \{\Phi\}_I = \{0\} \tag{6.160}$$

$[K]_{II}$ is the stiffness matrix found by internal nodes (without considering boundary nodes). Similarly, $[M]_{II}$ is the appropriate portion of $[M]^S$. $[\Phi]_N$ are general coordinates related to the natural modes of the substructure.

$[\Phi]_c$ the constrained modes of the substructure is physically explained as each column representing the values at the internal nodes for a unit value of one of the degrees of freedom at an interface boundary mode. They are obtained from the following equation

$$[K]_{II} \{x\}_I + [K]_{IB} \{x\}_B = \{0\} \tag{6.161}$$

which gives

$$\{x\}_I = -[K]_{II}^{-1} [K]_{IB} \{x\}_B = [\Phi]_C \{x\}_B \tag{6.162}$$

Both fixed interface and constrained modes are shown in Fig. 6.24

The energy expression for each substitution is given by

$$T_S = \frac{1}{2} \left((\{\dot{x}\}^S)^T [M]^S \{\dot{x}\}^S \right) \tag{6.163}$$

$$U_S = \frac{1}{2} \left((\{x\}^S)^T [K]^S \{x\}^S \right) \tag{6.164}$$

where superscript S denotes a substructure.

Using transformation given by Eq. (6.158) in Eqs. (6.163) and (6.164) yields

$$T_S = \frac{1}{2}\left(\{\xi\}^S\right)^T \left([T_F]^S\right)^T [M]^S [T_F]^S \{\dot{\xi}\}^S \tag{6.165}$$

$$U_S = \frac{1}{2}\left(\{\dot{\xi}\}^S\right)^T \left([T_F]^S\right)^T [K]^S\, [T_F]^S \{\xi\}^S \tag{6.166}$$

or,

$$T_S = \frac{1}{2}\left(\{\xi\}^S\right)^T [\bar{M}]^S \{\dot{\xi}\}^S \tag{6.167}$$

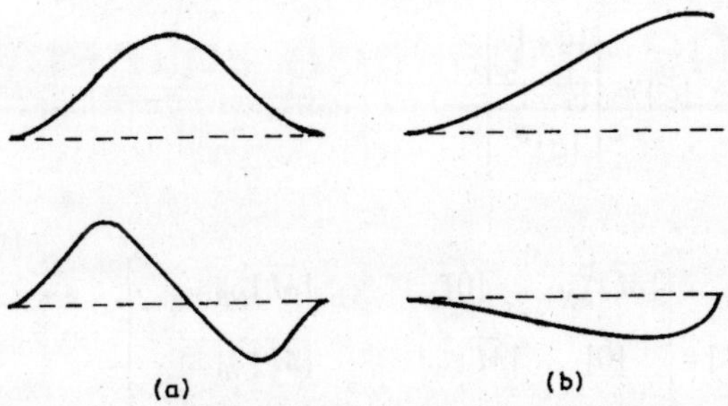

Fig. 6.24 (a) Fixed interface mode (b) Constrained mode

$$U_S = \frac{1}{2}\left(\{\xi\}^S\right)^T [\bar{K}]^S \{\xi\}^S \tag{6.168}$$

where

$$[\bar{M}]^S = \left([T_F]^S\right)^T [M]^S [T_F]^S \tag{6.169}$$

$$[\bar{K}]^S = \left([T_F]^S\right)^T [K]^S [T_F]^S \tag{6.170}$$

$[\bar{M}]^S$ and $[\bar{K}]^S$ can be reduced to the following form with the help of Eq. (6.158).

$$[\bar{M}]^S = \begin{bmatrix} [\bar{M}]_{NN} & [\bar{M}]_{NB} \\ [\bar{M}]_{BN} & [\bar{M}]_{BB} \end{bmatrix} \tag{6.171}$$

and

$$[\bar{K}]^S = \begin{bmatrix} [\bar{K}]_{NN} & [0] \\ [0] & [\bar{K}]_{BB} \end{bmatrix} \tag{6.172}$$

Now,

$$[\bar{M}]_{NN} = [\Phi]_N^T [M]_{II} [\Phi]_N \tag{6.173}$$

$$[\bar{K}]_{NN} = [\Phi]_N^T [K]_{II} [\Phi]_N \tag{6.174}$$

Diagonal matrices can be formed accordingly.

$[K]$ matrix can be positioned as follows

$$[K]^S = \begin{bmatrix} [K]_{II} & [K]_{IB} \\ [K]_{BI} & [K]_{BB} \end{bmatrix} \tag{6.175}$$

It can be shown that

$$[\bar{K}]_{BB} = [K]_{BB} - [K]_{BI}\left([K]_{II}\right)^{-1} [K]_{IB} \tag{6.176}$$

Eq. (6.176) indicates that internal nodes have been eliminated by static condensation.

If the contributions of two substructures shown in Fig. 6.24 are added, it yields

$$T = \frac{1}{2}\{\dot{\xi}\}^T [\overline{M}]\{\dot{\xi}\} \tag{6.177}$$

$$U = \frac{1}{2}\{\xi\}^T [\overline{K}]\{\xi\} \tag{6.178}$$

where

$$\{\xi\} = \begin{Bmatrix} \{\xi_N\}^1 \\ \{\xi_N\}^2 \\ \{\xi\}^B \end{Bmatrix} \tag{6.179}$$

$$[M] = \begin{bmatrix} [\overline{M}]^1_{NN} & [0] & [\overline{M}]^1_{NB} \\ [0] & [\overline{M}]^2_{NN} & [\overline{M}]^2_{NB} \\ [\overline{M}]^1_{BN} & [\overline{M}]^2_{BN} & [\overline{M}]^1_{BB} + [\overline{M}]^2_{BB} \end{bmatrix} \tag{6.180}$$

and

$$[K] = \begin{bmatrix} [\overline{K}]^1_{NN} & [0] & [0] \\ [0] & [\overline{K}]^2_{NN} & [0] \\ [0] & [0] & [\overline{K}]^1_{BB} + [\overline{K}]^2_{BB} \end{bmatrix} \tag{6.181}$$

Subscripts 1 and 2 indicate the substructure number.

The equation of motion of the complete structure is

$$[M]\{\ddot{\xi}\} + [K]\{\xi\} = \{0\} \tag{6.182}$$

Once Eq. (6.182) is solved, the displacements of each substructure can then be found from Eq. (6.1). The method is very elegant and possesses good convergence characteristic with the increase of number of component modes.

6.21.2 Free Interface Method

In the free interface method, the following transformation is used for reducing the number of degrees of freedom in a substructure

$$\{x\}^S = [\Phi]_N \{\xi_N\} \tag{6.184}$$

$[\Phi]_N$ is obtained from the solution of the eigenvalue problem of the substructure

$$\left([K]^S - p^2[M]^S\right)\{\phi\} = \{0\} \tag{6.185}$$

Free interface modes of substructure 1 of Fig.6.23 are shown in Fig. 6.25. $[\phi]_N$ are modes of the substructure with interface boundary.

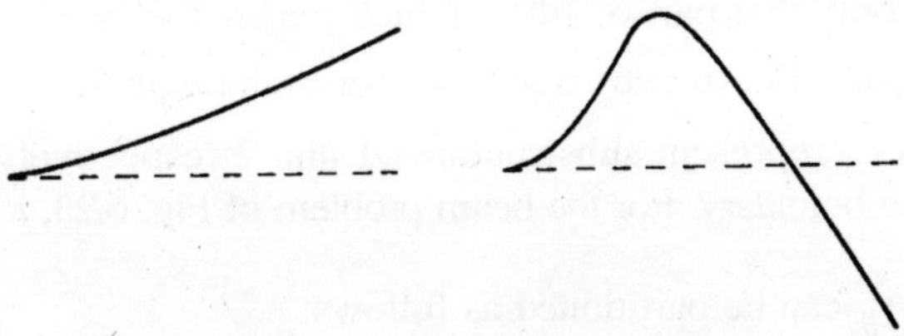

Fig. 6.25 Free interface nodes

Substructures having both ends free will result in a rigid body mode which is to be treated as a mode with zero frequency.

Substituting Eq. (6.184) into Eq. (6.162), yields

$$T_S = \frac{1}{2}\left(\{\dot{\xi}_N\}^S\right)^T [\overline{M}]^S \{\dot{\xi}_N\}^S \tag{6.186}$$

$$U_S = \frac{1}{2}\left(\{\xi_N\}^S\right)^T [\overline{K}]^S \{\xi_N\}^S \tag{6.187}$$

where $$[\overline{M}]^S = [\phi]_N^T [M]^S [\phi]_N \tag{6.188}$$

and $$[\overline{K}]^S = [\phi]_N^T [K]^S [\phi]_N \tag{6.189}$$

Both $[\overline{M}]^S$ and $[\overline{K}]^S$ are diagonal matrices.

Putting

$$[\overline{M}]_S = [I] \tag{6.190}$$

and $$[\overline{K}]_S = \left[1/p_r^2\right]^S \tag{6.191}$$

If the contributions of two substructures of Fig. 6.21 are added, then

$$\{\xi\} = \begin{Bmatrix} \{\xi_N\}^1 \\ \{\xi_N\}^2 \end{Bmatrix} \tag{6.192}$$

$$[M] = \begin{bmatrix} [I] & [0] \\ [0] & [I] \end{bmatrix} \tag{6.193}$$

and $$[K] = \begin{bmatrix} [1/p^2]^1 & [0] \\ [0] & [1/p^2]^2 \end{bmatrix} \tag{6.194}$$

The constraints at the interface of two substructure boundaries are next to be applied. At the interface both of them will have the same displacement

$$\{x\}_B^1 = \{x_B\}_B^2 \tag{6.195}$$

Substituting Eq. (6.184) in Eq. (6.195), yields

$$[\Phi]_B^1 \{\xi_N\}^1 = [\Phi]_B^2 \{\xi_N\}^2 \tag{6.196}$$

where $[\Phi]_B$ contains only that part of $[\Phi]_N$ which relates the interface degrees of freedom which are at the boundary. Eq. (6.196) is a set of constraint equations.

Let n^1 and n^2 modes represent substructures 1 and 2 respectively. n_B is the degrees of freedom at the interface boundary. For the beam problem of Fig. 6.23, $n_B = 2$.

If $n^1 > n_B$, the $[\phi]_B^1$ can be partitioned as follows:

$$\underset{(n_B \times n^1)}{[\Phi]_B^1} = \left\lfloor \underset{(n_B \times n_B)}{[\Phi]_{B1}^1} \quad \underset{(n_B \times (n^1 - n_B))}{[\phi]_{B2}^1} \right\rfloor \tag{6.197}$$

Eq. (6.196) becomes

$$[[\Phi]_{B1}^1 \, [\Phi]_{B2}^1] \begin{Bmatrix} \{\xi_{N1}\}^1 \\ \{\xi_{N2}\}^1 \end{Bmatrix} = [\Phi]_B^2 \, \{\xi_N\}^2 \tag{6.198}$$

From Eq. (6.198), we get

$$\{\xi_{N1}\}^1 = -\,([\Phi]_{B1}^1)^{-1} [\Phi]_{B2}^1 \{\xi_{N2}\}^1 + ([\Phi]_{B1}^1)^{-1} [\Phi]_B^2 \{\xi_N\}^2 \tag{6.199}$$

Therefore,

$$\{\xi\} = \begin{Bmatrix} \{\xi_N\}^1 \\ \{\xi_N\}^2 \end{Bmatrix} = \begin{Bmatrix} \{\xi_{N1}\}^1 \\ \{\xi_{N2}\}^1 \\ \{\xi_N\}^2 \end{Bmatrix} = [T]_C \{\bar{\xi}\} \tag{6.200}$$

where

$$\{\bar{\xi}\} = \begin{Bmatrix} \{\xi_{N2}\}^1 \\ \{\xi_N\}^2 \end{Bmatrix} \tag{6.201}$$

and

$$[T]_C = \begin{bmatrix} -\,([\Phi]_{B1}^1)^{1} [\Phi]_{B2}^1 & ([\Phi]_{B1}^1)^{-1} [\Phi]_B^2 \\ [I] & 0 \\ [0] & [I] \end{bmatrix} \tag{6.202}$$

Substituting Eq. (6.200) into Eq. (6.177), and then combining with Eq. (6.193) and (6.194), yields

$$T = \frac{1}{2} \{\dot{\bar{\xi}}\}^T [M]_R \, [\dot{\bar{\xi}}\} \tag{6.203}$$

$$U = \frac{1}{2} \{\bar{\xi}\}^T [K]_R \{\bar{\xi}\} \tag{6.204}$$

where

$$[M]_R = [T]_C^T \, [T_C] \tag{6.205}$$

and

$$[K]_R = [T]_C^T \, [K] \, [T]_C \tag{6.206}$$

The equation of motion of the complete structure is

$$[M]_R \{\ddot{\xi}\} + [K]_R \{\xi\} = \{0\} \tag{6.207}$$

After solving this equation, the displacements for the complete structure are calculated using

$$\begin{Bmatrix} \{x\}^1 \\ \{x\}^2 \end{Bmatrix} = \begin{bmatrix} [\Phi]_N^1 & [0] \\ [0] & [\Phi]_N^2 \end{bmatrix} [T]_C \{\xi\} \tag{6.208}$$

6.22 LAGRANGE'S EQUATION

A general formulation of the equation of motion of a dynamical system can often be derived in terms of general coordinate by Lagrange's equations [6.36]. Both linear and nonlinear systems can be dealt with this method. The equations are formulated on the basis of energies of the system.

Lagrange's equation is given by

$$\frac{d}{dt}\left(\frac{\partial t}{\partial \dot{q}_j}\right) - \frac{\partial T}{\partial q_j} + \frac{\partial V}{\partial V_j} = Q_j \tag{6.209}$$

where $\dot{q}_j = \frac{\partial q_j}{\partial t}$ is the generalised velocity.

Q_j is the nonconservative generalised force corresponding to the coordinate q_j.

For a conservative system $Q_j = 0$ and Eq. (6.209) then becomes

$$\frac{d}{dt}\left(\frac{\partial T}{\partial \dot{q}_j}\right) - \frac{\partial T}{\partial q_j} + \frac{\partial V}{\partial q_j} = 0 \qquad j = 1, 2, \ldots\ldots, n \tag{6.210}$$

Corresponding to *j*-th generalised coordinate, we get equations (6.209) or (6.210) as the case may be. Thus for n generalised systems we get n equations.

Examples 6.15

A spring-mass system is shown in Fig. 6.26. Derive the equations of motion using Lagrange's equations.

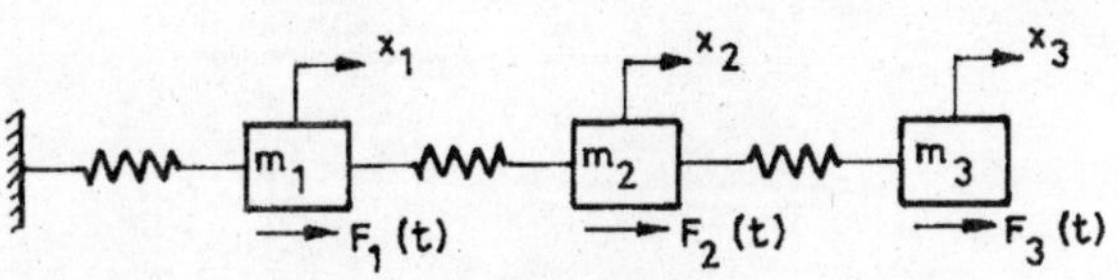

Fig. 6.26

In this case $q_1 = x_1$, $q_2 = x_2$ and $q_3 = x_3$, and the kinetic energy of the system is given by

$$T = \frac{1}{2} m_1 x_1^2 + \frac{1}{2} m_2 x_2^2 + \frac{1}{2} m_3 x_3^2 \tag{a}$$

The potential energy of the spring-mass system is given by

$$V = \frac{1}{2} k_1 x_1^2 + \frac{1}{2} k_2 (x_2 - x_1)^2 + \frac{1}{2} k_3 (x_3 - x_2)^2 \tag{b}$$

The external forces applied to the masses, yields

$$Q_j^{(1)} = F_1(t)\frac{\partial x_1}{\partial x_1} + F_2(t)\frac{\partial x_2}{\partial x_1} + F_3(t)\frac{\partial x_3}{\partial x_1} = F_1(t) \tag{c}$$

Similarly,

$$Q_j^{(2)} = F_2(t), \quad Q_i^{(3)} = F_3(t) \tag{d}$$

Combining Eqs. (a), (b), (c) and (d), yields

$$\left.\begin{aligned} m_1\ddot{x}_1 + (k_1 + k_2)\,x_1 - k_2x_2 = F_1(t) \\ m_2\ddot{x}_2 + (k_2 + k_3)\,x_2 - k_1x_1 - k_3x_3 = F_2(t) \\ m_3\ddot{x}_3 + k_3x_3 + k_3x_2 = F_3(t) \end{aligned}\right\} \tag{e}$$

Eq. (e) written in matrix form becomes

$$\begin{bmatrix} m_1 & 0 & 0 \\ 0 & m_1 & 0 \\ 0 & 0 & m_1 \end{bmatrix}\begin{Bmatrix} \ddot{x}_1 \\ \ddot{x}_2 \\ \ddot{x}_3 \end{Bmatrix} + \begin{bmatrix} k_1 + k_2 & -k_2 & 0 \\ -k_2 & k_2 + k_3 & -k_3 \\ 0 & -k_3 & k_3 \end{bmatrix}\begin{Bmatrix} x_1 \\ x_2 \\ x_3 \end{Bmatrix} = \begin{Bmatrix} F_1(t) \\ F_2(t) \\ F_3(t) \end{Bmatrix} \tag{f}$$

References

6.1 E.C. Pestel and E.A. Leckie, Matrix Methods in Elasto-mechanics, McGraw Hill, 1963

6.2 A.K. Chopra, Dynamics of Structures, Prentice-Hall of India Pvt. Ltd., New Delhi, 1995

6.3 A. Jennings, Eigenvalue methods for vibration analysis, Shock and Vibration Digest, V.12, No. 2, 1980, pp. 3-16.

6.4 H. Rutihauser, Deflection der Bandmatrizen, ZAMP, V.10, 1959, pp. 314-319.

6.5 G.W. Stewart, The economical storage of plane rotations, Numerical Mathematics, V. 25, 1976, pp. 137-138.

6.6 R.H. Anderson, Fundamentals of Vibrations, McMillan Co., New York, 1967.

6.7 N.Y. Myklestad, Vibration Analysis, McGraw Hill, New York, 1965.

6.8 G.L. Rogers, Dynamics of Framed Structures, John Wiley, 1959.

6.9 K.J. Bathe and E.L. Wilson, Large eigenvalue problem in dynamic analysis, Journal of Engineering Mechanics Division, ASCE, V.98, 1972, pp. 1471-1485.

6.10 K.J. Bathe, Finite Element Procedures in Engineering Analysis, Prentice-Hall, 1996.

6.11 K.J. Bathe and E.L. Wilson, Solution methods for eigenvalue problems in structural mechanics, International Journal for Numerical Methods in Engineering, V.6, 1973, pp. 213-226.

6.12 R.B. Corr and A. Jennings, A simultaneous iteration algorithm for symmetric eigenvalue problems, International Journal for Numerical Methods in Engineering, V.10, 1976, pp. 657-663.

6.13 R.J. Guyan, Reduction of stiffness and mass matrices, AIAA Journal, V.3, No.2, Feb, 1965, p-380.

6.14 B.M. Irons, Structural eigenvalue problems: elimination of unwanted variables, AIAA Journal, V.3, No.5, 1965, pp. 961-962.

6.15 W.C. Hurty, Vibrations of structural systems by component mode synthesis, Journal of Engineering Mechanics Division, ASCE, V.86, August 1960, pp. 57-69.

6.16 S.N. Hou, Review of modal synthesis techniques and a new approach, Shock and Vibration Bulletin, Vol.40, No.4, 1965, pp. 25-39.

6.17 R.D. Henshell and J.H.Ong, Automatic masters of eigenvalue economisation, International Journal for Earthquake Engineering and Structural Dynamics, V.3, 1975, pp. 375-383.

6.18 V.N.Sha and R.Raymund, Analytical selection of masters for reduced eigenvalue problem, International Journal for Numerical Methods in Engineering, V.18, 1982, pp. 89-98.

6.19 G.C.Wright and G.A. Miles, An economical method for determining the smallest eigenvalue of large linear system, International Journal for Numerical Methods in Engineering, V.3, 1971, pp. 25-34.

6.20 M.Paz, Dynamic condensation, AIAA Journal, V.22, No.5, May 1985, pp. 724-727.

EXERCISE 6

6.1 A three spring-mass system is shown in the figure. All the masses are subjected to dynamic forces. Derive the equations of motion in terms of displacements x_1, x_2 and x_3 of the masses along the axis of the springs.

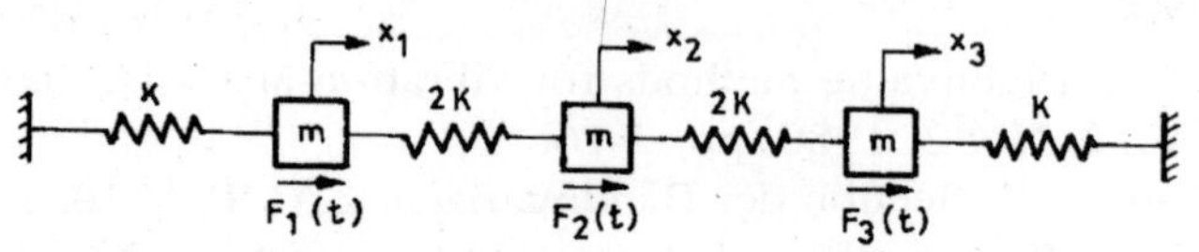

Prob. 6.1

6.2 Using Dunkerley's method, determine the fundamental frequency of a uniformly loaded cantilever beam with a concentrated mass M at the end, equal to the mass of the uniform beam having flexural rigidity EI. The frequency of the beam due to uniform load is

$$p_1 = 3.515^2\left(\frac{EI}{ML^3}\right)$$

6.3 Determine the fundamental frequency of the lumped mass simply supported on an uniform beam shown in the figure by Dunkerley's method.

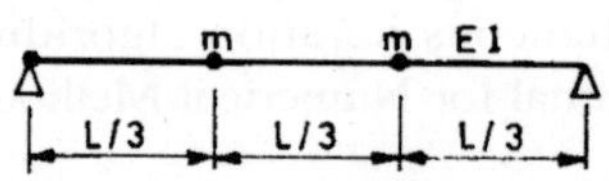

Prob. 6.3

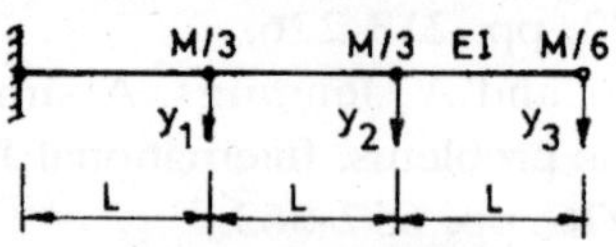

Prob. 6.4

6.4 For the cantilever uniform beam having lumped masses as shown, write the equations of motion and apply matrix iteration technique to obtain the frequency and the mode shape for the fundamental mode.

6.5 For the spring- mass system shown in the figure, determine all the natural frequencies and mode shapes by Stodola's method.

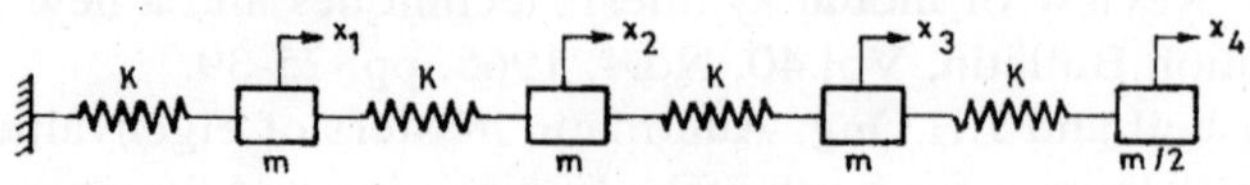

Prob. 6.5

6.6 Treat the 3-degree spring-mass system shown in the figure as a periodic structure. Solve for frequencies and modes of vibration.

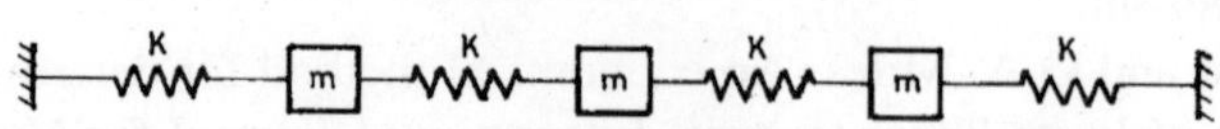

Prob. 6.6

6.7 A massless cantilever beam of length L supports two lumped masses $mL/2$ and $mL/4$ at the midpoint and free end as shown in the figure (a). The degrees of freedom and applied forces are shown in figure (b). Formulate the equations of motion of the system.

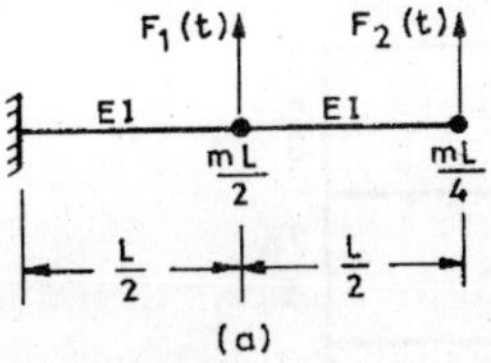

Prob. 6.7 (a)

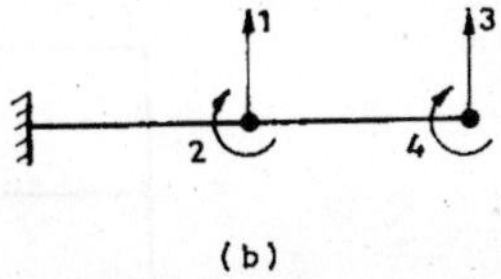

Prob. 6.7 (b)

6.8 The fundamental frequency of a uniform cantilever beam of mass M and length L is $3.515\sqrt{EI/ML^3}$. If a lumped mass m_0 is attached to the beam at a distance $L/3$ from the fixed end, determine the new fundamental frequency.

6.9 Determine the natural frequencies and modes of the system shown in the figure

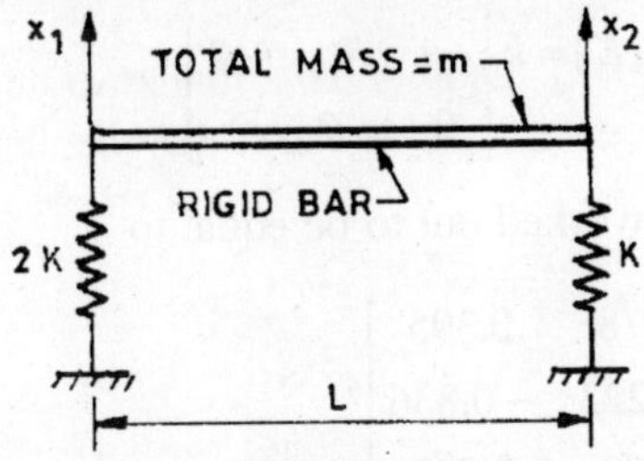

Prob. 6.9

6.10 Determine the natural frequencies and modes for the system of figure (Prob. 6.4). Show that the modes satisfy orthogonality properties.

6.11 Determine the natural frequencies and modes of the system shown in the figure for the two-storey frame idealised as a shear building. Normalise the modes.

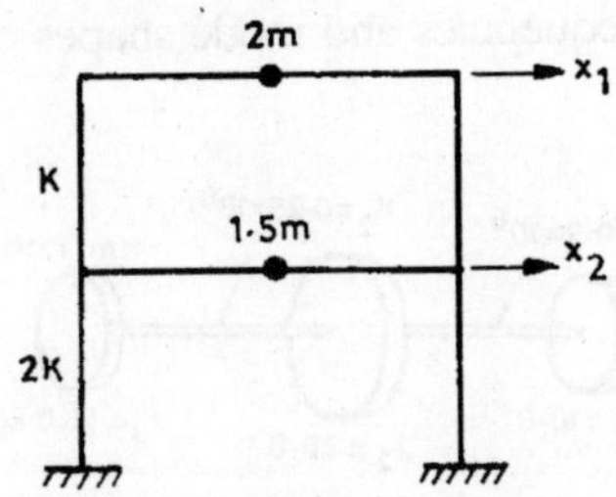

Prob. 6.11

6.12 The properties of a three-storey shear building are given in the figure. Derive the Rayleigh damping matrix such that the damping ratio is 5% for the first and third modes. Compute the damping ratio for the second mode.

$$w = 100 \quad k = 168$$

$$\{p\}^T = \{12.01 \quad 25.47 \quad 38.90\} \quad rad/s$$

$$[\Phi] = \begin{bmatrix} 0.6375 & 0.95527 & 1.5778 \\ 1.2750 & 0.9829 & -1.1270 \\ 1.9125 & -1.9642 & 0.4508 \end{bmatrix}$$

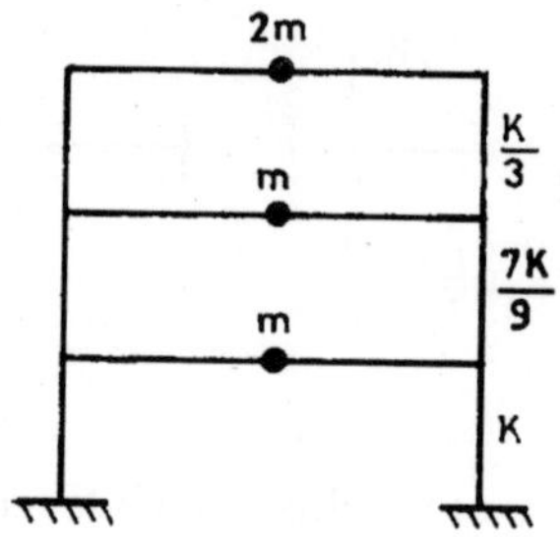

Prob. 6.12

6.13 The mass and stiffness matrix of a MDF system is given by

$$[M]=m\begin{bmatrix}1&0&0\\0&1&0\\0&0&2\end{bmatrix},\quad [K]=k\begin{bmatrix}2&-1&0\\-1&3&-2\\0&-2&2\end{bmatrix}$$

The modal matrix has worked out to be equal to

$$[\Phi]=\sqrt{m}\begin{bmatrix}0.269&0.878&0.395\\0.501&-0.223&-0.836\\0.582&0.299&0.269\end{bmatrix}$$

Show that when used as a transformation matrix, [F] diagonalizes [*M*] and [*K*] simultaneously.

6.14 Estimate the fundamental frequency by means of Rayleigh's quotient using the data of Prob. 6.13

6.15 Determine the natural frequencies and mode shapes of the system shown in the figure by Holzer's method.

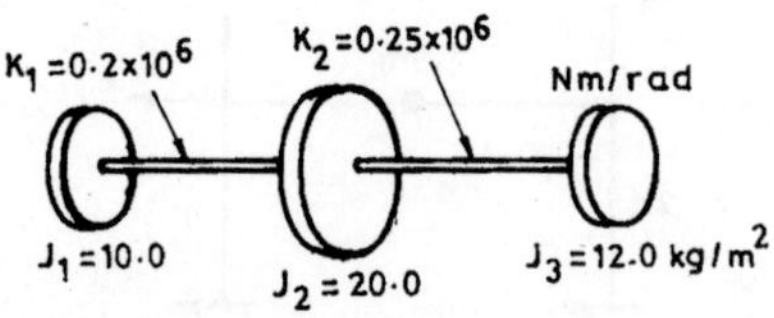

Prob. 6.15

6.16 For the spring-mass system shown in the figure, determine the natural frequencies and mode shapes for the vertical vibration.

Prob. 6.16

6.17 Determine the fundamental frequency and mode shape for the cantilever beam shown in the figure of Prob. 6.3 by Myklestad's method.

6.18 For the system shown in the figure, determine the characteristic polynomial. Determine the roots. Determine the mode shape corresponding to the third frequency.

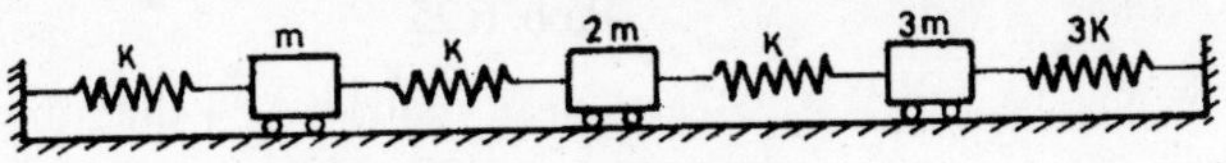

Prob. 6.18.

6.19 From the solution of Prob. 6.5, show that the normal modes are mutually orthogonal.

6.20 For the frame shown in the figure, determine the natural frequencies and mode shapes.

6.21 Determine the eigenvalues and eigenvectors for Prob. 6.4 by transfer matrix method.

6.22 Fundamental mode shape for the beam of Prob. 6.4 can be approximated as $\{\phi\}^T = \{0.175 \quad 0.566 \quad 1.000\}$. Obtain the estimate of fundamental frequency by Rayleigh's method.

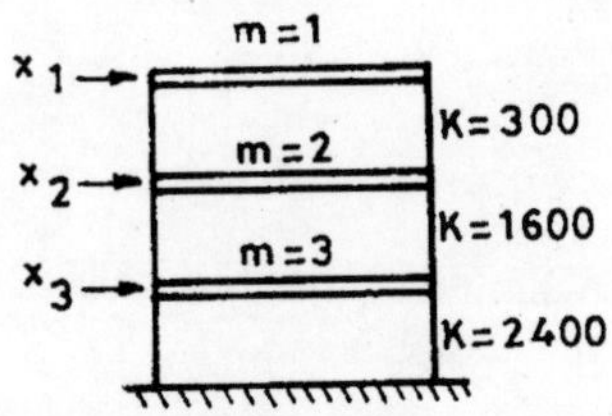

Prob. 6.20

6.23 For the free torsional vibration problem of four discs connected at different locations of a cantilever shaft, the first two mode shapes are approximated as

$$[\Phi] = \begin{bmatrix} 0.25 & 0.06 \\ 0.50 & 0.25 \\ 0.75 & 0.56 \\ 1.00 & 1.00 \end{bmatrix}$$

The stiffness and the mass matrices are

$$[M]=I\begin{bmatrix}1&0&0&0\\0&1&0&0\\0&0&1&0\\0&0&0&0.5\end{bmatrix},\quad [K]=k\begin{bmatrix}2&-1&0&0\\-1&2&-1&0\\0&-1&2&-1\\0&0&-1&1\end{bmatrix}$$

Determine the first two natural frequencies by Rayleigh-Ritz method.

6.24 Determine the natural frequencies and location of nodes by transfer matrix method.

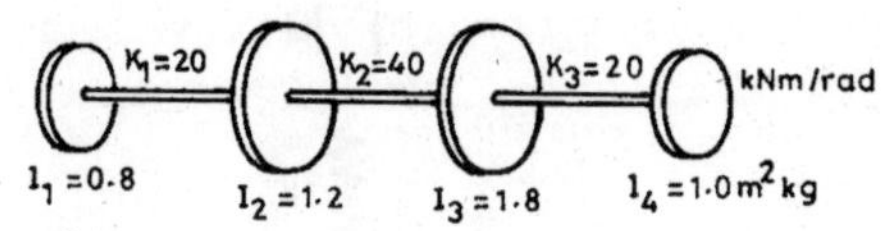

Prob. 6.24

6.25 Determine the natural frequencies of lumped mass system by Myklestad's method.

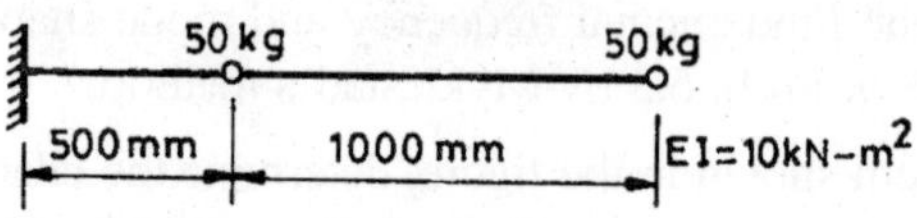

Prob. 6.25

Chapter 7

Forced Vibration Analysis of Multiple Degrees of Freedom System

7.1 INTRODUCTION

Most of the structures represented as multiple degrees of freedom system are subjected to dynamic loads, for which they are to be analysed. The equations that result for these systems are all coupled. The equations of motion are transformed for facilitating the analysis. The method which is adopted for this purpose is known as the mode superposition method. There is yet another method developed for this purpose, which is termed as the mode acceleration method. The mode acceleration method may be treated as an extension of mode superposition method, the main difference between them lies in obtaining the final response of the system. The mode acceleration method gives a better convergence characteristics. The equations of motion for a MDF system may also be directly evaluated. For many problems, particularly when external loading is complicated, the direct integration is performed numerically. Different methods for the forced vibration analysis of MDF systems are presented in this chapter.

7.2 MODE SUPERPOSITION METHOD FOR THE DETERMINATION OF RESPONSE OF UNDAMPED MDF SYSTEM

The equations of motion for an undamped MDF system is given by

$$[M]\{\ddot{x}\}+[K]\{x\}=\{F(t)\} \tag{7.1}$$

where $[M]$, $[K]$ and $\{F(t)\}$ are the mass, stiffness and the loading matrix. $\{x\}$ and $\{\ddot{x}\}$ are the displacement and acceleration matrices. For a n-degree of freedom system, Eq. (7.1) possesses n unknowns and they are to be obtained from simultaneous solution of n equations.

In mode superposition method, which is also termed as normal mode method, the equations are uncoupled with the help of normal coordinates [7.1-7.5]. But before we start to evaluate the response, the results of natural frequencies and mode shapes must be available. The displacements are expressed as a linear function of mode shapes [same as Eq. (6.26)].

$$\{x\}=[\Phi]\{\xi\} \tag{7.2}$$

Substituting $\{x\}$and $\{\ddot{x}\}$ from Eq. (7.2) into Eq. (7.1), and pre-multiplying by $[\phi]^T$, we get

$$[\Phi]^T[M][\Phi]\{\ddot{\xi}\}+[\Phi]^T[K][\Phi]\{\xi\}=[\Phi]^T\{F(t)\} \tag{7.3}$$

Combining Eqs. (6.29) and (7.3), we get

$$[\Phi]^T [M][\Phi](\{\ddot{\xi}\}+[p^2]\{\xi\}) = [\Phi]^T \{F(t)\} \tag{7.4}$$

If [*M*] is a diagonal matrix, then Eq. (7.4) will result in *n* uncoupled equations. The *r*th equation is given by

$$\ddot{\xi}_r + p_r^2 \xi_r = \frac{\bar{f}_r}{\bar{m}_r} \tag{7.5}$$

where
$$\bar{f}_r = \{\phi^{(r)}\}^T \{F(t)\} = \sum_{i=1}^{n} \phi_i^{(r)} F_i(t) \tag{7.6}$$

and
$$\bar{m}_r = \{\phi^{(r)}\}^T [M]\{\phi^{(r)}\} = \sum_{i=1}^{n} m_i \{\phi_i^{(r)}\}^2 \tag{7.7}$$

Equation (7.5) is identical to the SDF system forced vibration equation. The solution of this equation is [see Eq. (3.69)]

$$\xi_r = \frac{1}{p_r}\int_o^t \frac{\bar{f}_r}{\bar{m}_r} \sin p_r (t-\tau)\, d\tau \tag{7.8}$$

The original coordinate $\{x\}$ is related to the transformed coordinate and is given by

$$x_i(t) = \sum_{r=1}^{n} \phi_i^{(r)} \xi_r \tag{7.9}$$

As per the discussion above, it may appear that all the n modes for a *n*-degree of freedom system have been considered in the analysis. For most practical problems, it may be noted that only a few lower modes may be considered in the analysis. For such cases, inclusion of all modes will not yield much better accuracy, as the effect of higher modes on the total response is nominal. Further, this will entail in much more computation and hence more computer time will be required without attainment of any significant accuracy. Truncation of the modes for such cases is highly desirable and to what extent this is to be done depends on the forcing function and free vibration characteristics. The problem has been investigated in the following example.

Example 7.1

A three storey frame shown in Fig. 7.1 is subjected to an excitation force of $P \cos\omega t$ at the top level due to steady state vibration. Determine the response at the top level, on the basis of consideration of

(a) First mode only

(b) First two modes only

(c) All three modes

for $\omega = 0$, $\omega = 0.5p_1$ and $\omega = 1.3\, p_2$.

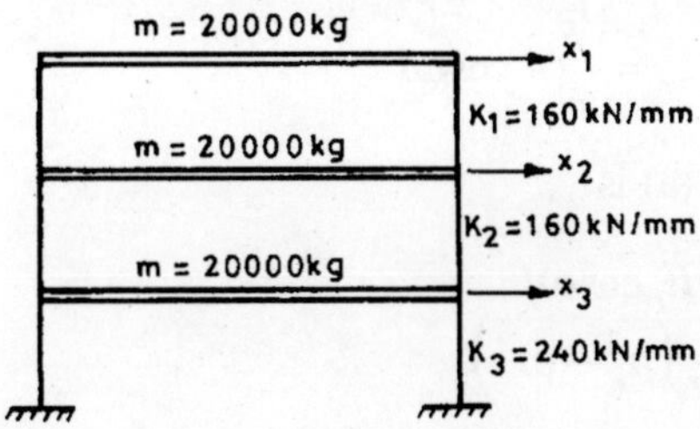

Fig. 7.1 Example 7.1

For this problem,

$$[M] = 20000\begin{bmatrix}1 & 0 & 0\\0 & 1 & 0\\0 & 0 & 1\end{bmatrix}, \quad [K] = 8\times10^7\times\begin{bmatrix}2 & -2 & 0\\-2 & 4 & -2\\0 & -2 & 5\end{bmatrix}$$

$$\{p\} = \begin{Bmatrix}43.872\\120.155\\167.00\end{Bmatrix} \quad \text{and} \quad \{p^2\} = \begin{Bmatrix}1924.733\\14437.20\\27889.00\end{Bmatrix}$$

$$[\Phi] = \begin{bmatrix}1.0000 & 1.0000 & 1.000\\0.7594 & -0.8047 & -2.427\\0.3361 & -1.1572 & 2.512\end{bmatrix}$$

$$[\Phi]^T[M][\Phi] = \begin{bmatrix}1.000 & 0.7594 & 0.3361\\1.000 & -0.8047 & -1.1572\\1.000 & -2.427 & 2.512\end{bmatrix}$$

$$\times 20000\begin{bmatrix}1 & 0 & 0\\0 & 1 & 0\\0 & 0 & 1\end{bmatrix}\begin{bmatrix}1.000 & 1.000 & 1.000\\0.7594 & -0.8047 & -2.427\\0.3361 & -1.1572 & 2.512\end{bmatrix}$$

$$= \begin{bmatrix}1.6896 & 0 & 0\\0 & 2.9867 & 0\\0 & 0 & 13.200\end{bmatrix}\times 20000$$

$$[\Phi]^T\{F(t)\} = \begin{bmatrix}1.000 & 0.7594 & 0.3361\\1.000 & -0.8047 & -1.1572\\1.000 & -2.427 & 2.512\end{bmatrix}\begin{Bmatrix}P\cos\omega t\\0\\0\end{Bmatrix}$$

$$= \begin{Bmatrix}P\cos\omega t\\P\cos\omega t\\P\cos\omega t\end{Bmatrix} = \{\bar{F}\}\cos\omega t$$

The *r*th equation for this harmonic excitation is [from Eq. (7.5)]

$$\ddot{\xi}_r + p_r^2 \xi_r = \frac{\bar{F}_r}{\bar{m}_r} \cos \omega t \qquad \text{(a)}$$

Steady state solution of Eq. (a) is

$$\xi_r = \frac{\bar{F}_r \cos \omega t}{\bar{m}_r \left(p_r^2 - \omega^2\right)} \qquad \text{(b)}$$

The displacement at the top level is [Eq. (7.9)]

$$x_1 = \sum_{r=1}^{n} \phi_1^{(r)} \xi_r \qquad \text{(c)}$$

Let us first write the expression for $x_1(t)$ for all the modes and then we shall take out the terms for the necessary truncation

$$x_1(t) = \left.\left.\left.\frac{1.0(\text{P}\cos\omega t)}{20000 \times 1.6896 \times \left(1924.733 - \omega^2\right)}\right]_{r=1} + \frac{1.0(\text{P}\cos\omega t)}{20000 \times 2.9867\left(14437.2 - \omega^2\right)}\right]_{r=2} + \frac{1.0(\text{P}\cos\omega t)}{20000 \times 13.2\left(27889 - \omega^2\right)}\right]_{r=3}$$

Say $$C_o = \frac{x_1(t)}{P\cos\omega t}$$

For different values of ω, the values of C_0 has been shown in the following table for different values of *r*.

Values of C_0

	r = 1	r = 2	r = 3
$\omega = 0$	1.5375×10^{-8}	1.6535×10^{-8}	1.667×10^{-8}
$\omega = 0.5\ p_1$	2.059×10^{-8}	2.17×10^{-8}	2.1838×10^{-8}
$\omega = 1.3\ p_2$	-1.3167×10^{-9}	-2.9972×10^{-9}	-1.9119×10^{-9}

For $\omega = 0$ and $\omega = 0.5\, p_1$, a two-mode solution is reasonably accurate. But for $\omega = 1.3\, p_2$, for obtaining correct solution, all modes must be considered. In this case, the frequency of the external force ω is between the second and third natural frequency of the system. As such, the contribution of both the second and the third mode is very important in this case.

From this, we may generalise that if the frequency of the external forcing function lies between two particular natural frequencies of the system, the modes in the vicinity of that frequency are important. When the truncation of mode shapes is to be considered, this aspect should be borne in mind.

7.3 MODE-ACCELERATION METHOD FOR THE DETERMINATION OF RESPONSE OF MDF SYSTEM

In mode superposition method, many modes are needed to obtain an accurate solution. Total number of modes needed for desired degree of accuracy can be further reduced by the application of mode-acceleration method, as the method possesses better convergence characteristics.

The equations of motion of an undamped MDF system is given by

$$[M]\{\ddot{x}\}+[K]\{x\}=\{F(t)\} \tag{7.1}$$

The mode superposition solution $\{x\}$ is given by Eq. (7.9).

$$x_i(t)=\sum_{r=1}^{N}\phi_i^{(r)}\xi_r \tag{7.9}$$

It may be noted in the above equation that modes are truncated up to N and all modes beyond $(N + 1)$ to n for a n-degree of freedom system are not considered at all.

Equation (7. 1) is modified to the following form

$$\{x\}=[K]^{-1}(\{F(t)\}-[M]\{\ddot{x}\}) \tag{7.10}$$

Combining Eqs. (7.9) and (7.10), we get

$$\{x\}=[K]^{-1}(\{F(t)\})-[K]^{-1}[M][\Phi]\{\ddot{\xi}\}) \tag{7.11}$$

Incorporating Eq. (6.12) into Eq. (7.11), we get

$$\{x\}=[K]^{-1}\{F(t)\}-\left[1/p^2\right][\Phi]\{\ddot{\xi}\} \tag{7.12}$$

The first term in Eq. (7.12) is the pseudo-static response. This method is called mode-acceleration method, due to the presence of the second term and as p^2 is present in the denominator, values associated with this term reduce with higher frequencies.

Example 7.2

Solve the problem of Example 7.1 by the mode-acceleration method.

$[K]^{-1}$ is to be determined first.

$$[K]^{-1}=\frac{1}{48\times10^7}\begin{bmatrix}8 & 5 & 2\\5 & 5 & 2\\2 & 2 & 2\end{bmatrix}=[F]$$

From Eq. (7.12)

$$x_i(t)=f_{i1}\,P\cos\omega t-\sum_{r=1}^{N}\left(\frac{1}{p_r^2}\right)\phi_i^{(r)}\ddot{\xi}_r \tag{a}$$

Combining Eq. (b) of Example 7.1 with the above Eq. (a),

$$x_i(t) = f_{i1} P \cos\omega t + \sum_{r=1}^{N} \frac{\omega^2}{p_r^2} \phi_i^{(r)} \xi_r \quad \text{(b)}$$

Therefore

$$\left.\left.\left.\begin{aligned} x_1(t) &= \frac{1}{48\times10^7} \times 8P\cos\omega t \\ &+ \frac{\omega^2}{1924.733} \frac{(1.0)\,P\cos\omega t}{20000\times1.6896\times(1924.733-\omega^2)} \end{aligned}\right] N=1 \\ + \frac{\omega^2}{14437.20} \frac{(1.0)\,P\cos\omega t}{20000\times2.9867\times(14437.20-\omega^2)} \right] N=2 \\ + \frac{\omega^2}{27889} \frac{(1.0)\,P\cos\omega t}{20000\times13.2\times(27889-\omega^2)} \right] N=3 \quad \text{(c)}$$

Let $C_o = \dfrac{x_1(t)}{P\cos\omega t}$

For different values of ω, the values of C_0 has been shown in the following table for different values of *r*.

Values of C_0

	N = 1	N = 2	N = 3
$\omega = 0$	1.667×10^{-8}	1.667×10^{-8}	1.667×10^{-8}
$\omega = 0.5\, p_1$	2.1793×10^{-8}	2.1833×10^{-8}	2.1835×10^{-8}
$\omega = 1.3\, p_2$	-2.5098×10^{-9}	-2.8652×10^{-9}	-1.916×10^{-9}

When a comparison of the values of the above Table is made with that in Example 7.1, it is seen that they are nearly same for $N = 3$. We can draw certain conclusions based on the above Table.

For the first two cases, that is, $\omega = 0$ and $\omega = 0.5\, p_1$, the first mode solution is accurate enough. However, when $\omega = 1.3\, p_2$, the mode acceleration method does not perform better than the mode superposition method for the truncated solution. Further, it may be noted that at $\omega = 0$, there is no contribution from the normal modes and the solution obtained is same as that of static case.

7.4 RESPONSE OF MDF SYSTEMS UNDER THE ACTION OF TRANSIENT FORCES

Let us rewrite the *r*th modal equation [Eq. (7.5)] for an undamped MDF system

$$\bar{m}_r \ddot{\xi}_r + \bar{m}_r p_r^2 \xi_r = \bar{f}_r \quad (7.13)$$

$\overline{m}_r$ is called the generalised mass or modal mass or equivalent mass and from Eq. (7.6) and (7.7) we can write

$$\left.\begin{aligned}\overline{m}_r &= \sum_{i=1}^{n} m_i \left(\phi_i^{(r)}\right)^2 \\ \bar{f}_r &= \sum_{i=1}^{n} \phi_i^{(r)} F_i(t)\end{aligned}\right\} \tag{7.14}$$

Similarly,

Equation (7.13) is basically an equation exactly in the same form as a SDF system. Modal static deflection is given by

$$\xi_{rst} = \frac{\sum_{i=1}^{n} \overline{F}_i \phi_i^{(r)}}{p_r^2 \sum_{i=1}^{n} m_i \left(\phi^{(r)}\right)^2} \tag{7.15}$$

where $$F_i(t) = \overline{F}_i\, f_i(t) \tag{7.16}$$

$\overline{F}_i$ represents the amplitude of the external force. Solution of Eq. (7.13) is given by [similar to Eq. (3.144)]

$$\xi_r(t) = \xi_{rst}(DLF)_r \tag{7.17}$$

where $(DLF)_r$ depends only on $f(t)$ and p_r.

Therefore

$$[\xi_r(t)]_{\max} = \xi_{rst}(DLF)_{r,\max} \tag{7.18}$$

All the charts concerning DLF of the type of Figs. 3.22 and 3.23 applied to SDF systems for different pulses can be utilised for MDF systems as well.

Total deflection of the ith mass is obtained by superimposing the modes

$$x_i(t) = \sum_{i=1}^{n} \phi_i^{(r)} \xi_r$$

or $$x_1(t) = \sum_{i=1}^{n} \phi_i^{(r)} \xi_{rst}\,(DLF)_r \tag{7.19}$$

Example 7.3

For the structure of Fig. 7.2, m_1 = 20,000 kg, m_2 = 20,000 kg and m_3 = 20,000 kg. The stiffnesses are k_1 = 16 × 107 kN/m, k_2 = 16 × 10^7 kN/m and k_3 = 24 × 10^7 kN/m. The pulses to which the masses are subjected, have been shown in the figure. Find out the deflection at the level of second mass. What is the maximum force developed just below the second mass?

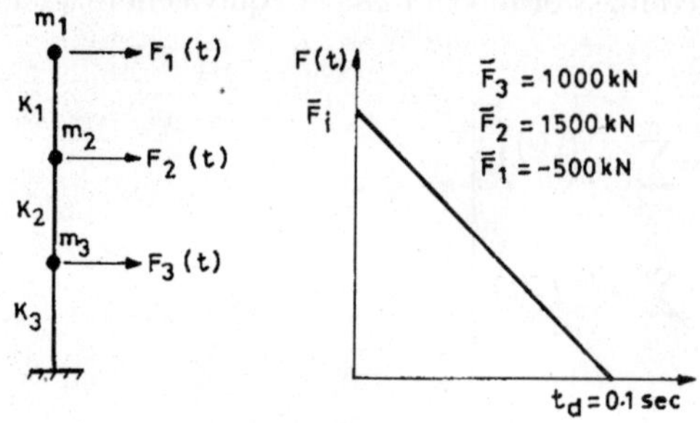

Fig. 7.2 Example 7.3

First step obviously is to calculate the natural frequencies and the mode shapes. They are

$$\{p\}=\begin{Bmatrix}43.872\\120.155\\167.00\end{Bmatrix} \text{ and } [\Phi]=\begin{bmatrix}1.000 & 1.000 & 1.000\\0.7594 & -0.8047 & -2.427\\0.3361 & -1.1572 & 2.512\end{bmatrix}$$

Let us compute the modal static displacement in a tabular manner.

We know that the solution of Eq. (7.13) in the form of Duhamel's integral [Eq. (3.136)] is

$$\xi_r = \xi_{rst}\, p_r \int_o^t f(\tau)\sin p(t-\tau)d\tau \qquad \text{(a)}$$

and

$$(DLF)_r = p_r \int_o^t f(\tau)\sin p(t-\tau)d\tau \qquad \text{(b)}$$

For the present problem

$$f(\tau) = 1-\frac{\tau}{t_d} \quad \text{for} \quad t<t_d \qquad \text{(c)}$$

$$\left.\begin{aligned}(DLF)_r &= 1-\cos p_r t + \frac{\sin p_r t}{p_r t_d} - \frac{t}{t_d} \quad \text{for} \quad t \le t_d\\ (DLF)_r &= \frac{1}{p_r t_d}\left[\sin p_r t_d - \sin p_r (t-t_d)\right] - \cos p_r t \quad \text{for} \quad t \ge t_d\end{aligned}\right\} \qquad \text{(d)}$$

Though internal force acting at the three mass levels have the same time function, $(DLF)_r$ for different modes will have different values. The horizontal deflection of the second mass is given by

$$x_2(t) = \sum_{r=1}^{n} \phi_2^{(r)} \xi_{rst} (DLF)_r$$

$= (0.7594)\,(0.0150)\,(DLF)_1 + (-0.8047)\,(-0.003775)\,(DLF)_2$

$+ (-2.427)\,(-0.00022)\,(DLF)_3$

$= 0.01139\,(DLF)_1 + 0.00304\,(DLF)_2 + 0.0000534\,(DLF)_3$ (e)

In order to calculate maximum value of $x_2(t)$, the expression given in Eq. (e) is to be differentiated. However, the process is indeed tedious. One may prefer to evaluate the value of $x_2(t)$ at different time intervals and then picking up the maximum value. Alternately, one may plot the displacement for the various modes and evaluate the values from the graph. A typical plot for the response at the second mass as given by Eq. (e) is shown in Fig. 7.3.

Mode	Mass point	$\phi_i^{(r)}$	$\overline{F}_i$	$\overline{F}_i\phi_i^{(r)}$	m_i	$\left(m_{i\phi_i^{(r)}}\right)^2$	ξ_{rst}
1	1	1.000	-5×10^5	-5×10^3	2×10^4	2×10^4	$p_1^2 = 1936$
	2	0.7594	15×10^5	-11.391×10^3	2×10^4	1.1533×10^4	$\xi_{1st} = \dfrac{9.752\times10^5}{1924.733\times3.3792\times10^4}$
	3	0.3361	10×10^5	3.361×10^3	2×10^4	0.2259×10^4	
	Σ			9.752×10^3		3.3792×10^4	$= 0.015$ m.
2	1	1.000	-5×10^5	-5×10^5	2×10^4	2×10^4	$p_2^2 = 14400$
	2	−0.8047	15×10^5	-12.071×10^5	2×10^4	1.2951×10^4	$\xi_{2st} = \dfrac{-28.6425\times10^5}{14400\times5.2684\times10^4}$
	3	−1.1572	10×10^5	-11.572×10^5	2×10^4	2.6782×10^4	
	Σ			-28.6425×10^5		5.2684×10^4	$= 0.003775$ m
3	1	1.000	-5×10^5	-5×10^5	2×10^4	2×10^4	$p_3^2 = 27889$
	2	−2.427	15×10^5	-36.405×10^5	2×10^4	11.78×10^4	$\xi_{3st} = \dfrac{-16.285\times10^5}{27889\times26.4\times10^4}$
	3	2.512	10×10^5	25.12×10^5	2×10^4	12.62×10^4	
	Σ			-16.285×10^5	2×10^4	26.4×10^4	$= -0.00022$ m

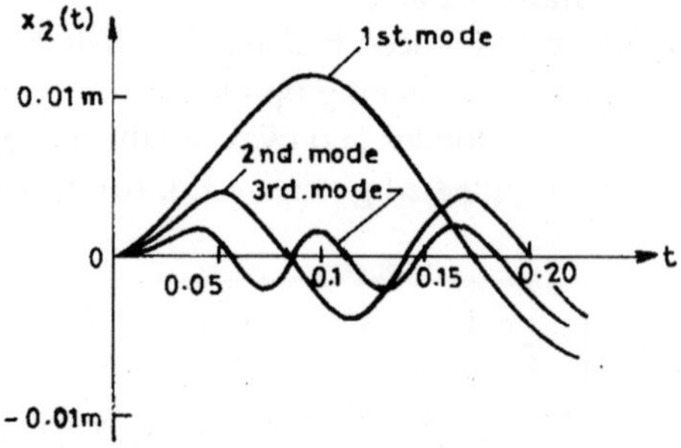

Fig. 7.3 Variation of displacement

The maximum force developed at the first mass is

$$P_1 = k_1(x_1 - x_2)$$

$$= k_1 \sum \left(\phi_1^{(r)} - \phi_2^{(r)}\right) \xi_{rst} (DLF)_r$$

$$= k_1\Big[\left(\phi_1^{(1)} - \phi_2^{(1)}\right)\xi_{1st}(DLF)_1 + \left(\phi_1^{(2)} - \phi_2^{(2)}\right)\xi_{2st}(DLF)_2$$

$$+ \left(\phi_1^{(3)} - \phi_2^{(3)}\right)\xi_{3st}(DLF)_3\Big]$$

$$= k_1\big[(0.2406)(0.0150)(DLF)_1 + (1.8047)(-0.003775)(DLF)_2 + (3.427)(-0.00022)(DLF)_3\big]$$

or, $$P_1 = k_1\left[(-0.00361)(DLF)_1 - (0.00681)(DLF)_2 - (0.00075)(DLF)_3\right]$$

In order to determine the maximum value of P_1, the same procedure as mentioned for the determination of maximum $x_2(t)$ may be followed. An upper bound of maximum force at the first mass is given by

$$(P_1)_{max} \le 16\times10^7\,\big[(0.00361)(DLF)_{1,\,max} - (0.00681)(DLF)_{2,\,max} + (0.00075)(DLF)_{3,\,max}\big]\ \text{kN}$$

For the given loading function

$(DLF)_{1,\,max} = 1.12$, $(DLF)_{2,\,max} = 1.53$ and $(DLF)_{3,\,max} = 1.68$

Therefore

$$(P_1)_{max} \le 16\times10^7\,[(0.00361)(1.12) + (0.00681)(1.53) + (0.00075)(1.68)]$$
$$\le 2{,}247{,}400\ kN$$

A root mean square of the maximum values for each mode may be more realistic. The root mean square value for this particular problem corresponding to maximum (*DLF*) in each mode is rms of

$$P_1 = 16 \times 10^7 \sqrt{(0.0036 \times 1.12)^2 + (0.0068 \times 1.53)^2 + (0.0007 \times 1.68)^2}$$
$$= 1{,}795{,}163 \text{ kN}$$

7.5 DAMPING IN MDF SYSTEMS

It is very difficult to know the nature of damping by any theoretical means. Experimental investigation has not done much to throw sufficient light on damping of different systems. The arrangement of dampers and the coefficients that are to be considered are perplexing problems even today. As such, treatment of damping is always based on simplifying assumptions.

Equations of motion for a damped MDF system are given by

$$[M]\{\ddot{x}\} + [C]\{\dot{x}\} + [K]\{x\} = \{F(t)\} \tag{7.20}$$

The displacements are expressed in terms of normal coordinates

$$\{x\} = [\Phi]\{\xi\} \tag{7.21}$$

Substituting $\{x\}$ and its derivatives into Eq. (7.20) and then pre-multiplying by $[\Phi]^T$, we get

$$[\Phi]^T[M][\Phi]\{\ddot{\xi}\} + [\Phi]^T[C][\Phi]\{\dot{\xi}\} + [\Phi]^T[K][\Phi]\{\xi\} = [\Phi]^T\{F(t)\} \tag{7.22}$$

It is assumed that the frequencies and mode shapes obtained from undamped free vibration analysis is valid, even when damping is present in the system. This is reasonably true for systems having small values of damping. Using the orthogonality relation of Eqs. (6.21) and (6.29), Eq. (7.22) can be written as

$$[\Phi]^T[M][\Phi]\{\ddot{\xi}\} + [\Phi]^T[C][\Phi]\{\dot{\xi}\} + [\Phi]^T[M][\Phi][p^2]\{\xi\} = [\Phi]^T\{F(t)\} \tag{7.23}$$

$[\Phi]^T[M][\Phi]$ is a diagonal matrix. In order to uncouple Eq. (7.23), it is assumed that the orthogonality condition applies to damping as well, i.e.

$$\{\phi^{(r)}\}^T[C]\{\phi^{(s)}\} = \{0\} \tag{7.24}$$

and

$$[\Phi]^T[C][\Phi] = [\Phi]^T[\mathrm{M}][\Phi][2p\zeta] \tag{7.25}$$

Combining Eqs. (7.23) and (7.25), we get

$$[\Phi]^T[M][\Phi]\left(\{\ddot{\xi}\} + [2p\zeta]\{\dot{\xi}\} + [p^2]\{\xi\}\right) = [\Phi]^T\{F(t)\} \tag{7.26}$$

The *r*th equation of Eq. (7.26) is given by

$$\overline{m}_r\left(\ddot{\xi}_r + 2p_r\zeta_r\dot{\xi}_r + p_r^2\xi_r\right) = \bar{f}_r \tag{7.27}$$

Converting the damping matrix into diagonal form suggests that a reasonable percentage of critical damping is assumed in each mode. It is not only convenient, but also physically reasonable to assume damping in this form, than trying to evaluate the elements of [*C*] matrix.

The final values of the displacements are obtained by substituting the solution of ζ_r from Eq. (7.27) into Eq. (7.21).

Example 7.4

For the two-storeyed frame with viscous damping shown in Fig. 7.4, compute the displacement at the top storey.

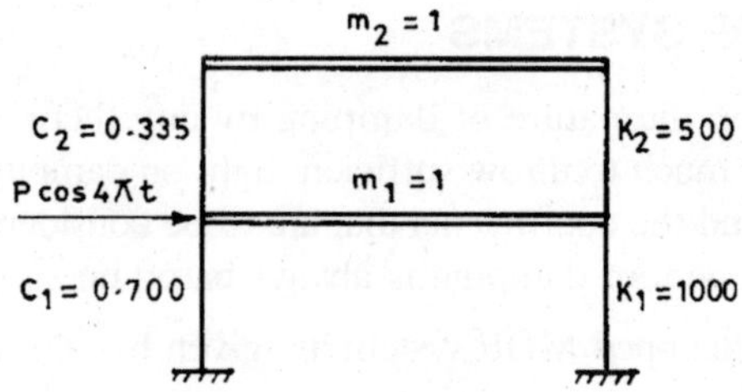

Fig. 7.4 Example 7.4

The natural frequencies and mode shapes for the free undamped vibration is to be obtained.

$$[M]=\begin{bmatrix} 1.0 & 0 \\ 0 & 1.0 \end{bmatrix} \quad \text{and} \quad [K]=\begin{bmatrix} 1500 & -500 \\ -500 & 500 \end{bmatrix}$$

The free vibration equation is

$$\left(500\begin{bmatrix} 3 & -1 \\ -1 & 1 \end{bmatrix} - p^2\begin{bmatrix} 1 & 0 \\ 0 & 1 \end{bmatrix}\right)\begin{Bmatrix} \phi_1 \\ \phi_2 \end{Bmatrix} = \begin{Bmatrix} 0 \\ 0 \end{Bmatrix}$$

The frequencies and mode shapes are found to be

$$p_1 = 17.11 \text{ and } p_2 = 41.32$$

$$[\Phi] = \begin{bmatrix} 1.000 & 1.000 \\ 2.414 & -0.414 \end{bmatrix}$$

$$[\Phi]^T [M] [\Phi] = \begin{bmatrix} 1.000 & 1.000 \\ 2.414 & -0.414 \end{bmatrix}^T \begin{bmatrix} 1 & 0 \\ 0 & 1 \end{bmatrix} \begin{bmatrix} 1.000 & 1.000 \\ 2.414 & -0.414 \end{bmatrix}$$

$$= \begin{bmatrix} 6.827 & 0 \\ 0 & 1.171 \end{bmatrix}$$

$$[C] = \begin{bmatrix} 1.035 & -0.335 \\ -0.335 & 0.335 \end{bmatrix}$$

$$[\Phi]^T [C][\Phi] = \begin{bmatrix} 1.000 & 2.414 \\ 1.000 & -0.414 \end{bmatrix} \begin{bmatrix} 1.035 & -0.335 \\ -0.335 & 0.335 \end{bmatrix} \begin{bmatrix} 1.000 & 1.000 \\ 2.414 & -0.414 \end{bmatrix}$$

$$= \begin{bmatrix} 1.370 & 0 \\ 0 & 1.370 \end{bmatrix}$$

$$\{F(t)\} = \begin{Bmatrix} P\cos 4\pi t \\ 0 \end{Bmatrix}$$

$$[\Phi]\{F(t)\} = \begin{bmatrix} 1.000 & 2.414 \\ 1.000 & -0.414 \end{bmatrix} \begin{Bmatrix} P\cos 4\pi t \\ 0 \end{Bmatrix}$$

$$= P\cos 4\pi t \begin{Bmatrix} 1 \\ 1 \end{Bmatrix}$$

Therefore, Eq. (7.27) for this problem for $r = 1$ is as follows:

$$6.827\ddot{\xi}_1 + 1.37\dot{\xi}_1 + 292.75 \times 6.827\xi_1 = P\cos 4\pi t \qquad \text{(a)}$$

Before, we proceed further, let us check the damping ratio for this mode

$$\zeta = \frac{1.37}{2 \times 6.827 \times 17.11} = 0.00586$$

Equation (a) is written after dividing both sides by 6.827 as

$$\ddot{\xi}_1 + 0.2\dot{\xi}_1 + 292.75\xi_1 = \frac{P}{6.827} \cos 4\pi t \qquad \text{(b)}$$

Therefore, for the steady-state vibration

$$\xi_1 = \frac{\dfrac{P}{6.827}}{\sqrt{\left(292.75 - (4\pi)^2\right)^2 + (2 \times 0.1 \times 4\pi)^2}} \cos 4\pi t$$

$$= \frac{P}{1737.67} \cos 4\pi$$

Similarly

$$1.171\ddot{\xi}_2 + 1.37\dot{\xi}_2 + 1707.34 \times 1.171\xi_2 = \mathrm{P}\cos 4\pi t$$

or

$$\ddot{\xi}_2 + 1.17\dot{\xi}_2 + 1707.34\xi_2 = \frac{\mathrm{P}\cos 4\pi t}{1.171} \qquad \text{(c)}$$

For the steady-state vibration

$$\xi_2 = \frac{\dfrac{P}{1.171}}{\sqrt{\left(1707.34 - (4\pi)^2\right)^2 + (2 \times 1.17 \times 4\pi)^2}} \cos 4\pi t$$

$$= \frac{P}{2200} \cos 4\pi t$$

Therefore, the displacement at the top storey level

$$x_1 = \phi_1^{(1)}\xi_1 + \phi_1^{(2)}\xi_2$$

$$= \frac{P}{1737.67} \cos 4\pi t + \frac{P}{2200} \cos 4\pi t$$

$$= \frac{P}{970.85} \cos 4\pi t$$

7.5.1 Conditions for Damping Uncoupling

In the derivation of uncoupled damped equation, Eq. (7.20), it is assumed that the normal coordinate transformation, Eq. (7.21) that has been used to uncouple the inertial and elastic forces also uncouples the damping forces, Eq. (7.23). This results in transforming the damping matrix in terms of modal damping ratio. However, there are situations where the principle of superposition cannot be applied for the dynamic response analysis so that damping matrix cannot be expressed by damping ratios, rather the damping matrix expressed explicitly is needed in such cases. The cases are (a) nonlinear responses, for which the mode shapes are not fixed, but are changing with changes of stiffness and (b) analysis of linear systems having nonproportional damping.

The easiest approach to formulate a proportional damping matrix is to treat it as proportional to either the mass matrix or the stiffness matrix, because the undamped mode shapes are orthogonal with respect to each other. Thus, the damping matrix can be expressed as

$$[C] = \alpha[M], \quad [C] = \beta[K] \tag{7.28 a, b}$$

in which proportionality constants α and β have units of $\sec^{-1}$ and sec, respectively. The left hand equation of Eq. (7.28) is mass proportional damping and the right hand equation of Eq. (7.28) is stiffness proportional damping

For Eq. (7.28 a)

$$\bar{c}_r = \alpha\{\phi^{(r)}\}^T [M]\{\phi^{(r)}\}$$

or,

$$2\, p_r\, \zeta_r\, \overline{m}_r = \alpha\, \overline{m}_r$$

or,

$$\zeta_r = \frac{\alpha}{2\, p_r} \tag{7.29 a}$$

Similarly, for Eq. (7.28 b)

$$\bar{c}_r = \beta\{\phi^{(r)}\}^T [K]\{\phi^{(r)}\} \tag{7.29 b}$$

From Eq. (7.29)

$$\bar{c}_r = \beta\, p_r^2\, \overline{m}_r$$

or,

$$2\, p_r\, \zeta_r\, \overline{m}_r = \beta\, p_r^2\, \overline{m}_r$$

or,

$$\zeta_r = \frac{\beta\, p_r}{2} \tag{7.29 c}$$

Typical variations of mass and stiffness-proportional damping ratios are shown in Fig. 7.5. In practice it has been found that mass proportional damping can represent friction damping whilst stiffness proportional damping can represent internal material damping.

Some typical values of modal damping ratios are 0.01 for small diameter piping systems to 0.07 for bolted joints and reinforced concrete structures. If all modal damping ratios can be estimated, it is not necessary to form the damping matrix. The values of ζ_r are substituted into Eq. (7.27)

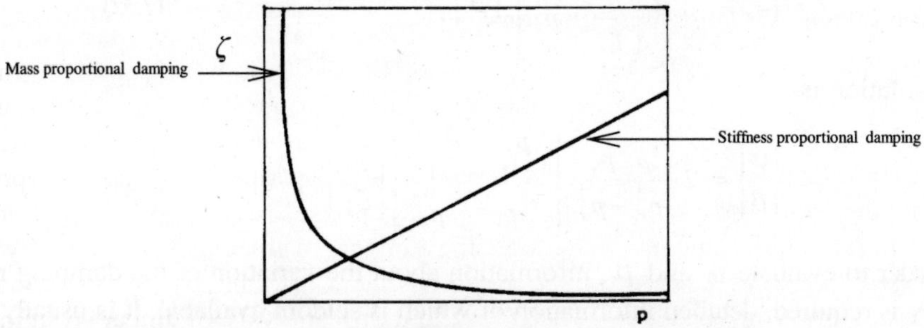

Fig. 7.5 Mass and stiffness-proportional damping

Eq. (7.29 a) and (7.29c) reveal that for mass proportional damping, the damping ratio is inversely proportional to the frequency, while for stiffness proportional damping it is directly in proportion to the frequency.

In dynamic analysis, contribution of all *n* modes are involved, though only a few modes are included in the uncoupled equations of motion. Thus when the significant modes span over a wide range, neither of the above types are suitable for the dynamic analysis of MDF systems, as the relative amplitudes of different modes will be seriously distorted by the inappropriate damping ratios.

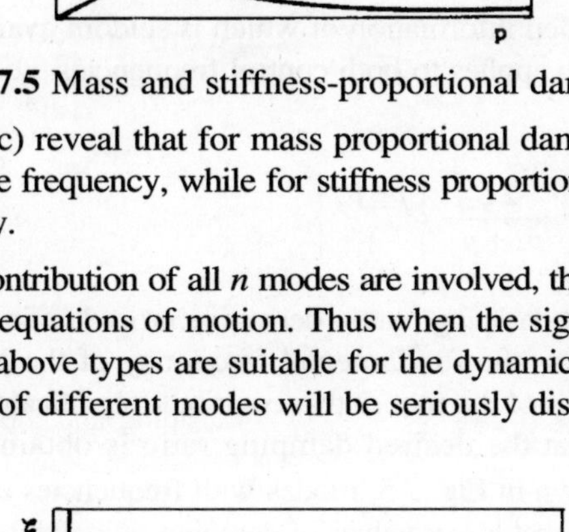

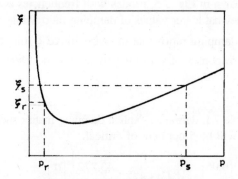

Fig. 7.6 Rayleigh damping

An improvement is obvious if the damping is assumed to be proportional to a combination of the mass and the stiffness matrices and is given by

$$[C] = \alpha[M] + \beta[K] \tag{7.30}$$

This is called Rayleigh damping [Fig. 7.6]. From Eq. (7.29), it is evident that Rayleigh damping leads to the following relationship between damping ratio and frequency,

$$\zeta_r = \frac{\alpha}{2\,p_r} + \frac{\beta\,p_r}{2} \tag{7.31}$$

Now, it is apparent that the two Rayleigh damping factors, α and β can be evaluated from the solution of a pair of simultaneous equations if the damping ratios ζ_r and ζ_s associated with two specific frequencies p_r and p_s are known. Writing Eq. (7.31) for each of these two cases and expressing the two equations in matrix form leads to

$$\begin{Bmatrix} \zeta_r \\ \zeta_s \end{Bmatrix} = \frac{1}{2} \begin{bmatrix} \dfrac{1}{p_r} & p_r \\ \dfrac{1}{p_s} & p_s \end{bmatrix} \begin{bmatrix} \alpha \\ \beta \end{bmatrix} \tag{7.32}$$

and the solution is

$$\begin{Bmatrix} \alpha \\ \beta \end{Bmatrix} = 2 \frac{p_r\, p_s}{p_r^2 - p_s^2} \begin{bmatrix} p_r & -p_s \\ -\dfrac{1}{p_r} & \dfrac{1}{p_s} \end{bmatrix} \begin{bmatrix} \zeta_r \\ \zeta_s \end{bmatrix} \tag{7.33}$$

In order to evaluate α and β, information about the variation of the damping ratio with frequency is required, detailed information of which is seldom available. It is usually assumed that the same damping ratio applies to both control frequencies, i.e., $\zeta_r = \zeta_s = \zeta$. In this case, then

$$\begin{Bmatrix} \alpha \\ \beta \end{Bmatrix} = \frac{2\zeta}{p_r + p_s} \begin{Bmatrix} p_r\, p_s \\ 1 \end{Bmatrix} \tag{7.34}$$

In applying this proportional damping matrix derivative procedure in practice, it is recommended that p_r be generally taken as the fundamental frequency of the MDF system and p_s be set among the higher frequencies of the modes that contribute significantly to the dynamic response. The derivation ensures that the desired damping ratio is obtained from these two modes ($\zeta_r = \zeta_s = \zeta$), then as shown in Fig. 7.5, modes with frequencies between these two specified frequencies will have somewhat lower values of damping ratio, while all modes with frequencies greater than p_s will have damping ratios that increase above ζ monotonically with frequency. The end result is that the responses of very high frequency modes are effectively eliminated by their high damping ratios.

Example 7.5

For the structure of Example 7.1, define an explicit damping matrix such that the damping ratio in the first and third modes will be 5 per cent of critical.

$$\begin{Bmatrix} \zeta_1 \\ \zeta_3 \end{Bmatrix} = \begin{Bmatrix} 0.05 \\ 0.05 \end{Bmatrix} = \frac{1}{2} \begin{bmatrix} \dfrac{1}{43.872} & 43.872 \\ \dfrac{1}{167} & 167 \end{bmatrix} \begin{Bmatrix} \alpha \\ \beta \end{Bmatrix}$$

or,

$$\begin{Bmatrix} \alpha \\ \beta \end{Bmatrix} = \begin{Bmatrix} 3.474 \\ 4.74 \times 10^{-4} \end{Bmatrix}$$

Hence,

$$[C] = 3.474 \begin{bmatrix} 1 & 0 & 0 \\ 0 & 1 & 0 \\ 0 & 0 & 1 \end{bmatrix} \times 20000 + 4.74 \times 10^{-4} \times 8 \times 10^{7} \times \begin{bmatrix} 2 & -2 & 0 \\ -2 & 4 & -2 \\ 0 & -2 & 5 \end{bmatrix}$$

$$= \begin{bmatrix} 145320 & -75840 & 0 \\ -75840 & 221160 & -75840 \\ 0 & -75840 & 259080 \end{bmatrix} \text{ N s/m}$$

It is interesting to note that the damping ratio of this matrix will yield the second mode

$$\zeta_2 = \frac{1}{2}\begin{bmatrix} \frac{1}{120.155} & 120.155 \end{bmatrix}\begin{Bmatrix} \alpha \\ \beta \end{Bmatrix}$$

$$= 0.0429 = 4.29\%$$

Hence, even though the first and third damping ratios were specified, the resulting damping ratio of the second mode will have a reasonable value.

7.5.2 Extended Rayleigh Damping

The mass and stiffness matrices used to formulate Rayleigh damping are not the only matrices to which the free-vibration mode-shape orthogonality conditions apply. However, there are other matrices formed from mass and stiffness matrices which also satisfy the orthogonality conditions. In general, the damping matrix may be of the form

$$[C] = [M]\sum_i a_i\,([M]^{-i}[K])^i \tag{7.35}$$

where i can be anywhere in the range $-\alpha < i < \alpha$ and the summation may include as many terms as desired. The damping matrix of Eq. (7.30) can be obtained as a special case of Eq. (7.35). By taking two terms corresponding to $i = 0$ and $i = 1$ in Eq. (7.35), we obtain the damping matrix given by Eq. (7.30). With this form of damping matrix it is possible to compute the damping coefficients necessary to provide uncoupling of a system having any desired degree of damping ratios in any specific number of modes. For any mode r, the modal damping matrix is

$$\bar{c}_r = \{\phi^{(r)}\}^T [C]\{\phi^{(r)}\} = 2\zeta_r\, p_r\, \bar{m}_r \tag{7.36}$$

If $[C]$ given by Eq. (7.35) is substituted in the expression for $\bar{c}_r$, we obtain

$$\bar{c}_r = \{\phi^{(r)}\}^T [M]\sum_i a_i\left([M]^{-1}[K]\right)^i \{\phi^{(r)}\} \tag{7.37}$$

Now, using $\left([K]\{\phi^{(r)}\} = p_r^2\,[M]\{\phi^{(r)}\}\right)$ and performing several algebraic operations, we can show that the damping coefficient associated with any mode r may be written as

$$\bar{c}_r = \sum_i a_i\, p_r^{2i}\, \bar{m}_r = 2\zeta_r\, p_r\, \bar{m}_r \tag{7.38}$$

from which

$$\zeta_r = \frac{1}{2p_r}\sum_i a_i\, p_e^{2i} \tag{7.39}$$

Eq. (7.39) may be used to determine constants a_i for any desired value of modal damping ratios corresponding to any specified number of modes. For example, to evaluate these constants specifying the first four modal damping ratios ζ_1, ζ_2, ζ_3 and ζ_4, we may choose i = 1, 2, 3, 4. In this case, Eq. (7.39) gives the following system of equations

$$\begin{Bmatrix} \zeta_1 \\ \zeta_2 \\ \zeta_3 \\ \zeta_4 \end{Bmatrix} = \frac{1}{2} \begin{bmatrix} p_1 & p_1^3 & p_1^5 & p_1^7 \\ p_2 & p_2^3 & p_2^5 & p_2^7 \\ p_3 & p_3^3 & p_3^5 & p_3^7 \\ p_4 & p_4^3 & p_4^5 & p_4^7 \end{bmatrix} \begin{Bmatrix} a_1 \\ a_2 \\ a_3 \\ a_4 \end{Bmatrix} \tag{7.40}$$

In general, Eq. (7.40) may be written symbolically as

$$\{\zeta\} = \frac{1}{2}[Q]\{a\} \tag{7.41}$$

where $[Q]$ is a square matrix having different powers of the natural frequencies. On solving Eq. (7.41), yields

$$\{a\} = 2[Q]^{-1}\{\zeta\} \tag{7.42}$$

Finally, the damping matrix is obtained after the substitution of Eq. (7.42) into Eq. (7.35)

Fig. 7.7 illustrates the relationship between damping ratio and frequency that would result from this matrix. To simplify the figure, it has been assumed that the same damping ratio $\bar{\zeta}$ was specified for all four frequencies; however each of the damping ratios could have been specified arbitrarily. p_1 is the fundamental frequency and p_4 is intended to approximate the frequency of the highest mode that contributes significantly to the response, while p_2 and p_3 are spaced about equally within the frequency range. It is evident in Fig. 7.7 that damping ratio remains close to the desired value $\bar{\zeta}$ throughout the frequency range, being exact at four specified frequencies and ranging slightly above or below at other frequencies. It is important to note however, that the damping increases monotonically with frequency for frequencies above p_4. This has the effect of excluding any significant contribution from any modes with frequencies much greater than p_4. This has the effect of excluding any significant contribution from any mode with frequencies much greater than p_4. Thus these modes need not be included in the response superposition.

An even more important point to note is the consequence of including only three terms in the derivation of the viscous damping matrix in Eq. (7.35). In that case three simultaneous equations equivalent to Eq. (7.40) would be obtained and the resulting damping ratio-frequency relation obtained after solution is shown in Fig. 7.7(b). As required by the solution of the simultaneous equation, the desired damping ratio is obtained exactly at the three specified frequencies and is approximated well at intermediate frequencies. However, the serious defect of this result is that the damping decreases monotonically with frequencies increasing above p_3 and negative damping is indicated for all highest modal frequencies. This is an unacceptable result because the contribution of the negatively damped modes would tend to increase without limit in the analysis, but certainly would not do so in actuality. The general implication of the observation is that extended Rayleigh damping may be used effectively only if an even number of terms is included in the series expression.

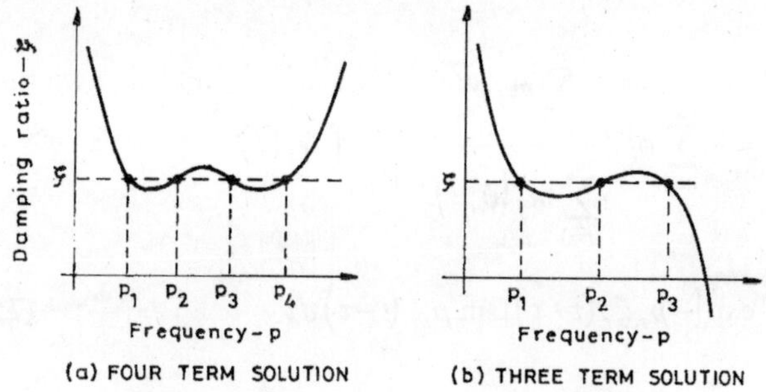

Fig. 7.7 Extended Rayleigh damping

7.6 RESPONSE OF MDF SYSTEMS TO SUPPORT MOTION

The equations which have been developed for the response of SDF systems to ground motion can be extended to MDF systems. The equations of motion for a structure subjected to ground motion is given by

$$[M]\{\ddot{z}\}+[C]\{\dot{z}\}+[K]\{z\}= -[M]\{\ddot{x}_s\} \tag{7.43}$$

where z represents the relative displacement of the mass with respect to the ground and $\ddot{x}_s$ is the ground acceleration. Let us assume the solution in terms of normal coordinates as usual.

$$\{z\}=[\Phi]\{\xi\} \tag{7.44}$$

Substituting $\{z\}$ from Eq. (7.44) into Eq. (7.43) and pre-multiplying both sides by $[\phi]_T$, it can be shown as demonstrated in the previous sections that it results in a series of uncoupled differential equations. For the response due to support motion, the rth equation will be

$$\ddot{\xi}_r + 2p_r\zeta_r\dot{\xi}_r + p_r^2\xi_r = -\frac{\bar{f}_{yr}}{\overline{m}_r} \tag{7.45}$$

where
$$\bar{f}_{yr} = \sum_{j=1}^{n}\phi_j^{(r)} m_j \ddot{x}_s \tag{7.46}$$

and
$$\overline{m}_r = \sum_{j=1}^{n} m_j \left(\phi_j^{(r)}\right)^2 \tag{7.47}$$

From Eq. (7.44), the displacement of the ith mass is

$$z_i = \sum_{r=1}^{n}\phi_i^{(r)}\xi_r \tag{7.48}$$

Solution of Eq. (7.45) is

$$\xi_r = -\int_o^t \frac{\bar{f}_{yr}}{\overline{m}_r\, p_{dr}} \exp\left[-p_r\zeta_r(t-\tau)\right]\sin p_{dr}(t-\tau)\,d\tau \tag{7.49}$$

where p_{dr} is the damped natural frequency in the rth mode. Substituting ξ_r from Eq. (7.44), and $\bar{f}_{yr}$ and $\overline{m}_r$ from Eqs. (7.46) and (7.47) respectively, into Eq. (7.48), we get

$$z_i = -\sum_{r=1}^{n} \phi_i^{(r)} \frac{\sum_{j=1}^{n} m_j \phi_j^{(r)}}{\sum_{j=1}^{n} m_j \left(\phi_j^{(r)}\right)^2} \frac{1}{p_{dr}} \int_o^t \ddot{x}_s(\tau)$$

$$\exp\left[-p_r \zeta_r (t-\tau)\right] \sin p_{dr}(t-\tau)\, d\tau \qquad (7.50)$$

The quantity $B_r = \dfrac{\sum_{j=1}^{n} m_j \phi_j^{(r)}}{\sum_{j=1}^{n} m_j \left(\phi_j^{(r)}\right)^2}$ is called the mode participation factor.

From Eqs. (3.153) and (3.157), it is known that the spectral displacement of the rth mode is

$$z_{\max}^{(r)} = S_d^{(r)} = \left| \frac{1}{p_{dr}} \int_o^t \ddot{x}_s(\tau) \exp\left(-p_r \zeta_r (t-\tau)\right) \sin p_{dr}(t-\tau)\, d\tau \right|_{\max} \qquad (7.51)$$

where $\ddot{x}_{s0}$ is the maximum amplitude of the ground motion. *Thus, Eq. (7.50) is written as*

$$| z_i^{(r)} |_{\max} = \phi_i^{(r)} B_r S_d^{(r)} \qquad (7.51a)$$

Example 7.6

A two storeyed bent is shown in Fig. 7.8. Compute the displacements at the floor levels due to the motion of the support as shown. The natural frequencies for this bent are 9.00 rad/s and 23.5 rad/s respectively. The mode shape matrix is

$$[\phi] = \begin{bmatrix} 1.00 & 1.00 \\ 1.57 & -1.06 \end{bmatrix}$$

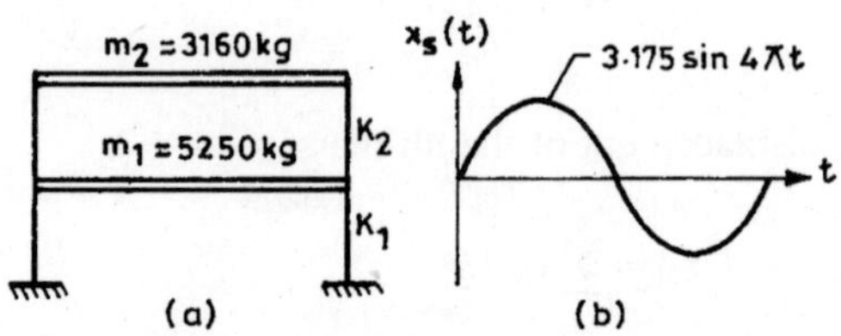

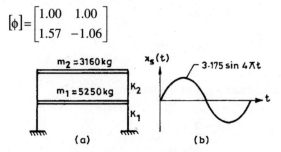

Fig. 7.8 Example 7.6

Let us next compute the mode participation factors. They are

$$B_1 = \frac{\sum_{j=1}^{n} m_j \phi_j^{(1)}}{\sum_{j=1}^{n} m_j \left(\phi_j^{(1)}\right)^2} = \frac{(5250)\times 1.00 + (3160)\times 1.57}{5250\times(1.00)^2 + (3160)\times(1.57)^2} = 0.783$$

Similarly

$$B_2 = \frac{(5250)(1.00)+(3160)(-1.06)}{(5250)(1.00)^2+(3160)(-1.06)^2} = 0.216$$

The dynamic load factor is determined as follows:

$$(DLF)_{a,r} = p_r \int_o^t f(\tau)\sin p_r(t-\tau)\,d\tau$$

$$= p_r \int_o^t \sin\omega\tau \, \sin p_r(t-\tau)\,d\tau$$

$$= p_r^2\left[\frac{\sin\omega t}{p_r^2-\omega^2} - \frac{\omega}{p_r}\frac{\sin p_r t}{p_r^2-\omega^2}\right]$$

$$= \frac{1}{1-\frac{\omega^2}{p_r^2}}\left(\sin\omega t - \frac{\omega}{p_r}\sin p_r t\right)$$

For the first normal mode, when $r = 1$

$$(DLF)_{a,1} = \frac{1}{1-\frac{(4\pi)^2}{9^2}}\left(\sin 4\pi t - \frac{4\pi}{9}\sin 9t\right)$$

$$= -1.053(\sin 4\pi t - 0.535\sin 23.5t)$$

Similarly,

$$(DLF)_{a,2} = 1.400(\sin 4\pi t - 0.535\sin 23.5t)$$

Therefore,

$$z_1 = \phi_1^{(1)} B_1 \frac{\ddot{x}_{so}}{p_1^2}(DLF)_{a,1} + \phi_1^{(2)} B_2 \frac{\ddot{x}_{so}}{p_2^2}(DLF)_{a,2}$$

$$= 1\times 0.783\times\frac{3.175}{81}(DLF)_{a,1} + 1.00\times 0.216\times\frac{3.175}{23.5^2}(DLF)_{a,2}$$

$$= 0.307(DLF)_{a,1} + 0.00124(DLF)_{a,2}$$

$$= 0.0307\left[-1.053(\sin 4\pi t - 1.396\sin 9t)\right]$$

$$+ 0.00124\left[1.40(\sin 4\pi t - 0.535\sin 23.5t)\right]$$

$$= -0.0306\sin 4\pi t - 0.0428\sin 9t - 0.00066\sin 23.5t$$

Similarly,

$$z_2 = (1.57)(0.783)\frac{3.175}{81}(DLF)_{a,1} + (-1.06)(0.216)\frac{3.175}{23.5^2}(DLF)_{a,2}$$

$$=0.0482(DLF)_{a,1}-0.00131(DLF)_{a,2}$$

7.7 EARTHQUAKE SPECTRUM ANALYSIS OF STRUCTURES HAVING MDF SYSTEM

Earthquake spectrum analysis, which has been applied to SDF system in Chapter 3, is extended here to MDF system [7.2, 7.3, 7.6, 7.7].

Let us rewrite Eq. (7.51) by placing

$$\ddot{x}_s(\tau)=\ddot{x}_{so}f_a(\tau)$$

$$S_d^{(r)}=\left|\frac{1}{p_{dr}}\int_o^t \ddot{x}_{so}f_a(\tau)\exp\left(-p_r\zeta_r(t-\tau)\right)\sin p_{dr}(t-\tau)\,d\tau\right|_{\max} \tag{7.52}$$

This equation is similar to Eqs. (3.153) and (3.157). So for the rth equation, the same response spectrum diagram can be used. But these response spectrum diagrams give only the maximum response. The maximum values from all the equations will not occur at the same time instant. The total displacement of any mass is obtained by the superimposition of displacements for each mode. As the maximum values are only obtained from the response spectrum diagrams, its sum would give excessively conservative values. So in order to get a reasonable value, IS 1893 suggests the adoption of the root mean square of the modal maximum. This is based on the assumption that the modal components are random variables, which are consistent with the random nature of the input. The accuracy of this approach increases with the increase of number of degrees of freedom.

The maximum relative displacement for the rth mode from Eq. (7.51) can be written as

$$\left|z_i^{(r)}\right|_{\max}=\phi_i^{(r)}B_rS_d^{(r)} \tag{7.53}$$

The maximum storey shear at any level in the rth mode is given by

$$V_i^{(r)}=K_i\left\{\phi_i^{(r)}-\phi_{i-1}^{(r)}\right\}B_r\,S_d^{(r)} \tag{7.54a}$$

The maximum load acting at any floor level i due to rth mode of vibration is given by

$$P_i^{(r)}=m_ip_{dr}^2\left|z_i^{(r)}\right|_{\max} \tag{7.54b}$$

Substituting $z_i^{(r)}$ from Eq. (7.50) and replacing the necessary quantities by the notation B_r and $S_d^{(r)}$ into Eq. (7.54a), we get

$$P_i^{(r)}=m_ip_{dr}^2\phi_i^{(r)}B_rS_d^{(r)} \tag{7.55}$$

From Eqs. (3.122) and (3.128), we get

$$S_a^{(r)}=p_{dr}^2S_d^{(r)} \tag{7.56}$$

Combining Eqs. (7.55) and (7.56), we get

$$P_i^{(r)}=m_i\phi_i^{(r)}B_rS_a^{(r)} \tag{7.57}$$

IS Code 1893 has specified the equation for the load acting at each floor level in the same form as Eq. (7.57).

The maximum storey shear at any level i in the rth mode then, is

$$V_i^{(r)} = \sum_{j=i}^{n} P_j = B_r S_a^{(r)} \sum_{j=i}^{n} m_j \phi_j^{(r)} \tag{7.58}$$

where n represents the topmost storey and i is the storey under consideration.

Though the expressions of Eqs. (7.54) and (7.58) appear to be different, they are in fact identical. This is proved as follows.

Consider a three-storeyed shear building of Fig. 7.9. The equations of motion for the masses in free vibration for the rth mode can be written as

$$-p_r^2 \begin{bmatrix} m_1 & 0 & 0 \\ 0 & m_2 & 0 \\ 0 & 0 & m_3 \end{bmatrix} \begin{Bmatrix} \phi_1^{(r)} \\ \phi_2^{(r)} \\ \phi_3^{(r)} \end{Bmatrix} + \begin{bmatrix} k_1 + k_2 & -k_2 & 0 \\ -k_2 & k_2 + k_3 & -k_3 \\ 0 & -k_3 & k_3 \end{bmatrix} \begin{Bmatrix} \phi_1^{(r)} \\ \phi_2^{(r)} \\ \phi_3^{(r)} \end{Bmatrix} = \begin{Bmatrix} 0 \\ 0 \\ 0 \end{Bmatrix} \tag{7.59}$$

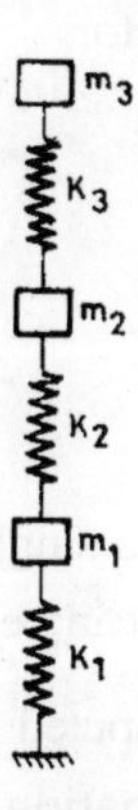

Fig. 7.9 Shear binding

Adding the three equations of Eq. (7.59), we get

$$\sum_{j=1}^{3} m_j p_r^2 \phi_j^{(r)} = k_1 \phi_1^{(r)} \tag{7.60}$$

Multiplying both sides of Eq. (7.60) by $B_r S_d^{(r)}$, we get

$$\sum_{j=1}^{3} m_j p_r^2 \phi_j^{(r)} B_r S_d^{(r)} = k_1 \phi_1^{(r)} B_r S_d^{(r)} \tag{7.61a}$$

Combining Eqs. (7.56) and (7.61 a), we get

$$\sum_{j=1}^{3} m_j \phi_j^{(r)} \mathrm{B}_r S_a^{(r)} = k_1 \phi_1^{(r)} B_r S_d^{(r)} = \text{Base shear} = V_3^{(r)} \tag{7.61b}$$

Similarly, adding last two equations of Eq. (7.59), multiplying both sides by $B_r S_d^{(r)}$ and substituting the relation given by Eq. (7.56), we get

$$\sum_{j=2}^{3} m_j \phi_j^{(r)} B_r S_a^{(r)} = k_2 \left\{\phi_2^{(r)} - \phi_1^{(r)}\right\} B_r S_d^{(r)} = V_2^{(r)} \tag{7.62}$$

Equations (7.61 a) and (7.62) conclusively prove that Eqs. (7.54) and (7.58) are identical.

7.8 USE OF RESPONSE SPECTRA FOR DESIGNING MDF SYSTEMS

The various steps for designing MDF systems by using response spectra of the type shown in Fig. 7.3 are as follows:

(1) Compute the lowest few modes and natural periods of the structural system. For multistoried buildings, first four modes are usually sufficient. First two modes are predominant in the case of masonry dams. First five or six modes may be necessary for the analysis of earth dams.

Let the time periods be T_1, T_2, T_3, ..., T_N and the modes be denoted by for the $\phi_i^{(1)}, \phi_i^{(2)}, \phi_i^{(3)}, \ldots \phi_i^{(N)}$ *i*th mass for first *N* modes,.

(2) Compute the mode participation factors

$$B_r = \frac{\sum_{i=1}^{n} m_i \phi_i^{(r)}}{\sum_{i=1}^{n} m_i \left(\phi_i^{(r)}\right)^2}$$

where n is the total number of masses.

(3) Now, each mode can be treated as a single degree of freedom system. Corresponding to the time period in each mode, $S_a^{(r)}$ or $S_d^{(r)}$ can be obtained from the figure similar to Fig. 3.38.

(4) Then $V_i^{(r)}$ or $P_i^{(r)}$, either can be computed for each mode, based on Eq. (7.54) or (7.57).

(5) The last step involves the combination of various modal quantities. Usually, the maximum values of them are of interest in design. To obtain the theoretical exact value, time wise integrals of the ground motion are to be superimposed with proper sign. The maximum value thus obtained may occur at any time instant and should be picked up as the design value. The computation involved in this case is too tedious and a simpler procedure is adopted in design.

The maximum values of the modal quantities are calculated following steps 1 to 4. These maximum values will not occur at the same instant of time, nor their sign will be same for each mode. As discussed in Art. 7.4, the assumption of root mean square value of this quantity is more realistic.

Thus

$$V_i = \left[\left(V_i^{(1)}\right)^2 + \left(V_i^{(2)}\right)^2 + \ldots + \left(V_i^{(N)}\right)^2\right]^{\frac{1}{2}} \tag{7.63}$$

Example 7.7

The three storeyed shear building with pertinent data is shown in Fig. 7.10. The system has 5 percent damping. Calculate the maximum storey shears based on the spectral diagrams of Fig. 3.35. Also find the maximum displacement of each mass.

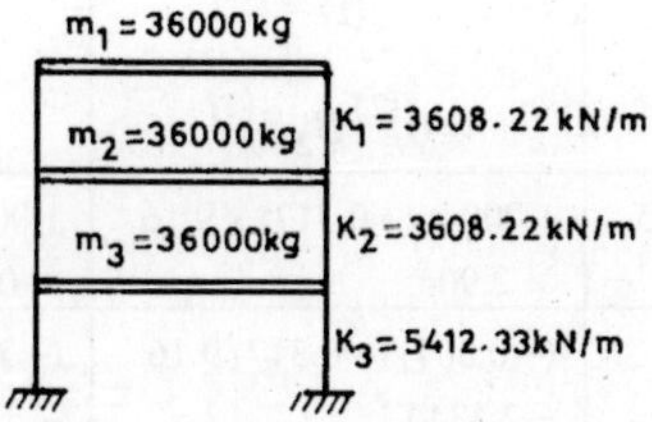

Fig. 7.10 Example 7.7

The natural frequencies and mode shapes are to be calculated first. The values obtained for the problem are indicated in the Table below.

	Mode		
	1	2	3
p_r	4.92 rad/s	13.45 rad/s	18.7 rad/s
T_r	1.227 s	0.467 s	0.336 s
$\phi_1^{(r)}$	1.000	1.000	1.000
$\phi_2^{(r)}$	0.759	– 0.804	– 2.462
$\phi_3^{(r)}$	0.336	– 1.157	2.580

The mode participation factors are then computed in the following Table.

Mass	m_j^*	1st mode		2nd mode		3rd mode	
		$m_j\phi_j^{(1)}$	$m_j\left(\phi_j^{(1)}\right)^2$	$m_j\phi_j^{(2)}$	$m_j\left(\phi_j^{(2)}\right)^2$	$m_j\phi_j^{(3)}$	$m_j\left(\phi_j^{(3)}\right)^2$
1	1	1.000	1.000	1.000	1.000	1.000	1.000
2	1	0.759	0.576	– 0.804	0.646	– 2.462	6.061
3	1	0.336	0.113	– 1.157	1.339	2.580	6.656
	Σ	2.095	1.689	– 0.961	3.027	1.118	13.717
	$B_1 = \frac{2.095}{1.689} = 1.240$		$B_2 = -\frac{0.961}{3.027} = -0.317$		$B_3 = \frac{1.118}{13.717} = 0.082$		

**As all masses are same and there are same masses in both the numerator and the denominator, which cancel out, they are taken as unity in this table for simplicity.*

The values of spectral ground acceleration and spectral ground displacement are then determined from Fig. 3.38, considering 5 per cent damping in each mode.

Mode	T_r	$S_a^{(r)}$ mm/s²	$S_d^{(r)}$ mm
1	1.227	1150	47.50
2	0.467	1650	9.16
3	0.336	1900	5.44

Applying Eq. (7.53), the relative displacement with respect to the ground for each storey is determined.

Mass	$z_i^{(1)} = \phi_i^{(1)} B_1 S_d^{(1)}$	$z_i^{(2)} = \phi_i^{(2)} B_2 S_d^{(2)}$	$z_i^{(3)} = \phi_i^{(3)} B_3 S_d^{(3)}$	z_i
1	1.000 × 1.240 ×47.5 = 58.9	1.000 × (– 0.317) × 9.16 = 2.904	1.000 × 0.082 × 5.55 = 0.446	58.976
2	0.759 × 1.240 × 47.5 = 44.705	(– 0.804) (– 0.317) 9.16 = 2.3346	(– 2.462) (0.082) 5.44 = – 1.098	44.782
3	0.336 × 1.24 × 47.5 = 19.790	(– 1.157) (– 0.317) 9.16 = 3.3596	(2.580) (0.082) 5.44 = 1.15	20.15

The values of z_i given in the last column have been calculated on the basis of root mean square value of each mode.

Let us calculate the storey shear on the basis of Eq. (7.54) in the following Table :

Mode	Top storey		Second storey		1st story	
	$\phi_1^{(r)} - \phi_2^{(r)}$	$V_1^{(r)}$ kN	$\phi_2^{(r)} - \phi_3^{(r)}$	$V_2^{(r)}$ kN	$\phi_3^{(r)}$	$V_3^{(r)}$ kN
1	0.241	51.218	0.423	89.898	0.336	107.112
2	1.804	– 18.89	0.353	– 3.901	– 1.157	18.184
3	– 3.462	– 5.572	– 5.042	– 8.115	2.580	6.229
rms of maximum storey shear (kN)		54.767		90.360		108.930

A look into the tables will immediately reveal, that both in the case of displacements and the storey shear, the first mode always plays the most dominant part.

The base shear = 108.930 kN

Let us check the base shear from Eq. (7.58)

$$V_3^{(1)} = B_1 S_a^{(1)} \left[m_1\phi_1^{(1)} + m_2\phi_2^{(1)} + m_3\phi_3^{(1)}\right]$$

$$= 1.24 \times 1.15 \times [1.00 + 0.759 + 0.336] \times 36000$$

$$= 107.548 \quad \text{kN}$$

$$V_3^{(2)} = B_2 S_a^{(2)} \left[m_1\phi_1^{(2)} + m_2\phi_2^{(2)} + m_3\phi_3^{(2)}\right]$$

$$= -0.317 \times 1.65 \times [1.000 - 0.804 - 1.157] \times 36000$$

$$= 18.095 \quad \text{kN}$$

$$V_3^{(3)} = B_3 S_a^{(3)} \left[m_1\phi_1^{(3)} + m_2\phi_2^{(3)} + m_3\phi_3^{(3)}\right]$$

$$= 0.082 \times 1.9 \times [1.000 - 2.462 + 2.58] \times 36000$$

$= 6.271\,kN$

Thus, the base shear in each mode calculated from Eq. (7.58) agreed exactly with Eq. (7.54) [the difference between them is due to round off errors].

7.9 DIRECT INTEGRATION FOR DETERMINING RESPONSE OF MDF SYSTEMS

So far the forced vibration analysis of MDF systems has been made by the mode superposition method, in that the unknown displacements are expressed in terms of normal coordinates. However, the solution of equations of motion can also be obtained directly. But a closed bound solution for such cases is a very difficult proposition. The problem can be tackled in a more elegant manner, by using numerical techniques. We have already looked into the principles of some of the numerical methods applied to SDF systems. All these methods can be extended to MDF systems. In this section, we limit ourselves only to Newmark's method of direct integration. As mentioned earlier, the method is very popular and can be applied to nonlinear systems as well.

Time *T*, over which one is interested in knowing the response is divided into *n* intervals, each of duration Δt. Knowing the displacement, velocity and acceleration at the *i*th time instant, one can obtain those values at $(i + 1)$th time instant.

Let $\{x\}$, $\{\dot{x}\}$ and $\{\ddot{x}\}$ be the displacement, velocity and acceleration matrices of the masses. Equations (4.20) and (4.21) are extended to MDF systems.

$$\{\dot{x}\}_{i+1} = \{\dot{x}\}_i + \left[(1-\delta)\{\ddot{x}\}_i + \delta\{\ddot{x}\}_{i+1}\right]\Delta t \qquad (7.64a)$$

$$\{x\}_{i+1} = \{x\}_i + \{\dot{x}\}_i\,\Delta t + \left[\left(\frac{1}{2}-\alpha\right)\{\ddot{x}\}_i + \alpha\{\ddot{x}\}_{i+1}\right]\Delta t^2 \qquad (7.64b)$$

Discussion has been made in section 4.2.3 regarding the choice of the values of α and δ. To evaluate $\{x\}$ and $\{\dot{x}\}$ at $(i+1)$th time instant, the information of acceleration at $(i+1)$th time instant is needed. It can be obtained from the equation of motion at $(i+1)$th time instant.

$$[M]\{\ddot{x}\}_{i+1} + [C]\{\dot{x}\}_{i+1} + [K]\{x\}_{i+1} = \{F\}_{i+1} \qquad (7.65)$$

An iterative scheme may be adopted for the solution of necessary quantities at $(i+1)$th time instant from three sets of simultaneous equations given by Eqs. (7.64) to (7.65). They, however, can be explicitly determined on the basis of the solution of equations. The algorithm for Newmark's method as given by Bathe and Wilson [7.8] is appended below.

(1) The initial conditions of the systems are always specified. In most cases, the system starts at rest.

(2) Based on the material, geometry and damping properties of the structure, the stiffness matrix [*K*], mass matrix [*M*] and damping matrix [*C*] may be formed.

(3) Based on the chosen values of α and δ and the time interval Δt, the integration constants may be determined. They are

$$a_o = \frac{1}{\alpha \Delta t^2}; \quad a_1 = \frac{\delta}{\alpha \Delta t}; \quad a_2 = \frac{1}{\alpha \Delta t}; \quad a_3 = \frac{1}{2\alpha} - 1;$$

$$a_4 = \frac{\delta}{\alpha} - 1; \quad a_5 = \frac{\Delta t}{2}\left(\frac{\delta}{\alpha} - 2\right); \quad a_6 = \Delta t(1-\delta); \quad a_7 = \delta \Delta t$$

(4) The effective stiffness matrix may be formed. It is given by

$$[\bar{K}] = [K] + a_o[M] + a_1[C] \quad (7.66)$$

(5) $[\bar{K}]$ matrix is inverted.

For a given problem, steps 1 through 5 are to be carried out once only. The following steps are to be performed for each time interval.

(1) The effective load at $(i+1)$th time instant is given by

$$\{\bar{F}\}_{i+1} = \{F\}_{i+1} + [M]\left(a_o\{x\}_i + a_2\{\dot{x}\}_i + a_3\{\ddot{x}\}_i\right)$$
$$+ [C]\left(a_1\{x\}_i + a_4\{\dot{x}\}_i + a_5\{\ddot{x}\}_i\right) \quad (7.67)$$

(2) Solve for displacements at $(i+1)$th time instant

$$\{x\}_{i+1} = [\bar{K}]^{-1}\{\bar{F}\}_{i+1} \quad (7.68)$$

(3) Calculate accelerations and velocities at $(i+1)$th time instant

$$\{\ddot{x}\}_{i+1} = a_o\left(\{x\}_{i+1} - \{x\}_i\right) - a_2\{\dot{x}\}_i - a_3\{\ddot{x}\}_i \quad (7.69)$$

$$\{\dot{x}\}_{i+1} = \{\dot{x}\}_i + a_6\{\ddot{x}\}_i + a_7\{\ddot{x}\}_{i+1} \quad (7.70)$$

The method has been demonstrated in the following example

Example 7.8

Calculate the displacement response of a 2-degree of freedom system having the following data:

$$[K] = \begin{bmatrix} 72 & -72 \\ -72 & 144 \end{bmatrix}; \quad [M] = \begin{bmatrix} 8 & 0 \\ 0 & 8 \end{bmatrix}$$

$$\{F(t)\} = \begin{Bmatrix} 0 \\ 480t \end{Bmatrix}; \quad \{x\}_o = \begin{Bmatrix} 0 \\ 0 \end{Bmatrix}_i; \quad \{\dot{x}\}_o = \begin{Bmatrix} 0 \\ 0 \end{Bmatrix}$$

$\alpha = \frac{1}{4}$ and $\delta = \frac{1}{2}$

$$\Delta t = 0.2\text{s}$$

Choose

The constants of integration are calculated first. They are:

$$a_o = 100.0, \quad a_1 = 10.0, \quad a_2 = 20.0, \quad a_3 = 1.00$$
$$a_4 = 1.00, \quad a_5 = 0.00, \quad a_6 = 0.1, \quad a_7 = 0.1$$

The effective stiffness matrix is

$$[\bar{K}]=\begin{bmatrix}72 & -72\\ -72 & 144\end{bmatrix}+100\begin{bmatrix}8 & 0\\ 0 & 8\end{bmatrix}=\begin{bmatrix}872 & -72\\ -72 & 944\end{bmatrix}=8\begin{bmatrix}109 & -9\\ -9 & 118\end{bmatrix}$$

$$[\bar{K}]^{-1}=\frac{1}{8\times 12781}\begin{bmatrix}118 & -9\\ -9 & 109\end{bmatrix}$$

For every time interval, the following calculation is to be performed.

$$\{\bar{F}\}_{i+1}=\begin{Bmatrix}0\\ 480t\end{Bmatrix}_{i+1}+\begin{bmatrix}8 & 0\\ 0 & 8\end{bmatrix}(100\{x\}_i+20\{\dot{x}\}_i+\{\ddot{x}\}_i)$$

From Eq. (7.65), it is seen that as $\{x\}_0=\{\dot{x}\}_0=\{F\}_0=\{0\}$

Therefore, $\{\ddot{x}\}_0=\{0\}$

$$\{\bar{F}\}_1=\begin{Bmatrix}0\\ 96\end{Bmatrix}+\begin{Bmatrix}0\\ 0\end{Bmatrix}=\begin{Bmatrix}0\\ 96\end{Bmatrix}$$

From Eq. (7.68), we get

$$\{x\}_1=[\bar{K}]^{-1}(\bar{F}\}_1$$

or

$$\{x\}_1=\frac{1}{8\times 12781}\begin{bmatrix}118 & -9\\ -9 & 109\end{bmatrix}\begin{Bmatrix}0\\ 96\end{Bmatrix}=\begin{Bmatrix}-0.00845\\ 0.10234\end{Bmatrix}$$

From Eqs. (7.69) and (7.70), we get

$$\{\ddot{x}\}_1=100(\{x\}_1-\{x\}_0)-20\{\dot{x}\}_0-1.0\{\ddot{x}\}_0$$

$$=100\begin{Bmatrix}-0.00845\\ 0.10234\end{Bmatrix}-20\begin{Bmatrix}0\\ 0\end{Bmatrix}-1.0\begin{Bmatrix}0\\ 0\end{Bmatrix}=\begin{Bmatrix}-0.845\\ 10.234\end{Bmatrix}$$

$$\{\dot{x}\}_1=\{\dot{x}\}_0+0.1\{x\}_0+0.1\{\ddot{x}\}_1=0.1\begin{Bmatrix}-0.845\\ 10.234\end{Bmatrix}$$

$$=\begin{Bmatrix}-0.0845\\ 1.0234\end{Bmatrix}$$

Let us pass on to the next time interval and repeat the above steps

$$\{\bar{F}\}_2=\begin{Bmatrix}0\\ 192\end{Bmatrix}+\begin{bmatrix}8 & 0\\ 0 & 8\end{bmatrix}(100\{x\}_1+20\{\dot{x}\}_1+\{\ddot{x}\}_1)$$

$$=\begin{Bmatrix}0\\ 192\end{Bmatrix}+\begin{bmatrix}8 & 0\\ 0 & 8\end{bmatrix}\left(100\begin{Bmatrix}-0.00845\\ 0.10234\end{Bmatrix}+20\begin{Bmatrix}-0.0845\\ 1.0234\end{Bmatrix}+\begin{Bmatrix}-0.845\\ 10.234\end{Bmatrix}\right)$$

$$=\begin{Bmatrix}-27.040\\ 519.488\end{Bmatrix}$$

$$\{x\}_2 = \frac{1}{8\times 12781}\begin{bmatrix} 118 & -9 \\ -9 & 109 \end{bmatrix}\begin{Bmatrix} -27.04 \\ 519.488 \end{Bmatrix} = \begin{Bmatrix} -0.0765 \\ 0.5562 \end{Bmatrix}$$

$$\{\ddot{x}\}_2 = 100\left(\begin{Bmatrix} -0.0765 \\ 0.5562 \end{Bmatrix} - \begin{Bmatrix} -0.00845 \\ 0.10234 \end{Bmatrix}\right) - 20\begin{Bmatrix} -0.0845 \\ 1.0234 \end{Bmatrix}$$

$$-1.0\begin{Bmatrix} -0.845 \\ 10.234 \end{Bmatrix} = \begin{Bmatrix} -4.274 \\ 14.681 \end{Bmatrix}$$

$$\{\dot{x}\}_2 = \begin{Bmatrix} -0.0845 \\ 1.0234 \end{Bmatrix} + 0.1\begin{Bmatrix} -0.00845 \\ 0.10234 \end{Bmatrix} + 0.1\begin{Bmatrix} -4.274 \\ 14.681 \end{Bmatrix}$$

$$= \begin{Bmatrix} -0.5127 \\ 2.5018 \end{Bmatrix}$$

The above procedure can thus be repeated for other time steps.

Instead of working directly on the coupled equations of the MDF system, the equations may be decoupled by using normal coordinates. The equation in each mode can then be solved by Newmark's method. The final result of the displacements for such cases will be obtained by superimposing the effects of all the modes.

7.10 COMPLEX MATRIX INVERSION METHOD FOR FORCED VIBRATION ANALYSIS OF MDF SYSTEMS

The external force acting on a system is of the following form:

$$\{F(t)\} = \{\bar{F}\}e^{i\omega t} \tag{7.71}$$

The equations of motion for MDF system with damping then becomes

$$[M]\{\ddot{x}\} + [C]\{\dot{x}\} + [K]\{x\} = \{\bar{F}\}e^{i\omega t} \tag{7.72}$$

The steady-state solution of Eq. (7.72) is

$$\{x\} = \{X\}e^{i\omega t} \tag{7.73}$$

Therefore

$$\{\dot{x}\} = i\omega\{X\}e^{i\omega t} = i\omega\{x\} \tag{7.74}$$

and

$$\{\ddot{x}\} = -\omega^2\{x\} \tag{7.75}$$

Combining Eqs. (7.72), to (7.75). we get

$$\left(-\omega^2[M] + i\omega[C] + [K]\{x\}\right) = \{F\}e^{i\omega t} \tag{7.76}$$

Equation (7.76) can be written as

$$[A]\{x\} = \{F\}e^{i\omega t} \tag{7.77}$$

where matrix $[A]$ is a complex matrix of order $n \times n$.

Let $$[A] = [A'] + i\,[A''] \tag{7.78}$$

where $[A']$ and $[A'']$ are real square matrices of same order as that of $[A]$.

In the present case

$$[A'] = \text{Real}\,[A] = [K] - \omega^2 [M] \tag{7.79}$$

$$[A''] = \text{Im}[A] = \omega [C] \tag{7.80}$$

Solution of Eq. (7.77) is

$$\{x\} = [A]^{-1}\{\bar{F}\}e^{i\omega t} = [B]\{\bar{F}\}e^{i\omega t} \tag{7.81}$$

$[B]$ is also a complex matrix of order $n \times n$ and can be split into real and imaginary parts

$$[B] = [B'] + i[B''] \tag{7.82}$$

where $$[B'] = ([A''] + [A'][A'']^{-1}[A']^{-1}[A'][A''])^{-1} \tag{7.83}$$

and $$[B''] = [A'']^{-1}[A'][B'] - [A'']^{-1} \tag{7.84}$$

The inversion of complex matrix has been obtained by operating on the real parts.

7.11 FREQUENCY DOMAIN ANALYSIS OF MDF SYSTEMS BY MODAL SUPERPOSITION FOR HARMONIC LOADS

The equation of motion of a viscously damped MDF system is given by

$$[M]\{\ddot{x}\} + [C]\{\dot{x}\} + [K]\{x\} = \{F(t)\} \tag{7.85}$$

In mode superposition method we assume normal mode solution as usual [Eq. (6.26)]

$$\{x\} = [\Phi]\{\xi(t)\} \tag{7.86}$$

Following steps given in Section 7.1, we get n uncoupled equations. The rth equation is given by

$$\ddot{\xi}_r + 2p_r\,\zeta_r\,\dot{\xi}_r + p_r^2\,\xi_r = \frac{\bar{f}_r}{\bar{m}_r} \tag{7.87}$$

with the same notations as adopted in Chapter 6.

It may be noted that if the damping ratio for each eigenfrequency is known, then it is not essential to establish damping matrix $[C]$. It is also not necessary to generate $[K]$ or $[\Phi]^T[K][\Phi]$, as p_r is obtained from the eigenvalue solution of the MDF system. It has been shown in Section 6.6 that $[\Phi]^T[M][\Phi]$ can be reduced to a identity matrix. As such $\bar{m}_r = 1$, obtained by adjustment of $[\Phi]$ matrix. However, $\bar{f}_r$ quantities are to be calculated.

In the frequency response method, the solution is supposed to be complex and so is the load. The load is therefore extended to a complex load vector to obtain a simple solution of the differential equation. The dynamic response of the structure will then be given by the real part of the solution.

The harmonic load in d.o.f. r is $F_r(t)$ which is

$$F_r(t) = \bar{F}_r \cos(\omega t + \alpha_r) \tag{7.88}$$

To facilitate the solution of the differential equation, the load vector is expanded into a complex quantity :

$$F_r(t) = \bar{F}_r [\cos(\omega t + \alpha_r) + i \sin(\omega t + \alpha_r)] \tag{7.89}$$

which is expressed as

$$F_r(t) = \bar{F}_r e^{i(\omega t + \alpha_r)} \tag{7.90}$$

Eq. (7.90) can be expressed as follows

$$F_r(t) = \left(\bar{F}_R^{(r)} + i \bar{F}_I^{(r)}\right) e^{i\omega t} \tag{7.91}$$

where $\bar{F}_R^{(r)}$ is the real part and $\bar{F}_I^{(r)}$ is the imaginary part.

Therefore, from Eqs. (7.6) and (7.91), we get

$$\bar{f}_r = \left[\sum_{j=1}^{n} \phi_j^{(r)} F_r(t) \right]$$

or,

$$\bar{f}_r = \left[\sum_{j=1}^{n} \phi_j^{(r)} \bar{F}_R^{(r)} + i \sum_{j=1}^{n} \phi_j^{(r)} \bar{F}_I^{(r)} \right] e^{i\omega t} \tag{7.92}$$

In Eq. (7.92), we indicate the real terms by $\bar{f}_R^{(r)}$ and imaginary terms by $\bar{f}_I^{(r)}$ and the resultant complex vector of Eq. (7.92) is given by

$$\bar{f}_r = \left(\bar{f}_R^{(r)} + i \bar{f}_I^{(r)}\right) e^{i\omega t} = \bar{f}_r^* e^{i\omega t} \tag{7.93}$$

The solution of equation (7.87) is assumed to be of the following form

$$\xi_r(t) = \bar{\xi}_r e^{i\omega t} = \left(\bar{\xi}_R^{(r)} + i \bar{\xi}_I^{(r)}\right) e^{i\omega t} \tag{7.94}$$

Therefore,

$$\dot{\xi}_r(t) = i\omega \bar{\xi}_r e^{i\omega t} \tag{7.95}$$

$$\ddot{\xi}_r(t) = -\omega^2 \bar{\xi}_r e^{i\omega t} \tag{7.96}$$

Combining Eqs. (7.94), (7.95), (7.96) and (7.87), yields

$$-\omega^2 \bar{\xi}_r + 2i\omega p_r \zeta_r \bar{\xi}_r + p_r^2 \bar{\xi}_r = \bar{f}_r^*$$

or,

$$\left(-\omega^2 + 2i\omega p_r \zeta_r + p_r^2\right) \bar{\xi}_r = \bar{f}_r^* \tag{7.97}$$

Eq. (7.97) combining with Eq. (7.94), yields

$$\left(-\omega^2 + 2i\omega p_r \zeta_r + p_r^2\right)\left(\bar{\xi}_R^{(r)} + i\bar{\xi}_I^{(r)}\right) = \bar{f}_R^{(r)} + i\,\bar{f}_I^{(r)} \tag{7.98}$$

Equating the real parts and imaginary parts separately in Eq. (7.98), yields the following two equations

$$\left(-\omega^2 + p_r^2\right)\bar{\xi}_R^{(r)} - 2\omega\, p_r \zeta_r \bar{\xi}_I^{(r)} = \bar{f}_R^{(r)} \tag{7.99}$$

$$2\omega\, p_r \zeta_r \bar{\xi}_R^{(r)} + \left(-\omega^2 + p_r^2\right)\bar{\xi}_I^{(r)} = \bar{f}_I^{(r)} \tag{7.100}$$

From the solution of Eqs. (7.99) and (7.100), real and imaginary parts of the modal displacement amplitude are obtained. They are

$$\bar{\xi}_R^{(r)} = \frac{\left(-\omega^2 + p_r^2\right)\bar{f}_R^{(r)} + 2\omega\, p_r \zeta_r \bar{f}_{\mathrm{I}}^{(r)}}{\left(-\omega^2 + p_r^2\right)^2 + \left(2\omega\, p_r \zeta_r\right)^2} \tag{7.101a}$$

$$\bar{\xi}_{\mathrm{I}}^{(r)} = \frac{\left(-\omega^2 + p_r^2\right)^2 \bar{f}_R^{(r)} - 2\omega\, p_r \zeta_r \bar{f}_{\mathrm{I}}^{(r)}}{\left(-\omega^2 + p_r^2\right)^2 + \left(2\omega\, p_r \zeta_r\right)^2} \tag{7.101b}$$

Substituting Eq. (7.94) into Eq. (7.86), we get a complex vector of displacement $\{x\}$. It is to be noted that the solution of the displacements is the real part of the time dependent solution. From Eq. (7.9), we find

$$x_r(t) = \sum_{r=1}^{n} \phi_i^{(r)} \xi_r \tag{7.102}$$

Substitution of ξ_r from Eq. (7.94) into Eq. (7.102), yields

$$x_r(t) = \sum_{r=1}^{n} \phi_i^{(r)}\left(\bar{\xi}_R^{(r)} + i\,\bar{\xi}_I^{(r)}\right)e^{i\omega t} \tag{7.103}$$

or,

$$x_r(t) = \left(x_R^{*(r)} + i\,x_{\mathrm{I}}^{*(r)}\right)e^{i\omega t} = x_r^{*(r)}\,e^{i\omega t} \tag{7.104}$$

where

$$x_R^{(r)} = \sum_{r=1}^{n} \phi_i^{(r)} \bar{\xi}_R^{(r)}, \quad x_I^{(r)} = \sum_{r=1}^{n} \phi_i^{(r)} \bar{\xi}_I^{(r)} \tag{7.105}$$

Eq. (7.104) is written as

$$\begin{aligned} x_r(t) &= x_0^{*(r)}\left(\cos\theta_r + i\sin\theta_r\right)e^{i\omega t} \\ &= x_0^{*(r)}\,e^{i(\omega t + \theta_r)} \\ &= x_0^{*(r)}\left[\cos(\omega t + \theta_r) + i\sin(\omega t + \theta_r)\right] \end{aligned} \tag{7.106}$$

The solution of $x_r(t)$ is the real part of Eq. (7.106), that is,

$$x_r(t) = x_0^{*(r)}\cos(\omega t + \theta_r) \tag{7.107}$$

7.12 FREQUENCY DOMAIN ANALYSIS OF DIRECT FREQUENCY RESPONSE METHOD

The equation of motion of the MDF system is

$$[M]\{\ddot{x}\}+[C]\{\dot{x}\}+[K]\{x\}=\{F(t)\} \tag{7.108}$$

The harmonic load in d.o.f '*r*' is given by

$$F_r=\left(F_R^{(r)}+i\,F_I^{(r)}\right)e^{i\omega t}=F_r^*\,e^{i\omega t}$$

$$=F_0^{(r)}\left(\cos\alpha_r+i\sin\alpha_r\right)e^{i\omega t}$$

$$=F_0^{(r)}\,e^{i(\omega t+\alpha_r)}$$

or,

$$F_r=F_0^{(r)}\left[\cos(\omega t+\alpha_r)+i\sin(\omega t+\alpha_r)\right] \tag{7.109}$$

Now,

$$\{F(t)\}=\{F^*\}e^{i\omega t} \tag{7.110}$$

The solution of the differential equation has a steady state solution, which is a complex number. The complex solution vector for the differential equation is [from Eq. (7.110)],

$$\{x(t)\}=\{x^*\}e^{i\omega t} \tag{7.111}$$

The dynamic response of the structure in d.o.f '*r*' will now be real part of the above solution given by eqn (7.112).

$$x_r(t)=x_0^{*(r)}\cos(\omega t+\theta_r) \tag{7.107}$$

Differentiating $x_r(t)$ once and twice with repeat to *t*, yields [from Eq. (7.111)].

$$\dot{x}_r(t)=i\omega\,x^{*(r)}\,e^{i\omega t}\,(=i\omega\,x_r(t)) \tag{7.112 a}$$

$$\ddot{x}_r(t)=-\omega^2\,x^{*(r)}\,e^{i\omega t}\left(=-\omega^2\,x_r(t)\right) \tag{7.112 b}$$

Substituting Eqs. (7.110), (7.112 a) and (7.112 b) into Eq. (7.108), yields

$$\left(-p_r^2\,[\mathrm{M}]+i\,p_r\,[C]+[\mathrm{K}]\right)\{x^*\}=\{F^*\} \tag{7.113}$$

Let

$$[H]=-p_r^2\,[M]+i\,p_r\,[C]+[\mathrm{K}] \tag{7.114}$$

Therefore, from Eqs (7.113) and (7.114), we can write

$$[H]\{x^*\}=\{F^*\} \tag{7.115}$$

or,

$$\{x^*\}=[H]^{-1}\{F^*\} \tag{7.116}$$

Splitting the load into real and imaginary parts gives (Eq. 7.109)

$$\{x^*\}=[H]^{-1}\{F_R^{(r)}\}+i[H]^{-1}\{F_I^{(r)}\} \tag{7.117}$$

If we assume that the real part of the load vector has the solution, then

$$\left\{x_R^{*(r)}\right\}^R + i\left\{x_I^{*(r)}\right\}^R = [H]^{-1}\left\{F_R^{(r)}\right\} \tag{7.118}$$

and imaginary part of the load vector has the solution,

$$\left\{x_R^{*(r)}\right\}^I + i\left\{x_I^{*(r)}\right\}^I = [H]^{-1}\left\{F_I^{(r)}\right\} \tag{7.119}$$

Combining Eqs. (7.117), (7.118) and (7.119) indicates that the solution for a complex load vector may be found from a combination of two separate solutions of the complex equation with the real part and imaginary part of the load vector as real right hand sides of the complex equation

$$\left\{x^*\right\} = \left\{x_R^{(r)}\right\}^R + i\left\{x_I^{(r)}\right\}^R + i\left[\left\{x_R^{(r)}\right\}^I + \left\{x_I^{(r)}\right\}^I\right]$$

$$= \left\{x_R^{(r)}\right\}^R - \left\{x_I^{(r)}\right\}^I + \left(i\left\{x_R^{(r)}\right\}^I + \left\{x_I^{(r)}\right\}^R\right) \tag{7.120}$$

References

7.1 R.R. Craig, Jr, Structural Dynamics, John Wiley & Sons, 1981

7.2 R.W. Clough and J. Penzien, Dynamics of Structures, 2nd Edition McGraw Hill, New York, 1993.

7.3 A.K. Chopra, Dynamics of Structures, Practice-Hall of India, New Delhi, 1995

7.4 L. Meirovitch , Elements of Vibration Analysis, McGraw Hill, New York, 1975

7.5 G.L. Rogers, Dynamics of Framed Structures, John Wiley, 1959

7.6 J.H. Biggs, Introduction to Structural Dynamics, McGraw-Hill, New York, 1964

7.7 D.G. Fertis, Dynamics and Vibrations of Structures, John Wiley and Sons, New York, 1973

7.8 K.J. Bathe, Finite Element Procedures in Engineering Analysis, Prentice Hall of India Ltd., 1996.

EXERCISE 7

7.1 An unsymmetrical frame is shown in the figure. The members *AB* and *DC* are constrained to move in the vertical direction only. E is the modulus of elasticity. *L* is the length of *AB* and *DC*, θ is the rotation of *BC* about *G*, *m* is the mass of *BC* and *J* is the moment of inertia of *BC* about the horizontal axis through *G*. Determine the natural frequencies of the system.

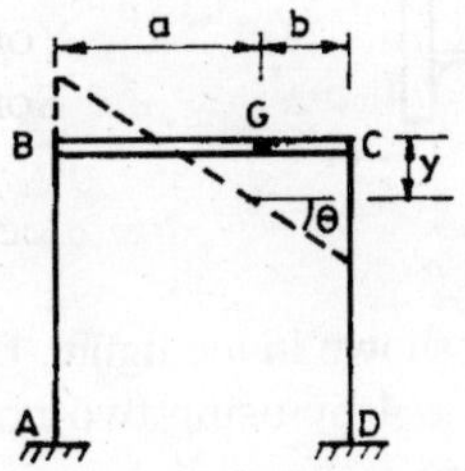

Prob. 7.1

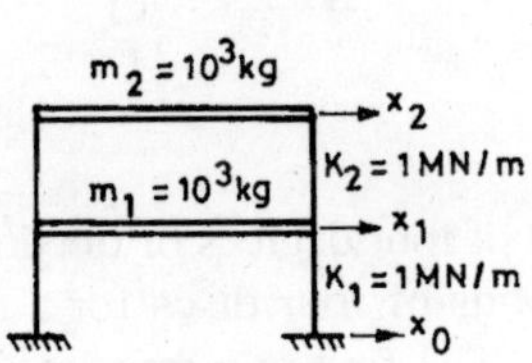

Prob. 7.3

7.2 Determine the solution of the forced vibration problem shown in the figure. Initial conditions are at $t=0,\ x_1=\dot{x}_1=x_2=\dot{x}_2=0$

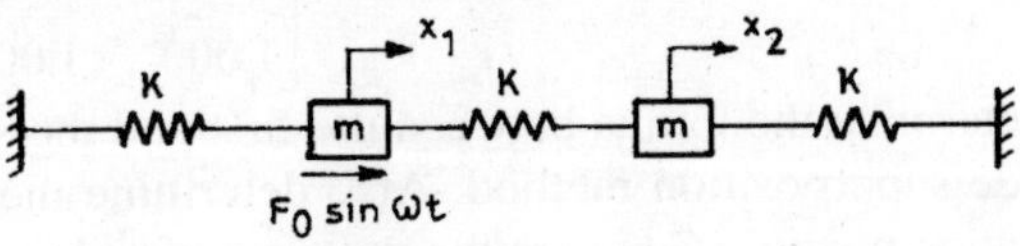

Prob. 7.2

7.3 The base of the frame shown in the figure is subjected to a horizontal displacement

$$\left.\begin{array}{ll} x_o = 0.1\sin \pi t / t_d\ (m) & 0 \le t \le t_d \\ x_o = 0 & t \ge t_d \end{array}\right\}$$

Neglecting damping and assuming t_d = 0.1s, determine the horizontal displacements of the masses. What are their maximum values?

7.4 If in Prob 7.3, instead of the base motion, if the mass m_1 is subjected to a horizontal force *F*(*t*) = 20 kN, determine the displacements of masses m_1 and m_2.

7.5 Calculate the displacement response by Newmark's method of a two-degree freedom system with the following data:

$$[M]=\begin{bmatrix}2 & 0\\ 0 & 1\end{bmatrix},\quad [K]=\begin{bmatrix}6 & -2\\ -2 & 4\end{bmatrix},\quad \{F(t)\}=\begin{Bmatrix}0\\ 10\end{Bmatrix}$$

$$\alpha=0.25 \text{ and } \delta=0.5,\quad \Delta t = 0.1\text{s}$$

7.6 Calculate the displacement response of system, whose data are given below by Newmark's method. Use $\alpha=1/6$ and $\delta=1/2\ (\Delta t=0.2\text{s})$

$$[K]=10\begin{bmatrix}12 & -4\\ -4 & 1\end{bmatrix} \quad [M]=\begin{bmatrix}8 & 0\\ 0 & 4\end{bmatrix}$$

$$\{x\}_0=\{\dot{x}\}_0=\{0\} \qquad \{F(t)\}=\begin{Bmatrix}0\\ 10\end{Bmatrix}$$

Calculation should be carried out up to five time steps

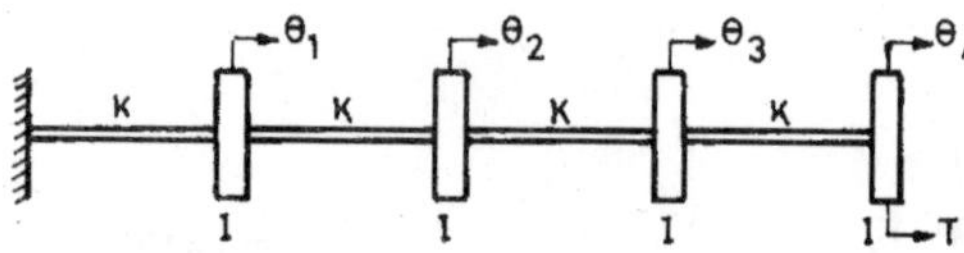

Prob. 7.7

7.7 The discs of four degrees of freedom torsional problem is shown in the figure. Determine the rotations of four discs for a torque T applied to disc 4 by using two normal co-ordinates. The first two frequencies and mode shapes are

$$\{p^2\}=\begin{Bmatrix}0.1530\\ 1.502\end{Bmatrix}\frac{K}{I} \quad \text{and} \quad [\Phi]=\begin{bmatrix}0.406 & -0.475\\ 0.707 & -0.467\\ 0.904 & 0.024\\ 1.000 & 1.000\end{bmatrix}$$

Also determine the torque between the first and the second discs. Solve the problem by mode superposition method. Also determine the necessary values using mode-acceleration method. Compare the values obtained from both the methods.

7.8 Figure shows a structural steel beam with $E = 205 \times 109$ N/m^2, $I = 400$ cm^4, $L = 400$ cm and $m = 10$ kg/m. Determine the displacement response of the system to a suddenly applied force of 400kN applied at the left mass.

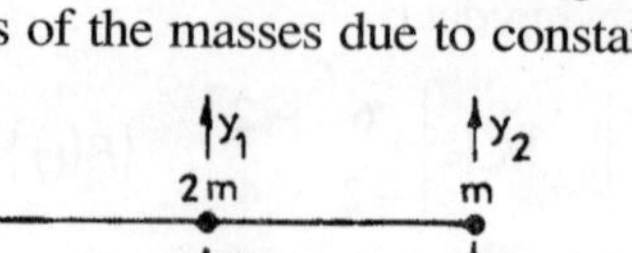

Prob. 7.8

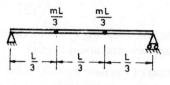

7.9 Lumped mass representation of a uniform beam having flexural rigidity El is shown. Determine the displacements of the masses due to constant force P acting on them.

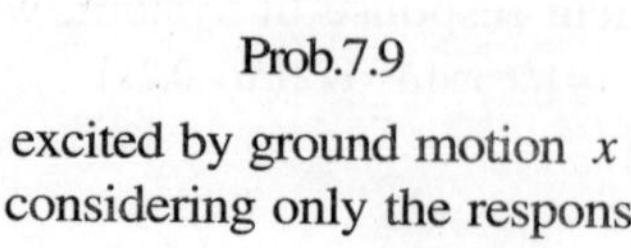

Prob.7.9

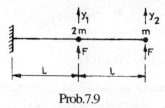

7.10 An idealised building is excited by ground motion $x = X \cos \omega t$. Find the steady state shear force at the base, considering only the response in the fundamental mode.

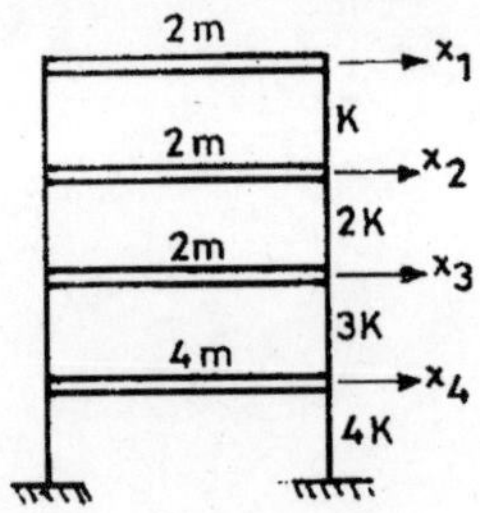

Prob. 7.10

7.11 A reinforced concrete chimney idealised as lumped mass cantilever is subjected at the top level to a step force $F(t)$ = 4500 kN with m = 7000 kg-sec.^{2}/cm. $EI = 2.305 \times 10^{10}$ kN-m^2. Solve by any numerical technique the equations of motion after transforming them to the first two modes by the lower acceleration method with $\Delta t = 0.1$ s.

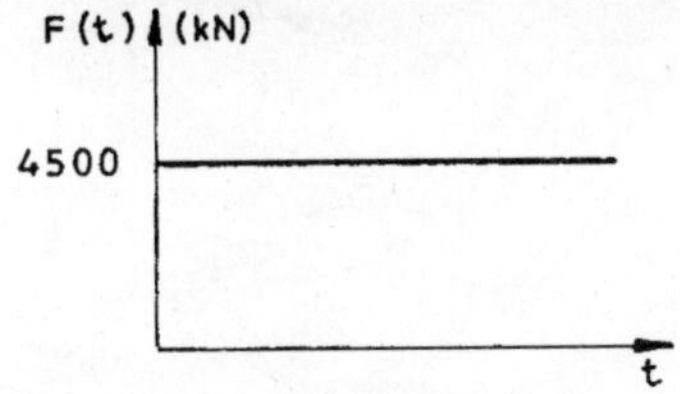

Prob. 7.11

7.12 For the system whose spring-mass-damper representation is shown in the figure, the different quantities have the following values:

$$F_1^{(t)} = \text{P}_1 \cos \omega t, \quad k_1 = 1000, \quad k_2 = 500, \quad m = 1$$
$$c_1 = 0.5, \qquad c_2 = 0.05$$

Determine the response of the masses.

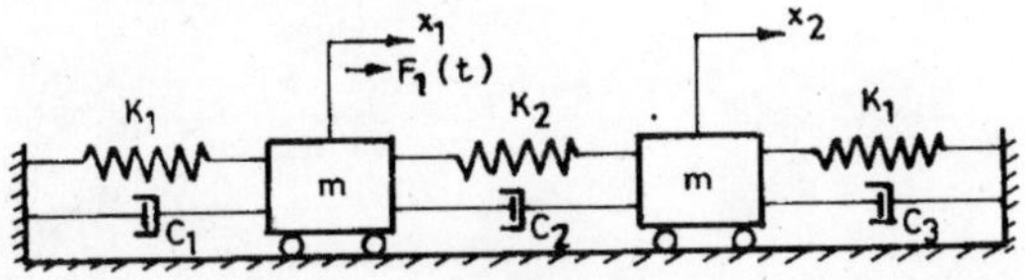

Prob. 7.12

7.13 Find the steady-state response of the system shown in the figure.

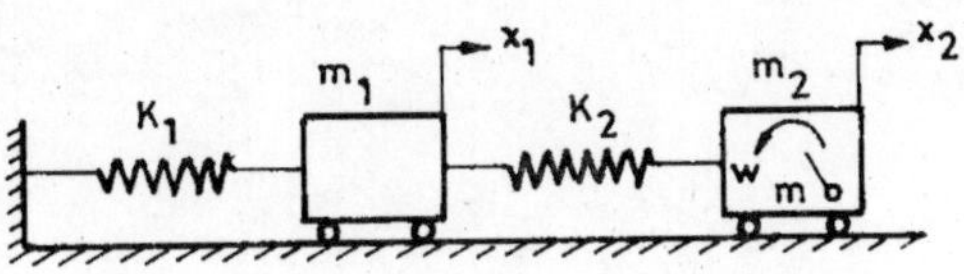

Prob. 7.13

7.14 For the Prob. 7.3, determine the base shear by using spectral diagrams of Fig. 3.35. Assume 2 percent critical damping.

Chapter 8

Free Vibration Analysis of Continuous Systems

8.1 INTRODUCTION

Structures analysed so far have been treated as discrete systems. Structures have been idealised and for convenience of computation, simplifying assumptions have been introduced and as such results obtained can only be treated as approximate. But they are, however, sufficiently accurate for most practical purposes. Increase of structural discretisation will entail an increase of degrees of freedom and this will improve the accuracy of the results.

For all systems, the mass of the members is continuously distributed. As such, specifying the displacement at every point in the system will require infinite number of coordinates. The system in this case is assumed to have infinite degrees of freedom. For such cases the mass is inseparable from the elasticity of the system. Continuous models of vibrating systems are indeed more realistic because structural properties are distributed rather than concentrated at discrete points. The price however to be paid for increased realism is the increased complexity.

The reason as to why the practical structures are reduced to discrete systems, is due to the fact that the analysis of continuous system is much more involved. The argument is similar to that applied to the static analysis, in which for a somewhat redundant structure, solutions obtained by classical methods are very tedious and for such cases, matrix or the finite element analysis is preferred. However, there are certain applications where modelling the system on the basis of distributed parameter may be justified. In this chapter, the free vibration analysis of continuous systems has been discussed. The equation of motion is a partial differential equation for continuous systems. In the treatment of the entire chapter, it has been assumed that the material is elastic, homogeneous and isotropic.

8.2 VIBRATION OF STRINGS

Study of vibration of strings will be helpful in understanding the dynamic behaviour of strings.

The free body diagram for a section of the displaced string is shown in Fig. 8.1. In this figure $T(x)$ is the tension in the string, $f(x, t)$ is the applied force per unit length and $m(x)$ is the mass per unit length. Considering the equilibrium of vertical forces (for small displacements, where $\sin\theta \approx \theta$), we get

$$\left(T(x) + \frac{\partial T(x)}{\partial x}dx\right)\left(\frac{\partial y}{\partial x} + \frac{\partial^2 y}{\partial x^2}dx\right) + f(x, t)\,dx - T(x)\frac{\partial y}{\partial x} = m(x)\,dx\,\frac{\partial^2 y}{\partial t^2} \qquad (8.1)$$

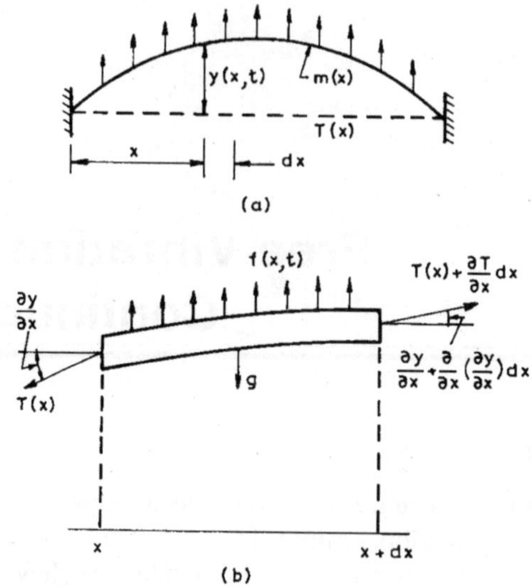

Fig. 8.1 Free-body diagram of the string

Ignoring second order effects, Eq. (8.1) reduces to

$$\frac{\partial}{dx}\left[T(x)\frac{\partial y}{\partial x}\right]+f(x,t)=m(x)\frac{\partial^2 y}{\partial t^2} \tag{8.2}$$

If the string is uniform and the tension is constant, Eq. (8.2) reduces to

$$T(x)\frac{\partial^2 y}{\partial x^2}+f(x,t)=m(x)\frac{\partial^2 y}{\partial t^2} \tag{8.3}$$

If $f(x,t)=0$, we get the free vibration equation

$$T\frac{\partial^2 y}{\partial x^2}=m\frac{\partial^2 y}{\partial t^2} \tag{8.4}$$

This can be written as wave equation.

$$\frac{\partial^2 y}{\partial x^2}=\frac{1}{c^2}\frac{\partial^2 y}{\partial t^2} \tag{8.5}$$

where $c=\sqrt{T/m}$ = velocity of wave propagation.

In order to find natural modes of vibration of a string, the free vibration Eq. (8.5) is to be solved by separation of variables as

$$y(x,t)=Y(x)\,q(t) \tag{8.6}$$

Substituting Eq. (8.6) into Eq. (8.2) and setting $f(x,t)=0$, yields

$$\frac{d}{dx}[T(x)\,Y'(x)\,q(t)] = m(x)\,Y(x)\,\ddot{q}(t) \tag{8.7}$$

or,

$$\frac{1}{m(x)\,Y(x)}\,\frac{d}{dx}[T(x)\,Y'(x)] = \frac{\ddot{q}}{q} = -\,p^2 \tag{8.8}$$

If we carefully look into Eq. (8.8), we will notice that the space dependent variables are on one side of the equation and the time dependent variables are on the other side. Eq. (8.8) can be written as

$$-\frac{d}{dx}[T(x)\,Y'(x)] = p^2 m(x)\,Y(x) \tag{8.9}$$

When the string is uniform and the tension is constant, Eq. (8.9) becomes

$$\frac{d^2Y}{dx^2} + \frac{p^2}{c^2}Y = 0 \tag{8.10}$$

$$\frac{d^2q}{dt^2} + p^2 q = 0 \tag{8.11}$$

The solution of Eqs. (8.10) and (8.11) are

$$Y(x) = A\cos\frac{px}{c} + B\sin\frac{px}{c} \tag{8.12}$$

$$q(t) = C_1\cos pt + C_2\sin pt \tag{8.13}$$

Therefore, the general solution of the equation is

$$y(x,t) = \left(A\cos\frac{px}{c} + B\sin\frac{px}{c}\right)(C_1\cos pt + C_2\sin pt) \tag{8.14}$$

where, A and B depend on the boundary conditions of the problem and C_1 and C_2 on the initial conditions of the problem.

Let us consider the string having both ends fixed. Satisfaction of the boundary conditions requires that the following two equations are solved

$$Y(0) = 0 = C_2 \tag{8.15a}$$

$$Y(L) = 0 = B\sin\frac{p}{c}L \tag{8.15b}$$

Since B cannot be zero for a non-trivial solution, we have

$$\sin\frac{pL}{c} = 0 \tag{8.16}$$

The solution of Eq. (8.16) is

$$\frac{p_nL}{c} = n\pi \qquad n = 1, 2, \ldots\ldots \tag{8.17}$$

Therefore,

$$p_n = n\pi \sqrt{\frac{T}{mL^2}} \qquad n = 1, 2 \ldots\ldots \tag{8.18}$$

The mode shape is given by

$$Y_n(x) = B_n \sin \frac{n\pi x}{L} \tag{8.19}$$

Therefore, the general solution of Eq. (8.5) is

$$y(x, t) = \sum_{n=1}^{\infty} y_n(x, t)$$

$$= \sum_{n=1}^{\infty} B_n \sin \frac{n\pi x}{L} \left[C_{1n} \cos \frac{nc\pi t}{L} + C_{2n} \frac{nc\pi t}{L} \right] \tag{8.20}$$

The mode shapes of the string are shown in Fig. 8.2.

8.2.1 Wave Propagation Solution

For structures having infinite or semi-infinite dimensions, the solution based on the principles of wave propagation are particularly useful. Finite structures can also be studied using these principles. Modelling of ground vibration in earthquake engineering forms a typical example. [8.1, 8.2]

The classical wave equation governs the case of a string undergoing small transverse vibration having constant tension T and a constant mass per unit length m and is given by

$$\frac{\partial^2 y}{\partial x^2} = \frac{1}{c^2} \frac{\partial^2 y}{\partial t^2} \tag{8.21}$$

where $c = \sqrt{T/m}$

The general solution is based on the assumption that the response $y(x, t)$ equals the sum of two different waves travelling in opposite directions,

$$y(x, t) = F_1(x - ct) + F_2(x + ct) \tag{8.22}$$

where F_1 and F_2 are arbitrary functions that can be differentiated twice with respect to x and t.

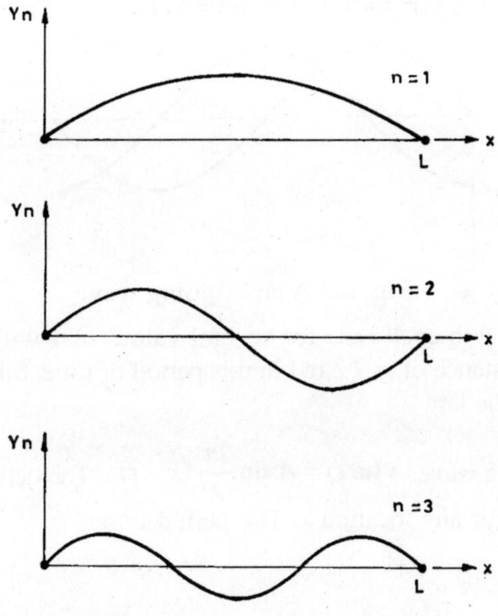

Fig. 8.2 Mode shapes of a string

The following derivatives are obtained

$$\frac{\partial^2 y}{\partial t^2} = c^2 \frac{\partial F_1}{\partial t^2} + c^2 \frac{\partial^2 F_2}{\partial t^2} \tag{8.23}$$

$$\frac{\partial^2 y}{\partial x^2} = \frac{\partial^2 F_1}{\partial x^2} + \frac{\partial^2 F_2}{\partial x^2} \tag{8.24}$$

Substituting the above values in Eq. (8.21), yields

$$\frac{\partial^2 F_1}{\partial x^2} + \frac{\partial^2 F_2}{\partial x^2} = \frac{1}{c^2}\left(c^2 \frac{\partial^2 F_1}{\partial t^2} + c^2 \frac{\partial^2 F_2}{\partial t^2} \right) \tag{8.25}$$

to find that it is satisfied for any two different functions of $x \mp ct$.

In order to demonstrate this, a harmonic solution is assumed

$$y(x,t) = A \sin \frac{2\pi}{\lambda}(x - ct) + B \cos \frac{2\pi}{\lambda}(x + ct) \tag{8.26}$$

Eq. (8.26) satisfies the wave equation given by Eq. (8.21). In order to understand the terms of Eq. (8.26), $y(x,t) = \sin(x - t)$ is considered at two time instances, $t = 0$ sec and $t = \frac{\pi}{2}$ sec.

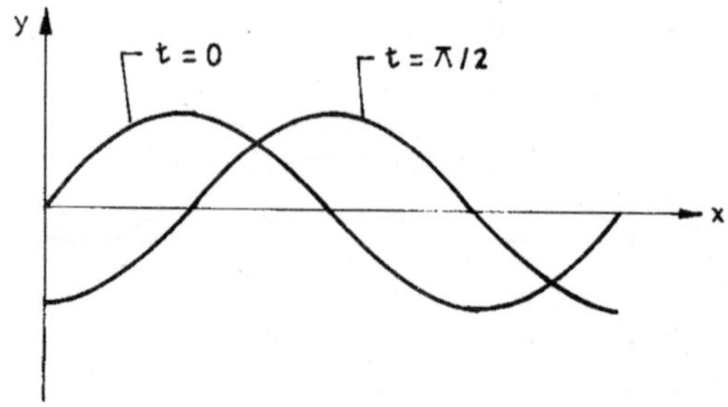

Fig. 8.3 A right-going wave

Picking up the value of y in each case for several values of x will show that the wave has travelled to the right a distance of $\pi/2$ rad in this period of time. Similarly, it can be seen that $\sin(x + t)$ will travel to the left.

Consider a right-going wave, $y(x, t) = A \sin \frac{2\pi}{\lambda}(x - ct)$. The definition of period is time for the value of y to repeat for any location x. The period is

$$\tau = \frac{2\pi}{p} \tag{8.27}$$

For the right-going wave

$$p = \frac{2\pi c}{\lambda} \tag{8.28}$$

or,
$$\tau = \frac{\lambda}{c} \tag{8.29}$$

or,
$$\lambda = \tau c \tag{8.30}$$

The number of waves per unit length, known as wave number d, equals the inverse of the wavelength, $\alpha = 1/\lambda$. With these new definitions, the right-going wave can be written as

$$y(x, t) = A \sin(2\pi\alpha x - pt) \tag{8.31}$$

If two waves of equal amplitude A and frequency p travel in opposite directions, the solution of the wave propagation will be the summation of two waves

$$y(x, t) = A \sin(2\pi\alpha x - pt) + A \cos(2\pi\alpha x + pt) = 2A \sin 2\pi\alpha x \cos pt \tag{8.32}$$

Eq. (8.32) represents a standing wave, that is, it is oscillating rather than propagating.

$$y(0, t) = 0 \tag{8.33}$$

$$y(L, t) = 0 = 2A \sin 2\pi\alpha L \cos pt \tag{8.34}$$

From Eq. (8.34), we get

$$\sin 2\pi\alpha L = 0 = \sin n\pi \qquad n = 1, 2 \,\ldots\ldots$$

or,
$$\alpha = \frac{n}{2L} \tag{8.35}$$

But

$$\alpha = \frac{1}{\lambda} = \frac{p}{2\pi c} \tag{8.36}$$

From Eq. (8.35) and (8.36), we get

$$p_n = \frac{n\pi c}{L} = \frac{n\pi}{L}\sqrt{\frac{T}{m}} \qquad n = 1, 2 \ldots\ldots \tag{8.37}$$

Values of p_n given by Eq. (8.37) are the allowable vibration frequencies of the string, for the given boundary conditions.

The purpose of this section is to show alternative approaches of solution which can sometimes prove to be useful.

8.3 FREE LONGITUDINAL VIBRATION OF A BAR

Let us consider the longitudinal vibration of a slender, straight elastic bar. It is assumed that the cross-sections, which are initially plane and perpendicular to the axis of the bar remain plane and perpendicular to the axis at all stages of the vibratory motion.

Let the bar be uniform and has a cross-sectional area A. E is the modulus of elasticity of the material of the bar, L is its length and ρ is the mass of the material per unit volume.

An element of this bar of length dx is shown in Fig. 8.4. u is the longitudinal displacement at x and $u + \frac{\partial u}{\partial x}dx$ is the displacement at $x + dx$.

The change of length of the element is $\frac{\partial u}{\partial x}dx$ and the strain at x is therefore $\frac{\partial u}{\partial x}$. The strain ε is therefore given by

$$\varepsilon = \frac{\partial u}{\partial x} \tag{8.38}$$

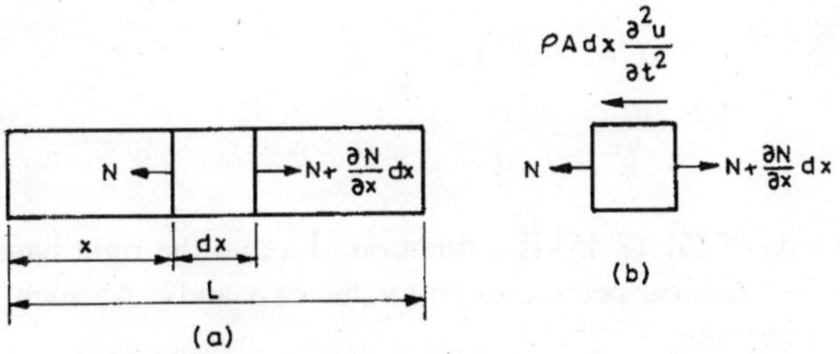

Fig. 8.4 A bar and its freebody diagram

Applying Hooke's law, we can write

$$\frac{N}{AE} = \frac{\partial u}{\partial x} \tag{8.39}$$

Considering the dynamic equilibrium of the element [Fig. 8.4(b)] and applying D'Alembert's principle for the equilibrium of the forces in the x-direction, we get

$$\frac{\partial N}{\partial x}\,dx = \rho A\,dx\,\frac{\partial^2 u}{\partial t^2}$$

or
$$\frac{\partial N}{\partial x} = \rho A\,\frac{\partial^2 u}{\partial t^2} \tag{8.40}$$

Assuming A to be constant and differentiating both sides of Eq. (8.39), we get

$$\frac{1}{AE}\frac{\partial N}{\partial x} = \frac{\partial^2 u}{\partial x^2} \tag{8.41}$$

Combining Eqs. (8.40) and (8.41), we get

$$\frac{\partial^2 u}{\partial t^2} = \frac{E}{\rho}\frac{\partial^2 u}{\partial x^2} \tag{8.42}$$

or
$$\frac{\partial^2 u}{\partial t^2} = a^2\,\frac{\partial^2 u}{\partial x^2} \tag{8.43}$$

where $a = \sqrt{\dfrac{E}{\rho}}$

Let us assume the solution of Eq. (8.43) as

$$u(x,t) = U(x)\,q(t) \tag{8.44}$$

Substituting $u(x,t)$ and its derivatives from Eq. (8.44) into Eq. (8.43), we get

$$U\,\frac{d^2 q}{dt^2} = a^2 q\,\frac{d^2 U}{dx^2}$$

or
$$a^2\,\frac{\dfrac{d^2 U}{dx^2}}{U} = \frac{\dfrac{d^2 q}{dt^2}}{q} \tag{8.45}$$

The left hand side of Eq. (8.45) is a function of x and the right hand side is a function of t and this equation is valid irrespective of any value of x and t. As such, it is a constant which is chosen as $-p^2$. Therefore

$$\frac{\dfrac{d^2 q}{dt^2}}{q} = -\,p^2 \tag{8.46}$$

or
$$\frac{d^2 q}{dt^2} + p^2 q = 0 \tag{8.47}$$

The solution of Eq. (8.47) is

$$q = C_1 \cos pt + C_2 \sin pt \tag{8.48}$$

where C_1 and C_2 are constants. Similarly

$$a^2 \frac{\frac{d^2U}{dx^2}}{U} = -p^2 \tag{8.49}$$

or

$$a^2 \frac{d^2U}{dx^2} + p^2U = 0 \tag{8.50}$$

The solution of Eq. (8.50) is

$$U = A_1 \cos \frac{px}{a} + A_2 \sin \frac{px}{a} \tag{8.51}$$

Therefore, the general solution is

$$u(x, t) = \left(A_1 \cos \frac{px}{a} + A_2 \sin \frac{px}{a} \right) (C_1 \cos pt + C_2 \sin pt) \tag{8.52}$$

where A_1 and A_2 are to be determined from the boundary conditions of the problem and C_1 and C_2 from the initial conditions of the problem. Further, it may be noted that the solution given by Eq. (8.52) is independent of the cross-sectional area of the bar.

There can be two end conditions for the problem. They are:

(a) *Clamped end*

In this case, the axial displacement is restrained at the end. Therefore, at a clamped end

$$U = 0 \tag{8.53}$$

(b) *Free end*

The end being free, no axial force can be developed. As such, at a free end

$$N = 0 \tag{8.54}$$

Using the relation of N from Eq. (8.39), we can write

$$\frac{dU}{dx} = 0 \tag{8.55}$$

8.3.1 Free longitudinal vibration of a bar clamped at $x = 0$ and free at $x = L$

The boundary condition for this particular bar is

at $$x = 0, \quad U = 0 \tag{8.56}$$

and at $$x = L, \quad \frac{dU}{dx} = 0 \tag{8.57}$$

Substituting the values of U and x from Eq. (8.56) into Eq. (8.51), gives

$$A_1 = 0 \tag{8.58}$$

Differentiation with respect to x of both sides of Eq. (8.51) gives

$$\frac{dU}{dx} = \frac{p}{a}\left(-A \sin \frac{px}{a} + B \cos \frac{px}{a}\right) \tag{8.59}$$

Substituting the end condition given in Eq. (8.57) we get

$$A_2 \cos \frac{pL}{a} = 0 \tag{8.60}$$

$\triangle_2$ cannot be zero, otherwise there will be no vibration in the system. Therefore,

$$\cos \frac{pL}{a} = 0 \tag{8.61}$$

Thus $$\frac{pL}{a} = \frac{(2n-1)\pi}{2}, \qquad n = 1, 2, 3 \ldots\ldots$$

The natural frequencies are given by

$$p_n = \frac{(2n-1)\pi}{2L}\sqrt{\frac{E}{\rho}}, \qquad n = 1, 2, 3 \ldots\ldots \tag{8.62}$$

The mode shapes are given by

$$U_n = \sin \frac{(2n-1)\pi x}{2L}, \qquad n = 1, 2, 3 \ldots\ldots \tag{8.63}$$

A few lower mode shapes have been plotted in Fig. 8.5. The complete solution of the equation is

$$u(x, t) = \sum B_n \sin \frac{(2n-1)\pi x}{2L}\left(C_1 \cos \frac{(2n-1)\pi}{2L}\sqrt{\frac{E}{\rho}}\,t + C_2 \sin \frac{(2n-1)\pi}{2L}\sqrt{\frac{E}{\rho}}\,t\right) \tag{8.64}$$

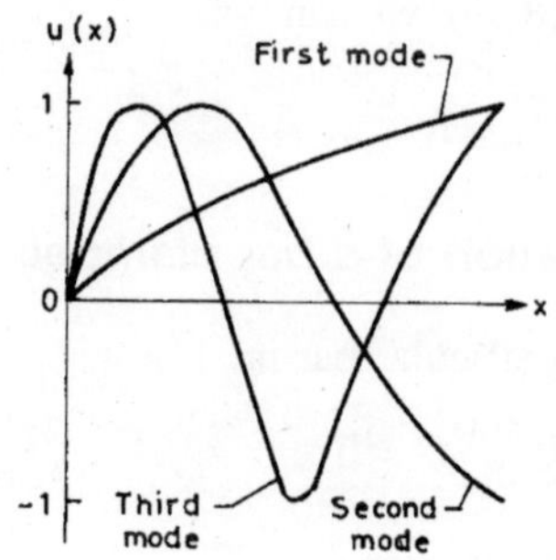

Fig. 8.5 Mode shapes of the bar

8.4 FREE TORSIONAL VIBRATION OF THE SHAFT

The initial derivation will be made by considering the cross-section of the bar to be circular. It is assumed that the cross-section of the shaft during torsional vibrations remain plane and the radii of these cross-sections remain straight. From the knowledge of strength of materials, we know

$$T = GJ \frac{\partial \theta}{\partial x} \tag{8.65}$$

where G is the shear modulus of elasticity,

J is the polar moment of inertia, and

θ is the angle of twist.

If ρ is the mass per unit volume, then applying D'Alembert's principle to the dynamic equilibrium of the elemental length dx [Fig. 8.6]

$$\left(T + \frac{\partial T}{\partial x}\,dx\right) - T = \rho J\,dx\,\frac{\partial^2 \theta}{\partial t^2}$$

or

$$\frac{\partial T}{\partial x} = \rho J \frac{\partial^2 \theta}{\partial t^2} \tag{8.66}$$

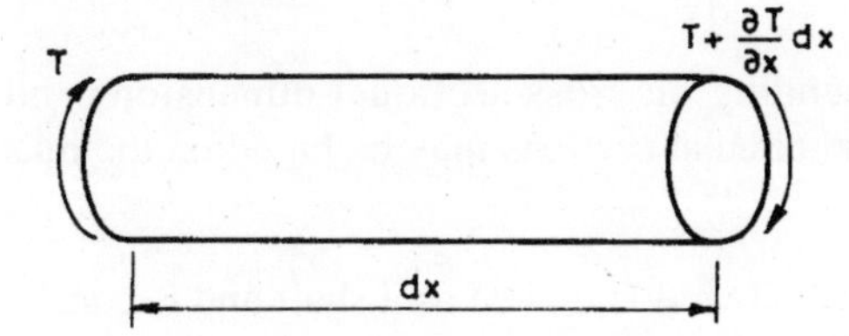

Fig. 8.6 A shaft in torsion

Combining Eqs. (8.65) and (8.66), we get

$$\frac{\partial^2 \theta}{\partial t^2} = a^2 \frac{\partial^2 \theta}{\partial x^2} \tag{8.67}$$

where $a^2 = G/\rho$.

Equation (8.67) is identical to Eq. (8.43) and its solution is therefore equal to

$$\theta(x, t) = \left(A \cos \frac{px}{a} + B \sin \frac{px}{a}\right) C_1 \sin(pt - \alpha) \tag{8.68}$$

At the free end of the bar, $T = 0$ and hence $\frac{\partial \theta}{\partial x} = 0$.

At the clamped end of the bar, $\theta = 0$.

One end of the shaft may be supported by a torsional spring of rotational stiffness k_0. If the torsional spring is provided as the left hand support, then

$$k_0 \theta = GJ \frac{\partial \theta}{\partial x} \tag{8.69}$$

If the torsional spring is provided as the right hand support, then

$$k_0 \theta = - GJ \frac{\partial \theta}{\partial x} \tag{8.70}$$

One end of the shaft may have a disc of polar moment of inertia J_0.

If the disc is placed on the left hand end, then the end condition is

$$J_0 \frac{\partial^2 \theta}{\partial t^2} = GJ \frac{\partial \theta}{\partial x} \tag{8.71}$$

If the disc is placed on the right hand end, then the end condition is

$$J_0 \frac{\partial^2 \theta}{\partial t^2} = - GJ \frac{\partial \theta}{\partial x} \tag{8.72}$$

The natural frequencies and mode shapes can be determined in the same way as indicated in Art. 8.3.1.

In order to treat torsional vibrations of non-circular sections, the torque-twist relation given by Eq. (8.65) is to be modified. It becomes

$$T = \mu GJ \frac{\partial \theta}{\partial x} \tag{8.73}$$

where μ is a factor depending on cross-sectional dimensions and the type of the section. Further derivation for non-circular sections may be based on the relation given by Eq. (8.73).

Example 8.1

A drill pipe of an offshore oil well is treated as a shaft and is considered fixed to the platform at the top. The pipe is 1000 m long. It is terminated at the lower end to a drill bit 10 m long. The dimensions of the drill pipe and the drill bit, which is modelled as a disc (Fig. 8.7) are as follows:

Drill pipe:	outside diameter	= 110 mm.
	inside diameter	= 80 mm.
Drill bit:	outside diameter	= 160 mm.
	inside diameter	= 50 mm.

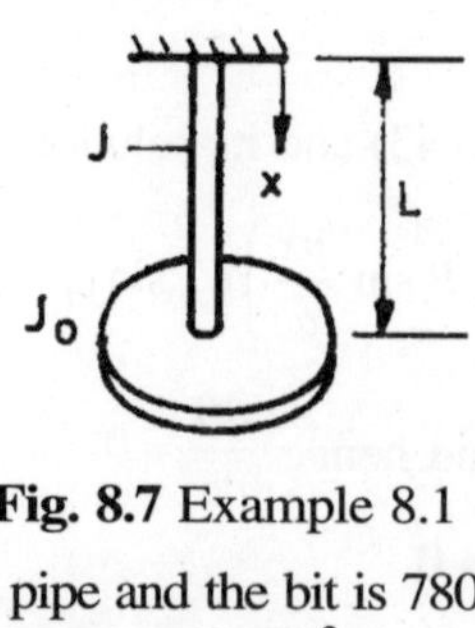

Fig. 8.7 Example 8.1

The density of the material of the pipe and the bit is 7800 kg/m^3. Determine the fundamental natural frequency of the system. $G = 84 \times 1010$ N/m^2.

The general solution for the torsional vibration problem is

$$\theta(x, t) = \left(A \cos \frac{px}{a} + B \sin \frac{px}{a} \right) C_1 \sin(pt - \alpha) \tag{a}$$

where $a = \sqrt{G/\rho}$

The boundary condition for the problem at $x = 0$ is $\theta = 0$. Putting this condition in Eq. (a) gives

$$A = 0 \tag{b}$$

At $x=L$
$$J_0 \frac{\partial^2 \theta}{\partial t^2} = -GJ \frac{\partial \theta}{\partial x} \tag{c}$$

Combining Eqs. (a), (b) and (c), we get

$$-BJ_0 p^2 \sin \frac{pL}{a} C_1 \sin(pt - \alpha) = -GJB \frac{p}{a} \cos \frac{pL}{a} C_1 \sin(pt - \alpha)$$

or
$$\tan \frac{pL}{a} = \frac{GJ}{aJ_0 p} = \frac{GJ}{J_0 p}\sqrt{\frac{\rho}{G}} = \frac{\rho JL}{pJ_0 L}\sqrt{\frac{G}{\rho}} = \frac{I_p a}{J_0 pL} \tag{d}$$

I_p = mass moment of inertia of the pipe.

Therefore
$$\frac{pL}{a} \tan \frac{pL}{a} = \frac{I_p}{J_0} \tag{e}$$

For the data given

$$J = \frac{\pi}{32}(0.11^4 - 0.08^4) = 1.03525 \times 10^{-5} \text{m}^4$$

$$I_p = \rho JL = 7800 \times 1.03525 \times 10^{-5} \times 1000 = 80.75 \text{kgms}^2$$

$$J_0 = 7800 \times \frac{\pi}{32}(0.16^4 - 0.05^5) \times 10 = 4.971 \text{kgms}^2$$

Therefore, equation to be solved is

$$\frac{pL}{a} \tan \frac{pL}{a} = \frac{80.75}{4.971} = 16.244$$

The first root of Eq. (e) is

$$\frac{pL}{a} = 1.48$$

or
$$p = \frac{1.48}{L}\sqrt{\frac{G}{\rho}} = \frac{1.48}{1000}\sqrt{\frac{8.4 \times 10^{10}}{7800}} = 4.85 \text{ rad/s}$$

8.5 FREE FLEXURAL VIBRATION OF BEAMS

Let us consider a beam sufficiently long in comparison to its cross section, so that the shear deformation is ignored. The effect of rotary inertia, which has been discussed later has not been considered in this derivation. The freebody diagram of a beam segment of elemental length *dx* has been shown in Fig. 8.8.

Let A be the cross-sectional area of the beam,

I be the second moment of area of beam cross-section,

ρ be the mass density for the material of the beam,

$w(x, t)$ be the intensity of the external loading, acting on the beam per unit length, and

V, M are the shear force and bending moment at a section, respectively.

Considering the dynamic equilibrium of the system [Fig. 8.8(a)] and applying D'Alembert's principle for the equilibrium of the vertical forces, the equation becomes

$$\left(V + \frac{\partial V}{\partial x}\,dx\right) - V + w\,dx + \rho A\,dx\,\frac{\partial^2 y}{\partial t^2} = 0$$

or

$$\frac{\partial V}{\partial x} = -\,w - \rho A\,\frac{\partial^2 y}{\partial t^2} \tag{8.74}$$

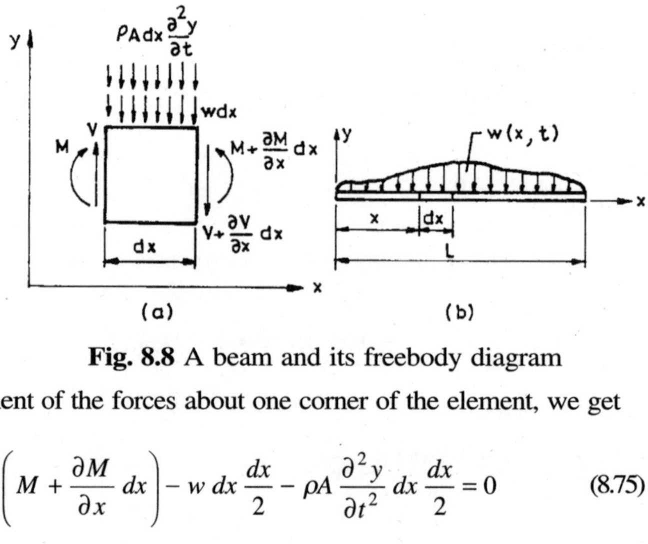

Fig. 8.8 A beam and its freebody diagram

Taking moment of the forces about one corner of the element, we get

$$V\,dx + M - \left(M + \frac{\partial M}{\partial x}\,dx\right) - w\,dx\,\frac{dx}{2} - \rho A\,\frac{\partial^2 y}{\partial t^2}\,dx\,\frac{dx}{2} = 0 \tag{8.75}$$

Neglecting products of small quantities in Eq. (8.75), we get

$$\frac{\partial M}{\partial x} = V \tag{8.76}$$

Again, from our knowledge of the strength of materials, we know that

$$M = EI\,\frac{\partial^2 y}{\partial x^2} \tag{8.77}$$

Combining Eqs. (8.75), (8.76) and (8.77), we get

$$\frac{\partial^2}{\partial x^2}\left(EI\,\frac{\partial^2 y}{\partial x^2}\right) = -\,w - \rho A\,\frac{\partial^2 y}{\partial t^2} \tag{8.78}$$

For free vibration, $w = 0$. Therefore Eq. (8.78) becomes

$$\frac{\partial^2}{\partial x^2}\left(EI\,\frac{\partial^2 y}{\partial x^2}\right) + \rho A\,\frac{\partial^2 y}{\partial t^2} = 0 \tag{8.79}$$

If the beam is uniform, then Eq. (8.79) becomes

$$EI\,\frac{\partial^4 y}{\partial x^4} + \rho A\,\frac{\partial^2 y}{\partial t^2} = 0 \tag{8.80}$$

In free vibration, $y(x, t)$ is a harmonic function of time, so that

$$y(x, t) = Y(x)\,(A_1 \cos pt + A_2 \sin pt) \tag{8.81}$$

where $Y(x)$ is a function of x only and represents the mode shape.
Substituting $y(x, t)$ from Eq. (8.81) into Eq. (8.80) for the uniform beam, we get

$$EI\,\frac{d^4Y}{dx^4} - \rho A p^2 Y = 0 \tag{8.82}$$

or

$$\frac{d^4Y}{dx^4} - \lambda^4 Y = 0 \tag{8.83}$$

where

$$\lambda^4 = \frac{\rho A p^2}{EI} \tag{8.84}$$

The general solution of Eq. (8.84) is

$$Y = C_1 \sin \lambda x + C_2 \cos \lambda x + C_3 \sinh \lambda x + C_4 \cosh \lambda x \tag{8.85}$$

where C_1, C_2, C_3 and C_4 are constants and they depend upon the boundary conditions for the problem.

8.6 FREE FLEXURAL VIBRATION OF THE SIMPLY SUPPORTED BEAM

The beam has hinged supports at both ends. The end conditions of the beam (Fig. 8.9) are

At $$x = 0,\quad Y = 0 \quad\text{and}\quad \frac{d^2Y}{dx^2} = 0 \tag{8.86a}$$

At $$x = L,\quad Y = 0 \quad\text{and}\quad \frac{d^2Y}{dx^2} = 0 \tag{8.86b}$$

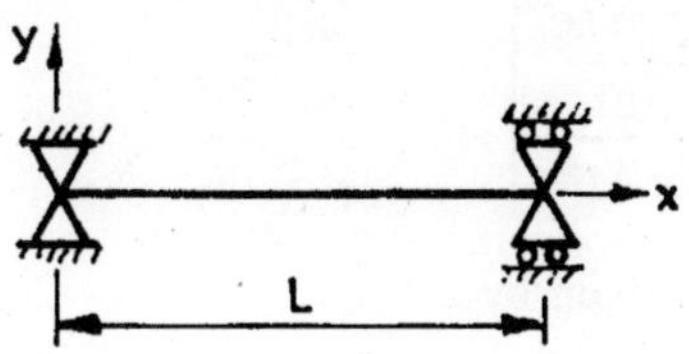

Fig. 8.9 A beam with simple supports

Applying the end conditions given by Eq. (8.86a), we get

$$\left.\begin{aligned} C_2 + C_4 &= 0 \\ -\lambda^2 C_2 + \lambda^2 C_4 &= 0 \end{aligned}\right\} \tag{8.87}$$

Solving Eq. (8.87), we get

$$C_2 = C_4 = 0 \tag{8.88}$$

Substituting the end conditions given by Eq. (8.86b), we get

$$\left.\begin{aligned} C_1 \sin \lambda L + C_3 \sinh \lambda L &= 0 \\ \lambda^2 (-C_1 \sin \lambda L + C_3 \sinh \lambda L) &= 0 \end{aligned}\right\} \tag{8.89}$$

which gives

and

$$\left.\begin{aligned} C_3 \sinh \lambda L &= 0 \\ C_1 \sin \lambda L &= 0 \end{aligned}\right\} \tag{8.90}$$

As λ is not zero, $\sinh \lambda L \neq 0$ for all values of λ. Therefore, from Eq. (8.90)

$$C_3 = 0 \tag{8.91}$$

If we assume $C_1 = 0$ in Eq. (8.89), then there cannot be any vibration of the system. Therefore

$$\sin \lambda L = 0 \tag{8.92}$$

which leads to

$$\lambda L = r\pi, \quad r = 1, 2, 3, ... \tag{8.93}$$

or

$$\lambda = \frac{r\pi}{L} \tag{8.94}$$

Substituting the value of A from Eq. (8.84) to Eq. (8.94), we get

$$p_r = r^2 \pi^2 \sqrt{\frac{EI}{\rho A L^4}}, \quad r = 1, 2,3, ... \tag{8.95}$$

Various natural frequencies of the beam are obtained by substituting the values of r. For example

$$\left.\begin{aligned} p_1 &= \pi^2 \sqrt{\frac{EI}{\rho A L^4}} \\ p_2 &= 4\pi^2 \sqrt{\frac{EI}{\rho A L^4}} \\ p_3 &= 9\pi^2 \sqrt{\frac{EI}{\rho A L^4}} \end{aligned}\right\} \tag{8.96}$$

The mode shape of the beam is given by

$$Y_r = \sin \frac{r\pi x}{L} \tag{8.97}$$

The general solution of the problem as given by Eq. (8.81) is

$$y(x, t) = \sum_{r=1}^{\infty} \sin \frac{r\pi x}{L} (A_r \cos p_r t + B_r \sin p_r t) \tag{8.98}$$

The first four modes of vibration have been shown in Fig. 8.10

8.7 FREE FLEXURAL VIBRATION OF BEAMS WITH OTHER END CONDITIONS

Beams may have a variety of end conditions. Depending on the type of supports at the two ends, the natural frequencies and mode shapes are to be determined.

8.7.1 Uniform Beam Having Both Ends Free

Structures such as aeroplanes and ships, if treated as a beam have both the ends free. The end conditions for this case are

$$\text{At } x = 0, \quad EI \frac{d^2Y}{dx^2} = 0 \text{ and } \frac{d}{dx}\left(EI \frac{d^2Y}{dx^2} \right) = 0 \tag{8.99}$$

$$\text{At } x = L \quad EI \frac{d^2Y}{dx^2} = 0 \text{ and } \frac{d}{dx}\left(EI \frac{d^2Y}{dx^2} \right) = 0 \tag{8.100}$$

The derivatives which are required from the end conditions are obtained from Eq. (8.85) as

$$\frac{d^2Y}{dx^2} = \lambda^2 (- C_1 \sin \lambda x - C_2 \cos \lambda x + C_3 \sinh \lambda x + C_4 \cosh \lambda x) \tag{8.101}$$

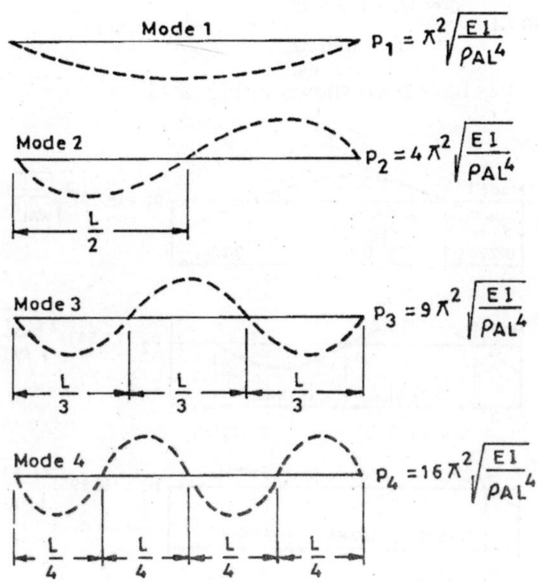

Fig. 8.10 Mode shapes of a uniform simply supported beam

$$\frac{d^3Y}{dx^3} = \lambda^3 (-C_1 \cos \lambda x + C_2 \sin \lambda x + C_3 \cosh \lambda x + C_4 \sinh \lambda x) \tag{8.102}$$

Putting the boundary conditions given by Eq. (8.99) into Eq. (8.101) after considering the beam to be uniform, we get

$$\left.\begin{aligned} -C_2 + C_4 &= 0 \\ -C_1 + C_3 &= 0 \end{aligned}\right\} \tag{8.103}$$

or

$$\left.\begin{aligned} C_2 &= C_4 \\ C_1 &= C_3 \end{aligned}\right\} \tag{8.104}$$

Substituting the end conditions given by Eq. (8.100) into Eq. (8.102), and noting the relation given by Eq. (8.104), we get

$$\left.\begin{aligned} C_2 (-\cos \lambda L + \cosh \lambda L) &= C_1 (\sin \lambda L - \sinh \lambda L) \\ C_2 (\sin \lambda L + \sinh \lambda L) &= C_1 (\cos \lambda L - \cosh \lambda L) \end{aligned}\right\} \tag{8.105}$$

Combining the two equations given above in Eq. (8.105), we get the frequency equation as

$$(\sin \lambda L + \sinh \lambda L)(\sin \lambda L - \sinh \lambda L) = (\cos \lambda L - \cosh \lambda L) \times (-\cos \lambda L + \cosh \lambda L)$$

or

$$\cos \lambda L \cosh \lambda L = 1 \tag{8.106}$$

Equation (8.106) is a transcendental equation. The roots of this equation are

$$\lambda L = 4.730,\ 7.853,\ 10.996,\ 14.137,\ 17.219,\ \ldots \tag{8.107}$$

The mode shape for the free beam is given by

$$Y = \cos \lambda x + \cosh \lambda x + \frac{\cos \lambda L - \cosh \lambda L}{\sin \lambda L + \sinh \lambda L} (\sin \lambda x + \sinh \lambda x) \tag{8.108}$$

The first three mode shapes have been shown in Fig. 8.11

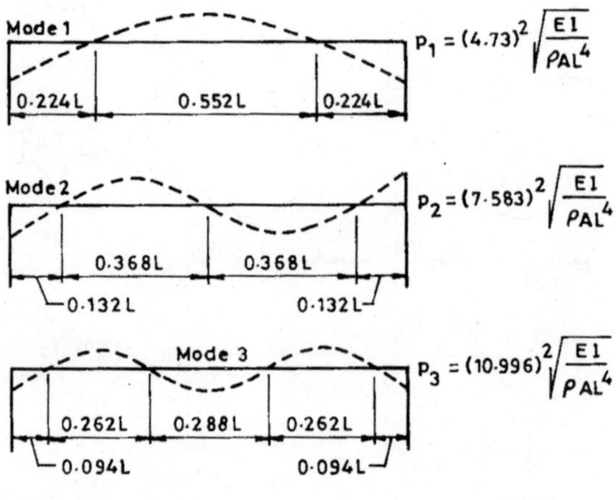

Fig. 8.11 Mode shapes of a free-free beam

8.8 FREE FLEXURAL VIBRATION OF BEAMS WITH GENERAL END CONDITIONS

A beam end may not have only classical boundary conditions, such as hinged support, clamped support or free support. The support at one end may be elastically restrained in translation, or rotation, or both. In addition, the end may have a concentrated mass.

Let the coefficients of elastic restraint in translation and rotation of the left hand support be T_L and R_L, respectively. Also let the coefficients of elastic restraint in translation and rotation of the right hand support be T_R and R_R, respectively.

If the concentrated mass on the left hand support be M_L and that of the right hand support be M_R (Fig. 8.12), then the boundary conditions on the left hand end are

at $$x = 0, \quad EI\left(\frac{\partial^2 y}{\partial x^2}\right) = R_L\left(\frac{\partial y}{\partial x}\right) \tag{8.109}$$

and $$-EI\left(\frac{\partial^3 y}{\partial x^3}\right) = T_L y + M_L\left(\frac{\partial^2 y}{\partial t^2}\right) \tag{8.110}$$

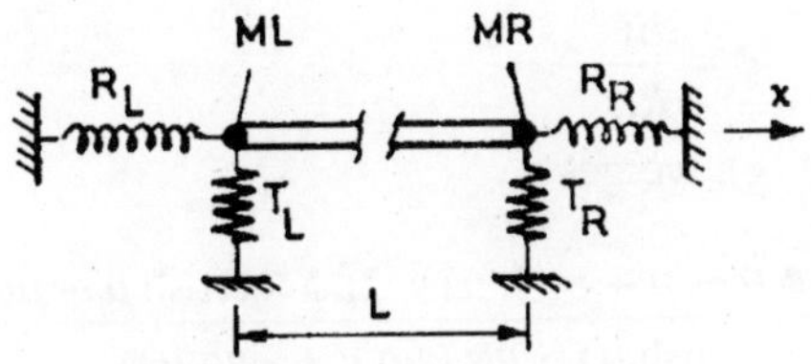

Fig. 8.12 Restrained supports

At $x = L$

$$EI\left(\frac{\partial^2 y}{\partial x^2}\right) = R_R\left(\frac{\partial y}{\partial x}\right) \tag{8.111}$$

and $$-EI\left(\frac{\partial^3 y}{\partial x^3}\right) = -T_R y - M_R\left(\frac{\partial^2 y}{\partial t^2}\right) \tag{8.112}$$

The frequency equation can be derived by substituting the conditions given by Eqs. (8.109) – (8.112) into Eq. (8.85).

Problem 8.2

Determine the natural frequencies and mode shapes for a uniform beam having both ends elastically restrained in rotation, the coefficient of elastic restraint being R.

The end conditions for this problem are

At $$x = 0, \quad Y = 0 \quad \text{and} \quad R\frac{dY}{dx} = EI\frac{d^2Y}{dx^2} \tag{a}$$

and at $$x = L, \quad Y = 0 \quad \text{and} \quad R\frac{dY}{dx} = -EI\frac{d^2Y}{dx^2} \tag{b}$$

Substituting the conditions given by Eq. (a) in Eq. (8.85), we get

$$C_2 + C_4 = 0 \tag{c}$$

$$C_1 + C_3 = \frac{EI\lambda}{R}[-C_2 + C_4] \tag{d}$$

$$C_1 \sin\mu + C_2 \cos\mu + C_3 \sinh\mu + C_4 \cosh\mu = 0 \tag{e}$$

where

$$\mu = \lambda L \tag{f}$$

and

$$R\,[C_1 \cos\mu - C_2 \sin\mu + C_3 \cosh\mu + C_4 \sinh\mu]$$

$$= -\frac{EI\mu}{L}[-C_1 \sin\mu - C_2 \cos\mu + C_3 \sinh\mu + C_4 \cosh\mu] \tag{g}$$

Combining Eqs. (c), (d) and (e), we get

$$\frac{C_1}{C_2} = \frac{\cosh\mu - \cos\mu + 2\phi\,\mu\sinh\mu}{\sin\mu - \sinh\mu} \tag{h}$$

where

$$\phi = \frac{EI}{RL} \tag{i}$$

Combining Eqs. (c), (d) and (g), we get

$$\frac{C_1}{C_2} = \frac{[(\sin\mu + \sinh\mu) + \mu\phi(\cos\mu + \cosh\mu) + 2\mu\phi \times (\cosh\mu + \mu\phi\sinh\mu)]}{(\cos\mu - \cosh\mu) - \mu\phi(\sin\mu + \sinh\mu)} \tag{j}$$

From Eqs. (h) and (j), one obtains the following transcendental equation

$$\frac{\cosh\mu - \cos\mu + 2\phi\mu\sinh\mu}{\sin\mu - \sinh\mu}$$

$$= \frac{(\sin\mu + \sinh\mu) + \mu\phi(\cos\mu + \cosh\mu) + 2\mu\phi(\cosh\mu + \phi\sinh\mu)}{(\cos\mu - \cosh\mu) - \mu\phi(\sin\mu + \sinh\mu)} \tag{k}$$

From Eq. (k), depending on the magnitude of coefficient of restraint, different values of μ can be obtained.

Values of μ_m for a beam for various values of the degrees of elastic restraint against rotation, which are same at both ends have been shown in Table 8.1.

The mode shape equation for this case is

$$Y_m = \left[\sin\left(\frac{\mu_m x}{L}\right) - \sinh\left(\frac{\mu_m x}{L}\right)\right] + \alpha_m\left[\cos\left(\frac{\mu_m x}{L}\right)\right.$$

$$\left. - 2\mu_m\phi\sinh\left(\frac{\mu_m x}{L}\right) - \cosh\left(\frac{\mu_m x}{L}\right)\right] \tag{l}$$

Table 8.1. Values of μ_m for a beam having the same degree of elastic restraint against rotation at both ends

ϕ	m					
	1	2	3	4	5	6
0.01	4.642	7.711	10.802	13.895	16.991	20.089
0.05	4.374	7.329	10.339	13.375	16.431	19.592
0.10	4.156	7.069	10.066	13.106	16.172	19.257
0.50	3.577	6.547	9.613	12.712	15.827	18.950
1.00	3.399	6.428	9.525	12.643	15.770	18.902
10.00	3.173	6.399	9.436	12.575	15.715	18.855

where

$$\alpha_m = \frac{\sin \mu_m - \sinh \mu_m}{\cosh \mu_m - \cos \mu_m + 2\phi\mu_m \sinh \mu_m} \tag{m}$$

From Table 8.1, for $\phi = 0.1$, $\mu_1 = 4.462$

Therefore $\quad \lambda_1 L = 4.462$

or

$$p_1 = (4.462)^2 \sqrt{\frac{EI}{\rho A L^4}}$$

The natural frequencies and mode shapes for various end conditions for a beam are given below [note that $\mu_m = \lambda_m L$]:

(1) Left hand end simply supported, right hand end clamped

$$Y_m = \sin \frac{\mu_m x}{L} - \alpha_m \sinh \frac{\mu_m x}{L} \tag{8.113}$$

where

$$\mu_m = \frac{4m+1}{4}\pi \qquad m = 1, 2, 3, \ldots.. \tag{8.114}$$

$$\alpha_m = \frac{\sin \mu_m}{\sinh \mu_m} \tag{8.115}$$

(2) Both ends clamped

$$Y_m = \sin \frac{\mu_m x}{L} - \sinh \frac{\mu_m x}{L} - \alpha_m \left(\cos \frac{\mu_m x}{L} - \cosh \frac{\mu_m x}{L} \right) \tag{8.116}$$

$$\mu_m = \frac{2m+1}{2}\pi \qquad m = 1, 2, 3, \ldots.. \tag{8.117}$$

where

$$\alpha_m = \frac{\sin \mu_m - \sinh \mu_m}{\cos \mu_m - \cosh \mu_m} \tag{8.118}$$

(3) Both ends free

$$Y_m = \sin \frac{\mu_m x}{L} + \sinh \frac{\mu_m x}{L} - \alpha_m \left(\cos \frac{\mu_m x}{L} + \cosh \frac{\mu_m x}{L} \right) \tag{8.119}$$

$$\mu_m = \frac{2m+1}{2}\pi \qquad m = 1, 2, 3, \ldots.. \tag{8.120}$$

$$\alpha_m = \frac{\sin \mu_m - \sinh \mu_m}{\cos \mu_m - \cosh \mu_m} \tag{8.121}$$

(4) Left hand end clamped, right hand end free

$$Y_m = \sin \frac{\mu_m x}{L} - \sinh \frac{\mu_m x}{L} - \alpha_m \left(\cos \frac{\mu_m x}{L} - \cosh \frac{\mu_m x}{L} \right) \tag{8.122}$$

$$\mu_1 = 1.875, \quad \mu_2 = 4.694, \quad \mu_3 = 7.855$$

$$m \geq 4, \quad \mu_m = \frac{2m-1}{2}\pi \tag{8.123}$$

$$\alpha_m = \frac{\sin \mu_m + \sinh \mu_m}{\cos \mu_m + \cosh \mu_m} \tag{8.124}$$

(5) Left hand end simply supported, right hand end free

$$Y_m = \sin \frac{\mu_m x}{L} + \alpha_m \sinh \frac{\mu_m x}{L} \tag{8.125}$$

where
$$\mu_m = \frac{4m-3}{4}\pi, \qquad m = 2, 3, 4, \ldots \tag{8.126}$$

$$\alpha_m = \frac{\sin \mu_m - \sinh \mu_m}{\cos \mu_m - \cosh \mu_m} \tag{8.127}$$

(6) Left hand end clamped, the right hand end elastically restrained in rotation

$$Y_m = \sin \frac{\mu_m x}{L} - \sinh \frac{\mu_m x}{L} + \alpha_m \left(\cos \frac{\mu_m x}{L} - \cosh \frac{\mu_m x}{L} \right) \tag{8.128}$$

where
$$\alpha_m = \frac{\sin \mu_m - \sinh \mu_m}{\cos \mu_m - \cosh \mu_m}$$

Different values of μ_m for various values of $\phi = EI/RL$, where R is the coefficient of elastic restraint and EI is the flexural rigidity of the beam, are given in Table 8.2.

Table 8.2. Values of μ_m for a beam having one edge clamped and the other end elastically restrained against rotation

ϕ	*m*					
	1	2	3	4	5	6
0.01	4.686	7.782	10.898	14.015	17.134	20.254
0.05	4.547	7.585	10.661	13.751	16.850	19.958
0.10	4.431	7.450	10.552	13.615	16.720	19.834
1.00	4.042	7.134	10.257	13.388	16.523	19.660
1000.00	3.927	7.069	10.211	13.352	16.494	19.635

(7) Left hand end simply supported, right hand end elastically restrained against rotation

$$Y_m = \sin \frac{\mu_m x}{L} - \alpha_m \sinh \frac{\mu_m x}{L} \tag{8.130}$$

where
$$\alpha_m = \frac{\sin \mu_m}{\sinh \mu_m} \tag{8.131}$$

Different values of μ_m for various values of $\phi = EI/RL$, where R is the coefficient of elastic restraint and EI is the flexural rigidity of the beam, are given in Table 8.3.

Table 8.3. Values of μ_m for one edge simply supported and the other edge elastically restrained against rotation

ϕ	*m*					
	1	2	3	4	5	6
0.01	3.890	7.004	10.119	13.256	16.354	19.474
0.05	3.770	6.820	9.891	12.977	16.075	19.180
0.10	3.665	6.688	9.752	12.840	15.943	19.055
0.50	3.367	6.418	9.520	12.640	15.768	18.900
1.00	3.274	6.356	9.475	12.605	15.739	18.876
1000.00	3.142	6.284	9.425	12.567	15.708	18.850

Natural frequencies of uniform beams having various boundary conditions are given in Table 8.4 where

$$p_n = c_n \sqrt{\frac{EI}{mL^4}}$$

p_n is given in cycles/s.

8.9 ORTHOGONALITY PROPERTIES OF NORMAL MODES

Let us consider free flexural vibration of beams. For the *r*th mode of vibration, we can write from Eq. (8.82), the following equation for a non-uniform beam:

$$\frac{d^2}{dx^2}\left(EI \frac{d^2 Y_r}{dx^2} \right) = \rho A p_r^2 \, Y_r \tag{8.132}$$

Let us multiply both sides of Eq. (8.132) by Y_s, the mode shape function for the *s*th mode and integrate with respect to x over the length of the beam

$$\int_0^L Y_s \frac{d^2}{dx^2}\left(EI \frac{d^2 Y_r}{dx^2} \right) dx = \int_o^L \rho A p_r^2 \, Y_s \, Y_r \, dx \tag{8.133}$$

After integrating by parts, the left hand side of Eq. (8.133) results in

$$Y_s \frac{d}{dx}\left(EI \frac{d^2 Y_r}{dx^2} \right)\Bigg|_0^L - EI \frac{d^2 Y_r}{dx^2} \cdot \frac{dY_s}{dx}\Bigg|_0^L + \int_0^L EI \frac{d^2 Y_r}{dx^2} \frac{d^2 Y_s}{dx^2} dx$$

$$= \int_0^L \rho A p_r^2 Y_s Y_r \, dx \tag{8.134}$$

The first two terms of Eq. (8.134) are zero for any type of boundary conditions. The mode shapes conform to the boundary conditions of the problem, and one can verify instantaneously by substituting any classical boundary conditions in the first two terms, which will be zero.

This is, however, valid for non-classical boundary conditions as well. Therefore, Eq. (8.134) can be written as

$$\int_0^L EI \frac{d^2Y_r}{dx^2} \frac{d^2Y_s}{dx^2} dx = p_r^2 \int_0^L \rho A Y_s Y_r \, dx \tag{8.135}$$

Repeating the above procedure by starting with sth mode in Eq. (8.134), then integrating both sides after multiplying by Y_r and substituting the boundary conditions, we get

$$\int_0^L EI \frac{d^2Y_r}{dx^2} \frac{d^2Y_s}{dx^2} dx = p_s^2 \int_0^L \rho A Y_r Y_s \, dx \tag{8.136}$$

Subtracting Eq. (8.136) from Eq. (8.135), we get

$$(p_r^2 - p_s^2) \int_0^L \rho A Y_r Y_s \, dx = 0$$

or

$$\int_0^L \rho A Y_r Y_s \, dx = 0 \tag{8.137}$$

CANTILEVER	C = .560	.226 C = 3.51	.4999 .132 C = 9.82	.644 .356 .094 C = 19.2	.721 .500 .277 .0725 C = 31.8
SIMPLY SUPPORTED ENDS	C = 1.57	.500 C = 6.28	.333 .333 C = 14.1	.75 .50 .25 C = 25.2	.80 .60 .40 .20 C = 39.4
FIXED ENDS	C = 3.56	.500 C = 9.82	.641 .359 C = 19.2	.722 .5 .278 C = 31.8	.773 .591 .409 .227 C = 47.5
FREE ENDS	.776 .224 C = 3.56	.868 .500 .132 C = 9.82	.906 .644 .356 .094 C = 19.2	.927 .723 .500 .277 .034 C = 31.8	.940 .774 .591 .409 .227 .060 C = 47.5
FIXED-HINGED	C = 2.45	.440 C = 7.95	.616 .308 C = 16.6	.706 .471 .235 C = 28.4	.762 .571 .381 .190 C = 43.3
HINGED - FREE	.264 C = 2.45	.554 .147 C = 7.96	.692 .384 .102 C = 16.6	.765 .529 .233 .0777 C = 28.4	.810 .619 .429 .237 .063 C = 43.3

Table 8.4 Natural frequencies of unform beams

Combining Eq. (8.137) with either Eq. (8.135) or Eq. (8.136), we get

$$\int_0^L EI \frac{d^2Y_r}{dx^2}\frac{d^2Y_s}{dx^2}\,dx = 0 \tag{8.138}$$

Equations (8.137) and (8.138) are known as orthogonality conditions for normal modes for flexural vibrations of beams. If the beam is uniform, Eqs.(8.137) and (8.138) reduce to

$$\left.\begin{aligned}&\int_0^L Y_r Y_s\,dx = 0\\ &\int_0^L \frac{d^2Y_r}{dx^2}\frac{d^2Y_s}{dx^2}\,dx = 0\end{aligned}\right\} \tag{8.139}$$

Example 8.3

Find the general solution of free vibration of a simply supported beam, if all the masses of the beam are given a uniform velocity *v*.

The general solution of the vibrating beam is

$$y(x,t) = \sum_{m=1}^{\infty} \sin\frac{m\pi x}{L}(A_m \cos p_m t + B_m \sin p_m t) \tag{a}$$

The initial conditions of the problem are

at $$t = 0,\quad y(x,t) = 0,\quad \frac{\partial y}{\partial t} = v \tag{b}$$

Let us substitute the first condition from Eq. (b) into Eq. (a), to give

$$0 = \sum_{m=1}^{\infty} \sin\frac{m\pi x}{L}(A_m)$$

Multiplying both sides by $\sin\frac{n\pi x}{L}$ and integrating with respect to x throughout the length of the beam, we get

$$\sum_{m=1}^{\infty} A_m \int_0^L \sin\frac{m\pi x}{L}\sin\frac{n\pi x}{L}\,dx = 0 \tag{c}$$

Using orthogonality relationship given by Eq. (8.137), we get

$$A_n \int_0^L \left(\sin\frac{n\pi x}{L}\right)^2 dx = 0$$

or $$A_n \cdot \frac{L}{2} = 0$$

or $$A_n = 0 \tag{d}$$

Substituting second condition from Eq. (b) into Eq. (a), we get

$$v = \sum p_m (B_m) \sin \frac{m\pi x}{L} \tag{e}$$

Multiplying both sides of Eq. (e) by $\sin \frac{n\pi x}{L}$ and integrating with respect to x throughout the length of the beam, we get

$$\int_0^L v \sin \frac{n\pi x}{L}\, dx = \sum_m p_m B_m \int_0^L \sin \frac{m\pi x}{L} \sin \frac{n\pi x}{L}\, dx$$

Using the orthogonality relationship given by Eq. (8.137), we get

$$\frac{2}{n\pi} v = p_n B_n \frac{L}{2} \quad \text{for} \quad n = 1, 2, 3, ...$$

or

$$B_n = \frac{4}{n\pi} \frac{v}{p_n} \quad \text{for} \quad n = 1, 2, 3, ... \tag{f}$$

Therefore, the general solution for transverse displacement is

$$y(x, t) = 4v \sum_{m=1,3,5,...}^{\infty} \frac{1}{m\pi p_m} \sin \frac{m\pi x}{L} \sin p_m t \tag{g}$$

Now

$$p_m = (m\pi)^2 \sqrt{\frac{EI}{\rho A L^4}}$$

Therefore

$$y(x, t) = 4v \sqrt{\frac{\rho A L^4}{EI}} \sum_{m=1,3,5,...}^{\infty} \frac{1}{(m\pi)^3} \sin \frac{m\pi x}{L} \sin p_m t \tag{h}$$

8.10 EFFECT OF ROTARY INERTIA ON THE FREE FLEXURAL VIBRATION OF BEAMS

The displacement of a fibre located at a distance z from the axis of the beam is given by [Fig. 8.13]

$$u = -z \frac{\partial y}{\partial x} \tag{8.140}$$

The displacement u is a function of y. As y varies with time, u also varies with time. Further, u also varies with the distance of the fibre from the axis. As such, the axial fibres will have acceleration $\ddot{u} (= \partial^2 u / \partial t^2)$. Therefore, a cross-sectional area dA of length dx will have an inertia force equal to $\rho dA\, dx\, \ddot{u}$. On substitution of the value of u from Eq.(8.140)

$$\text{Inertia force} = \rho dA\, dx\, z \frac{\partial^3 y}{\partial x\, \partial t^2}$$

Just as we have in the case of beam bending, along with the normal force in the fibre at a particular distance z above the neutral axis, there will be an equal normal force but in the opposite direction, at the same distance below the neutral axis. As this normal force is time dependent, this will give rise to a time dependent moment, which is given by

$$M_R = \int_A \rho \, dx \, z^2 \frac{\partial^3 y}{\partial x \, \partial t^2} dA \tag{8.141}$$

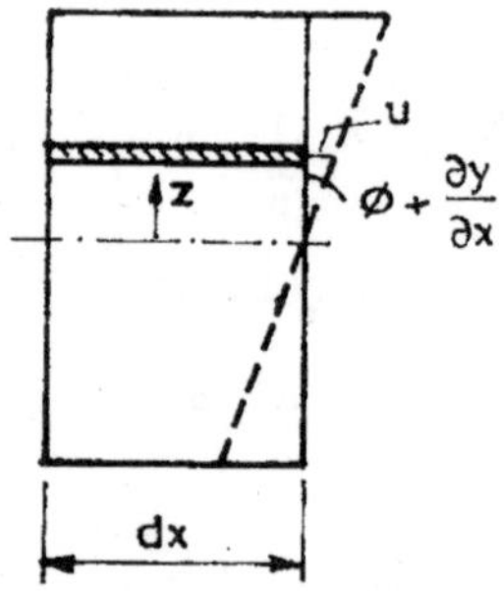

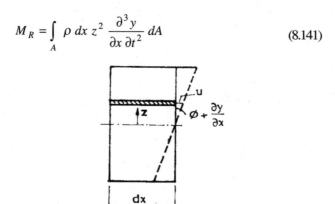

Fig. 8.13 Rotation of an elemental beam

The second moment of the area is

$$I = \int_A z^2 \, dA \tag{8.142}$$

Therefore

$$M_R = \rho I \frac{\partial^3 y}{\partial x \, \partial t^2} dx \tag{8.143}$$

The effect of this couple is included in Eq. (8.76) and it becomes

$$\frac{\partial M}{\partial x} = V + \rho I \frac{\partial^3 y}{\partial x \, \partial t^2} \tag{8.144}$$

Substituting M from Eq. (8.77) into Eq. (8.144), and then differentiating both sides with respect to x, we get

$$\frac{\partial^2}{\partial x^2}\left(EI \frac{\partial^2 y}{\partial x^2} \right) = \frac{\partial V}{\partial x} + \rho I \frac{\partial^4 y}{\partial x^2 \, \partial t^2} \tag{8.145}$$

Combining Eqs.(8.74) and (8.145) and assuming $w = 0$, we get

$$\frac{\partial^2}{\partial x^2}\left(EI \frac{\partial^2 y}{\partial x^2} \right) = - \rho A \frac{\partial^2 y}{\partial t^2} + \rho I \frac{\partial^4 y}{\partial x^2 \, \partial t^2} \tag{8.146}$$

Equation (8.146) is written as

$$\frac{\partial^2}{\partial x^2}\left(EI \frac{\partial^2 y}{\partial x^2} \right) = - \rho A \left(\frac{\partial^2 y}{\partial t^2} - r^2 \frac{\partial^4 y}{\partial x^2 \, \partial t^2} \right) \tag{8.147}$$

where $r^2 = \dfrac{I}{A}$, r is the radius of gyration of the cross-section.

The additional term introduced in Eq. (8.147) due to the effect of time-dependent couple of the axial fibres of the beam is referred to as rotary or rotatory inertia.

For free vibration, we assume

$$y(x, t) = Y(x) \sin(pt - \alpha) \tag{8.148}$$

Substituting y from Eq. (8.148) into Eq. (8.147), and assuming the beam to be uniform, we get

$$EI \frac{d^4Y}{dx^4} = \rho A p^2 \left(Y - r^2 \frac{d^2Y}{dx^2} \right) \tag{8.149}$$

Equation (8.149) is rewritten as

$$\frac{d^4Y}{dx^4} = \lambda^4 \left(Y - r^2 \frac{d^2Y}{dx^2} \right) \tag{8.150}$$

$$\text{where } \lambda^4 = \frac{\rho A p^2}{EI} \tag{8.151}$$

The solution of Eq. (8.150) is as follows:

$$Y = C_1 \sin kx + C_2 \cos kx + C_3 \sinh k'x + C_4 \cosh k'x \tag{8.152}$$

Substituting Y from Eq.(8.152) into Eq.(8.150) yields the following relationship

$$\left.\begin{aligned} k^4 - r^2 \lambda^4 k^2 &= \lambda^4 \\ k'^4 + r^2 \lambda^4 k'^2 &= \lambda^4 \end{aligned}\right\} \tag{8.153}$$

Example 8.4

A freely vibrating simply supported uniform beam is considered. Determine the effects of rotary inertia.

Let the span of the beam be L and the flexural rigidity of the beam be EI. The boundary conditions of the beam are:

$$\text{at } x = 0, \quad Y = 0 \quad \text{and} \quad \frac{d^2Y}{dx^2} = 0 \tag{a}$$

and

$$\text{at } x = L, \quad Y = 0 \quad \text{and} \quad \frac{d^2Y}{dx^2} = 0 \tag{b}$$

Substituting the conditions given by Eq. (a) in Eq. (8.152), we get

$$C_2 = C_4 = 0 \tag{c}$$

and on substitution of condition given by Eq. (b) into Eq. (8.152), finally gives the frequency equation as

$$\sin kL = 0 \tag{d}$$

which gives

$$k = \frac{n\pi}{L}, \quad n = 1, 2, 3, ... \tag{e}$$

Replacing the value of k from Eq. (e) into the first of Eq. (8.153) gives

$$\lambda^4 = \frac{k^4}{1 + r^2 k^2} \tag{f}$$

or

$$p^2 = \frac{EI}{\rho A}\left(\frac{k^4}{1 + r^2 k^2}\right)$$

We know that the angular frequencies of an elastic simply supported beam without rotary inertia effect is [Eq. (8.95)]

$$\bar{p}_n^2 = \frac{n^4 \pi^4 EI}{\rho A L^4} \tag{g}$$

Now, combining Eqs. (e), (f) and (g), we get

$$p^2 = \bar{p}_n^2 \left(1 - \frac{n^2 r^2 \pi^2}{L^2} + \frac{n^4 r^4 \pi^4}{L^4} - \cdots\right) \tag{h}$$

The rotary inertia is thus a function of r/L. Its effect is most prominent in short and stocky beams and in higher modes of vibration.

8.11 FREE VIBRATION OF THE SHEAR BEAM

For short and sturdy beams, the contribution of the shear force towards the total deflection of the beam is not negligible. In this section, let us limit our study only to the transverse shear effect. Though the shear force is the rate of change of bending moment, the effect of bending moment is not considered in the following analysis. Similarly, the rotary inertia effect is also ignored. The beam which is analysed on the basis of transverse shear only, is referred to as shear beam.

Consider a shear beam of length L, shown in Fig. 8.14.

Let us consider an infinitesimal element of length dx at a distance x from the origin. Let y_s be the transverse deflection at that section and V is the shear force at the left hand end.

Then, we may write

$$V = \mu AG \frac{\partial y_s}{\partial x} \tag{8.154}$$

where μ is called the shape factor and it depends on the shape of the cross-section. A is the cross-sectional area and G is the shear modulus of elasticity.

If the beam is vibrating due to its own mass, then considering the dynamic equilibrium, the following equation results [Fig. 8.14(b)]

$$\frac{\partial V}{\partial x} = \rho A \frac{\partial^2 y_s}{\partial t^2} \tag{8.155}$$

where ρ is the mass density of the material of the beam and A is the cross-sectional area.

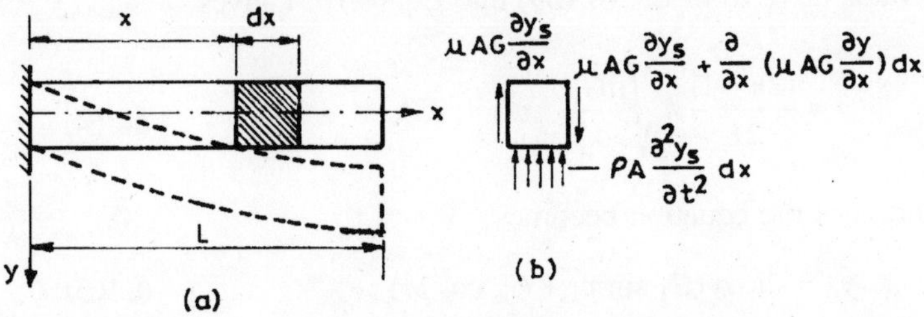

Fig. 8.14 (a) A shear beam and (b) its freebody diagram

Combining Eqs. (8.154) and (8.155) for a uniform shear beam, we get

$$\mu AG \frac{\partial^2 y_s}{\partial x^2} = \rho A \frac{\partial^2 y_s}{\partial t^2}$$

or

$$\lambda^2 \frac{\partial^2 y_s}{\partial x^2} = \frac{\partial^2 y_s}{\partial t^2} \tag{8.156}$$

where

$$\lambda^2 = \frac{\mu G}{\rho} \tag{8.157}$$

Equation (8.156) is of the same form as Eq. (8.43), derived for free longitudinal vibration of bars. The general solution of Eq. (8.156) is

$$y_s(x, t) = (C_1 \sin \alpha x + C_2 \cos \alpha x)(C_3 \sin pt + C_4 \cos pt) \tag{8.158}$$

where C_1 and C_2 are constants, which depend on the end conditions of the beam. C_3 and C_4 are also constants, but they depend on the initial conditions.

$$\alpha^2 = \frac{p^2}{\lambda^2} \tag{8.159}$$

For a cantilever uniform shear beam, the end conditions are

$$\left.\begin{aligned} &\text{at} \quad x = 0, \quad y_s = 0 \\ &\text{at} \quad x = l, \quad \partial y_s / \partial x = 0 \end{aligned}\right\} \tag{8.160}$$

Substituting Y_s from Eq. (8.158) into the conditions given by Eq. (8.160), we get

$$C_2 = 0 \tag{8.161}$$

and the frequency equation becomes

$$\cos \alpha L = 0 \tag{8.162}$$

Solution of Eq. (8.162) gives

$$\alpha = \frac{(2n-1)}{2L}\pi, \quad n = 1, 2, 3, \ldots \tag{8.163}$$

Substituting the value of α from Eq. (8.159) into Eq. (8.163), gives

$$p = \frac{(2n-1)}{2L}\pi\sqrt{\frac{\mu G}{\rho}}, \quad n = 1, 2, 3, \ldots \tag{8.164}$$

The general solution of the equation becomes

$$y_s = \sin \alpha x\,(C_3 \sin pt + C_4 \cos pt) \tag{8.165}$$

8.12 EFFECT OF AXIAL FORCE ON THE FREE FLEXURAL VIBRATION OF BEAMS

In practical problems, a beam undergoing flexural vibration may be subjected to an axial tension or compression.

Freebody diagram or an element of length dx of a beam has been shown in Fig. 8.15. The beam is subjected to an axial tension N, which is supposed to be constant throughout the length for small deflection of the beam. The shear deformation and rotary inertia effects have not been considered in the present derivation.

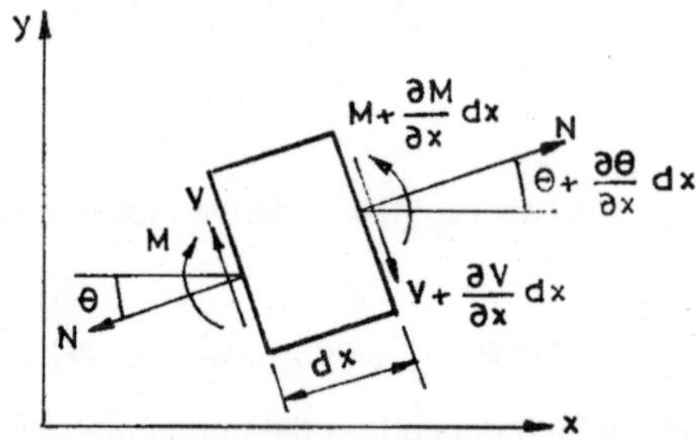

Fig. 8.15 Freebody diagram of a beam

Considering the dynamic equilibrium of the forces acting on the element the following equation results

$$\rho A \frac{\partial^2 y}{\partial t^2} dx = -\left(V + \frac{dV}{\partial x} dx\right) + V + N\left(\theta + \frac{\partial \theta}{\partial x} dx\right) - N\theta$$

or

$$\rho A \frac{\partial^2 y}{\partial t^2} = -\frac{\partial V}{\partial x} + N \frac{\partial \theta_x}{\partial x}$$

or

$$\rho A \frac{\partial^2 y}{\partial t^2} = -\frac{\partial V}{\partial x} + N \frac{\partial^2 y}{\partial x^2} \tag{8.166}$$

Combining Eqs. (8.76), (8.77) and (8.166), we get

$$\rho A \frac{\partial^2 y}{\partial t^2} = -\frac{\partial^2}{\partial x^2}\left(EI \frac{\partial^2 y}{\partial x^2} \right) + N \frac{\partial^2 y}{\partial x^2} \tag{8.167}$$

For a uniform beam, Eq. (8.167) becomes

$$\rho A \frac{\partial^2 y}{\partial t^2} = -EI \frac{\partial^4 y}{\partial x^4} + N \frac{\partial^2 y}{\partial x^2}$$

or

$$\rho A \frac{\partial^2 y}{\partial t^2} + EI \frac{\partial^4 y}{\partial x^4} - N \frac{\partial^2 y}{\partial x^2} = 0 \tag{8.168}$$

The sign of N will have to be reversed, if axial compression is applied.

Example 8.5

Determine the natural frequency of a uniform, simply supported beam subjected to an axial compression.

The solution is assumed as

$$y_n(x, t) = C \sin \frac{n\pi x}{L} \sin(p_n t - \alpha) \tag{a}$$

Substituting solution given by Eq. (a) in Eq. (8.157), we get

$$EI\left(\frac{n\pi}{L}\right)^4 - N\left(\frac{n\pi}{L}\right)^2 - \rho A p_n^2 = 0$$

or

$$p_n = (n\pi)^2 \sqrt{\left(1 - \frac{NL^2}{n^2\pi^2 EI}\right)\frac{EI}{\rho AL^2}} \tag{b}$$

$$n = 1, 2, 3, \ldots$$

when $N = 0$, Eq. (b) reduces to the case of a simply supported beam. The effect of axial compression is to reduce the natural frequency, and the axial tension will increase the natural frequency of the beam.

The fundamental frequency is

$$p_1 = \pi \sqrt{\left(1 - \frac{N}{N_{cr}}\right)\frac{EI}{\rho AL^4}} \tag{c}$$

where $N_{cr} = \dfrac{\pi^2 EI}{L^2}$ is the critical buckling load for the member.

8.13 FREE VIBRATION OF BEAMS INCLUDING SHEAR DEFORMATION AND ROTARY INERTIA EFFECTS

Let y_b, y_s and y be the bending, shear and total deflection of the beam, then

$$y = y_b + y_s \tag{8.169}$$

and

$$\frac{\partial y}{\partial x} = \frac{\partial y_b}{\partial x} + \frac{\partial y_s}{\partial x} \tag{8.170}$$

The bending and shear deformation effects have been considered separately and the total effect is considered to be the summation of the two.

Some of the equations derived earlier are written as

$$V = \mu AG \frac{\partial y_s}{\partial x} \tag{8.154}$$

$$\frac{\partial V}{\partial x} = \rho A \frac{\partial^2 y}{\partial t^2} \tag{8.155}$$

$$M = - EI \frac{\partial^2 y_b}{\partial x^2} \tag{8.77}$$

Negative sign has been put in Eq. (8.77), because positive direction of y is downwards. Again

$$\frac{\partial M}{\partial x} = V + \rho I \frac{\partial^3 y_b}{\partial x \, \partial t^2} \tag{8.144}$$

Combining Eqs. (8.155), (8.169) and (8.170) for a uniform beam, we get

$$-\mu AG \frac{\partial^2 y_s}{\partial x^2} = \rho A \left(\frac{\partial^2 y_b}{\partial t^2} + \frac{\partial^2 y_s}{\partial t^2} \right) \tag{8.171}$$

or

$$\rho A \frac{\partial^2 y_b}{\partial t^2} + \rho A \frac{\partial^2 y_s}{\partial t^2} = \mu AG \frac{\partial^2 y_s}{\partial x^2} \tag{8.172}$$

Combining Eqs. (8.169), (8.154), (8.77) and (8.144), we get

$$EI \frac{\partial^3 y_b}{\partial x^3} + \mu AG \frac{\partial y_s}{\partial x} - \rho I \frac{\partial^3 y_b}{\partial x \, \partial t^2} = 0 \tag{8.173}$$

Equations (8.172) and (8.173) form a pair of coupled equations in y_b and y_s.

They can be combined into a single equation in y

$$\frac{\partial^4 y}{\partial x^4} + \frac{\rho A}{EI} \frac{\partial^2 y}{\partial t^2} - \left(\frac{\rho}{E} + \frac{\rho}{\mu G} \right) \frac{\partial^4 y}{\partial x^2 \, \partial t^2} + \frac{\rho^2}{\mu EG} \frac{\partial^4 y}{\partial t^4} = 0 \tag{8.174}$$

Equation (8.174), which includes the shear deformation and rotary inertia effects is known as Timoshenko beam equation.

The equations are sometimes derived in terms of total deflection y and the bending slope ψ. Eq. (8.173) can be rewritten as follows

$$EI\frac{\partial^2\psi}{\partial x^2} + \mu AG\left(\frac{\partial y}{\partial x} - \psi\right) = \rho I\frac{\partial^2\psi}{\partial t^2} \tag{8.175}$$

Equation (8.172) is rewritten as

$$\mu AG\left(\frac{\partial^2 y}{\partial x^2} - \frac{\partial\psi}{\partial x}\right) = \rho A\frac{\partial^2 y}{\partial t^2} \tag{8.176}$$

Eliminating ψ from Eqs. (8.175) and (8.176) results in Eq. (8.174). Eliminating y from Eqs. (8.175) and (8.176) results in

$$EI\frac{\partial^4\psi}{\partial x^4} + \rho A\frac{\partial^2\psi}{\partial t^2} - \left(\rho I + \frac{\rho EI}{\mu G}\right)\frac{\partial^4\psi}{\partial x^2\,\partial t^2} + \frac{\rho^2 I}{\mu G}\frac{\partial^4\psi}{\partial t^4} = 0 \tag{8.177}$$

Example 8.6

Determine the natural frequencies of a beam simply supported at both ends, by including the shear deformation and rotary inertia effects.

The solution of Timoshenko beams with arbitrary boundary conditions is somewhat difficult. The only beam problem which can be solved somewhat easily, is the one having both ends simply supported.

The boundary conditions are same as those used in elementary beam theory that is

at $\quad x = 0, \quad y = 0 \;$ and $\; \dfrac{\partial^2 y}{\partial x^2} = 0$

and at $\quad x = L, \quad y = 0 \;$ and $\; \dfrac{\partial^2 y}{\partial x^2} = 0$

After eliminating the time function from Eq. (8.176) by considering the total solution as a product of two functions of space and time, we get

$$\frac{d^4Y}{dx^4} - \lambda^4 Y + \lambda^4 r^2\left(1 + \frac{E}{\mu G}\right)\frac{d^2Y}{dx^2} + \lambda^8 r^4\left(\frac{E}{\mu G}\right)Y = 0 \tag{a}$$

where $\lambda^4 = \dfrac{mp^2}{EI}$ and $r^2 = \dfrac{I}{A}$.

The simply supported beam mode shape is assumed as follows:

$$Y_n(x) = A\sin\frac{n\pi x}{L} \tag{b}$$

Equation (b) satisfies all the boundary conditions of the problem. Therefore, the shear deformation and the rotary inertia do not have any effect on the mode shape of the uniform simply supported beam. Substituting Y_n from Eq.(b) into Eq. (a), we get

$$\left(\frac{n\pi}{L}\right)^4 - \lambda^4 - \lambda^4 n^2 \left(\frac{n\pi}{L}\right)^2 \left(1+\frac{E}{\mu G}\right) + \lambda^8 r^4 \left(\frac{E}{\mu G}\right) = 0 \tag{c}$$

Substituting the values of λ and r in Eq. (c), we get

$$p^4 - \frac{1}{\rho}\left[\frac{(E+\mu G)n^2\pi^2}{L^2} + \frac{\mu GA}{I}\right]p^2 + \frac{\mu G E n^4 \pi^4}{\rho^2 L^4} = 0 \tag{d}$$

Depending on the geometrical and material properties of the beam, n frequencies can be calculated from Eq. (d). The natural frequencies have been plotted for three values of $E/\mu G$ in Fig. 8.16. It is seen in the figure, that the correction due to shear deformation and rotary inertia increases with mode number and decreases with the increase of slenderness ratio.

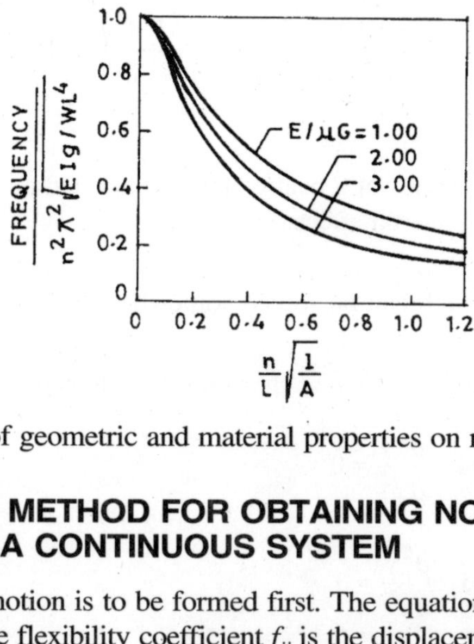

Fig. 8.16 Effect of geometric and material properties on natural frequencies

8.14 COLLOCATION METHOD FOR OBTAINING NORMAL MODES OF VIBRATION OF A CONTINUOUS SYSTEM

The integral equation of motion is to be formed first. The equation is derived on the basis of the flexibility function. The flexibility coefficient f_{ij} is the displacement at point i due to a unit load placed at point j. In the case of continuous systems, the term coefficient is replaced by function, as here we deal with infinite number of points. The flexibility function $f(x, \xi)$ represents the displacement at x due to a unit load applied at ξ.

Consider an element of the beam of infinitesimal length $d\xi$ at a distance ξ from the origin (Fig. 8.17). For free vibration case, it is the inertia force which only acts and its magnitude is equal to $-\rho A \ddot{y}\, d\xi$, where ρ is the density and A is the cross-sectional area. Therefore, the displacement dy at x resulting from the inertia force of this infinitesimal element is

$$dy(x,t) = -f(x,\xi)\,\rho A(\xi)\,\ddot{y}(\xi,t)\,d\xi \tag{8.178}$$

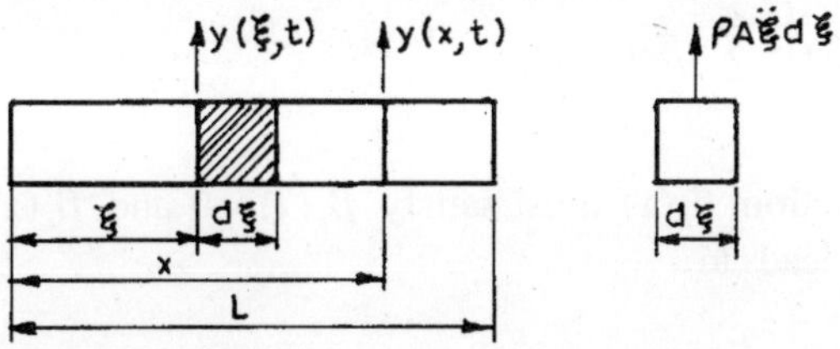

Fig. 8.17 An infinite beam element

Now, considering all the infinitesimal elements and their contributions towards total displacement at x is

$$y(x,t) = -\int_0^L f(x,\xi)\,\rho A(\xi)\,\ddot{y}(\xi,t)\,d\xi \tag{8.179}$$

The influence function should satisfy the end conditions of the problem.

Assuming that variables can be separated in the solution

$$y(x,t) = Y(x)\,q(t) \tag{8.180}$$

Substituting $y(x,t)$ from Eq. (8.180) into Eq. (8.178), we get

$$\frac{Y(x)}{\int_0^L f(x,\xi)\,\rho A(\xi) Y(\xi)\,d\xi} = \frac{\ddot{q}(t)}{q(t)} \tag{8.181}$$

If certain function of x is equal to a certain function of t, then both will be equal to a constant. Choosing p^2 as the constant

$$Y(x) = p^2 \int_0^L f(x,\xi)\,\rho A(\xi) Y(\xi)\,d\xi \tag{8.182}$$

and

$$\ddot{q} + p^2 q = 0 \tag{8.183}$$

Equation (8.183) is equivalent to the differential equation [Eq.(8.82)]. The solution will consist of infinite sets of eigenvalues and the corresponding set of eigenfunctions $Y(x)$.

In the method of collocation, we reduce the problem to a finite number of equations of motion.

Let us divide the beam into $(n + 1)$ stations, marked as $x_0, x_1, x_2, ..., x_n$. We shall satisfy the equations of motions at n points; between the points, however, the equation of motion in general will not be satisfied.

Let us take the case of a cantilever beam, as shown in Fig. 8.18. At $x = 0$, $Y(0) = 0$ and $f(0, \xi) = 0$. Substituting them in Eq. (8.182), gives $0 = 0$ in generalised coordinates; δ_s is described by

$$Y(x) = \sum_{i=1}^{n} \beta_i(x) \cdot \delta_i \tag{8.184}$$

in which the shape function $\beta_i(x)$ must satisfy $\beta_i(x) = 0$ and $\beta_i'(x) = 0$. Substitution of Eq. (8.184) into Eq. (8.182) leads to

$$\sum_{i=1}^{n} \delta_i \left[\beta_i(x_i) - p^2 \int_0^L f(x, \xi)\, \rho\, A(\xi)\, \beta_i(\xi)\, d\xi \right] = 0 \tag{8.185}$$

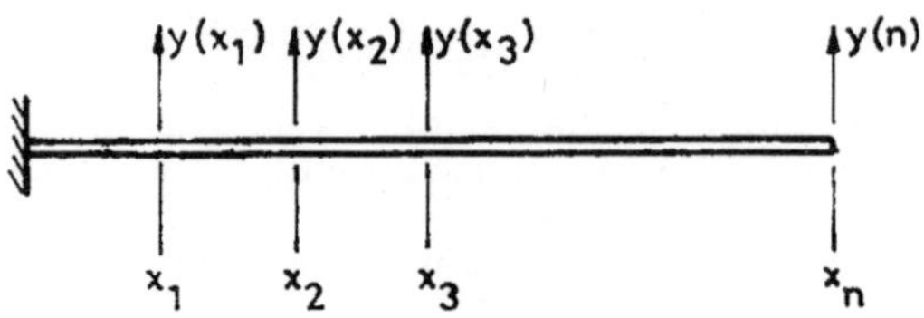

Fig. 8.18 A collocating beam

There will be n equations based on Eq. (8.183). The equations will be homogeneous. As discussed for the case of lumped mass system, n eigenvalues and n eigenvectors can be calculated.

Let f_{ij} be the flexibility coefficient for the station $x_i, x_j[f(x_i, x_j)]$. The integrand on the left hand side is approximated as follows:

$$\int_0^L f(x_i, \xi)\, \rho A(\xi) Y(\xi)\, d\xi = \sum_{j=1}^{n} w_j\, f_{ij}\, \rho A_j\, Y_j \tag{8.186}$$

where $A(x_j) = A_j$ and $Y(x_j) = Y_j$

and w_j is a weighing factor.

Therefore, Eq. (8.186) can be written as

$$Y_i = p^2 \sum_{j=1}^{n} w_j\, f_{ij}\, \rho A_j Y_j \tag{8.187}$$

Equation (8.186) written in matrix form becomes

$$\begin{Bmatrix} Y_1 \\ Y_2 \\ \vdots \\ Y_n \end{Bmatrix} = p^2 \begin{bmatrix} f_{11} & f_{12} & \cdots & f_{1n} \\ f_{21} & f_{22} & \cdots & f_{2n} \\ \cdots & \cdots & \cdots & \cdots \\ f_{n1} & f_{n2} & \cdots & f_{nn} \end{bmatrix} \rho \begin{bmatrix} A_1 & 0 & \cdots & 0 & 0 \\ 0 & A_2 & \cdots & 0 & 0 \\ \cdots & \cdots & \cdots & \cdots & \cdots \\ 0 & 0 & \cdots & 0 & A_n \end{bmatrix}$$

$$\begin{bmatrix} w_1 & 0 & \cdots & 0 \\ 0 & w_2 & \cdots & \cdots \\ \vdots & \cdots & \cdots & \cdots \\ 0 & 0 & \cdots & w_n \end{bmatrix} \begin{Bmatrix} Y_1 \\ Y_2 \\ \vdots \\ Y_n \end{Bmatrix} \tag{8.188}$$

or

$$\{Y\} = p^2 [F] \rho [A][w]\{Y\} \tag{8.189}$$

Equation (8.189) is an eigenvalue problem, which on solution will give n natural frequencies and n mode shapes.

Example 8.7

Determine the fundamental frequency of a cantilever beam by dividing it into three equal parts, that is, by considering four stations by the method of collocation.

The flexibility influence coefficients associated with points 1, 2 and 3 are (Fig. 8.19)

$$[F] = \frac{L^3}{162EI}\begin{bmatrix} 2 & 5 & 8 \\ 5 & 16 & 28 \\ 8 & 28 & 54 \end{bmatrix}$$

Numerical integration is done by the trapezoidal rule. According to trapezoidal rule, the weighting numbers are

$$[w] = \frac{L}{3}\begin{bmatrix} 1 & 0 & 0 \\ 0 & 1 & 0 \\ 0 & 0 & 1/2 \end{bmatrix}$$

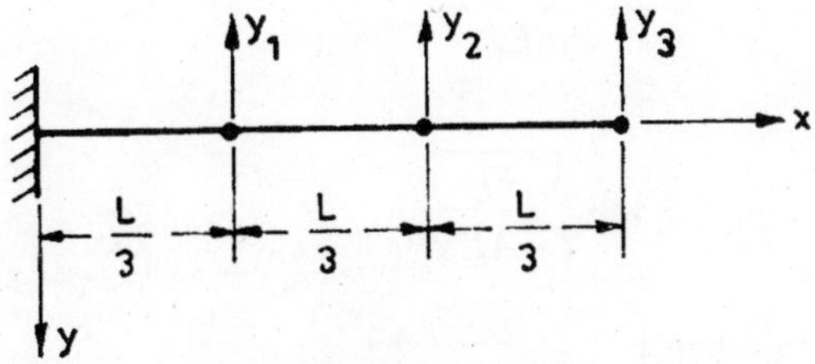

Fig. 8.19 Example 8.7

Since the beam is uniform, we can write

$$[A] = A\begin{bmatrix} 1 & 0 & 0 \\ 0 & 1 & 0 \\ 0 & 0 & 1 \end{bmatrix}$$

Therefore, from Eq. (8.188)

$$\begin{Bmatrix} Y_1 \\ Y_2 \\ Y_3 \end{Bmatrix} = p^2 \rho A \frac{L^3}{162EI} \begin{bmatrix} 2 & 5 & 8 \\ 5 & 16 & 28 \\ 8 & 28 & 54 \end{bmatrix} \begin{bmatrix} 1 & 0 & 0 \\ 0 & 1 & 0 \\ 0 & 0 & 1 \end{bmatrix} \frac{L}{3} \begin{bmatrix} 1 & 0 & 0 \\ 0 & 1 & 0 \\ 0 & 0 & 1/2 \end{bmatrix} \begin{Bmatrix} Y_1 \\ Y_2 \\ Y_3 \end{Bmatrix}$$

or,
$$\frac{1}{p^2} \begin{Bmatrix} Y_1 \\ Y_2 \\ Y_3 \end{Bmatrix} = \frac{\rho A L^4}{486EI} \begin{bmatrix} 2 & 5 & 4 \\ 5 & 16 & 14 \\ 8 & 28 & 27 \end{bmatrix} \begin{Bmatrix} Y_1 \\ Y_2 \\ Y_3 \end{Bmatrix}$$

The eigenvalues and eigenvectors can be solved by any suitable procedure. For determining the fundamental frequency, we adopt matrix iteration procedure.

$$\frac{\rho A L^4}{486EI} \begin{bmatrix} 2 & 5 & 4 \\ 5 & 16 & 14 \\ 8 & 28 & 27 \end{bmatrix} \begin{Bmatrix} 1 \\ 1 \\ 1 \end{Bmatrix} = \frac{\rho A L^4}{486EI} \begin{Bmatrix} 11 \\ 35 \\ 63 \end{Bmatrix} = \frac{11 \rho A L^4}{486EI} \begin{Bmatrix} 1.000 \\ 3.182 \\ 5.727 \end{Bmatrix}$$

$$\frac{\rho A L^4}{486EI} \begin{bmatrix} 2 & 5 & 4 \\ 5 & 16 & 14 \\ 8 & 28 & 27 \end{bmatrix} \begin{Bmatrix} 1.000 \\ 3.182 \\ 5.727 \end{Bmatrix} = \frac{\rho A L^4}{486EI} \begin{Bmatrix} 40.818 \\ 136.090 \\ 251.725 \end{Bmatrix} = \frac{40.81 \rho A L^4}{486EI} \begin{Bmatrix} 1.000 \\ 3.334 \\ 6.167 \end{Bmatrix}$$

or,

$$\frac{\rho A L^4}{486EI} \begin{bmatrix} 2 & 5 & 4 \\ 5 & 16 & 14 \\ 8 & 28 & 27 \end{bmatrix} \begin{Bmatrix} 1.000 \\ 3.334 \\ 6.167 \end{Bmatrix} = \frac{\rho A L^4}{486EI} \begin{Bmatrix} 43.338 \\ 144.682 \\ 267.861 \end{Bmatrix} = \frac{43.338 \rho A L^4}{486EI} \begin{Bmatrix} 1.000 \\ 3.338 \\ 6.180 \end{Bmatrix}$$

If we stop at this stage and determine the natural frequency, then

$$\frac{1}{p_1^2} = \frac{43.338 \rho A L^4}{486EI}$$

or
$$p_1 = 3.348 \sqrt{\frac{EI}{\rho A L^4}}$$

The exact value for the problem is

$$p_1 = 3.516 \sqrt{\frac{EI}{\rho A L^4}}$$

8.15 RAYLEIGH'S QUOTIENT FOR FUNDAMENTAL FREQUENCY

Rayleigh's method presented for lumped mass is extended here for continuous systems, for determination of fundamental frequency. It is based on the principle of conservation of energy, that is, the maximum potential energy is equal to the maximum kinetic energy. For a beam, the strain energy is given by

$$U_E = \frac{1}{2}\int_0^L EI\left(\frac{\partial^2 y}{\partial x^2}\right) dx \tag{8.190}$$

and the kinetic energy is given by

$$T_E = \frac{1}{2}\int_0^L \rho A\left(\frac{\partial y}{\partial t}\right)^2 dx \tag{8.191}$$

The deflection $y(x, t)$ is expressed as

$$y(x, t) = Y(x)\sin(pt - \alpha) \tag{8.192}$$

Therefore

$$U_E = \frac{1}{2}\int_0^L EI\left(\frac{d^2 Y}{dx^2}\right)^2 \sin^2(pt - \alpha)\, dx \tag{8.193}$$

and

$$T_E = -\frac{1}{2}p^2\int_0^L \rho A Y^2 \sin^2(pt - \alpha)\, dx \tag{8.194}$$

Equating the maximum values of U_E and T_E, we get

$$p^2 = \frac{\int_0^L EI\left(\frac{d^2 Y}{dx^2}\right)^2 dx}{\int_0^L \rho A Y^2\, dx} \tag{8.195}$$

If there are a number of springs and masses located at different points in the continuous beam, then

$$p^2 = \frac{\int_0^L EI\left(\frac{d^2 Y}{dx^2}\right)^2 dx + \sum_{i=1}^{n} k_i (Y_i)^2}{\int_0^L \rho A Y^2\, dx + \sum_{i=1}^{m} m_i (Y_i)^2} \tag{8.196}$$

If for a particular mode, the mode shape $Y(x)$ is known exactly, then Eq. (8.196) will give the exact natural frequency for that particular mode.

Example 8.8

Find the fundamental frequency for a uniform, simply supported beam by assuming the static deflection curve.

The static deflection curve for the simply supported beam due to its own weight is

$$Y = \frac{\rho A g}{24EI}(L^3 x - 2Lx^3 + x^4)$$

Therefore

$$\int_0^L \rho A(Y)^2\,dx = \rho A\left(\frac{\rho A g}{24EI}\right)^2 \int_0^L (L^3 x - 2Lx^3 + x^4)^2\,dx$$

$$= \rho A\left(\frac{\rho A g}{24EI}\right)^2 \frac{124L^9}{2520}$$

$$\frac{d^2Y}{dx^2} = \frac{\rho A g}{24EI}(-12Lx + 12x^2)$$

$$\int_0^L EI\left(\frac{d^2Y}{dx^2}\right)^2 dx = EI\left(\frac{\rho A g}{24EI}\right)^2 \int_0^L (-12Lx + 12x^2)^2\,dx$$

$$= 144\,EI\left(\frac{\rho A g}{24EI}\right)^2 \int_0^L (x^2 - Lx)^2\,dx$$

$$= \frac{144}{30}\,EI\left(\frac{\rho A g}{24EI}\right)^2 L^5$$

Therefore

$$p^2 = \frac{\frac{144}{30}\,EI\left(\frac{\rho A g}{24EI}\right)^2 L^5}{\rho A L^9\left(\frac{\rho A g}{24EI}\right)^2 \frac{124}{2520}} = \frac{144}{30}\times\frac{2520}{124}\,\frac{EI}{\rho A L^4} = 97.548\,\frac{EI}{\rho A L^4}$$

or

$$p = 9.876\sqrt{\frac{EI}{\rho A L^4}}$$

which is very close to the exact value of

$$p = 9.8696\sqrt{\frac{EI}{\rho A L^4}}$$

8.16 RAYLEIGH-RITZ METHOD FOR DETERMINING NATURAL FREQUENCIES OF CONTINUOUS SYSTEMS

In Rayleigh-Ritz method, the deflection curve $Y(x)$ is assumed in the form of a finite series

$$Y(x) = \sum_{i=1}^{n} C_i \phi_i(x) \tag{8.197}$$

where $\phi_i(x)$ is any function that satisfies the boundary conditions of the problem and C_i is a parameter.

Substituting Eq. (8.197) into Rayleigh's quotient in Eq. (8.195), we obtain

$$p^2 = \frac{\int_0^L EI \left\{ \sum_{i=1}^{n} C_i \frac{d^2\phi_i}{dx^2} \right\}^2 dx}{\int_0^L \rho A \left\{ \sum_{i=1}^{n} C_i \phi_i \right\}^2 dx}$$

$$= \frac{\sum_{i=1}^{n} \sum_{j=1}^{n} C_i C_j \left[\int_0^L EI \frac{d^2\phi_i}{dx^2} \cdot \frac{d^2\phi_j}{dx^2} dx \right]}{\sum_{i=1}^{n} \sum_{j=1}^{n} C_i C_j \left[\int_0^L \rho A \phi_i \phi_j \, dx \right]} \tag{8.198}$$

or

$$p^2 = \frac{\sum_{i=1}^{n} \sum_{j=1}^{n} k_{ij} C_i C_j}{\sum_{i=1}^{n} \sum_{j=1}^{n} m_{ij} C_i C_j} \tag{8.199}$$

where

$$k_{ij} = \int_0^L EI \frac{d^2\phi_i}{dx^2} \cdot \frac{d^2\phi_j}{dx^2} dx$$

$$m_{ij} = \int_0^L \rho A \phi_i \phi_j \, dx$$

The frequencies determined from Eq. (8.199) is a minimum. In order to do that, we differentiate p^2 with respect to each of these constants and put them equal to zero. Therefore

$$\frac{2 \sum_{j=1}^{n} k_{ij} C_j}{\sum_i \sum_j m_{ij} C_i C_j} - \frac{2 \left(\sum_{i=1}^{n} \sum_{j=1}^{n} k_{ij} C_i Cj \right) \left(\sum_{j=1}^{n} m_{ij} C_j \right)}{\left(\sum_{i=1}^{n} \sum_{j=1}^{n} m_{ij} C_i C_j \right)} = 0 \tag{8.200}$$

Combining Eq. (8.200) with Eq. (8.199), we get

$$\sum k_{ij} C_j - p^2 \sum m_{ij} C_j = 0 \tag{8.201}$$

Equation (8.201) can be written in matrix form as

$$([K] - p^2 [M]) \{\phi\} = 0 \tag{8.202}$$

Equation (8.202) represents an eigenvalue problem, which on solution yields natural frequencies and mode shapes.

Example 8.9

Determine the first two natural frequencies of a uniform cantilever beam by Rayleigh-Ritz method.

Let us assume

$$\phi(x) = C_1 x^2 + C_2 x^3$$

Therefore $$\frac{d\phi}{dx} = 2C_1 x + 3C_2 x^2 \text{ and } \frac{d^2\phi}{dx^2} = 2C_1 + 6C_2 x$$

Now

$$k_{11} = \int_0^L EI \cdot 2 \cdot 2 \, dx = 4EI \cdot L$$

$$k_{12} = k_{21} = \int_0^L EI \cdot 2 \cdot 6x \, dx = 6EIL^2$$

$$k_{22} = \int_0^L EI \cdot 6x \cdot 6x \, dx = 12EIL^3$$

$$m_{11} = \int_0^L \rho A x^2 \cdot x^2 \, dx = \frac{\rho A L^5}{5}$$

$$m_{22} = \int_0^L \rho A x^3 \cdot x^3 \, dx = \frac{\rho A L^7}{7}$$

$$m_{12} = m_{21} = \int_0^L \rho A x^2 \cdot x^3 \, dx = \frac{\rho A L^6}{6}$$

Substituting the above values in Eq. (8.202), we get

$$\left(-\rho A p^2 \begin{bmatrix} \dfrac{L^5}{5} & \dfrac{L^6}{6} \\ \dfrac{L^6}{6} & \dfrac{L^7}{7} \end{bmatrix} + \begin{bmatrix} 4EIL & 6EIL^2 \\ 6EIL^2 & 12EIL^3 \end{bmatrix} \right) \begin{Bmatrix} C_1 \\ C_2 \end{Bmatrix} = \begin{Bmatrix} 0 \\ 0 \end{Bmatrix}$$

For a nontrivial solution of the above equation

$$\begin{vmatrix} 4EIL - p^2 \rho A L^5/5 & 6EIL^2 - p^2 \rho A L^2/6 \\ 6EIL^2 - p^2 \rho A L^6/6 & 12EIL^3 - p^2 \rho A L^7/7 \end{vmatrix} = 0$$

The frequency equation becomes

$$p^4 - 1224 p^2 \left(\frac{EI}{\rho A L^4} \right) + 15.120 p^4 \left(\frac{EI}{\rho A L^4} \right)^2 = 0$$

The solution of the above equation yields

$$p_1 = 3.52 \sqrt{\frac{EI}{\rho A L^4}} \quad \text{and} \quad p_2 = 22.03 \sqrt{\frac{EI}{\rho A L^4}}$$

8.17 VIBRATION OF MEMBRANES

A membrane is a plate subjected to tensile loading. Entire loading is assumed to be inplane. There are plenty of application areas of the membranes, such as inflatable structures, parachutes, drumheads and biomedical components such as the valve of an artificial heart.

The freebody diagram of the rectangular membrane element is shown in Fig. 8.20. Let $f(x, y, t)$ be the pressure loading acting in the z-direction. Let T be the intensity of tension at a point. T remains constant throughout the membrane. On an elemental area $dx\,dy$, forces Tdx and Tdy act on the sides parallel to y and x axes respectively (Fig. 8.20).

The net forces in the z-direction due to these forces are

$$\left(T \frac{\partial^2 w}{\partial y^2} dx\,dy \right) \quad \text{and} \quad \left(T \frac{\partial^2 w}{\partial x^2} dx\,dy \right)$$

If m denotes the mass per unit area, the equation of motion of the forced transverse vibration of the membrane is

$$T\left(\frac{\partial^2 w}{\partial x^2} + \frac{\partial^2 w}{\partial y^2} \right) + f = m \frac{\partial^2 w}{\partial t^2} \tag{8.203}$$

The free vibration equation of the membrane is

$$T\left(\frac{\partial^2 w}{\partial x^2}+\frac{\partial^2 w}{\partial y^2}\right)=m\,\frac{\partial^2 w}{\partial t^2} \tag{8.204}$$

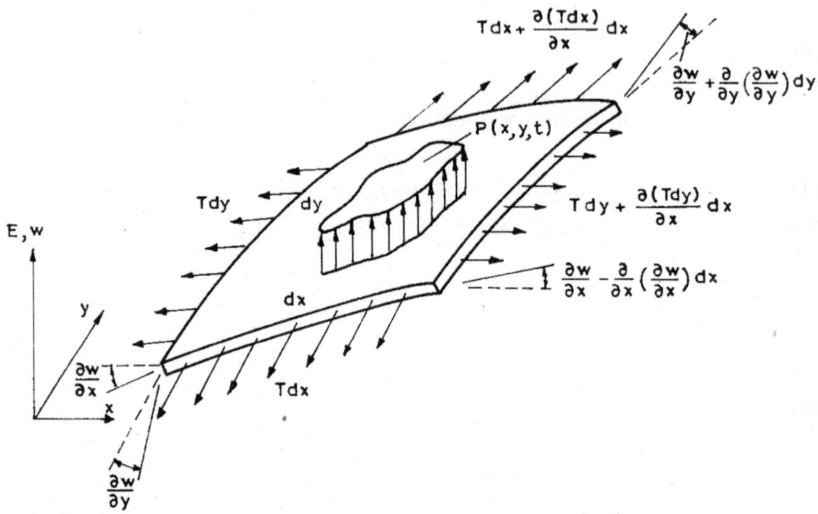

Fig. 8.20 Freebody diagram of a rectangular membrane

If $\dfrac{T}{m}=c^2$, then Eq. (8.204) reduces to

$$c^2\left(\frac{\partial^2 w}{\partial x^2}+\frac{\partial^2 w}{\partial y^2}\right)=\frac{\partial^2 w}{\partial t^2} \tag{8.205}$$

Assume a solution of Eq. (8.205) as

$$w(x,y,t)=W(x,y)\;q(t) \tag{8.206}$$

Substituting Eq. (8.206) into Eq. (8.205), yields

$$c^2\left[\frac{\partial^2 W}{\partial x^2}+\frac{\partial^2 W}{\partial y^2}\right]q=W\ddot{q} \tag{8.207}$$

The respective eigenvalue problem is

$$\nabla^2 W(x,y)+\beta^2\,W(x,y)=0 \tag{8.208}$$

where $$\nabla^2=\frac{\partial^2}{\partial x^2}+\frac{\partial^2}{\partial y^2} \tag{8.209}$$

$$\beta^2 = \left(\frac{p}{c}\right)^2 = p^2\left(\frac{m}{T}\right) \tag{8.210}$$

∇ is called the Laplacian operator.

We further substitute $W(x, y) = X(x)\,Y(y)$ into Eq. (8.208) to get

$$\frac{X''}{X} + \frac{Y''}{Y} + \beta^2 = 0 \tag{8.211}$$

Eq. (8.211) involves summation of functions x and y which add up to a constant. As such these functions of x and y separately equal to the constant.
As such

$$\frac{X''}{X} = -\frac{Y''}{Y} - \beta^2 = -\alpha^2 \tag{8.212}$$

The negative sign is chosen for the constant a^2 on physical grounds in order that the solution be harmonic in x and y. Therefore, following two equations result

$$X'' + \alpha^2 X = 0 \tag{8.213}$$

$$Y'' + \gamma^2 Y = 0 \tag{8.214}$$

where $\gamma^2 = \beta^2 - \alpha^2$.

Solutions of Eqs. (8.213) and (8.214) are

$$X(x) = A_1 \sin \alpha x + A_2 \cos \alpha x \tag{8.215}$$

$$Y(y) = A_3 \sin \gamma y + A_4 \cos \gamma y \tag{8.216}$$

Therefore, substituting Eqs. (8.215) and (8.216) into Eq. (8.206), yields

$$W(x, y) = C_1 \sin \alpha x \sin \gamma y + C_2 \sin \alpha x \cos \gamma y$$
$$+ C_3 \cos \alpha x \sin \gamma y + C_4 \cos \alpha x \cos \gamma y \tag{8.217}$$

The coefficients C_i are to be determined from boundary conditions. Let us consider a membrane simply supported on all four sides. The boundary conditions are

$W(0, y) = 0$, $w(a, y) = 0$, $W(x, 0) = 0$, $w(x, b) = 0$.

i) Along $x = 0$, $W(0, y) = C_3 \sin \gamma y + C_4 \cos \gamma y = 0$ which can be true for any value of y, if $C_3 = C_4 = 0$.

ii) Along $x = a$, $W(a, y) = C_1 \sin \alpha a \sin \gamma y + C_2 \sin \alpha a \cos \gamma y = 0$. One possible solution is $C_1 = C_2 = 0$ which is a trivial solution. Other possible solution is $\sin \alpha a = 0$.

iii) Along $y = 0$, $w(x, 0) = C_2 \sin \alpha x + C_4 \cos \alpha x = 0$ which can be true for any value of x, provided $C_2 = C_4 = 0$.

iv) Along $y = b$, $W(x, b) = C_1 \sin \alpha x \sin \alpha b + C_3 \cos \alpha x \sin \gamma b = 0$

One possible solution is $C_1 = C_3 = 0$ which is a trivial solution. Other possible solution is

$\sin \gamma b = 0$

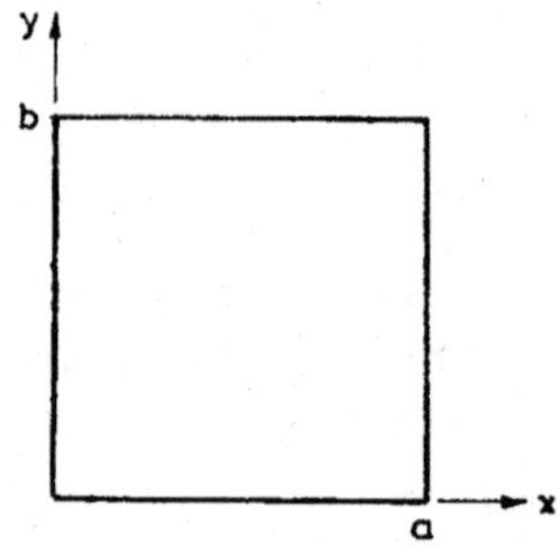

Fig. 8.21 A rectangular membrane

The only solution that do not satisfy all the above-mentioned conditions is that $C_1 \neq 0$ along with the following characteristic equations,

$$\sin \alpha a \rightarrow \alpha_j a = j\pi \tag{8.218}$$

$$\sin \gamma b \rightarrow \gamma_k b = k\pi \tag{8.219}$$

where $j, k = 1, 2, ...$

Therefore,

$$\beta_{jk} = (\alpha_j^2 + \gamma_k^2)^{1/2} = \pi\left[\left(\frac{j}{a}\right)^2 + \left(\frac{k}{b}\right)^2\right]^{1/2} \tag{8.220}$$

where $p_{jk} = C\beta_{jk} = \sqrt{\frac{T}{m}}\,\beta_{jk},\ j, k = 1, 2,......$

The mode shapes are given by

$$W_{jk}(x, y) = C_{jk}\sin\frac{j\pi x}{a}\sin\frac{k\pi y}{b} \quad j, k = 1, 2, ... \tag{8.221}$$

Two mode shapes have been presented for a square membrane in Fig. 8.22. For details of vibration of membranes, Ref. 8.4 and 8.5 may be looked into.

8.18 TRANSVERSE VIBRATION OF RECTANGULAR THIN PLATES

A body having its middle surface (the surface obtained by bisecting the upper and the lower face of the plate) in the form of a plane and whose thickness is sufficiently small compared to its other two dimensions, is called a thin plate. In deriving the theory, the following assumptions are made:

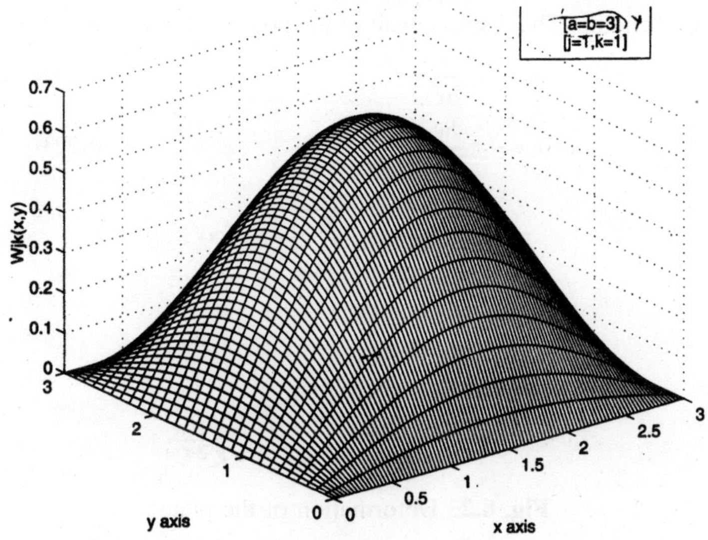

Fig. 8.22 (a) Example of mode for square membrane

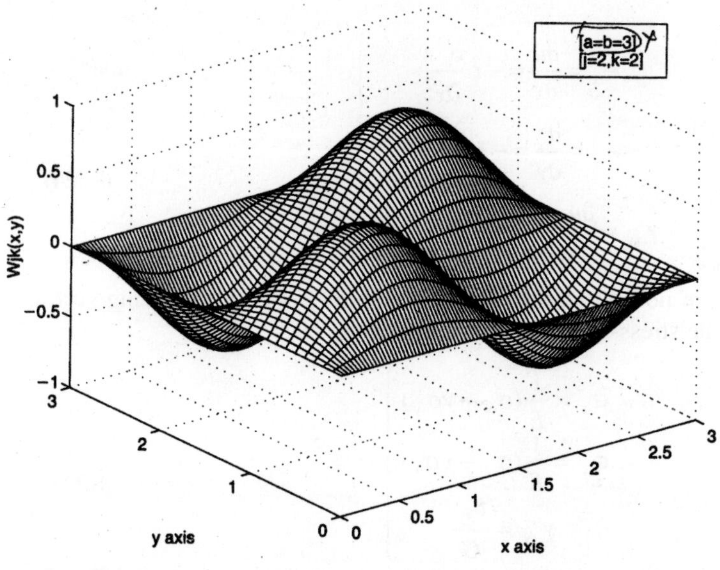

Fig. 8.22 (b) Example of mode for square membrane

(1) A normal to the middle surface of the plate before bending, transforms into a normal to the middle surface after bending.

(2) The middle surface remains unstrained after bending.

(3) The normal stress component perpendicular to the plane of the plate is small, compared to the other stress components and is neglected in the stress-strain relationship.

Figure 8.23 indicates the section of a plate parallel to *xz*-plane. During transverse vibration, a point P in the middle plane is deflected to p' with a deflection w. Another point Q, at a distance z from the undeformed middle plane is displaced to Q', which is on the normal to the middle plane after bending. The displacement of the point Q in the x-direction is given by

$$u = -z\frac{\partial w}{\partial x} \tag{8.222}$$

Similarly

$$v = -z\frac{\partial w}{\partial y} \tag{8.223}$$

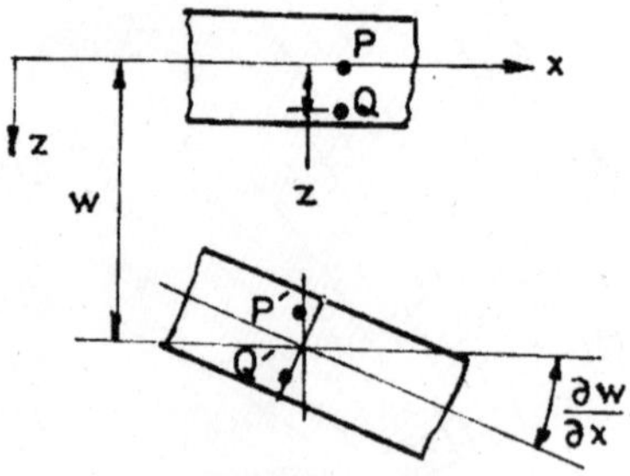

Fig. 8.23 Deformation of the plate

where, u, v and w are the displacements in the x, y and z directions.
The axial and shearing strains can be expressed in terms of w through Eqs. (8.222) and (8.223)

$$\left.\begin{aligned} \epsilon_x &= \frac{\partial u}{\partial x} = -z\frac{\partial^2 w}{\partial x^2} \\ \epsilon_y &= \frac{\partial v}{\partial y} = -z\frac{\partial^2 w}{\partial y^2} \\ \gamma_{xy} &= \frac{\partial u}{\partial y} + \frac{\partial v}{\partial x} = -2z\frac{\partial^2 w}{\partial x\,\partial y} \end{aligned}\right\} \tag{8.224}$$

where ϵ_x, ϵ_y are normal strain components and γ_{xy} is the shearing strain. As $\sigma_z = 0$ and $\tau_{xz} = \tau_{zx} = 0$, the stress-strain relation becomes

$$\left.\begin{aligned} \epsilon_x &= \frac{1}{E}(\sigma_x - \nu\sigma_y) \\ \epsilon_y &= \frac{1}{E}(\sigma_y - \nu\sigma_x) \\ \gamma_{xy} &= \frac{\tau_{xy}}{G} \end{aligned}\right\} \tag{8.225}$$

σ_x, σ_y and σ_z are the normal stresses in the x, y and z directions and τ_{xy}, τ_{yz} and τ_{zx} are the shearing stress components.

Equation (8.225) will yield the solution as

$$\left.\begin{aligned} \sigma_x &= \frac{E}{1-\nu^2}(\epsilon_x + \nu\epsilon_y) \\ \sigma_y &= \frac{E}{1-\nu^2}(\epsilon_y + \nu\epsilon_x) \\ \tau_{xy} &= \frac{E}{1-\nu^2}\left(\frac{1-\nu}{2}\gamma_{xy}\right) \end{aligned}\right\} \tag{8.226}$$

Combining Eqs. (8.224) and (8.226), we get

$$\left.\begin{aligned}\sigma_x &= -\frac{Ez}{1-\nu^2}\left(\frac{\partial^2 w}{\partial x^2}+\nu\frac{\partial^2 w}{\partial y^2}\right)\\ \sigma_y &= -\frac{Ez}{1-\nu^2}\left(\frac{\partial^2 w}{\partial y^2}+\nu\frac{\partial^2 w}{\partial x^2}\right)\\ \tau_{xy} &= \frac{Ez}{1-\nu^2}(1-\nu)\frac{\partial^2 w}{\partial x\,\partial y}\end{aligned}\right\} \tag{8.227}$$

The bending moment per unit width about *x*-axis is

$$M_x = \int_{-t/2}^{t/2} \sigma_x \, x\, dx \tag{8.228}$$

Substituting σ_x from the first of the Eq.(8.227) into Eq. (8.228), we get

$$M_x = -\frac{Et^3}{12(1-\nu^2)}\left(\frac{\partial^2 w}{\partial x^2}+\nu\frac{\partial^2 w}{\partial y^2}\right) \tag{8.229}$$

where $D = \dfrac{Et^3}{12(1-\nu^2)}$, the flexural rigidity of the plate.

Similarly, it can be shown that

$$M_y = -D\left(\frac{\partial^2 w}{\partial y^2}+\nu\frac{\partial^2 w}{d x^2}\right) \tag{8.230}$$

and

$$M_{xy} = -M_{yx} = D(1-\nu)\frac{\partial^2 w}{\partial x\,\partial y} \tag{8.231}$$

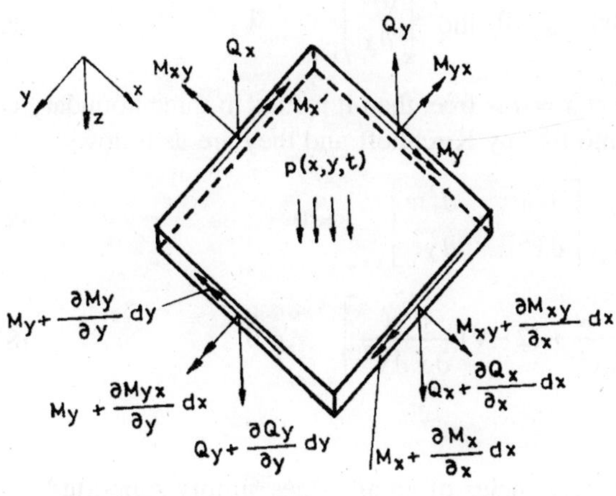

Fig. 8.24 Freebody diagram of a plate

The necessary forces acting on the element of the plate are shown in Fig. 8.24. The equilibrium equations obtained by resolving the forces in the z-direction and taking moments about y and x-axis, after dividing by dx dy are

$$\frac{\partial Q_x}{\partial x}+\frac{\partial Q_y}{\partial y}+q=\rho t\frac{\partial^2 W}{\partial t^2} \tag{8.232}$$

q(x,y,t)

$$\frac{\partial M_x}{\partial x}+\frac{\partial M_{yx}}{\partial y}-Q_x=0 \tag{8.233}$$

$$-\frac{\partial M_y}{\partial y}+\frac{\partial M_{xy}}{\partial x}+Q_y=0 \tag{8.234}$$

Eliminating Q_x and Q_y from Eqs. (8.232) to (8.234), we get

$$\frac{\partial^2 M_x}{\partial x^2}-2\frac{\partial^2 M_{xy}}{\partial x\partial y}+\frac{\partial^2 M_y}{\partial y^2}+q(x,y,t)=\rho t\frac{\partial^2 W}{\partial t^2} \tag{8.235}$$

Expressing the moments in terms of the curvatures as given in Eqs. (8.229) to (8.235), we get

$$D\left[\frac{\partial^4 w}{\partial x^4}+2\frac{\partial^4 w}{\partial x^2\,\partial y^2}+\frac{\partial^4 w}{\partial y^4}\right]+\rho t\frac{\partial^2 w}{\partial t^2}=p(x,y,t) \tag{8.236}$$

Equation (8.236) is the equation of motion for the vibrating plate. For free vibration, $q(x, y, t) = 0$. The value of $w(x, y, t)$ should be such that it must satisfy the boundary conditions at the edges of the plate.

If the plate has an edge x = a as simply supported, then

$$(w)_{x=a}=0 \text{ and } \left(\frac{\partial^2 w}{\partial x^2}\right)_{x=a}=0 \tag{8.237}$$

If an edge of the plate at $x = a$ is clamped, then

$$(w)_{x=a}=0 \text{ and } \left(\frac{\partial w}{\partial x}\right)_{x=a}=0 \tag{8.238}$$

If an edge of the plate at $x = a$ is free, then it results in three boundary conditions. Those three have been combined into two by Kirchhoff and they are as follows:

$$\left[\frac{\partial^2 w}{\partial x^2}+\nu\frac{\partial^2 w}{\partial y^2}\right]_{x=a}=0 \tag{8.239}$$

$$\left[\frac{\partial^3 w}{\partial x^3}+(2-\nu)\frac{\partial^2 w}{\partial x\,\partial y^2}\right]_{x=a}=0 \tag{8.240}$$

Example 8.10

Determine the natural frequencies of an all edges simply supported rectangular plate having side dimensions a and b and thickness t.

The deflection w is assumed as

$$w(x, y, t) = W(x, y)\sin(pt - \alpha) \tag{a}$$

Substituting Eq. (a) in Eq. (8.236) results in

$$\frac{\partial^4 W}{\partial x^4} + 2\frac{\partial^4 W}{\partial x^2 \partial y^2} + \frac{\partial^4 W}{\partial y^4} - \frac{\rho t p^2}{D} W = 0 \tag{b}$$

The deflection W is expressed in the following form

$$W = \sum_m \sum_n A_{mn} \sin\frac{m\pi x}{a} \sin\frac{n\pi y}{b} \tag{c}$$

Each term of the series in Eq. (c) satisfies the boundary conditions of the edges of the plate.

Substituting W from Eq.(c) into Eq.(b), gives

$$\frac{m^4\pi^4}{a^4} + 2\frac{m^2 n^2 \pi^4}{a^2 b^2} + \frac{n^4 \pi^4}{b^4} = \lambda^4 \tag{d}$$

where

$$\lambda^4 = \pi^4\left(\frac{m^2}{a^2} + \frac{n^2}{b^2}\right)^2 \tag{e}$$

Therefore

$$p(m, n) = \left(\frac{\pi^4 D}{\rho t b^4}\right)^{1/2}\left[m^2\left(\frac{b}{a}\right)^2 + n^2\right] \tag{f}$$

It is convenient to use the parameters (m, n) to describe the mode. The integers m and n are equal to half-sine waves in the x and y-directions, respectively.

References

8.1 H. Benaroya, Mechanical Vibration, Prentice-Hall, New Jersey, 1998.

8.2 J. B. Achenbach, Wave Propagation in Elastic Solids, North-Holland Publishing Co., Amsterdam, 1973.

8.3 M. S. Triantafyllou, Linear dynamics of cables and chains, Shock and Vibration Digest, V.16, March 1984, pp. 9-17.

8.4 N. Y. Oleer, General solution to the equation of vibrating membrane, Journal of Sound and Vibration, V.6, 1967, pp. 365-374.

8.5 J. Mazumdar, A review of approximate methods for determining the vibrational modes of membranes, Shock and Vibration Digest, V.14, Feb. 1982, pp. 11-17.

EXERCISE 8

8.1 Determine the velocity of wave propagation in an underwater cable of mass $m = 10$ kg/m and the applied tension $T = 6000$ N.

8.2 The cord of a musical instrument is fixed at both ends and has a length of 1 m, diameter $d = 0.5$ mm and density 7800 kg/m^3. Determine the tension needed in the cord so as to have a fundamental frequency of transverse vibration 200 Hz.

8.3 A flexible string of mass m per unit length is stretched under tension T between two fixed points placed at a distance L. Determine the natural frequencies of the string.

8.4 Determine the appropriate axial deformation. Boundary conditions at $x = 0$ for the two members shown in the figure.

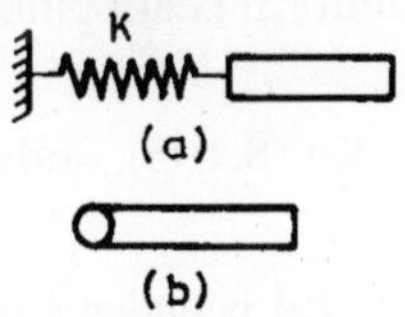

Prob. 8.4.

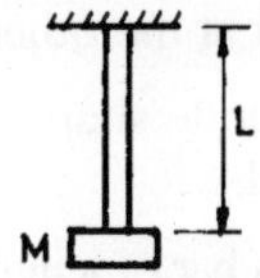

Prob. 8.7.

8.5 A uniform bar of length L having both ends clamped, is excited by a force in the axial direction, which is suddenly removed. Determine the displacement equation of the bar.

8.6 A uniform circular shaft of length L and polar moment of inertia I_p has two discs of moment of inertia I_0 fitted at two ends. Determine the natural frequency of the system and the general solution for the equation of motion.

8.7 A bar shown in the figure has a concentrated mass M attached at the free end. Determine its natural frequency. The bar is uniform and has a cross-sectional area A and density γ.

8.8 A prismatic bar having both ends free is 5.1 m long and weighs 27,000 N/m^3. The lowest natural frequency in longitudinal vibration of the rod is 500 cycles per second. Determine the modulus of elasticity of the material of the bar, if the area of the bar = 0.07 m^2.

8.9 A uniform bar of length L is clamped at one end and is attached to a torsional spring of stiffness k at the other end. Determine the frequency equation.

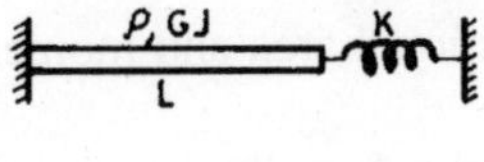

Prob. 8.9.

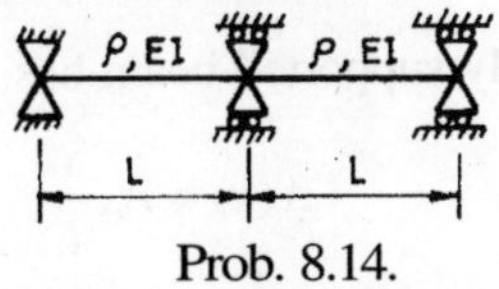

Prob. 8.14.

8.10 A shaft has a length of 2m, diameter 50 mm, modulus of rigidity 7.95×10^{11} N/m^2 and density 7800 kg/m^3. Both ends of the shaft are fixed. Determine the fundamental frequency of the torsional vibration of the shaft.

8.11 For the transverse vibration of a simply supported uniform beam, solve for free response if the initial conditions are given by

$$y(x, 0) = B\left(\frac{x}{L} - 3\frac{x^2}{L^2} + 2\frac{x^3}{L^3}\right), \quad y(x, 0) = 0$$

8.12 Derive an expression for natural frequencies of transverse vibration of a uniform fixed-fixed beam.

8.13 Determine the natural frequencies of flexural vibration of a cantilever beam.

8.14 Determine the natural frequencies of the flexural vibration of a two-span beam shown in the figure. Both the spans are identical.

8.15 Determine the solution of the free inplane vibrations of a rectangular membrane, having sides a and b. Assume the edges of the membrane to be clamped.

8.16 A beam of a particular material is supported at the left hand end and at a distance of $0.264L$ from the right hand end. The dimensions of beam in mm are $50 \times 50 \times 300$, its fundamental frequency is 1160 cps and its density is 2450 kg/m^3. Determine the modulus of elasticity of the beam.

8.17 Determine the frequency equation of a uniform beam pinned at one end and elastically supported at the other.

8.18 For the mode shape function given by Eq.(8.105), show that the mode shapes are orthogonal.

8.19 A uniform bar of length L and axial rigidity EA is undergoing axial vibrations. Determine the fundamental frequency of the bar by the method of collocation.

8.20 For a uniform beam simply supported at both ends, having span L and flexural rigidity EI, determine the natural frequencies of flexural vibrations due to its own mass.

8.21 Find the natural frequency of the wedge-shaped cantilever beam shown in the figure by Rayleigh-Ritz method. The beam is of constant width. Assume,

$$Y(x) = a_1 x^2 + a_2 x^3$$

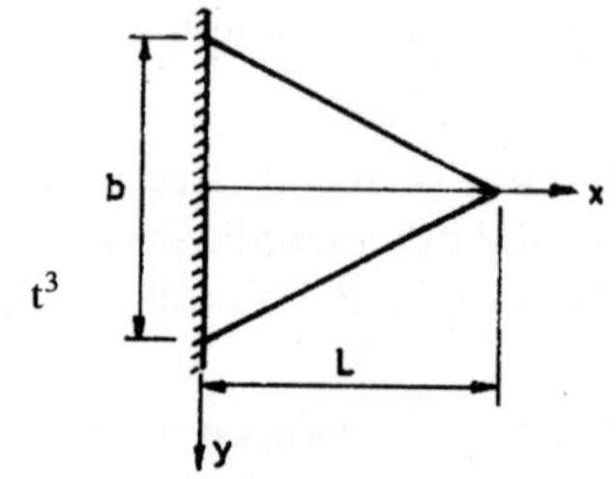

Prob. 8.21

8.22 A simply supported beam has cross-sectional area and moment of inertia varying as

$$A(x) = A_0 \frac{x}{L} \quad \text{and} \quad I(x) = I_0 \frac{x}{L}$$

Determine the fundamental frequency of the beam.

8.23 Determine the effect of rotary inertia on a freely vibrating uniform cantilever beam.

8.24 For a simply supported beam, determine the natural frequencies by considering it to be uniform. Consider the bending and shearing deformation and not rotary inertia.

8.25 Determine an approximate frequency of rectangular membrane supported along all edges by Rayleigh's method with

$$W(x, y) = C_1 \, xy(x-a)(y-b)$$

Chapter 9

Forced Vibration of Continuous Systems

9.1 INTRODUCTION

In the previous chapter, we have considered the free vibration analysis of continuous systems. We pass on to the forced vibration analysis of continuous systems in this chapter. Though we start with axial vibration problem, the major emphasis will be placed on flexural vibrations of beams. It might have been noted by the readers while going through the previous chapter that the natural frequencies associated with flexural vibrations are of much lower magnitude than those of torsional and axial vibrations. Hence, for practical problems they become much more important. The entire treatment of this chapter is based on mode summation procedure [9.1 - 9.4].

9.2 FORCED AXIAL VIBRATION OF BARS

Let us consider the bar of Fig. 9.1, which is fixed at one end, and at the free end an exciting force $P(t)$ is applied. As discussed in Art. 8.2, the forced vibration equation is given by

$$EA\frac{\partial^2 u}{\partial x^2} - \rho A\frac{\partial^2 u}{\partial t^2} = -P(t) \tag{9.1}$$

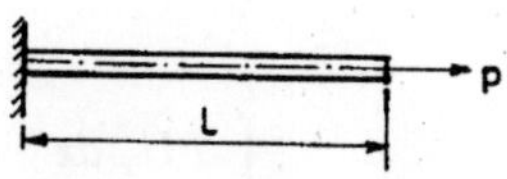

Fig. 9.1 A bar

Let us express the displacement u in terms of normal coordinates.

Therefore

$$u(x,t) = \sum U_r(x)\xi_r(t) \tag{9.2}$$

Let us substitute u from Eq. (9.2) into Eq. (9.1) which yields

$$\sum\left(EA\frac{d^2U_r}{dx^2}\xi_r - \rho A U_r \ddot{\xi}_r\right) = -P \tag{9.3}$$

From Eq. (8.13), we get

$$EA\frac{d^2U_r}{dx^2} = -\rho A p_r^2 U_r \tag{9.4}$$

Combining Eqs. (9.3) and (9.4), we get

$$\sum \rho A U_r (\ddot{\xi}_r + p^2 \xi_r) = P \tag{9.5}$$

Let

$$P = \sum U_s P_s \tag{9.6}$$

Multiplying both sides of Eq. (9.6) by U_r, and integrating from 0 to L, we get

$$\int_0^L PU_r\,dx = \int_0^L \sum U_r U_s\,dx\,P_s$$

or

$$\sum P_s \int_0^L U_r U_s\,dx = \int_0^L PU_r\,dx \tag{9.7}$$

From orthogonality relationship, we know that for a uniform bar

$$\int_0^L U_r U_s\,dx = 0 \text{ where } r \neq s \tag{9.8}$$

Therefore, making use of orthogonality relationship given by Eq. (9.8), Eq. (9.7) can be written as

$$P_r(t) = \frac{\int_0^L PU_r\,dx}{\int_0^L U_r^2\,dx} \tag{9.9}$$

In this case, P has been assumed to be constant and independent of time and acting at $x = L$. Combining Eqs. (9.5), (9.6) and (9.9), we get

$$\sum \rho A U_r (\ddot{\xi}_r + p_r^2 \xi_r) = \sum U_r \frac{\int_0^L PU_r\,dx}{\int_0^L U_r^2\,dx} \tag{9.10}$$

Therefore, rth equation of Eq. (9. 10) becomes

$$\ddot{\xi}_r + p_r^2 \xi_r = \frac{1}{\rho A}\frac{\int_0^L PU_r\,dx}{\int_0^L U_r^2\,dx} \tag{9.11}$$

From Eqs. (8.62) and (8.63), we know

$$U_r = \sin\frac{(2r-1)\pi x}{2L} \tag{9.12}$$

and
$$P_r = \frac{(2r-1)\pi}{2L}\sqrt{\frac{E}{\rho}} \tag{9.13}$$

Now
$$\int_0^L PU_r\,dx = \int_0^L P(x=L)\sin\frac{(2r-1)}{2L}\pi x\,dx = P.\sin\frac{(2r-1)\pi}{2} \tag{9.14}$$

and
$$\int_0^L U_r^2\,dx = \int_0^L \sin^2\frac{(2r-1)}{2L}\pi x\,dx = \frac{L}{2} \tag{9.15}$$

Combining Eqs. (9.11), (9.13), (9.14) and (9.15), we get

$$\ddot{\xi}_r + \frac{(2r-1)^2\pi^2}{4L^2}\frac{E}{\rho}\,\xi_r = \frac{2P}{\rho AL}\sin\frac{(2r-1)}{2}\pi \tag{9.16}$$

We know
$$a^2 = \frac{E}{\rho} \tag{9.17}$$

Substituting a^2 for $\frac{E}{\rho}$ in Eq. (9.16), we get

$$\ddot{\xi}_r + \frac{(2r-1)^2\pi^2 a^2}{4L^2}\,\xi_r = \frac{2P}{\rho AL}\sin\frac{(2r-1)}{L}\pi \tag{9.18}$$

Using Duhamel's integral, solution of Eq. (9.18) is

$$\xi_r = \frac{4P}{(2r-1)\rho A\pi a}\sin\frac{(2r-1)}{2}\pi\int_0^t \sin\frac{(2r-1)\pi a}{2L}(t-\tau)\,d\tau \tag{9.19}$$

assuming the structure to be at rest at $t = 0$.

Therefore, the displacement $u(x, t)$ is given by, after substituting U_r from Eq.(9.12) and ξ_r from Eq. (9.19) in Eq. (9.2)

$$u = \frac{4P}{\rho A\pi a}\sum\frac{1}{(2r-1)^2}\sin\frac{(2r-1)}{2}\pi\sin\frac{(2r-1)\pi x}{2L}\int_0^t \sin\frac{(2r-1)\pi a}{2L}(t-\tau)\,d\tau \tag{9.20}$$

If P is suddenly applied at the time $t = 0$, then Eq. (9.20) becomes

$$u = \frac{8PL}{\pi^2 a^2\rho A}\sum\frac{1}{(2r-1)^2}\sin\frac{(2r-1)\pi}{2}\sin\frac{(2r-1)\pi x}{2L}$$

$$\times\left[1-\cos\frac{(2r-1)\pi at}{2L}\right] \tag{9.21}$$

Maximum displacement will occur at the end of the bar, where $x = L$ and at time $t = 2L/a$. Noting that

$$\sum_{r=1,2,3...} \frac{1}{(2r-1)^2} = \frac{\pi^2}{8}$$

and substituting it in Eq. (9.21) for the maximum value, we get

$$(u)_{x=L} = \frac{2PL}{AE} \tag{9.22}$$

The suddenly applied load therefore produces twice the deflection than that one would obtain, if the load is applied gradually.

9.3 FORCED VIBRATION OF THE SHEAR BEAM UNDER GROUND MOTION EXCITATION

The equation of motion for the shear beam subjected to ground motion is given by

$$\frac{\partial}{\partial x}\left(\mu AG \frac{\partial z}{\partial x}\right) = \rho A \frac{\partial^2 z}{\partial t^2} + \rho A \ddot{y}_g(t) \tag{9.23}$$

where z is the relative displacement with respect to the base and $\ddot{y}_g(t)$ is the ground acceleration [Fig. 9.2].

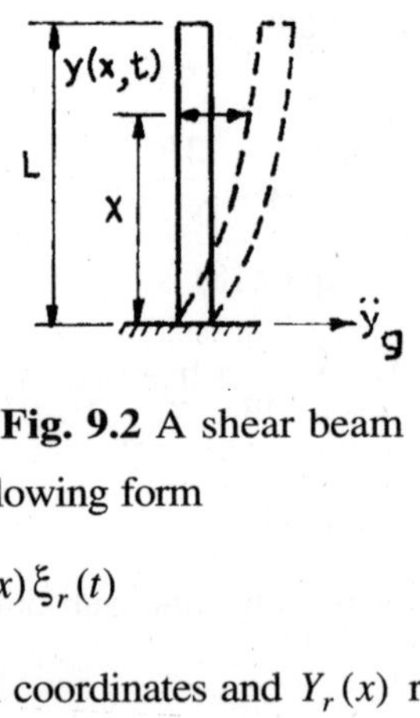

Fig. 9.2 A shear beam

Let us assume a solution of the following form

$$z(x,t) = \sum Y_r(x)\xi_r(t) \tag{9.24}$$

where $\xi_r(t)$ represents the normal coordinates and $Y_r(x)$ represents the mode shapes.

Substituting z from Eq. (9.24) into Eq. (9.23), we get

$$\sum \xi_r \left\{\frac{d}{dx}\left(\mu AG \frac{dY_r}{dx}\right)\right\} = \sum \rho A Y_r \ddot{\xi}_r + \rho A \ddot{y}_g \tag{9.25}$$

For the free vibration, it can be shown by substituting $y_r = Y_r \sin p_r t$ in Eq. (8.155), after modifying the equation for a beam having non-uniform cross-section that

$$\frac{d}{dx}\left(\mu AG \frac{dY_r}{dx}\right) = -\rho A p_r^2 Y_r \tag{9.26}$$

Combining Eqs. (9.25) and (9.26) we get

$$\sum \rho A Y_r (\ddot{\xi}_r + p_r^2 \xi_r) = -\rho A \ddot{y}_g \tag{9.27}$$

Let
$$\rho A \ddot{y}_g = \sum \rho A Y_r f_r(t) \tag{9.28}$$

Multiplying both sides of Eq. (9.28) by Y_s and integrating with respect to x from 0 to L, we get

$$\int_0^L \rho A \ddot{y}_g Y_s \, dx = \int_0^L \sum \rho A Y_r Y_s f_r(t) \, dx$$

or
$$\int_0^L \rho A \ddot{y}_g Y_s \, dx = \sum f_r(t) \int_0^L \rho A Y_r Y_s \, dx \tag{9.29}$$

Making use of the orthogonality relationship, we get

$$\int_0^L \rho A Y_r Y_s \, dx = 0 \text{ when } r \neq s \tag{9.30}$$

Equation (9.29) reduces to

$$f_r(t) = \frac{\int_0^L \rho A Y_r \, dx}{\int_0^L \rho A Y_r^2 \, dx} \ddot{y}_g \tag{9.31}$$

Combining Eqs. (9.27), (9.28) and (9.31), we get

$$\ddot{\xi}_r + p_r^2 \xi_r = -\ddot{y}_g \frac{\int_0^L \rho A Y_r \, dx}{\int_0^L \rho A Y_r^2 \, dx} \tag{9.32}$$

The solution of Eq. (9.32) can be written as

$$\xi_r = -\frac{\int_0^L \rho A Y_r \, dx}{\int_0^L \rho A Y_r^2 \, dx} \frac{1}{p_r} \int_0^t \ddot{y}_g(\tau) \sin p_r (t - \tau) d\tau \tag{9.33}$$

Substituting ξ_r from Eq. (9.33) into Eq. (9.24), we get

$$z(x,t) = -\sum \frac{Y_r(x) \int_0^L \rho A Y_r \, dx}{\int_0^L \rho A Y_r^2 \, dx} \frac{1}{p_r} \int_0^t \ddot{y}_g(\tau) \sin p_r (t - \tau) d\tau \tag{9.34}$$

The maximum relative displacement at any section x in the rth mode of vibration is

$$\left| Z_x^{(r)} \right|_{\max} = B_x^{(r)} \frac{1}{p_r} S_v^{(r)} \tag{9.35}$$

where $B_x^{(r)}$ = mode participation factor

$$B_x^{(r)} = \frac{Y_r(x) \int_0^L \rho A Y_r \, dx}{\int_0^L \rho A Y_r^2 \, dx} \tag{9.36}$$

and as given in Eq. (3.157)

$$S_v^{(r)} = \left| \int_0^t \ddot{y}(\tau) \sin p_r (t - \tau) \, dx \right|_{\max} \tag{9.37}$$

When damping is present in the system, Eq. (9.35) is modified as

$$\left| Z_x^{(r)} \right|_{\max} = B_x^{(r)} \frac{1}{p_{dr}} S_v^{(r)} \tag{9.38}$$

where $\quad p_{dr} = p_r \sqrt{1 - \zeta^2}$

$$S_v^{(r)} = \int_0^t \ddot{y}_g(\tau) \exp[-p_r \zeta_r (t - \tau)] \sin p_{dr} (t - \tau) d\tau \tag{9.39}$$

Therefore, the complete solution of $z(x, t)$ is therefore

$$z(x,t) = \sum B_x^{(r)} \frac{1}{p_{dr}} S_v^{(r)} \tag{9.40}$$

Using relation given by Eq. (3.128), Eq. (9.40) becomes

$$z(x,t) = \sum B_x^{(r)} S_d^{(r)} \tag{9.41}$$

9.4 FORCED VIBRATION OF FLEXURAL MEMBER

If an external force $P(t)$ per unit length acts transversely on a beam, the equation of motion for forced vibration is given by Eq. (8.78), which is reproduced below

$$\frac{\partial^2}{\partial x^2} \left(EI \frac{\partial^2 y}{\partial x^2} \right) + \rho A \frac{\partial^2 y}{\partial t^2} = P(t) \tag{9.42}$$

Let us assume the solution of y in the same form as done in the previous sections of this chapter, that is, considering it as a summation of modal components

$$y = \sum Y_r(x)\xi_r(t) \tag{9.43}$$

where Y_r is the mode shape for the rth mode and $\xi_r(t)$ represents the normal or principal coordinates.

Substituting y from Eq. (9.43) into Eq. (9.42), multiplying by Y_s and integrating with respect to x over the length of the beam

$$\int_0^L Y_s \frac{\partial^2}{\partial x^2}\left[EI\sum\frac{d^2Y_r}{dx^2}\xi_r\right]dx + \int_0^L \rho A Y_s \sum Y_r \ddot{\xi}_r\,dx$$

$$= \int_0^L P(t)Y_s\,dx \tag{9.44}$$

Using orthogonality relationships, from Eqs. (8.137) and (8.138) we get

$$\int_0^L \rho A Y_r Y_s\,dx = 0 \tag{9.45}$$

$$\int_0^L Y_s \frac{d^2}{dx^2}\left(EI\frac{d^2Y_r}{dx^2}\right)dx = 0 \tag{9.46}$$

when $r \neq s$.

Again when $r = s$, we can write from Eq. (8.133) that

$$\int_0^L Y_r \frac{d^2}{dx^2}\left[EI\frac{d^2Y_s}{dx^2}\right]dx = p_r^2\int_0^L \rho A Y_r^2\,dx \tag{9.47}$$

Combining Eqs. (9.44) and (9.47) we get

$$\int_0^L \rho A Y_r^2\left(\ddot{\xi}_r + p_r^2\xi_r\right)dx = \int_0^L P(t)Y_r\,dx \tag{9.48}$$

Now, if $P(t) = \overline{p}(x)f(t)$, then

$$\left(\ddot{\xi}_r + p_r^2\xi_r\right)\int_0^L \rho A Y_r^2\,dx = f(t)\int_0^L \overline{p}(x)Y_r\,dx$$

$$\ddot{\xi}_r + p_r^2\xi_r = \frac{f(t)\int_0^L \overline{p}(x)Y_r\,dx}{\int_0^L \rho A Y_r^2\,dx} \tag{9.49}$$

If $\bar{p}(x)$ and $f(t)$ are known, and based on known mode shapes for the problem, Eq. (9.49) can be evaluated. The general solution for Eq. (9.49) is

$$\xi_r(t) = A_r \cos p_r t + B_r \sin p_r t + \frac{\int_0^L \bar{p}(x) Y_r dx}{\int_0^L \bar{p}(x) Y_r^2 dx} \frac{1}{p_r} \int_0^t f(\tau) \sin p_r (t-\tau) d\tau \tag{9.50}$$

Substituting the value of ξ_r from Eq. (9.50) into Eq. (9.43), we get

$$y = \sum Y_r (A_r \cos p_r t + B_r \sin p_r t) + \sum \frac{Y_r \int_0^L \bar{p}(x) Y_r \, dx}{\int_0^L \bar{p}(x) Y_r^2 \, dx} \frac{1}{p_r} \int_0^t f(\tau) \sin p_r (t-\tau) d\tau \tag{9.51}$$

If the external force is applied to a structure at rest, then Eq. (9.51) becomes

$$y = \sum \frac{Y_r \int_0^L \bar{p}(x) Y_r \, dx}{\int_0^L \bar{p}(x) Y_r^2 \, dx} \frac{1}{p_r} \int_0^t f(t) \sin p_r (t-\tau) d\tau \tag{9.52}$$

The modal static deflection for the *r*th mode is given by

$$\xi_{rst} = \frac{\int_0^L \bar{p}(x) Y_r \, dx}{p_r^2 \int_0^L \rho A Y_r^2 \, dx} \tag{9.53}$$

and the dynamic load factor for the *r*th mode is given by

$$(DLF)_r = p_r \int_0^L f(\tau) \sin p_r (t-\tau) d\tau \tag{9.54}$$

Therefore, Eq.(9.52) can be written as

$$y(x,t) = \sum Y_r \xi_{rst} (DLF)_r \tag{9.55}$$

Example 9.1

A simply supported uniform beam having mass density ρ and cross-sectional area *A*, has a distributed load, whose variation with time is shown in Fig. 9.3. Derive the expression for the dynamic deflection of the beam.

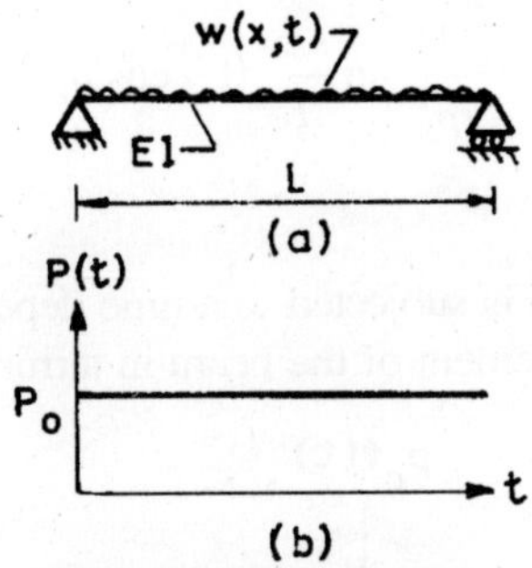

Fig. 9.3 Example 9.1

The normal modes of the beam are

$$Y_r(x) = \sin\frac{r\pi x}{L}$$

The natural frequencies of the beam are

$$p_r = (r\pi)^2\sqrt{\frac{EI}{\rho AL^3}}$$

Now,
$$\int_0^L \rho A Y_r^2\,dx = \int_0^L \rho A \sin^2\frac{r\pi x}{L}\,dx = \frac{\rho AL}{2}$$

and
$$\int_0^L \bar{p}(x)Y_r\,dx = \int_0^L P_0 \sin\frac{r\pi x}{L}\,dx = \frac{2P_0}{r\pi}, \quad r = 1,3,5,.....$$

$$f(t) = 1$$

Therefore, the equation of motion for the rth mode is

$$\ddot{\xi}_r + p_r^2\xi_r = \frac{2P_0\,L\cdot 2}{r\pi\rho AL} = \frac{4\,P_0}{\rho A\,r\pi}$$

which has the solution

$$\xi_r = \frac{4P_0}{\rho A\pi}\frac{1}{r\,p_r^2}(1-\cos p_r t)$$

Therefore, total dynamic displacement of the beam is expressed as

$$y(x,t) = \sum_{r=1,3,5,}^{\infty}\frac{4P_0}{\rho A\pi}\frac{1}{rp_r^2}\sin\frac{r\pi x}{L}(1-\cos p_r t)$$

$$= \frac{4P_0}{\rho A\pi} \sum_{r-1,3,5}^{\infty} \frac{1}{rp_r^2} \sin\frac{r\pi x}{L}(1-\cos p_r t)$$

Example 9.2

A simply supported uniform beam is subjected to a time dependent concentrated load at quarter span. Express the dynamic displacement of the beam in terms of dynamic load factor (Fig. 9.4).

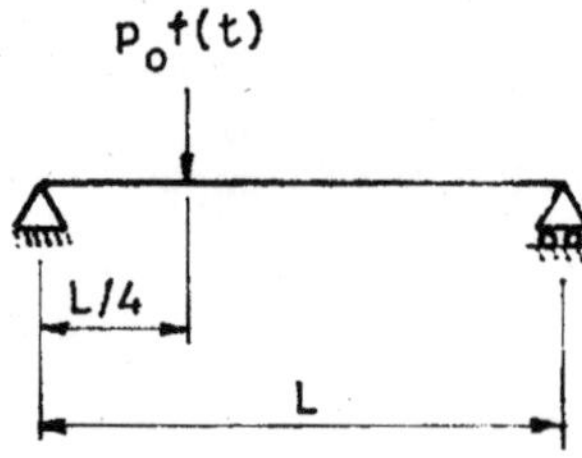

Fig. 9.4 Example 9.2

The normal modes of the beam are

$$Y_r(x) = \sin\frac{r\pi x}{L}$$

Now
$$\int_0^L \rho A Y_r^2\, dx = \int_0^L \rho A \sin^2\frac{r\pi x}{L}\, dx = \rho A\frac{L}{2}$$

$$\int_0^L \bar{p}(x) Y_r\, dx = P_0 \sin\frac{r\pi}{4}$$

Therefore, the equation of motion for the *r*th mode is

$$\ddot{\xi}_r + p_r^2\xi_r = \frac{2P_0}{\rho AL}\sin\frac{r\pi}{4} f(t)$$

which on solution yields

$$\xi_r = \frac{2P_0}{\rho AL}\sin\frac{r\pi}{4}\frac{1}{p_r}\int_0^t f(\tau)\sin p_r(t-\tau)d\tau$$

$$= \frac{2P_0}{\rho AL}\sin\frac{r\pi}{4}\frac{1}{p_r^2}(DLF)_r$$

Therefore

$$y(x,t) = \frac{2P_0}{\rho AL}\sum_{r=1,2,3}^{\infty}\frac{1}{p_r^2}\sin\frac{r\pi}{4}\sin\frac{r\pi x}{L}(DLF)_r$$

9.5 FORCED TRANSVERSE VIBRATION OF UNIFORM DAMPED BEAM

The equation of motion of a beam with viscous damping and external loading is given by

$$EI\frac{\partial^4 y}{\partial x^4}+c\frac{\partial y}{\partial t}+m\frac{\partial^2 y}{\partial t^2}=P(x,t) \tag{9.56}$$

The applied $P(x,t)$ force is expanded in terms of modes,

$$P(x,t)=\sum_{j=1}^{\infty}P_j(t)Y_j(x) \tag{9.57}$$

Both sides of Eq. (9.57) be multiplied by $m\,Y_k(x)$ and then integration is carried out over the span of the beam, yields

$$\int_0^L m\,P(x,t)\,Y_k(x)\,dx=\sum_{j=1}^{\infty}P_j(t)\int_0^L Y_j(x)Y_k(x)dx \tag{9.58}$$

While writing Eq. (9.58), orthogonality properties of modes have been used on the right hand side of the equation. The integral on the right hand side of the equation yield

$$\int_0^L m\,P(x,t)\,Y_k(x)\,dx=P_j(t) \tag{9.59}$$

Applying the same procedure for response

$$y(x,t)=\sum_{j=1}^{\infty}y_j(t)\,Y_j(x) \tag{9.60}$$

and then carrying out multiplication and integration as above

$$\int_0^L m\,y(x,t)Y_j(x)\,dx=y_j(t) \tag{9.61}$$

Substituting Eq. (9.57) and (9.60) into Eq. (9.56), yields

$$\sum_{j=1}^{\infty}\left[EI\,y_j(t)\frac{d^4Y_j}{dx^4}+c\frac{d\,y_j}{dt}Y_j+m\frac{d^2y_j}{dt^2}Y_j\right]=\sum_{j=1}^{\infty}P_j(t)Y_j(x) \tag{9.62}$$

Combining Eq.(8.82) and (9.62), yields

$$\sum_{j=1}^{\infty}\left[p^2my_j+c\dot{y}_j+m\ddot{y}_j-P_j(t)\right]Y_j(x)=0 \tag{9.63}$$

Since $Y_j(x)$ cannot be zero, the terms within the parenthesis must vanish,

$$\ddot{y}_j+\frac{c}{m}\dot{y}_j+p_j^2\,y_j=\frac{1}{m}P_j(t) \tag{9.64}$$

Eq. (9.64) can be written as

$$\ddot{y}_j + 2\zeta_j p_j \dot{y}_j + p_j^2 y_j = \frac{1}{m} P_j(t) \tag{9.65}$$

Solution of the convolution integral of Eq. (9.65) [using Eq.(3.68)] is

$$y_j(t) = \frac{1}{mp_j\sqrt{1-\zeta^2}} \int_0^t P_j(\tau)\, \exp[-\zeta\, p_j(t-\tau)]$$

$$\sin\left[p\sqrt{1-\zeta^2}\,(t-\tau)\right] d\tau \tag{9.66}$$

$y(x, t)$ can be obtained from Eq. (9.60) by substituting Eq. (9.66). Final solution is

$$y(x,t) = \sum_{j=1}^{\infty} Y_j(x) \int_0^t \left[\int_0^L p(\xi,\tau) Y_j(\xi)\, d\xi\right]$$

$$\frac{1}{mp_j\sqrt{1-\zeta^2}} \int_0^t \exp\left[-\zeta p_j(t-\tau)\right] \sin\left[p\sqrt{1-\zeta^2}\,(t-\tau)\right] d\tau \tag{9.67}$$

9.6 FORCED VIBRATION OF FLEXURAL MEMBER SUBJECTED TO GROUND MOTION EXCITATION

The equation of motion for a bending member subjected to ground motion excitation is given by

$$\frac{\partial^2}{\partial x^2}\left(EI\frac{\partial^2 z}{\partial x^2}\right) + \rho A\frac{\partial^2 z}{\partial t^2} = -\rho A \ddot{y}_g \tag{9.68}$$

where z is the relative displacement between the beam at a distance x and the ground. $\ddot{y}_g(t)$ represents the ground acceleration.

Equation (9.68) is similar to Eq. (9.42). Following procedures similar to that presented in Art. 9.4, it can be shown that

$$z(x,t) = -\sum \frac{Y_r(x)\int_0^L \rho A Y_r\, dx}{\int_0^L \rho A Y_r^2\, dx} \frac{1}{p_r} \int_0^L \ddot{y}_g(\tau) \sin p_r(t-\tau)\, d\tau \tag{9.69}$$

The maximum relative displacement at any section x in the rth mode of vibration is

$$\left|Z_x^{(r)}\right|_{\max} = B_x^{(r)} \frac{1}{p_r} S_v^{(r)} \tag{9.70}$$

where $B_x^{(r)}$ = mode participation factor

$$= \frac{Y_r \int_0^L \rho A Y_r \, dx}{\int_0^L \rho A Y_r^2 \, dx}$$

$$S_v^{(r)} = \left[\int_0^t \ddot{y}_g(\tau) \sin p_r (t-\tau) \, d\tau \right]_{\max}$$

Therefore

$$z(x,t) = \sum B_x^{(r)} \frac{1}{p_r} S_v^{(r)} \tag{9.71}$$

If damping is present in the system, Eq. (9.71) is to be modified as per equations given in Eqs. (9.38) to (9.41).

Example 9.3

The base of a cantilever beam undergoes a pulsating motion $\ddot{y}_{so} \sin \omega t$ Determine the response of the cantilever beam.

$$\int_0^t \ddot{y}_g(\tau) \sin p_r (t-\tau) \, d\tau$$

$$= \int_0^t \ddot{y}_{so} \sin \omega\tau \sin p_r (t-\tau) \, d\tau$$

$$= \ddot{y}_{so} \int_0^t \frac{1}{2} \left[\cos(\omega\tau - p_r t + p_r \tau) - \cos(\omega\tau + p_r t - p_r \tau) \right] d\tau$$

$$= \ddot{y}_{so} \frac{p_r}{p_r^2 - \omega^2} \sin \omega t$$

Let y_r be the mode shape for the rth mode and it is so scaled that

$$\int_0^L \rho A Y_r^2 \, dx = \rho A L$$

Therefore

$$z(x,t) = -\sum \frac{Y_r \int_0^L \rho A Y_r \, dx}{\int_0^L \rho A Y_r^2 \, dx} \frac{1}{p_r} \ddot{y}_{so} \frac{p_r}{p_r^2 - \omega^2} \sin \omega t$$

$$= -\sum \frac{Y_r \rho A \int_0^L Y_r \, dx}{\rho A L} \ddot{y}_{so} \frac{1}{p_r^2 - \omega^2} \sin \omega t$$

$$= -\frac{\ddot{y}_{so} \sin \omega t}{L} \sum \frac{Y_r \left(\int_0^L Y_r \, dx \right)}{p_r^2 - \omega^2}$$

9.7 RESPONSE OF BEAMS DUE TO MOVING LOADS

The problem of a load moving over a beam or girder is of much concern to structural engineers. The actual problem is somewhat complex. As such, it is approached with somewhat simplifying assumptions. A single concentrated load is assumed to move with a uniform velocity. The practical application for this problem occurs in bridges, aircraft carriers, etc.

Automobiles, aircrafts, etc., while moving over the deck produces a number of effects, which increase the stresses in the structural members. Some of them are the impact effect due to unbalanced parts of the vehicle, impact effect due to irregularities of the deck surface and the effect on the structure due to smooth running of the vehicle.

In this section, we limit ourselves to the live load effect of the vehicle on the deck girder. This effect lies between the following extreme cases:

(1) The mass of the moving load is large in comparison to the mass of the beam. In this case, therefore, the mass of the beam is neglected and the weight of the vehicle is considered as a concentrated weight, so that the problem reduces to a SDF system.

(2) In the other extreme, the mass of the moving load is considered to be small in comparison to the mass of the deck. This case is to be analysed as an elastic beam subjected to a moving load.

9.7.1 Response of the Beam when the Mass of the Vehicle is Large

In the following, we present the approximate solution given by Timoshenko *et al.* [9.5].

A simply supported beam is shown in Fig. 9.5, over which a load P is moving from left to right. Had the load P been placed statically at a distance x from left hand support, then the statical deflection would be

$$y = \frac{Px^2 (L - x)^2}{3EIL} \tag{9.72}$$

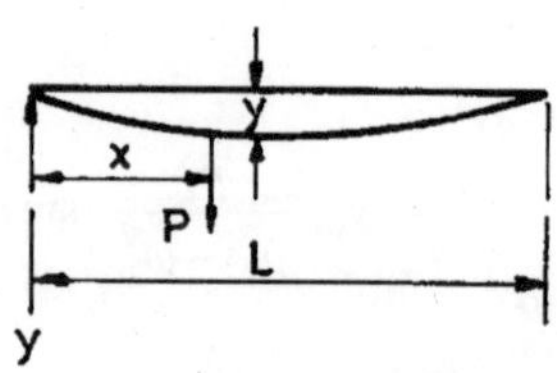

Fig. 9.5 A moving load over a simply supported beam

Due to application of P and its movement, the beam is vibrating. In addition to static load P, there is an inertia force associated with its vertical movement. Therefore, total force at the point of application of P is

$$R = P - \frac{\rho}{g}\frac{d^2 y}{dt^2} \tag{9.73}$$

If the load moves with a constant velocity v on the beam, then

$$\frac{dy}{dt} = v\frac{dy}{dx} \tag{9.74}$$

Differentiating both sides of Eq. (9.74) with respect to t, we get

$$\frac{d^2 y}{dt^2} = v\frac{d}{dt}\left(\frac{dy}{dx}\right) = v\frac{d}{dx}\left(\frac{dy}{dt}\right)$$

or

$$\frac{d^2 y}{dt^2} = v\frac{d}{dx}\left(v \cdot \frac{dy}{dx}\right) = v^2\frac{d^2 y}{dx^2} \tag{9.75}$$

Combining Eqs. (9.61) and (9.63), we get

$$R = P\left(1 - \frac{v^2}{g}\frac{d^2 y}{dx^2}\right) \tag{9.76}$$

Replacing P by R given in Eq. (9.76), Eq. (9.72) results in

$$y = P\left(1 - \frac{v^2}{g}\frac{d^2 y}{dx^2}\right)\frac{x^2(L-x)^2}{3EIL} \tag{9.77}$$

Eq.(9.77) indicates the path of contact of the rolling load with the beam. Timoshenko has approximated the solution of Eq. (9.77) by assuming $v = 0$, so that maximum deflection occurs at the centre and the maximum pressure is given by

$$R_{\max} = P\left(1 + \frac{v^2}{g}\frac{PL}{3EI}\right) \tag{9.78}$$

Therefore, the maximum deflection at the centre is

$$y_{\max} = y_{st}\left(1 + \frac{v^2}{g}\frac{PL}{3EI}\right) \tag{9.79}$$

9.7.2 Response of the Beam when the Mass of the Vehicle is Small

A harmonic force $P\sin(\omega t + \alpha)$ due to unbalanced parts of a vehicle moves with a uniform velocity v from left to right. At time $t = 0$, the load is assumed to remain at the left hand support.

Assuming the beam to possess uniform flexural rigidity *EI*, the equation of motion for the beam is (Fig. 9.6)

$$EI\frac{\partial^4 y}{dx^4} + \rho A\frac{\partial^2 y}{\partial t^2} = P\sin(\omega t + \alpha) \tag{9.80}$$

Assuming a solution in terms of normal modes

$$y = \sum \sqrt{2}\sin\frac{r\pi x}{L}\xi_r(t) \tag{9.81}$$

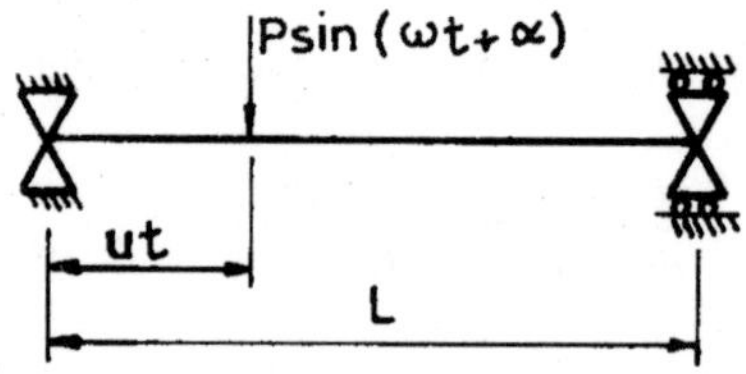

Fig. 9.6 A simply supported beam

The factor $\sqrt{2}$ is for normalising the mode shape and substituting it in Eq. (9.80) and following steps similar to that presented in Art. 9.4, we get the *r*th equation [Eq. (9.49)] as

$$\ddot{\xi}_r + p_r^2\xi_r = \frac{\sqrt{2}\,P\sin(\omega t + \alpha)}{\rho AL}\sin\frac{r\pi vt}{L} \tag{9.82}$$

The right hand side of Eq. (9.54) is

$$RHS = \frac{P}{\sqrt{2}\,\rho AL}\left[\cos\left(\omega t + \alpha - \frac{r\pi vt}{L}\right) - \cos\left(\omega t + \alpha + \frac{r\pi vt}{L}\right)\right] \tag{9.83}$$

The complete solution of Eq. (9.82) is

$$\xi_r = A_r \sin\omega_r t + B_r\cos\omega_r t$$

$$+\frac{P}{\sqrt{2}\,\rho AL}\left[\frac{\cos\left\{\omega t + \alpha - \dfrac{r\pi vt}{L}\right\}}{\omega_r^2 - \left\{\omega - \dfrac{r\pi v}{L}\right\}^2} - \frac{\cos\left\{\omega t + \alpha + \dfrac{r\pi vt}{L}\right\}}{\omega_r^2 - \left\{\omega + \dfrac{r\pi v}{L}\right\}^2}\right] \tag{9.84}$$

If the beam is at rest before the load is applied, then $\xi_r = \dot{\xi}_r = 0$ at $t = 0$ and the constants A_r and B_r can be evaluated from the initial conditions.

Equation (9.84) indicates that large amplitudes will be obtained when

$$\omega_r = \omega \pm \frac{r\pi v}{L} \tag{9.85}$$

In the derivation of this section, the mass of the load has been neglected.

References

9.1 G. L. Rogers, Dynamics of Framed Structures, John Wiley and Sons, Canada, 1959
9.2 G. B. Warburton, The Dynamical Behaviour of Structures, Pergamon Press, U. K., 1976
9.3 W.T. Thomson, Theory of Vibration with Application, 3rd Edition CBS Publishers and Distributors, New Delhi, 1988.
9.4 C. H. Norris, R.J. Hansen, M.J. Holley, J.M. Biggs, S. Namyet and J.K. Minami., Structural Design for Dynamic Loads, McGraw-Hill, New York, 1959
9.5 S.P. Timoshenko, D.A. Young and W. Weaver, Jr., Vibration Problems in Engineering, Fourth Edition, John Wiley & Sons, New York, 1974.

EXERCISE 9

9.1 A bar with built-in ends is acted upon by a constant axial force P, applied at quarter point, as shown in the figure. Write the equation of axial displacement of the bar, if the bar was unstressed before P was applied.

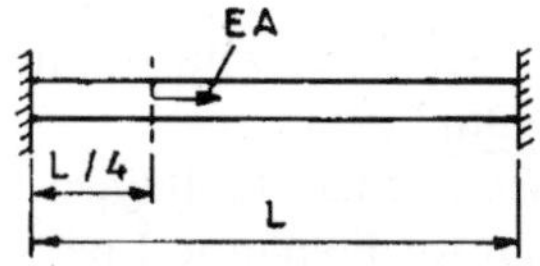

Prob. 9.1

9.2 A drill string is made of steel and is 1400 m long. Determine the fundamental mode of vibration by considering the string to be free at one end and clamped at the other. Find the displacement at the free end when $t = T/3$, where T is the time period due to a tensile stress $\sigma = P/A = 21$ N/mm^2 suddenly applied to this end. E = 2.1 × 10^{11} N/m^2 and $\rho = 7800$ kg/m^3.

9.3 A simply supported beam of mass M is suddenly loaded by a force shown in the figure. Determine the equation of deflection of the beam.

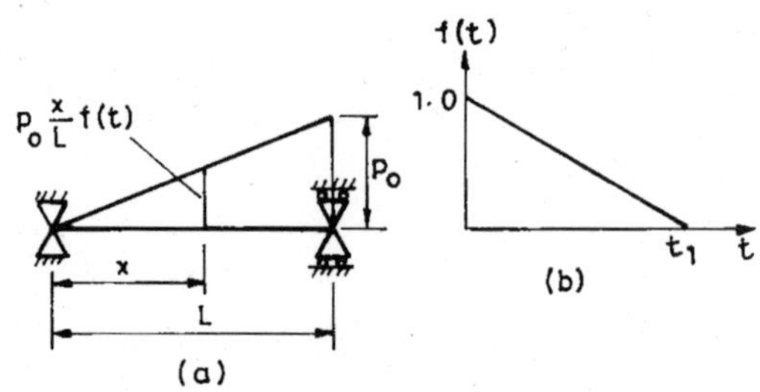

Prob. 9.3

9.4 A missile is excited longitudinally by the thrust $F(t)$ of its rocket engine, at the end $x = 0$ during its flight. Write a general expression for the displacement of the missile.

9.5 An earth dam 50 m high has on average a width of 90 m. For earth, $G = 9 \times 10^6$ N/m^2 and density 2000 kg/m^3. The dam is to be treated as a cantilever shear beam having uniform width of 90 m ($m = 0.833$).

Calculate the maximum deflection at the top, if the base is excited by a force shown in the figure.

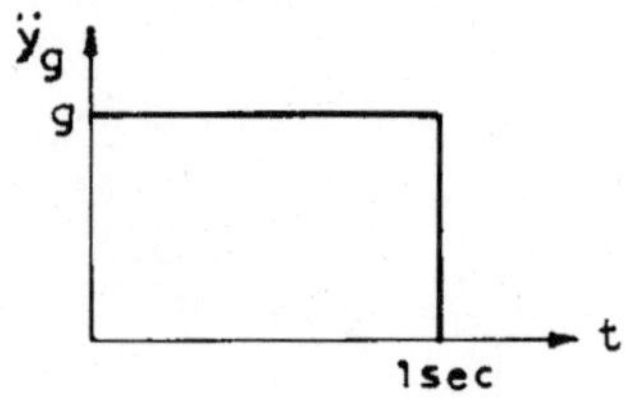

Prob. 9.5

9.6 A cantilever beam under the action of a force P applied at the free end deflects the end by 30 mm. What will be the amplitude of forced vibration produced by a pulsating force $P \sin \omega t$ applied at the free end, if the frequency w is equal to half the fundamental frequency of the beam?

9.7 A simply supported beam is subjected to pulsating loads as shown in the figure. Find the amplitude of forced vibration at the middle, if the frequency is ω equal to half the fundamental frequency of the beam.

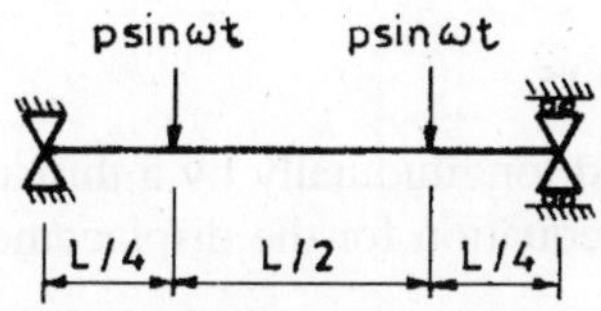

Prob. 9.7

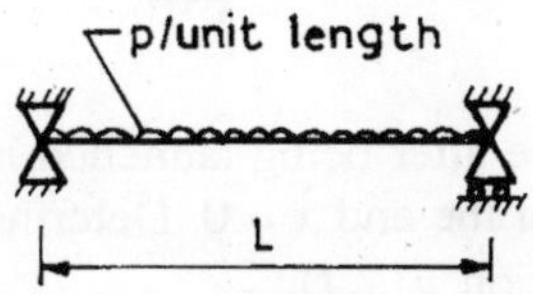

Prob. 9.8

9.8 Find the forced vibrations produced in the beam of figure by a distributed load

$$p = p_o \sin \frac{\pi x}{L} \sin \omega t$$

9.9 Deduce the equation of motion for forced flexural vibrations of a beam subjected to an applied force. Consider the material of the beam as viscoelastic, for which the stress-strain curve is

$$\sigma_n = E\left(\sigma_x + c\frac{\partial \varepsilon_x}{\partial t}\right)$$

9.10 For a uniform cantilever beam, determine the maximum deflection at the free end and maximum bending stress at the clamped end due to half-sine pulse.

9.11 A simply supported beam is subjected to a vertical motion of right support, of amount

$$y(L, t) = y_0 \sin \omega t$$

If at $t = 0$, the velocity and deflection at every point of the beam is zero, what will be the deflection equation for any time thereafter?

9.12 A vehicle having small mass exerts a pressure P on the deck beam. The magnitude of P remains constant as it crosses the span with uniform velocity v. Derive the equation of the deflection curve of the beam.

9.13 A simply supported bridge of span 20 m, $EI = 8$ GN m^2 and $\rho A = 20000$ kg/m is subjected to a load $P \sin 2\pi f t$, which moves with a velocity 20 m/s. If $f = 5$ Hz, what is the largest amplitude of vibration in the bridge?

9.14 Determine the expressions for the displacement and bending moment of a uniform simply supported beam to a step function force p_0 at a distance ξ from the left end.

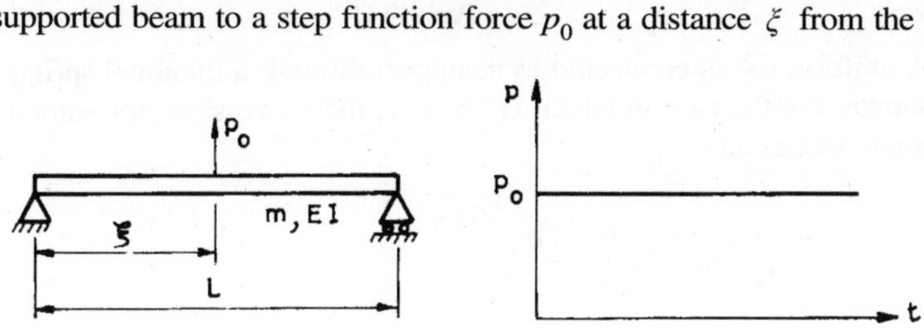

Prob. 9.14

9.15 Determine the response of a cantilever beam when its base is given a motion $y_b(t)$ normal to the beam axis as shown in the figure

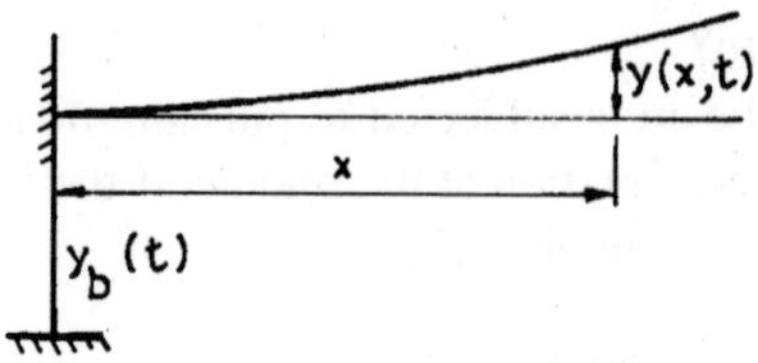

Prob. 9.15

9.16 A missile after being launched is excited longitudinally by a thrust $F(t)$ of its rocket engine at the end $x = 0$. Determine the equation for the displacement $u(x, t)$ and the acceleration $\ddot{u}(x, t)$.

9.17 A spring of stiffness k is attached to a uniform beam, as shown in the figure. Show that the one-mode approximation results in the frequency equation.

$$\left(\frac{p}{p_1}\right)^2 = 1 + 1.5\left(\frac{k}{m}\right)\left(\frac{ML^3}{\pi^4 EI}\right)$$

where
$$p_1^2 = \frac{\pi^4 EI}{ML^3}$$

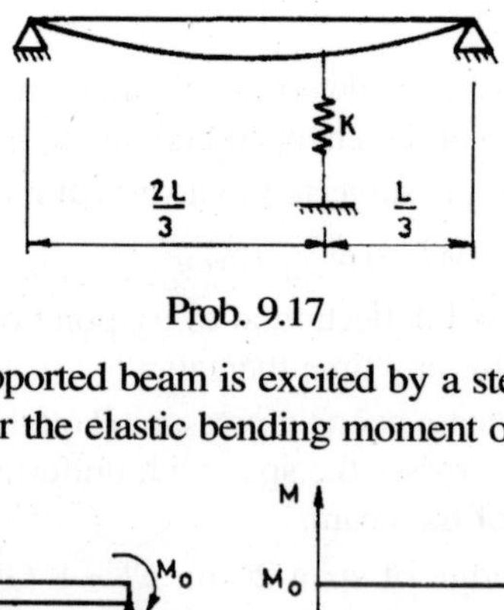

Prob. 9.17

9.18 A uniformly simply supported beam is excited by a step moment shown in the figure. Derive an expression for the elastic bending moment of the beam.

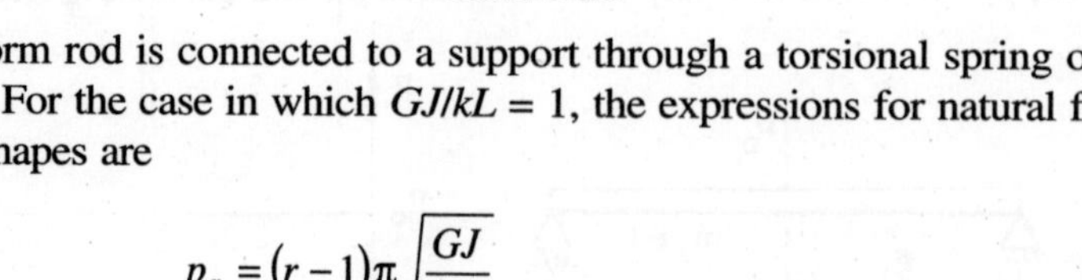

Prob. 9.18

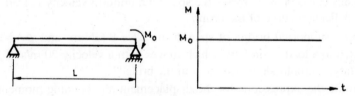

9.19 A uniform rod is connected to a support through a torsional spring of stiffness k as shown. For the case in which $GJ/kL = 1$, the expressions for natural frequencies and mode shapes are

$$p_r = (r-1)\pi\sqrt{\frac{GJ}{IL^2}}$$

$$\phi_r(x) = \cos(r-1)\frac{\pi x}{l}$$

The system is excited by a uniformly distributed torque $\tau(t)$ per unit length which acts over the outer half of the rod. Give the equations for the torsional motion in terms of the first two normal coordinates.

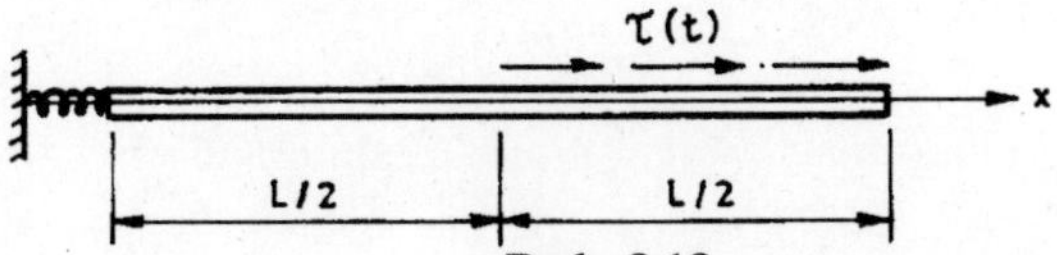

Prob. 9.19

Chapter 10

Dynamic Direct Stiffness Method

10.1 INTRODUCTION

The procedure presented in Chapter 8 for the determination of free vibration characteristics of the structural member, by considering it as a system having distributed mass, requires the evaluation of constants which are dependent on the boundary conditions. If the structure is made up of assemblage of structural members, then four constants are to be determined for each member. For a three-span continuous beam, the number of constants becomes 12, and with the increase in the number of members, it goes on increasing. The whole calculation procedure becomes tedious in this straight forward approach. A more methodical approach for carrying out the analysis for such cases, is performed by using dynamic direct stiffness approach. Once the free vibration characteristics is evaluated, the method can be applied to forced vibration analysis, in which case it becomes analogous to the displacement method used for statical problems. We first look into the classical approach for solving what is termed as statically indeterminate problems and then we shall pass on to the dynamic direct stiffness method [10.1 - 10.3].

10.2 CONTINUOUS BEAM

Consider a two-span continuous beam having uniform flexural rigidity *EI* throughout its length. Let us choose two coordinate systems for two spans-the origin of the left span be at left hand support and that of the right span be right hand support and positive directions of the axes are shown in Fig. 10.1.

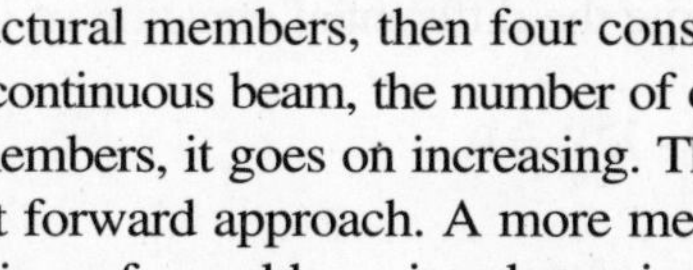

Fig. 10.1 A two-span beam

The boundary conditions of the beam are

$$y_1(0,t) = M_1(0,t) = y1(L_1,t) = y_2(L_2,t) = y_2(0,t) = M_2(0,t) = 0 \qquad (10.1)$$

Further, from the continuity condition of the central support, we have

$$M_1(L_1,t) = M_2(L_2,t) \qquad (10.2a)$$

and
$$\left(\frac{\partial y_1}{\partial y_2}\right)_{x_1=L_1} = -\left(\frac{\partial y_2}{\partial x_2}\right)_{x_2=L_2} \tag{10.2b}$$

The displacement functions are expressed as a product of space and time as follows,

$$y_1(x_1,t) = Y_1(x_1)q(t) \tag{10.3}$$

and
$$y_2(x_2,t) = Y_2(x_2)q(t) \tag{10.4}$$

Time function remains same for both spans, as the frequencies associated with them must be the same. Substituting Eqs. (10.3) and (10.4) into Eq. (10.1), we obtain

$$\begin{aligned} &Y_1(0) = Y_1''(0) = Y_1(L_1) = Y_2(L_2) = Y_2(0) = Y_2''(0) = 0 \\ &Y_1''(L_1) = Y_2''(L_2) \\ &Y_1'(L_1) = -Y_2'(L_2) \end{aligned} \tag{10.5}$$

Based on Eq. (8.46), we can write the differential equation of the two spans as given follows:

$$Y_1^{iv} - \lambda^4 Y_1 = 0 \tag{10.6}$$

$$Y_2^{iv} - \lambda^4 Y_2 = 0 \tag{10.7}$$

and
$$\ddot{q} + p^2 q = 0 \tag{10.8}$$

where
$$\lambda^4 = \rho \frac{Ap^2}{El}$$

Solution of Eqs. (10.6), (10.7) and (10.8) are

$$Y_1 = C_1 \sin \lambda x_1 + C_2 \cos \lambda x_1 + C_3 \sinh x_1 + C4 \cosh x_1 \tag{10.9}$$

and $Y_2 = D_1 \sin \lambda x_2 + D_2 \cos \lambda x_2 + D_3 \sinh \lambda x_2 + D_4 \cosh \lambda x_2$ (10.10)

Eight constants associated with Eqs. (10.9) and (10.10) are to be evaluated, and they can be done on the basis of boundary conditions given by Eq. (10.5).

Applying the end conditions at two extreme supports, that is, at $x_1 = x_2 = 0$ we obtain

$$C_2 = C_4 = D_2 = D_4 = 0 \tag{10.11}$$

Applying the condition at $x_1 = L_2, y_1 = 0$ and at $x_2 = L_2, y_2 = 0$, we obtain

$$C_1 \sin \lambda L_1 = -C_3 \sinh \lambda L_1 = C \tag{10.12}$$

and
$$D_1 \sin \lambda L_2 = -D_3 \sinh \lambda L_2 = D \tag{10.13}$$

Based on the constants evaluated as given by Eqs. (10.11) to (10.13), Eqs. (10.9) and (10.10) may be written as

$$Y_1 = C\left[\frac{\sin\lambda x_1}{\sin\lambda L_1} - \frac{\sinh\lambda x_1}{\sinh\lambda L_1}\right] \tag{10.14}$$

and
$$Y_2 = D\left[\frac{\sin\lambda x_2}{\sin\lambda L_2} - \frac{\sinh\lambda x_2}{\sinh\lambda L_2}\right] \tag{10.15}$$

Using the condition given by Eq. (10.2a), we get

$$C\left[\frac{\sin\lambda L_1}{\sin\lambda L_1} + \frac{\sinh\lambda L_1}{\sinh\lambda L_1}\right] = D\left[\frac{\sin\lambda L_2}{\sin\lambda L_2} + \frac{\sinh\lambda L_2}{\sinh\lambda L_2}\right]$$

or
$$C = D \tag{10.16}$$

Condition given by Eq, (10.2b) yields

$$C\left[\frac{\cos\lambda L_1}{\cos\lambda L_1} + \frac{\cosh\lambda L_1}{\cosh\lambda L_1}\right] = D\left[-\frac{\cos\lambda L_2}{\cos\lambda L_2} + \frac{\cosh\lambda L_2}{\cosh\lambda L_2}\right] \tag{10.17}$$

Combination of Eqs. (10.16) and (10.17) leads to the frequency equation as

$$\cot\lambda L_1 - \coth\lambda L_1 = \cot\lambda L_2 - \cot\lambda L_2 \tag{10.18}$$

When both spans are equal, that is $L_1 = L_2$, Eq. (10.18) becomes

$$\tan\lambda L = \tanh\lambda L \tag{10.19}$$

If $C = D$ and $L_1 = L_2 = L$, it can be shown from Eq. (10.17) that

$$\sin\lambda L \sinh\lambda L = 0$$

or
$$\sin\lambda L = 0 \tag{10.20}$$

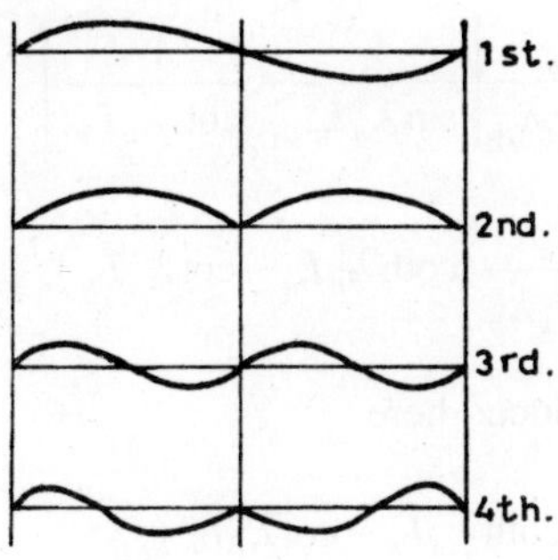

Fig. 10.2 Mode shapes of two-span beam

Therefore, for a two-span beam having equal span length, there are two frequency equations given by Eqs. (10.19) and (10.20), respectively. The lowest frequency is corresponding to the lowest root of Eq. (10.20), that is, $\lambda L = \pi$. Usually, the odd modes will correspond to the mode shape given by Eq. (10.20), that is, same as that of simply supported beam, and the even modes will correspond to Eq. (10.19). First four mode shapes have been shown in Fig. 10.2.

10.3 METHODS ANALOGOUS TO CLASSICAL METHODS IN STATICAL ANALYSIS

Free vibration analysis of indeterminate frames can be done by methods, which are similar to classical methods adopted in statical analysis. Hence, these methods have been given identical names as the classical methods, such as three-moment theorem, moment distribution method and so on. Here, we limit our discussion to the three-moment theorem only.

Let us consider a uniform simply supported beam of span L_n, mass m_n per unit length and flexural rigidity $E_n I_n$. Supports at two ends are indicated as $(n-1)$ and n as shown in Fig. 10.3. The end conditions for the beam are

$$y(0,t) = y(L_n,t) = M(0,t) = 0 \tag{10.21}$$

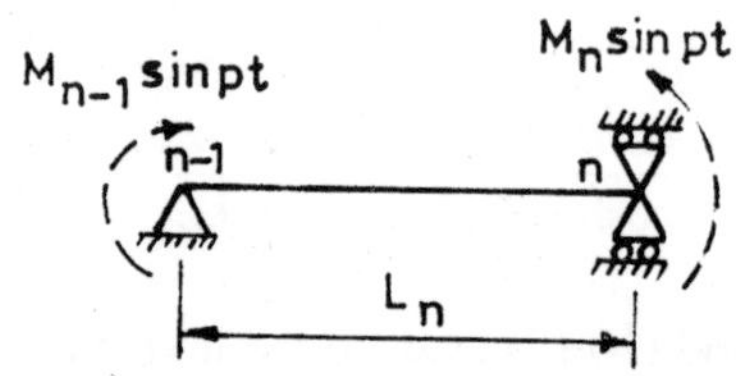

Fig. 10.3 A free-body diagram of a span

Let a pulsating moment $M_n \sin pt$ act on the support n, that is, the right support.

This gives $$M(L_n, t) = M_n \sin pt. \tag{10.22a}$$

It can be shown by following the usual procedure that the shape function for this beam is

$$Y = \frac{M_n}{2E_n I_n \lambda_n^2}\left[\frac{\sin\lambda_n x}{\sin\lambda_n L_n} - \frac{\sinh\lambda_n x}{\sinh\lambda_n L_n}\right] \tag{10.22b}$$

Based on the shape function of Eq. (10.22), the shapes at the two ends are

$$Y'(0) = \frac{M_n}{2E_n I_n \lambda_n}\left[\frac{1}{\sin\lambda_n L_n} - \frac{1}{\sinh\lambda_n L_n}\right] \tag{10.23}$$

and $$Y'(L_n) = \frac{M_n}{2E_n I_n \lambda_n}\left(\coth\lambda_n L_n - \cot\lambda_n L_n\right) \tag{10.24}$$

The following notations are introduced here

$$\theta_n = \frac{3}{2\lambda_n L_n}\left(\coth\lambda_n L_n - \cot\lambda_n L_n\right) \tag{10.25}$$

and $$\psi_n = \frac{3}{\lambda_n L_n}\left[\frac{1}{\sin\lambda_n L_n} - \frac{1}{\sinh\lambda_n L_n}\right] \tag{10.26}$$

On the basis of the above notations, Eqs. (10.23) and (10.24) are rewritten as

$$Y'(0) = \frac{M_n L_n \psi_n}{6E_n I_n} = M_n \beta_n \tag{10.27}$$

and
$$Y'(L_n) = -\frac{M_n L_n \theta_n}{3E_n I_n} = -M_n \beta_n \tag{10.28}$$

Now, in addition to the pulsating moment acting on the right end, if a pulsating moment $M_{n-1} \sin pt$ acts on the left end, then, we can write

$$Y'(0) = M_n \beta_n + M_{n-1} \alpha_n \tag{10.29}$$

and
$$Y'(L_n) = -M_n \alpha_n + M_{n-1} \beta_n \tag{10.30}$$

If we consider this span extending to the right from nth to $(n + 1)$th support, then the left end slope of this span is given by

$$Y'(0) = M_{n+1} \beta_n + M_n \alpha_{n+1} \tag{10.31}$$

For the continuity of the nth support, the slope to the left should be equal to the slope to the right at the support n. Therefore

$$Y'(L_n) \text{ at the left span} = Y'(0) \text{ at the right span} \tag{10.32}$$

Putting the respective values of Y_n and $Y'(0)$ from Eqs. (10.30) and (10.31), we get

$$M_{n+1} \beta_n + M_n (\alpha_n + \alpha_{n+1}) + M_n \beta_{n+1} = 0 \tag{10.33}$$

Equation (10.33) is the three-moment equation for the freely vibrating beam. In addition to the three bending moments at the three supports, it contains additional unknown λL associated with α and β terms.

Example 10.1

Determine the frequency equation of the continuous beam by using three-moment theorem. Three supports of Fig. 10.1are labeled as 1, 2 and 3 from the left.

We have

$$M_1 = M_3 = 0 \tag{a}$$

Three-moment equation for this problem reduces to

$$M_2 \alpha_2 + M_2 \alpha_3 = 0 \tag{b}$$

Substituting α_2 and α_3 into Eq. (b), we get

$$\frac{M_2 L_2 \theta_2}{3E_2 I_2} + \frac{M_2 L_3 \theta_3}{3E_3 I_3} = 0 \tag{c}$$

Now, $E_2 I_2 = E_3 I_3$, therefore, Eq. (c) becomes

$$L_2 \theta_2 + L_3 \theta_3 = 0 \tag{d}$$

Substituting θ_2 and θ_3 from Eq. (10.25) to Eq. (d), we get

$$\cot \lambda L_2 - \coth \lambda L_2 = \coth \lambda L_1 - \cot \lambda L_3 \tag{e}$$

Equation (e) agrees with Eq. (10.18).

Indeterminate frame problems can be formulated by other classical methods as well. However, we pass on to more formalised approach in the next section.

10.4 DYNAMIC STIFFNESS MATRIX IN BENDING

From the discussion of two-span continuous beam in the previous section, it is evident that as the number of spans goes on increasing, the number of constants to be evaluated also increase. As such, the problem tends to become more cumbersome. Dynamic stiffness matrix approach, which has a close resemblance with static direct stiffness matrix method, forms a more elegant method to the solution of these problems.

A beam segment with the prescribed displacements at its ends are shown in Fig. 10.4. The end displacements vary harmonically with time, frequency and phase difference, for all are same. As such, the end moments and shears also vary at the same frequency and the same phase relationship. The following derivation is mainly based on that given by Clough and Penzien [10.1].

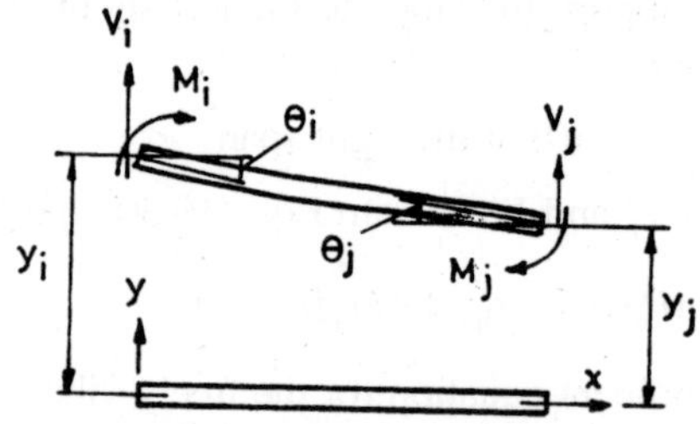

Fig. 10.4 A beam segment

For a uniform beam segment, the equation of motion for free flexural vibration is given by Eq. (8.43).

$$EI\frac{\partial^4 y}{\partial x^4} + m\frac{\partial^2 y}{\partial t^2} = 0 \tag{10.34}$$

where $m = \rho A$ = mass per unit length of the beam.

The boundary displacements vary harmonically, such that

$$y_i = y_{io} \sin pt \tag{10.35}$$

where y_{io} is the amplitude of boundary displacement at the end i. The displacement at any point with the span is

$$y(x,t) = Y(x)\sin(pt + e) \tag{10.36}$$

Substituting $y(x, t)$ from Eq. (10.36) into Eq. (10.34) gives

$$EI\frac{d^4Y}{dx^4} - mp^2Y = 0 \tag{10.37}$$

Equation (10.37) is of the same form as Eq. (8.45), and its solution is

$$Y(x) = C_1 \sin \lambda x + C_2 \cos \lambda x + C_3 \sinh \lambda x + C_4 \cosh \lambda x \tag{10.38}$$

where $\lambda^4 = \dfrac{mp^2}{EI}$

Expressing Eq. (10.38) and its derivatives in matrix form, we get

$$\begin{Bmatrix} Y \\ Y' \\ Y'' \\ Y''' \end{Bmatrix} = \begin{bmatrix} \sin \lambda x & \cos \lambda x & \sinh \lambda x & \cosh \lambda x \\ \lambda \cos \lambda x & -\lambda \sin \lambda x & \lambda \cosh \lambda x & \lambda \sinh \lambda x \\ -\lambda^2 \sin \lambda x & -\lambda^2 \cos \lambda x & \lambda^2 \sinh \lambda x & \lambda^2 \cosh \lambda x \\ -\lambda^3 \cos \lambda x & \lambda^3 \sin \lambda x & \lambda^3 \cosh \lambda x & \lambda^3 \sinh \lambda x \end{bmatrix} \begin{bmatrix} C_1 \\ C_2 \\ C_3 \\ C_4 \end{bmatrix} \tag{10.39}$$

The end deformations of the beam segment can be expressed in terms of four constants

$$\begin{Bmatrix} Y_i \\ Y_j \\ \theta_i \\ \theta_j \end{Bmatrix} = \begin{Bmatrix} Y \text{ at } x = 0 \\ Y \text{ at } x = L \\ -Y' \text{ at } x = 0 \\ Y' \text{ at } x = L \end{Bmatrix}$$

$$= \begin{bmatrix} 0 & 1 & 0 & 1 \\ \sin \lambda L & \cos \lambda L & \sinh \lambda L & \cosh \lambda L \\ -\lambda & 0 & -\lambda & 0 \\ -\lambda \cos \lambda L & \lambda \sin \lambda L & -\lambda \cosh \lambda L & -\lambda \sinh \lambda L \end{bmatrix} \begin{Bmatrix} C_1 \\ C_2 \\ C_3 \\ C_4 \end{Bmatrix} \tag{10.40}$$

In compact form, Eq. (10.40) can be written as

$$\{\delta\}_e = [G]\{C\} \tag{10.41}$$

The end forces acting on the member can be expressed in terms of constants

$$\begin{Bmatrix} V_i \\ V_j \\ M_i \\ M_j \end{Bmatrix} = \begin{Bmatrix} EIY''' \text{ at } x = 0 \\ -EIY''' \text{ at } x = L \\ EIY'' \text{ at } x = 0 \\ EIY'' \text{ at } x = L \end{Bmatrix}$$

$$= EI\lambda^2 \begin{bmatrix} -\lambda & 0 & \lambda & 0 \\ \lambda \cos \lambda L & -\lambda \sin \lambda L & -\lambda \cosh \lambda L & -\lambda \sinh \lambda L \\ 0 & -1 & 0 & 1 \\ \sin \lambda L & \cos \lambda L & -\sinh \lambda L & -\cosh \lambda L \end{bmatrix} \begin{Bmatrix} C_1 \\ C_2 \\ C_3 \\ C_4 \end{Bmatrix} \tag{10.42}$$

In compact form, Eq. (10.42) can be written as

$$\{p_e\} = [H]\{C\} \tag{10.43}$$

From Eq. (10.41), we get

$$\{C\} = [G]^{-1}\{\delta\}_e \tag{10.44}$$

Combining Eqs. (10.43) and (10.44) leads to

$$\{P_e\} = [H][G]^{-1}\{\delta\}_e \tag{10.45}$$

The dynamic stiffness matrix therefore is given by

$$[K(\lambda)] = [H][G]^{-1} \tag{10.46}$$

as the above Eq. (10.45) relates the nodal forces to the nodal displacements. It may be noted that the dynamic stiffness matrix is a function of the frequency parameter λ.

Equation (10.45) can be written in explicit form as follows.

$$\begin{Bmatrix} V_i \\ V_j \\ M_i \\ M_j \end{Bmatrix} = EI \begin{bmatrix} \frac{\alpha}{L^3} & -\frac{\bar{\alpha}}{L^3} & -\frac{\beta}{L^2} & -\frac{\bar{\beta}}{L^2} \\ -\frac{\bar{\alpha}}{L^3} & -\frac{\alpha}{L^3} & -\frac{\bar{\beta}}{L^2} & -\frac{\beta}{L^2} \\ -\frac{\beta}{L^2} & \frac{\bar{\beta}}{L^2} & \frac{\gamma}{L} & \frac{\bar{\gamma}}{L} \\ -\frac{\bar{\beta}}{L^2} & \frac{\beta}{L^2} & \frac{\bar{\gamma}}{L} & \frac{\gamma}{L} \end{bmatrix} \begin{Bmatrix} Y_i \\ Y_j \\ \theta_i \\ \theta_j \end{Bmatrix} \tag{10.47}$$

where

$$\alpha = \frac{\sin\lambda L\ \cosh\lambda L + \cos\lambda L\ \sinh\lambda L}{a} b^3,$$

$$\bar{\alpha} = \frac{\sinh\lambda L + \ \sin\lambda L}{a} b^3,$$

$$\beta = \frac{\sin\lambda L\ \ \sinh\lambda L}{a} b^2,$$

$$\bar{\beta} = \frac{\cosh\lambda L - \cos\lambda L}{a} b^2,$$

$$\gamma = \frac{\sin\lambda L\ \cosh\lambda L - \cos\lambda L\ \sinh\lambda L}{a} b,$$

$$\bar{\gamma} = \frac{\sinh\lambda L - \sin\lambda L}{a} b,$$

$a = 1 - \cos\lambda L \cosh\lambda L$ and

$b = \lambda L$

A plot of these coefficients with frequency parameter λL is shown in Fig. 10.5.

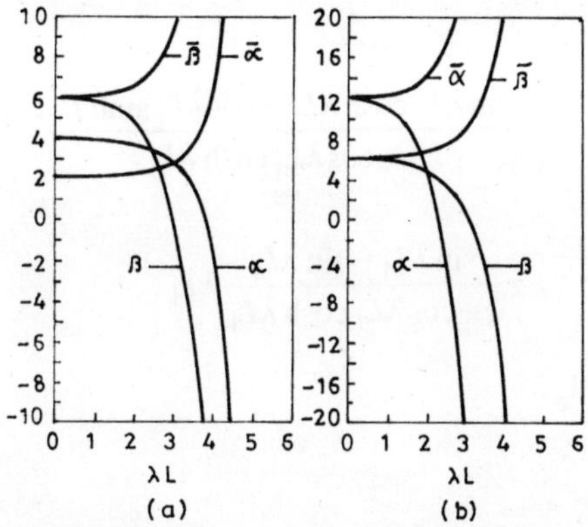

Fig. 10.5 Plot of coefficients with λL

If a structure is made up of a number of beam segments, then the overall dynamic stiffness matrix is formed by combining the dynamic stiffness matrix of the individual elements, exactly similar to that done in static analysis. The force-displacement relation for, the whole structure is given by

$$\{P\} = [K(\lambda)]\{\delta\} \tag{10.48}$$

Only the transverse vibrations have been considered here. In order to determine the free vibration of the structure, the external loading matrix {P} in Eq. (10.48) is a null matrix. So, for a non-trivial solution

$$\det [K(\lambda)] = 0 \tag{10.49}$$

The free vibration analysis for these cases has to be carried out by trial and error procedure, as the stiffnesses of the members depend on the unknown frequency.

When external forces are acting at the joints, {P} is no longer a null matrix and a forced vibration analysis can be performed on the basis of Eq. (10.48).

Example 10.2

For the continuous beam of Fig. 10.6, determine the fundamental frequency of the system by forming dynamic stiffness matrix.

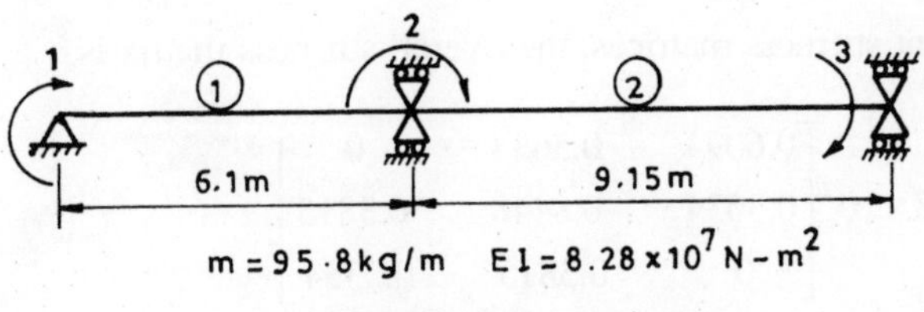

Fig. 10.6 Example 10.2

The continuous beam of Fig.10.6 consists of two spans. Both spans are considered as individual elements, with rotations at the two ends unknown. Arranging the stiffness matrix in the order of left hand rotation first and, right hand rotation next, the dynamic stiffness matrix for member 1 is

$$[K(\lambda)]_1 = \frac{1}{L}\begin{bmatrix} \gamma_1 & \bar{\gamma}_1 \\ \bar{\gamma}_1 & \gamma_1 \end{bmatrix} \tag{a}$$

where

$$\gamma_1 = \frac{\sin\lambda L_1 \cosh\lambda L_1 - \cos\lambda L_1 \sinh L_1}{1-\cos\lambda L_1 \cosh\lambda L_1}\lambda L_1$$

$$\bar{\gamma}_1 = \frac{\sin\lambda L_1 - \sin\lambda L_1}{1-\cos\lambda L_1 \cosh\lambda L_1}\lambda L_1$$

L_1 is the span of member 1.

Assuming $p = 133$ rad/s

and $$\gamma^4 = \frac{mp^2}{EI} = \frac{95.8*133^2}{8.28*10^7}\lambda = 0.3782$$

Therefore $$b_1 = \lambda L_1 = 0.3782*6.1 = 2.31$$

From Fig. 10.5, or by direct computation of γ and $\bar{\gamma}$ from the relations given by Eq. (10.47), we obtain γ = 3.7147 and $\bar{\gamma}$ = 2.2170. Substituting these values in Eq. (a) for the first member, we obtain

$$[K(\lambda)]_1 \begin{Bmatrix} \theta_1 \\ \theta_2 \end{Bmatrix} = \frac{8.28\times10^7}{6.1}\begin{bmatrix} 3.7174 & 2.2170 \\ 2.2170 & 3.7174 \end{bmatrix} \qquad \text{(b)}$$

Similarly, for the second member

$$b_2 = \lambda L_2 = 0.3782*9.15 = 3.465$$

For this value of b_2, γ = 2.1535 and $\bar{\gamma}$ = 3.4911

Therefore, for the second member

$$[K(\lambda)]_2 \begin{Bmatrix} \theta_2 \\ \theta_3 \end{Bmatrix} = \frac{8.28\times10^7}{9.15}\begin{bmatrix} 2.1535 & 3.4911 \\ 3.4911 & 2.1535 \end{bmatrix} \qquad \text{(c)}$$

Assembling two element stiffness matrices, the overall stiffness matrix is

$$[K(\lambda)] = 8.28\times10^7 \begin{bmatrix} 0.6094 & 0.3634 & 0 \\ 0.3634 & 0.8448 & 0.3815 \\ 0 & 0.3815 & 0.2354 \end{bmatrix}$$

Now, the determinant of the overall stiffness matrix is

$$|K(\lambda)| = 1.166 \times 10^5$$

Next, choose b_1 = 2.32 and following identical steps as above, we obtain the element stiffness matrix of the two spans as

$$[K(\lambda)]_1 = \frac{8.28\times10^7}{6.1}\begin{bmatrix}3.7094 & 2.2221\\2.2221 & 3.7094\end{bmatrix}$$

and
$$[K(\lambda)]_2 = \frac{8.28\times10^7}{9.15}\begin{bmatrix}2.1093 & 3.5289\\3.5289 & 2.1093\end{bmatrix}$$

The overall stiffness matrix resulting from the assembly of the beam element becomes

$$[K(\lambda)] = 8.28\times10^7\begin{bmatrix}0.6081 & 0.3643 & 0\\0.3643 & 0.8386 & 0.3857\\0 & 0.3857 & 0.2305\end{bmatrix}$$

The determinant of the overall stiffness matrix is

$$|K(\lambda)| = -2.9065*10^5$$

The value of b_1 for which the determinant will be zero can be obtained by linear interpolation, that is

$$b_1 = 2.31 + 0.01\left(\frac{1.166}{1.166+2.9065}\right) = 2.313$$

or
$$L_1\sqrt[4]{\frac{mp^2}{EI}} = 2.313$$

or
$$\frac{mp^2}{EI} = \left(\frac{2.313}{6.1}\right)^4$$

or
$$p = \left[\left(\frac{2.313}{6.1}\right)^4 \frac{8.28\times10^7}{95.8}\right]^{1/2} = 13367\,\text{rad/s}$$

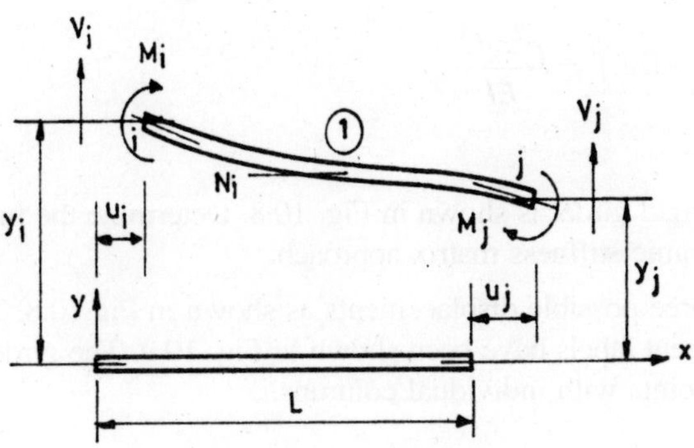

Fig. 10.7 A beam segment

10.5 DYNAMIC STIFFNESS MATRIX FOR FLEXURE AND RIGID AXIAL DISPLACEMENTS

For the analysis of the framed structures which are of moderate height, it is the usual practice to ignore the axial deformation. But certain members of the structure may undergo axial displacement as a rigid body, such as in a one-bay portal, where the frame undergoes sidesway movement involving the horizontal member to move as a rigid body in that direction. There is no contribution to the stiffness matrix due to the axial deformation of the member. As such, the element in the stiffness matrix due to the rigid axial displacement is represented by the inertial effect associated with its rigid body axial accelerations.

Referring to Fig. 10.7

$$N_i = -\overline{m}Lp^2u_i \tag{10.50}$$

where u_i is the rigid body axial displacement of member i and $\overline{m}$ is the mass of the member undergoing this displacement.

Incorporating this assitional term, the dynamic stiffness matrix is modified as

$$\begin{Bmatrix} V_i \\ V_j \\ M_i \\ M_j \\ N_i \end{Bmatrix}_e = EI \begin{bmatrix} \frac{\alpha}{L^3} & -\frac{\overline{\alpha}}{L^3} & -\frac{\beta}{L^2} & -\frac{\overline{\beta}}{L^2} & 0 \\ -\frac{\overline{\alpha}}{L^3} & \frac{\alpha}{L^3} & \frac{\overline{\beta}}{L^2} & \frac{\beta}{L^2} & 0 \\ -\frac{\beta}{L^2} & \frac{\overline{\beta}}{L^2} & \frac{\gamma}{L} & \frac{\overline{\gamma}}{L} & 0 \\ -\frac{\overline{\beta}}{L^2} & \frac{\beta}{L^2} & \frac{\overline{\gamma}}{L} & \frac{\gamma}{L} & 0 \\ 0 & 0 & 0 & 0 & -\frac{\mu^4}{L^3} \end{bmatrix} \begin{Bmatrix} y_i \\ y_j \\ \theta_i \\ \theta_j \\ u_i \end{Bmatrix}_e \tag{10.51}$$

where $u^4 = (\lambda L)^4 = \frac{p^2 mL^4}{EI}$

Example 10.3

A portal frame with rigid girder is shown in Fig. 10.8. Determine the fundamental frequency of the frame by the dynamic stiffness matrix approach.

The frame has three possible displacements as shown in Fig.10.8. The individual elements with global displacement labels have been shown in Fig. 10.9. The girder being rigid, there will be no rotation at its joints with individual columns.

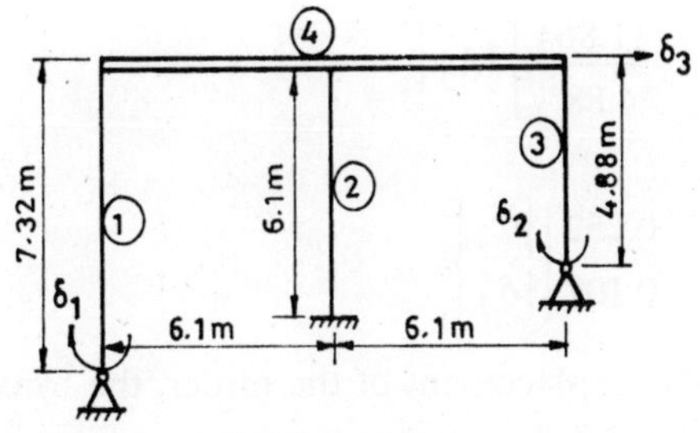

Fig. 10.8 The portal of Example 10.3

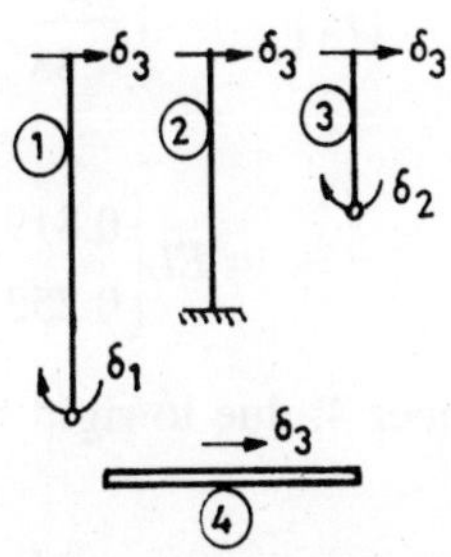

Fig 10.9 A free body diagram

Let $b_2 = \lambda L_2 = 0.97$ for the second member. For this value of b_2, $\alpha = 11.669$. Therefore, we can write from Eq. (10.51), that

$$\{P_3\}_2 = \frac{EI}{L^3} 11{,}669 \{\delta_3\}_2$$

or

$$\{P_3\}_2 = 0.0514 EI \{\delta_3\}_2 \qquad \text{(a)}$$

P_1, P_2 and P_3 are the forces corresponding to δ_1, δ_2 and δ_3. Similarly, for member 1,

$$b_1 = 0.97 \times \frac{7.32}{6.1} = 1.164$$

For this value of b_1, $\alpha = 11.315$, $\gamma = 3.9824$ and $\bar{\beta} = 6.0572$.

Therefore, for this member

$$\begin{Bmatrix} P_1 \\ P_3 \end{Bmatrix}_1 = EI \begin{bmatrix} \dfrac{3.9824}{7.32} & \dfrac{6.0572}{7.32^2} \\ \dfrac{6.0572}{7.32^2} & \dfrac{11.315}{7.32^3} \end{bmatrix} \begin{Bmatrix} \delta_1 \\ \delta_3 \end{Bmatrix}_1$$

$$= EI \begin{bmatrix} 0.544 & 0.113 \\ 0.113 & 0.029 \end{bmatrix} \begin{Bmatrix} \delta_1 \\ \delta_3 \end{Bmatrix}_1$$

For member 3, $b_3 = 0.97 \times \dfrac{4.88}{6.1} = 0.776.$

For this value of b_3, $\alpha = 11.864$, $\gamma = 3.997$ and $\bar{\beta} = 6.0103$.

Therefore

$$\begin{Bmatrix} P_1 \\ P_3 \end{Bmatrix}_1 = EI \begin{bmatrix} \dfrac{3.997}{4.88} & \dfrac{6.010}{4.88^2} \\ \dfrac{6.010}{4.88^2} & \dfrac{11.864}{4.88^3} \end{bmatrix} \begin{Bmatrix} \delta_2 \\ \delta_3 \end{Bmatrix}_1$$

$$= EI \begin{bmatrix} 0.819 & 0.252 \\ 0.252 & 0.102 \end{bmatrix} \begin{Bmatrix} \delta_2 \\ \delta_3 \end{Bmatrix}_1$$

Lastly, for member 4, due to rigid body displacement of the girder, the force-displacement relationship is

$$(P_3)_4 = \overline{m} L_4 p^2 (\delta_3)_4$$

or

$$(P_3)_4 = -\frac{\overline{m}}{m}\frac{mp^2}{EI} EIL_4 (\delta_3)_4$$

$$= -\frac{\overline{m}}{m}(\lambda L_4)^4 \frac{EI}{L_4^3}(\delta_3)_4$$

$$= -\frac{958}{95.8}(1.94)^4 \frac{EI}{12.2^3}(\delta_3)_4$$

$$= -0.078EI(\delta_3)_4$$

$$\left[b_4 = \lambda L_4 = 0.97 \times \frac{12.2}{6.1} = 1.94 \right]$$

Once the dynamic stiffness matrix for the individual element has been obtained, they can now be assembled to form the dynamic overall stiffness matrix for the frame

$$[K] = EI \begin{bmatrix} 0.544 & 0 & 0.113 \\ 0 & 0.819 & 0 \\ 0.113 & 0 & 0.05295 \end{bmatrix}$$

Therefore, the determinant of $[K]$ is

$$[K] = EI * 0.01313 = 8.28 * 10^7 * 0.01313 = 1.087 * 10^6$$

Calculations may be started with a fresh value of b_2 and the above steps are to be repeated. Natural frequency lies in between two values, where the determinant changes from positive to negative.

10.6 DYNAMIC STIFFNESS MATRIX OF A BAR UNDERGOING AXIAL DEFORMATION

We approach the problem in the same manner, as we have done in the case of bending member in Art. 10.4.

For a bar subjected to axial deformation due to its distributed mass, we can write the following equation based on Eq. (8.14) as

$$U = A_1 \sin \lambda_a x + A_2 \cos \lambda_a x \tag{10.52}$$

where
$$\lambda_a = \frac{p^2 \rho}{E} \tag{10.53}$$

Expressing, the axial displacement and its first derivative in matrix form, we get

$$\begin{Bmatrix} U(x) \\ U'(x) \end{Bmatrix} = \begin{bmatrix} \sin \lambda_a x & \cos \lambda_a x \\ \lambda_a \cos \lambda_a x & -\lambda \sin \lambda_a x \end{bmatrix} \begin{Bmatrix} A_1 \\ A_2 \end{Bmatrix} \tag{10.54}$$

(U_i and U_j are the displacements at the two nodes at $x = a$ and $x = L$ respectively (Fig. 10.10), then

$$\begin{Bmatrix} U_i \\ U_j \end{Bmatrix} = \begin{Bmatrix} U \text{ at } x=0 \\ U \text{ at } x=L \end{Bmatrix} = \begin{bmatrix} 0 & 1 \\ \sin \lambda_a L & \cos \lambda_a L \end{bmatrix} \begin{Bmatrix} A_1 \\ A_2 \end{Bmatrix} \tag{10.55}$$

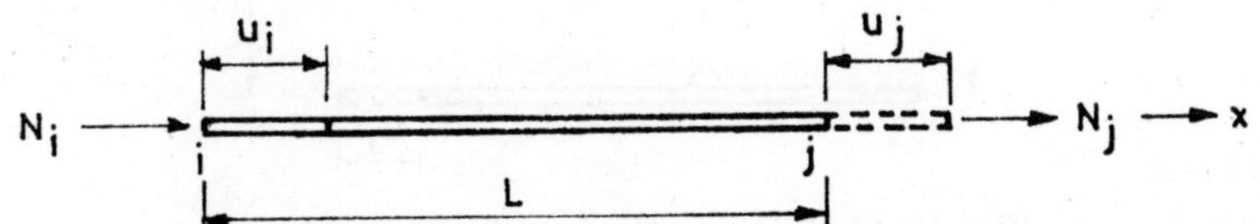

Fig. 10.10 A bar

Let N_i and N_j are the axial forces at the two ends, then

$$\begin{Bmatrix} N_i \\ N_j \end{Bmatrix} = EA \begin{Bmatrix} -U' \text{ at } x=0 \\ U' \text{ at } x=L \end{Bmatrix}$$

$$= EA \begin{bmatrix} -\lambda_a & 0 \\ -\lambda_a \cos\lambda_a L & -\lambda_a \sin\lambda_a L \end{bmatrix} \begin{Bmatrix} A_1 \\ A_2 \end{Bmatrix} \tag{10.56}$$

Combining Eqs. (10.55) and (10.56) and eliminating A_1 and A_2 between them, we get

$$\begin{Bmatrix} N_i \\ N_j \end{Bmatrix} = \frac{EA}{L} \begin{bmatrix} \psi_1 & -\psi_2 \\ -\psi_2 & \psi_1 \end{bmatrix} \begin{Bmatrix} U_i \\ U_j \end{Bmatrix} \tag{10.57}$$

where
$$\psi_1 = \frac{\lambda_a L \cos \lambda_a L}{\sin \lambda_a L} \quad \text{and} \quad \psi_2 = \frac{\lambda_a L}{\sin \lambda_a L}$$

Therefore, the dynamic stiffness matrix for a member subjected to axial vibration due to its own mass is

$$[K]_e - \frac{EA}{L} \begin{bmatrix} \psi_1 & -\psi_2 \\ -\psi_2 & \psi_2 \end{bmatrix} \tag{10.58}$$

The rest of the procedure is the same as that presented in Art. 10.4.

10.7 DYNAMIC STIFFNESS MATRIX OF A BAR SUBJECTED TO AXIAL AND BENDING DEFORMATIONS

When both axial and flexural deformations of a bar are considered, there are three degrees of freedom at each node, they are u_i, y_i and θ_i at node i and u_j, y_j and θ_j at node j (Fig.10.11). Two types of analyses are presented in the following. In the first case, it is assumed that the dynamic force effects are small enough, so that the axial and bending terms are not coupled. In the second category, we will consider the case where axial force has significant effect on bending, and as such the combined effect of axial force and bending is to be considered as coupled.

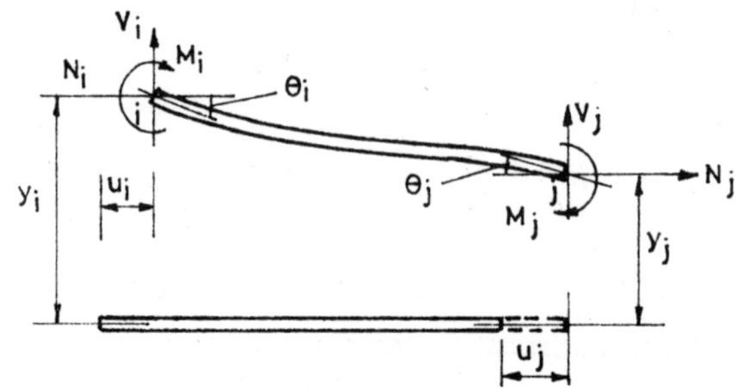

Fig. 10.11 A Bar with axial force and bending

10.7.1 Combined Uncoupled Axial and Bending Stiffness

As the dynamic bending and axial stiffness terms are uncoupled, the dynamic stiffness matrix for this case is simply a combination of axial stiffness matrix and bending stiffness matrix of Eqs. (10.58) and (10.47) respectively. It may be written as follows:

$$\begin{Bmatrix} V_i \\ V_j \\ M_i \\ M_j \\ N_i \\ N_j \end{Bmatrix} = \frac{EI}{L} \begin{bmatrix} \frac{\alpha}{L^3} & -\frac{\bar{\alpha}}{L^3} & -\frac{\beta}{L^2} & -\frac{\bar{\beta}}{L^2} & 0 & 0 \\ -\frac{\bar{\alpha}}{L^3} & \frac{\alpha}{L^3} & \frac{\bar{\beta}}{L^2} & \frac{\beta}{L^2} & 0 & 0 \\ -\frac{\beta}{L^2} & \frac{\bar{\beta}}{L^2} & \frac{\gamma}{L} & \frac{\bar{\gamma}}{L} & 0 & 0 \\ -\frac{\bar{\beta}}{L^2} & \frac{\beta}{L^2} & \frac{\bar{\gamma}}{L} & \frac{\gamma}{L} & 0 & 0 \\ 0 & 0 & 0 & 0 & \frac{\psi_1}{L^2}\left(\frac{L}{r}\right)^2 & \frac{\psi_2}{L^2}\left(\frac{L}{r}\right)^2 \\ 0 & 0 & 0 & 0 & \frac{\psi_2}{L^2}\left(\frac{L}{r}\right)^2 & \frac{\psi_1}{L^2}\left(\frac{L}{r}\right)^2 \end{bmatrix} \begin{Bmatrix} y_i \\ y_j \\ \theta_i \\ \theta_j \\ u_i \\ u_j \end{Bmatrix} \quad (10.59)$$

where $\quad r^2 = \frac{I}{A}$

It may be noted that the axial stiffness terms ψ_1 and ψ_2 in Eq. (10.59) are functions of axial frequency parameter λ_a, while the remaining terms depend on the flexural frequency parameter λ. We can compare the first axial frequency with the first flexural frequency of a cantilever beam. If p_a and p_f are the first axial and the first flexural frequencies of a cantilever beam, then

$$p_a - \frac{\pi L}{2r}\sqrt{\frac{El}{mL^4}} \tag{10.60}$$

and
$$p_f = 1.875^2\sqrt{\frac{El}{mL^4}} \tag{10.61}$$

Therefore
$$\frac{P_a}{P_f} = \frac{\pi L}{2*1.875^2 r} = \frac{L}{2.238r} \tag{10.62}$$

The fundamental frequency of axial vibration is always much higher than that of flexure.

10.7.2 Combined Coupled Axial and Bending Stiffness

In this part we assume that the dynamic axial force has a significant effect on the bending stiffness. However, if a constant axial force acts on the beam segment, its effect on transverse vibrations may not be negligible, and this should be accounted for in the analysis. In the following derivation, we assume a constant axial force acting over a uniform beam segment.

The freebody diagram of an infinitesimal length of the beam has been shown in Fig. 10.12. The moment equilibrium equation is

$$\left(M + \frac{\partial M}{\partial x}dx\right) - M - Vdx + N\frac{\partial y}{\partial x}dx = 0$$

or
$$\frac{\partial M}{\partial x} + N\frac{\partial y}{\partial x} = V \tag{10.63}$$

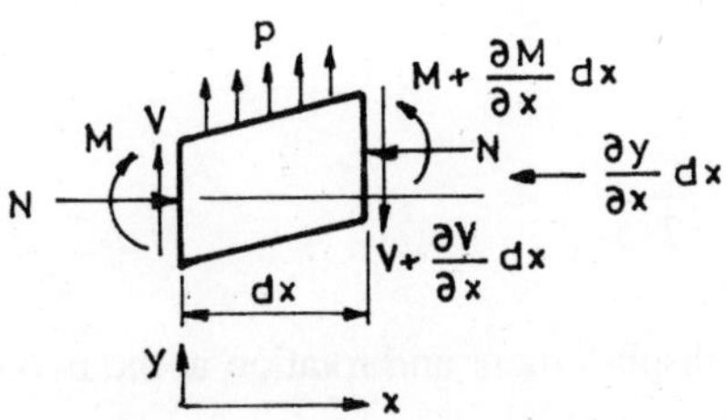

Fig. 10.12 A beam segment

Proceeding as per Art. 8.4, the equation of motion is given as

$$\frac{\partial^2}{\partial x^2}\left(El\frac{\partial^2 y}{\partial x^2}\right) + N\frac{\partial^2 y}{\partial x_2} + m\frac{\partial^2 y}{\partial x^2} = p \tag{10.64}$$

For free vibration of a uniform beam, Eq. (10.64) reduces to

$$El\frac{\partial^4 y}{\partial x^4} + N\frac{\partial^2 y}{\partial x^2} + m\frac{\partial^2 y}{\partial t^2} = 0 \tag{10.65}$$

Assuming a solution in the following form

$$y(x,t) = Y(x)q(t) \tag{10.66}$$

and substituting y from Eq. (10.66) into Eq. (10.65) we get

$$\frac{EIY^{iv}(x)}{Y(x)} + \frac{NY''(x)}{Y(x)} = -m\frac{\ddot{q}(t)}{q(t)} = C \tag{10.67}$$

From Eq. (10.67), we obtain the following two equations

$$\ddot{q}(t) + p^2 q(t) = 0 \tag{10.68}$$

and
$$EIY^{iv} + NY'' - mp^2 Y = 0 \tag{10.69}$$

Solution of Eq. (10.68) is

$$q(t) = A \cos pt + B \sin pt \tag{10.70}$$

Equation (10.69) is written as

$$Y^{iv} + g^2 Y'' - \lambda^4 Y = 0 \tag{10.71}$$

where $\lambda^4 = \dfrac{mp^2}{El}$ and $g^2 = \dfrac{N}{El}$

The solution of Eq. (10.71) is

$$y(x) = C_1 \sin \varepsilon x + C_2 \cos \varepsilon x + C_3 \sinh \phi x + C_4 \cosh \phi x \tag{10.72}$$

where $\varepsilon = \sqrt{\left(\lambda^4 + \dfrac{g^4}{4}\right)^{1/2} + \dfrac{g^2}{2}}$

and $\phi = \sqrt{\left(\lambda^4 + \dfrac{g^2}{4}\right)^{1/2} - \dfrac{g^2}{2}}$

Now, expressing the vertical displacement and rotation at the two nodes by means of Eq. (10.72) yields

$$\begin{Bmatrix} y_i \\ y_j \\ \theta_i \\ \theta_j \end{Bmatrix} = \begin{bmatrix} 0 & 1 & 0 & 1 \\ \sin \varepsilon L & \cos \varepsilon L & \sin \phi L & \cos \phi L \\ -\varepsilon & 0 & -\phi & 0 \\ -\varepsilon \cos \varepsilon L & \varepsilon \sin \varepsilon L & -\varepsilon \cosh \phi L & -\varepsilon \sinh \phi L \end{bmatrix} \begin{Bmatrix} C_1 \\ C_2 \\ C_3 \\ C_4 \end{Bmatrix} \tag{10.73}$$

or
$$\{\delta\}_e = [G][C] \tag{10.74}$$

or
$$\{C\} = [G]^{-1}\{\delta\}_e \tag{10.75}$$

Similarly, the nodal forces can be expressed through Eq. (10.58)

$$\begin{Bmatrix} V_i \\ V_j \\ M_i \\ M_j \end{Bmatrix} = EI \begin{Bmatrix} Y''' & \text{at } x=0 \\ -Y''' & \text{at } x=L \\ Y'' & \text{at } x=0 \\ Y'' & \text{at } x=L \end{Bmatrix}$$

$$= EI \begin{bmatrix} \varepsilon^3 & 0 & \varepsilon^3 & 0 \\ \varepsilon^3 \cos\varepsilon L & -\varepsilon^3 \sin\varepsilon L & -\varepsilon^3 \cos\phi L & -\varepsilon^3 \sin\phi L \\ 0 & -\varepsilon^2 & 0 & \varepsilon^2 \\ \varepsilon^2 \sin\varepsilon L & \varepsilon^2 \cos\varepsilon L & -\varepsilon^2 \sin\phi L & \varepsilon^2 \cos\phi L \end{bmatrix} \begin{Bmatrix} C_1 \\ C_2 \\ C_3 \\ C_4 \end{Bmatrix} \quad (10.76)$$

or

$$\{P\}_e = [H]\{C\} \quad (10.77)$$

Substituting $\{C\}$ from Eq. (10.75) into Eq. (10.77) results into

$$\{P\}_e = [H]\{G\}^{-1}\{\delta\}_e \quad (10.78)$$

Therefore, the dynamic stiffness matrix for this case is

$$[K(\varepsilon,\phi)]_e = [H][G]^{-1} \quad (10.79)$$

The dynamic stiffness matrix is thus dependent on both, the frequency and the axial force parameter. The stiffness matrix expression of Eq. (10.79) indicates the effect of the axial force on the transverse vibration of the beam. If the axial deformation due to time varying axial force as given in Art 10.6 is to be included in the stiffness matrix, it can be done exactly the same way as presented in Art 10.7.1. However, as suggested by Penzien and Clough, for complex framed structure, this approach will prove somewhat impractical and discrete parameter approach is preferred.

10.8 BEAM SEGMENTS WITH DISTRIBUTED MASS HAVING SHEAR DEFORMATION AND ROTARY INERTIA

Timoshenko beam equations in terms of y or ψ are given in Eq. (8.174) or Eq. (8.177). Solutions for Y given in Eq. (8.173) $[y(x,t) = Y(x)\sin(pt+\varepsilon)]$ and for $\overline{\psi}$ given in Eq. (8.176) $[\psi(x,t) = \overline{\psi}(x)\sin(pt+\varepsilon)]$ will take the following form[10.2].

$$Y = C_1 \cosh b\alpha\xi + C_2 \sinh b\alpha\xi + C_3 \cos b\beta\beta + C_4 \sin b\beta\beta \quad (10.80)$$

$$\overline{\psi} = C'_1 \sinh b\alpha\xi + C'_2 \cosh b\alpha\xi + C'_3 \sin b\beta\beta + C'_4 \cos b\beta\beta \quad (10.81)$$

in which

$$\frac{\alpha}{\beta} = \frac{1}{\sqrt{2}}\left\{ \mp\left(r^2+s^2\right) + \left[\left(r^2-s^2\right)^2 + \frac{4}{b^2}\right]^{1/2}\right\}^{1/2} \quad (10.82,83)$$

$$b^2 = \frac{1}{EI}\rho A L^4 p^2 \qquad (10.84)$$

$$r^2 = \frac{I}{AL^2} \qquad (10.85)$$

$$s^2 = \frac{EI}{\rho AGL^2} \qquad (10.86)$$

$$\xi = \frac{x}{L} \qquad (10.87)$$

$$C_1' = \frac{b}{L}\frac{\alpha^2 + s^2}{\alpha} C_1 = TC_1 \qquad (10.88)$$

$$C_2' = \frac{b}{L}\frac{\alpha^2 + s^2}{\alpha} C_2 = TC_2 \qquad (10.89)$$

$$C_3' = -\frac{b}{L}\frac{\beta^2 - s^2}{\beta} C_3 = -U'C_3 \qquad (10.90)$$

$$C_4' = \frac{b}{L}\frac{\beta^2 - s^2}{\beta} C_4 = -U'C_4 \qquad (10.91)$$

L is the length of the member and p is the angular frequency. Eqs. (10.80) and (10.81) when written in the matrix form become

$$\begin{Bmatrix} Y \\ \psi \end{Bmatrix} = \begin{bmatrix} \cosh b\alpha\xi & \sinh b\alpha\xi & \cos b\beta\xi & \sin b\beta\xi \\ T\sinh b\alpha\xi & T\cosh b\alpha\xi & -U'\sin b\beta\xi & U'\cos b\beta\xi \end{bmatrix} \begin{Bmatrix} C_1 \\ C_2 \\ C_3 \\ C_4 \end{Bmatrix} \qquad (10.92)$$

Equations (8.154) and (8.77) are rewritten in terms of Y and ψ

$$V = \mu AG\left(\frac{dY}{dx} - \psi\right) \qquad (10.93)$$

and

$$M = -EI\frac{d\psi}{dx} \qquad (10.94)$$

Substituting Y and $\bar{\psi}$ from Eqs. (10.80) and (10.81) into Eqs. (10.93) and (10.94) and using the relations (10.88) to (10.91), results in

$$\begin{Bmatrix} V \\ M \end{Bmatrix} = \begin{bmatrix} (\bar{u} - \mu AGT) & (\bar{u} - \mu AGT) & (-\eta + \mu AGU') & (\eta - \mu AGU') \\ \sinh b\alpha\xi & \cosh b\alpha\xi & \sin b\beta\xi & \cos b\beta\xi \\ -\Omega\cosh b\alpha\xi & -\Omega\sinh b\alpha\xi & \tau'\cos b\beta\xi & \tau'\sin b\beta\xi \end{bmatrix} \begin{Bmatrix} C_1 \\ C_2 \\ C_3 \\ C_4 \end{Bmatrix} \qquad (10.95)$$

in which

$$\bar{\mu} = \mu AG\frac{b\alpha}{L} \qquad (10.96)$$

$$\eta = \mu AG \frac{b\beta}{L} \qquad (10.97)$$

$$\Omega = El \frac{b\alpha}{L} T \qquad (10.98)$$

$$\tau' = El \frac{b\beta}{L} U' \qquad (10.99)$$

Let the displacements at the ith node of the beam segment be Y_i and $\overline{\psi}_i$ and at the *j*th node, they are Y_j and $\overline{\psi}_j$ (Fig.10.13) Similarly, the shear force and the bending moment at the *i*th end and *j*th end are V_i, M_i and V_j, M_j respectively. The displacement and the force matrices are given by

$$\{\delta\}_e = \begin{Bmatrix} \overline{\psi}_i \\ \overline{\psi}_j \\ Y_i \\ Y_j \end{Bmatrix} \qquad (10.100)$$

and

$$\{P\}_e = \begin{Bmatrix} M_i \\ M_j \\ V_i \\ V_j \end{Bmatrix} \qquad (10.101)$$

With the help of Eq. (10.92), the nodal displacements can be related to the constants. They are expressed in matrix form as

$$\{\delta\}_e = [G]\{C\} \qquad (10.102)$$

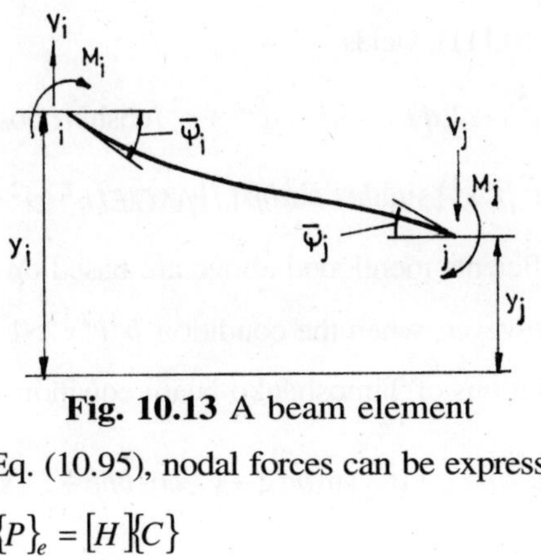

Fig. 10.13 A beam element

Similarly, with the help of Eq. (10.95), nodal forces can be expressed in terms of constants as

$$\{P\}_e = [H]\{C\} \qquad (10.103)$$

Eliminating $\{C\}$ between Eqs. (10.102) and (10.103), we get

$$\{P\}_e = [H][G]^{-1}\{\delta\}_e$$

or

$$\{P\}_e = [K]\{\delta\}_e \tag{10.104}$$

The dynamic stiffness matrix has been calculated explicitly by Cheng [7] and is given by

$$[K]_e = \begin{bmatrix} SM\overline{\psi}_1 & SM\overline{\psi}_2 & SMY_1 & SMY_2 \\ & SM\overline{\psi}_1 & SMY_2 & SMY_1 \\ & & SVY_1 & SVY_2 \\ Symmetrical & & & SVY_1 \end{bmatrix} \tag{10.105}$$

in which

$$SM\overline{\psi}_1 = \frac{L\left[-\alpha\left(\beta^2 - s^2\right)\sinh b\alpha\cos b\beta + \beta\left(\alpha^2 + s^2\right)\cosh b\alpha\sin b\beta\right]}{\mu AGbs^2\left(b^2 + \beta^2\right)D} \tag{10.106}$$

$$SM\overline{\psi}_2 = \frac{L\left[\alpha\left(\beta^2 - s^2\right)\sinh b\alpha - \beta\left(\alpha^2 + s^2\right)\sin b\beta\right]}{\mu AGbs^2\left(\alpha^2 + \beta^2\right)D} \tag{10.107}$$

$SMY_1 =$

$$\frac{L^2\left[\alpha\beta\left(2s^2 + \alpha^2 - \beta^2\right)\left(1 - \cosh b\alpha\cos b\beta\right) - \left(2\alpha^2\beta^2 - \alpha^2 s^2 + \beta^2 s^2\right)\sinh b\alpha\sin b\beta\right]}{Elb^2\left(\alpha^2 + \beta^2\right)\alpha\beta} \tag{10.108}$$

$$SMY_2 = \frac{(\beta^2 - s^2)(\alpha^2 + s^2)[-\cosh b\alpha + \cos b\beta]}{\mu AGbs^2(\alpha^2 + \beta^2)D} \tag{10.109}$$

$$SVY_1 = \frac{L[\alpha(\beta^2 - s^2)\cosh b\alpha\sin b\beta + \beta(\alpha^2 + s^2)\sinh b\alpha\cos b\beta]}{EIb\alpha\beta(\alpha^2 + \beta^2)D} \tag{10.110}$$

$$SVY_2 = \frac{L[\beta(\alpha^2 - s^2)\sinh b\alpha + \alpha(\beta^2 + s^2)\sin b\beta}{EIb\alpha\beta(\alpha^2 + \beta^2)D} \tag{10.111}$$

Combining Eqs.(10.106) to (10.111), yields

$$D = L^2[2\alpha\beta(\beta^2 - s^2)(\alpha^2 + s^2) - 2\alpha\beta(\beta^2 - s^2)(\alpha^2 + s^2)\cosh b\alpha\cos b\beta$$

$$\{(\alpha^2 - \beta^2)(\alpha^2\beta^2 - s^4) + 4\alpha^2\beta^2 s^2\}\sinh b\alpha\sin b\beta]/[\mu AGEIb^2(\alpha^2 + \beta^2)\alpha\beta s^2] \tag{10.112}$$

The dynamic stiffness coefficients mentioned above are based on Eqs. (10.82) and (10.83), by assuming that $b^2r^2s^2<1$ However, when the condition $b^2r^2s^2>1$ occurs, $\alpha' = i\alpha$ is to be used where $i = \sqrt{-1}$ then the solutions of Timoshenko beam equation will be

$$Y = C_1\cos b\alpha'\xi + iC_2\sin b\alpha'\xi + C_3\cos b\beta\beta + C_4\sin b\beta\beta \tag{10.113}$$

$$= iC'_1\sin b\alpha'\xi + iC'_2\cos b\alpha'\xi + C'_3\sin b\beta\beta + C'_4\cos b\beta\beta \tag{10.114}$$

The integration constants are, however, same as given by Eq (10.88) to Eq. (10.91). The dynamic stiffness coefficients for $b^2r^2s^2>1$ are

$$SM\overline{\psi}_1=\frac{1}{bD}[\alpha'(\beta^2-s^2)\sin b\alpha'\cos b\beta+\beta(\alpha^2+s^2)\sin b\beta\cos b\alpha']\frac{EI}{L} \tag{10.115}$$

$$SM\overline{\psi}_2=\frac{-1}{bD}[\alpha'(\beta^2-s^2)\sin b\alpha'+\beta(\alpha^2+s^2)\sin b\beta]\frac{EI}{L} \tag{10.116}$$

$$SMY_1=\frac{-1}{(\alpha^2+\beta^2)D}(\alpha^2+s^2)(\beta^2-s^2)\ [(s^2-r^2)-\frac{s^2(r^2+s^2)+2\beta^2\alpha^2}{\beta\alpha'} \tag{10.117}$$

$$\sin b\alpha'\sin b\beta]\frac{EI}{L^2}+\frac{(\alpha^2+s^2)(\beta^2-s^2)}{(\alpha^2+\beta^2)D}(r^2-s^2)\cos b\alpha'\cos b\beta\frac{EI}{L^2} \tag{10.118}$$

$$SMY_2=\frac{(\alpha^2+s^2)(\beta^2-\alpha^2)}{D}[\cos b\beta-\cos b\alpha']\frac{EI}{L^2} \tag{10.119}$$

$$SVY_1=\frac{1}{bD}\left[\frac{(\alpha^2+s^2)}{\alpha}\sin b\alpha'+\frac{(\beta^2-s^2)}{\beta}\sin b\beta\right]\frac{EI}{L^3} \tag{10.120}$$

$$SVY_2=\frac{1}{bD}\left[\frac{(\alpha^2+s^2)}{\alpha'}\cos b\beta\ \sin b\alpha'+\frac{(\beta^2-s^2)}{\beta}\sin b\beta\cos b\alpha'\right]\frac{EI}{L^3}$$

$$D=\frac{2(\alpha^2+s^2)(\beta^2-s^2)}{b^2(\alpha^2+\beta^2)}[1-\cos b\beta\cos b\alpha']+\frac{[\beta^2(\alpha^2+s^2)^2+\alpha'^2(\beta^2-s^2)^2]}{\alpha'\beta b^2(\alpha^2+\beta^2)}\sin b\beta\ \sin b\alpha' \tag{10.121}$$

References

10.1 R.W. Clough and J. Penzien, Dynamics of Structures, 2nd Edition McGraw-Hill, New York, 1993

10.2 M. Paz, Structural Dynamics, CBS Publishers and Distributors, New Delhi, 1987.

10.3 F.Y. Cheng, Vibration of Timoshenko beams and frameworks, Journal of Structural Division, ASCE, V 96, ST3, March 1973, pp. 551-557

EXERCISE 10

10.1 Determine the fundamental frequency of the frame shown in the figure by considering it as having distributed mass.

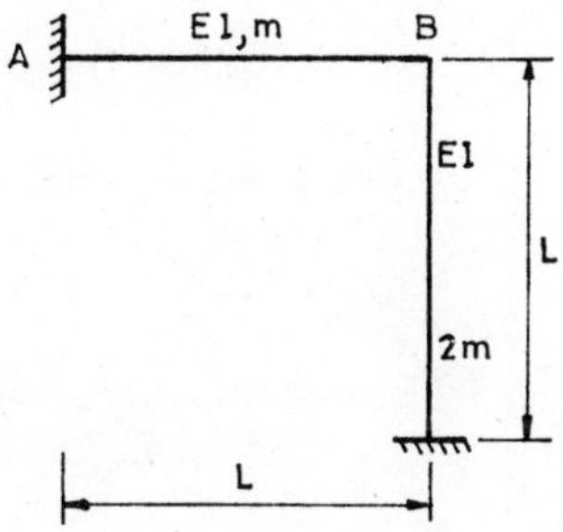

Prob. 10.1

10.2 If a clockwise moment M_0 acts at joint B, then determine the rotation at joint B for the frame of Prob 10.1.

10.3 Determine the fundamental frequency of the continuous beam of the figure by treating the whole system as having distributed mass.

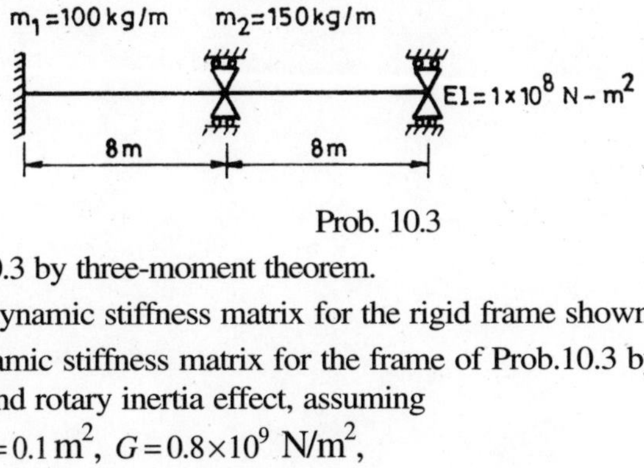

Prob. 10.3

10.4 Solve Prob. 10.3 by three-moment theorem.

10.5 Evaluate the dynamic stiffness matrix for the rigid frame shown in the figure.

10.6 Form the dynamic stiffness matrix for the frame of Prob.10.3 by considering the shear deformation and rotary inertia effect, assuming

$\mu = 0.8$, $A = 0.1\,\text{m}^2$, $G = 0.8\times10^9\ \text{N/m}^2$,

$E = 2\times10^9\ \text{N/m}^2$, and $\rho = 78.4\,\text{kg/m}^2$.

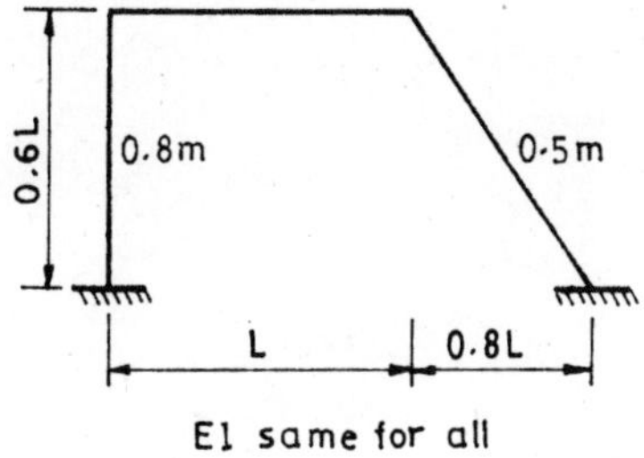

Prob. 10.5

Chapter 11

Vibration of Ship and Aircraft as a Beam

11.1 INTRODUCTION

Aircrafts and ships are complex structures. They have curved surfaces with various forms of intricate stiffening arrangements. For simplification of the analysis, they are sometimes treated as a beam having varying rigidity. This no doubt facilitates the analysis to a great extent, both for static and dynamic cases, but it introduces certain special problems, particularly in vibration analysis. Unlike a building, a bridge or a machine, the ship or the aircraft is unconstrained in space, that is, every point in the structure may undergo displacement. This introduces rigid body displacements. As a result of which, the stiffness matrix becomes singular. Various approaches have been sought to obviate these difficulties. Some of them are presented in the following [11.1-11.5]

11.2 SHIFT IN STIFFNESS MATRIX

Let us consider a multiple degrees of freedom system with lumped masses. The equations of motion in free vibration are given by

$$p^2[M]\{\phi\}=[K]\{\phi\} \tag{11.1}$$

For an unconstrained structure, $[K]$ is singular. Now to apply the general methods to the solution of the eigenvalue problem as presented in Chapter 6 for this case, the singularity of $[K]$ is to be disturbed. This is done as follows :

Let us add $\alpha[M]\{\phi\}$ to both sides of Eq. (11.1), where is an arbitrary constant.

$$[K]\{\phi\}+\alpha[M]\{\phi\}=p^2[M]\{\phi\}+\alpha[M]\{\phi\}$$

or

$$([K]+\alpha[M])\{\phi\}=(p^2+\alpha)[M]\{\phi\} \tag{11.2}$$

The shifted stiffness $([K]+\alpha[M])$ matrix will be non-singular, though $[K]$ is singular. It may be noted that α should be of the same order as p^2. The eigenvalues will be in terms of $(p^2 + \alpha)$ from where p can be evaluated. The shift approach has the advantage that the mode shapes are not changed.

11.3 ADDED MASS OF A SHIP

Ship's hull poses a typical problem. A ship's hull vibrates in a liquid medium with an air-water interface. 'The natural frequency of the hull, when calculated on the basis of the vessel's weight and elastic properties, results in a much higher value of the calculated frequency. In order to reconcile this difference, it is necessary to include the energy imparted to the surrounding water by the vibrating ship. The total mass thus considered is known as the virtual mass. The addition of mass over the hull weight due to the surrounding fluid is known as the added virtual mass or simply added mass.

The ship is divided into a number of sections. We consider the section as vibrating, by considering the motion of the fluid as two-dimensional. However, it is to be borne in mind that the actual motion of the fluid is three-dimensional. Therefore, the value of the added mass that has been obtained on the basis of two-dimensional motion is to be corrected for actual three-dimensional motion.

Considerable work has been done for evaluating added mass of ship sections. We shall not go into any detailed hydrodynamic theory, but will directly use the results.

Let us first consider the vertical vibration of ship's hull. If the ship is divided into n stations with the distance between each station being $\triangle L$, the added mass of water per unit length per section is given by

$$M_i = \frac{1}{2}\pi\rho B_i^2 C_i \Delta L J_i K_i \qquad (11.3)$$

where

- ρ is the density of water,
- B_i is the half-breadth of the section at water line,
- C_i is the shape factor depending on the relation between A_i / B_i and B_i / d_i (Fig. 11.1),
- A_i is the half-sectional area up to water line
- d_i is the section draft,
- J_i is the three-dimensional correction factor, which varies with each mode and is dependent of $L_{wl}/2Bi$ (Fig. 11.2).
- L_{wl} is the length of the ship at the waterline and
- K_i is a local correction factor (Fig. 11.3).

An interesting observation may be made at this stage. As the three-dimensional correction factor is dependent on the mode of vibration, Eq. (11.3) reveals that the added mass of water varies with each mode of vibration. Therefore, for ship's hull, one natural frequency is to be considered at a time. In vibration problems of other structures, once the stiffness matrix and mass matrix are formed, all the natural frequencies and mode shapes required can be obtained from that directly. However, in the case of ship's hull, the mass matrix is to be modified every time, after evaluating each natural frequency.

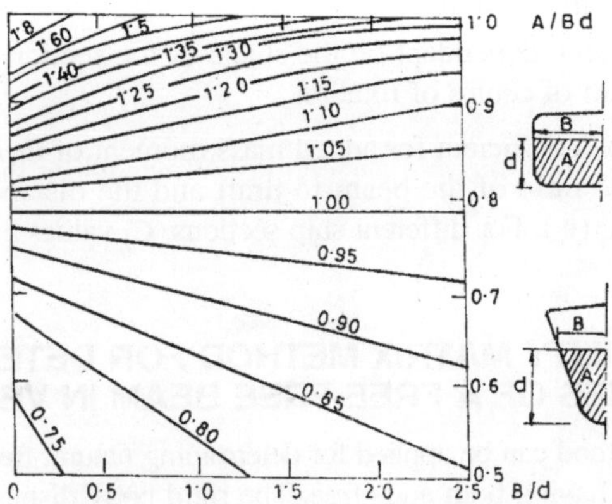

Fig. 11.1 Variation of *Ci with B/d* and *A/Bd*

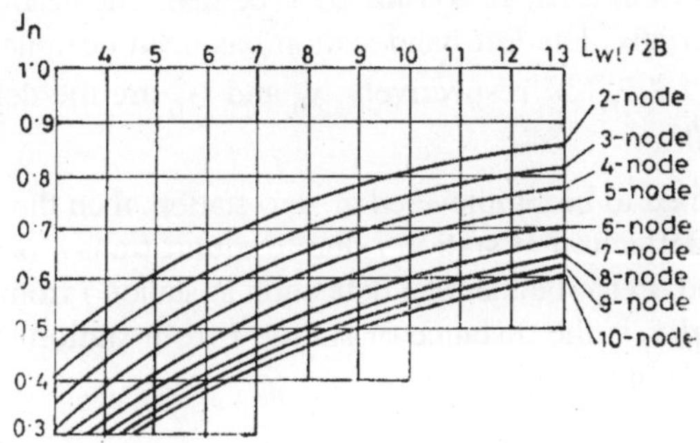

Fig. 11.2 Three-dimensional correction factor

In order to calculate added mass in horizontal vibration, the same curves given by Figs. 11.1 to 11.3 can be used, with *B*/*d* to be changed to *d*/2*B*, and *L*/2*B* to *L*/*d* for reading C_i from the graph for a particular section and for obtaining J_i.

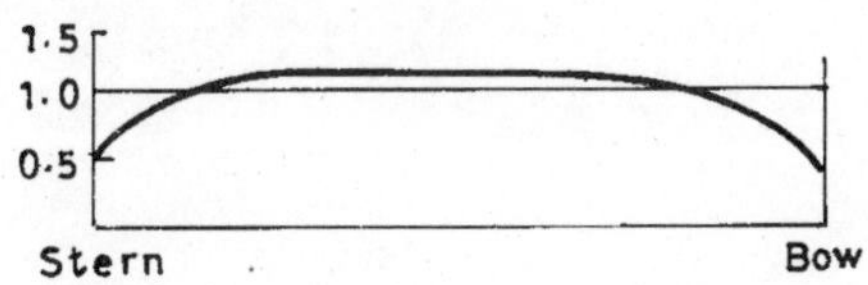

Fig. 11.3 Variation of local correction factor along length

Not much work has been done on added mass moment of inertia for torsional vibrations. Kumai [11.8] has made two-dimensional calculations of the added mass moment of inertia of ship sections, and has proposed the basic equation as

$$\Delta I = C_I \rho \pi d^4 \qquad (11.4)$$

where

$\triangle I$ = added mass moment of inertia per unit length,

ρ = density of the fluid,

d = draft of the section,

and C_I = coefficient, dependent on the shape of the section and on the location of centre of rotation.

Kumai found that the coefficient for added mass moment of inertia C_I depended upon the shape of the section, the ratio of the beam-to-draft and the distance between the centre of rotation and the waterline (y_o). For different ship sections, C_I values have been presented in Fig. 11.4. [11.6, 11.8]

11.4 THE FLEXIBILITY MATRIX METHOD FOR DETERMINING NATURAL FREQUENCIES OF A FREE-FREE BEAM IN VERTICAL VIBRATION

The flexibility matrix method can be applied for determining natural frequencies and mode shapes for a non-uniform beam having both ends free. The rigid body displacements which creep into the equations, are suitably eliminated with the help of dynamic equilibrium equations. The free vibration analysis of a free-free non-uniform beam is presented in the following.

Consider a beam of variable mass and section properties, which is representative of a ship or an aircraft (Fig. 11.5). Both ends are considered to be free. The beam is divided into a number of stations at equal intervals. The left hand station has been designated at 0 and succeeding stations to the right as 1, 2, 3, ..., n respectively. y_0 and θ_0 are the deflection and the angular rotation for the station 0.

If the beam is assumed to be cantilevered at zero station, then the flexibility coefficients a_{ij} (which represents the deflection at station i due to a unit transverse load at station j) at all stations can be calculated. Let y_i be the total deflection at station i from the equilibrium position of the elastic curve, and x_i is the distance of station i from station 0. Then the equations of motion for the vibrating beam are

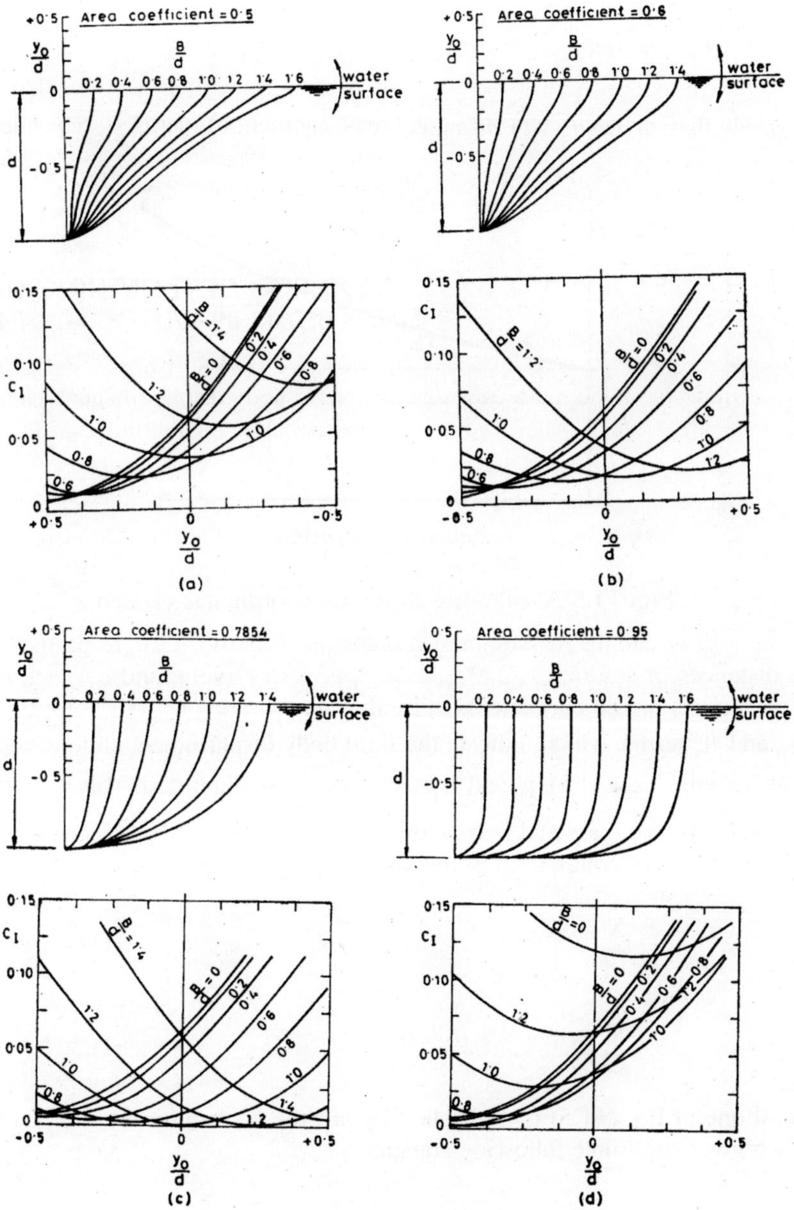

Fig. 11.4 C_I for different ship sections

$$\left.\begin{aligned} y_1 - y_0 - x_1\theta_0 &= \omega^2(m_1a_{11}y_1 + m_2a_{12}y_2 + ... + m_na_{1n}y_n) \\ y_2 - y_0 - x_2\theta_0 &= \omega^2(m_1a_{21}y_1 + m_2a_{22}y_2 + ... + m_na_{2n}y_n) \\ &\cdots\cdots\cdots\cdots\cdots\cdots \\ y_n - y_0 - x_n\theta_0 &= \omega^2(m_1a_{n1}y_1 + m_2a_{n2}y_2 + ... + m_na_{nn}y_n) \end{aligned}\right\} \tag{11.5}$$

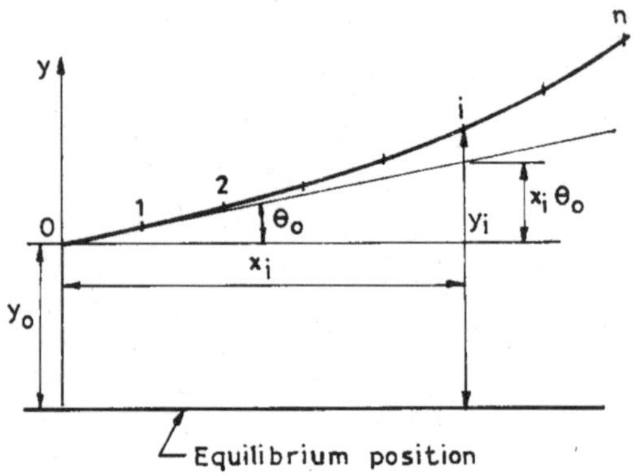

Fig. 11.5 A vibrating ship with coordinates chosen

where m_1, m_2, ….., m_n are the masses lumped at stations 1, 2, 3, …….n, respectively, x_1, x_2, …., x_n are the distances of stations 1, 2, 3, ………., n, respectively, and a_{11}, a_{12}, a_{13}, etc., are flexibility coefficients, which may cater both bending and shearing deformations. Equation (11.5) contains y_0 and θ_0 terms, which indicate the rigid body displacement and the rotation terms, respectively. To solve Eq. (11.5), y_0 and θ_0 terms are to be eliminated. This is done as follows.

Let m_0 and I_0 be the mass and mass moment of inertia respectively, of station zero. Then the dynamic equations of equilibrium of the vibrating beam are

$$\sum_{i=0}^{n} m_i y_i = 0 \tag{11.6}$$

and

$$\sum_{i=0}^{n} m_i x_i y_i + I_0\theta_0 = 0 \tag{11.7}$$

The first one of Eq. (11.5) is multiplied by m_1, the second by m_2, etc., and the results added, which gives rise to the following equation

$$\sum_{i=1}^{n} m_i y_i - \left(\sum_{i=1}^{n} m_i\right) y_0 - \left(\sum_{i=1}^{n} m_i x_i\right)\theta_0$$

$$= \omega^2(Q_1y_1 + Q_2y_2 + ... + Q_ny_n) \tag{11.8}$$

Adding and subtracting m_0y_0 on the left hand side gives

$$\sum_{i=0}^{n} m_i y_i - \left(\sum_{i=0}^{n} m_i\right) y_0 - \left(\sum_{i=0}^{n} m_i x_i\right) \theta_0$$

$$= \omega^2 \left(Q_1 y_1 + Q_2 y_2 + \ldots\ldots + Q_n y_n\right) \qquad (11.9)$$

Here $$Q_i = \sum_{j=1}^{n} m_i m_j a_{ji} \qquad (11.10)$$

Combining Eq. (11.6) with Eq. (11.10) gives

$$-M_t y_0 - S\theta_0 = \omega^2 \left(Q_1 y_1 + Q_2 y_2 + \ldots + Q_n y_n\right) \qquad (11.11)$$

where M_t = total mass = $\sum_{i=0}^{n} m_i$

$S = \sum m_i x_i$ = moment of the masses

Multiplying first one of Eq. (11.5) by $m_1 x_1$, the second by $m_2 x_2$ etc., and adding the results gives rise to the following equation

$$\sum_{i=1}^{n} m_i x_i y_i - \left(\sum_{i=1}^{n} m_i x_i\right) y_0 - \left(\sum_{i=1}^{n} m_i x_i^2\right) \theta_0$$

$$= \omega^2 \left(R_1 y_1 + R_2 y_2 + \ldots\ldots + R_n y_n\right) \qquad (11.12)$$

where $$R_i = \sum_{j=1}^{n} m_i x_j m_j a_{ji} \qquad (11.13)$$

Combining Eqs. (11.7) and (11.12), we get

$$-T\theta_0 - Sy_0 = \omega^2 (R_1 y_1 + \cdots + R_n y_n) \qquad (11.14)$$

where T is the pitching mass moment of inertia of the entire beam about station zero. Solving Eqs (11.11) and (11.14), we get

$$\theta_0 = \omega^2 \frac{1}{M_t T - S^2} \sum_{i=1}^{n} (SQ_i - M_t R_i)\, y_i \qquad (11.15)$$

$$y_0 = \omega^2 \frac{1}{M_t T - S^2} \sum_{i=1}^{n} (RQ_i - SR_i)\, y_i \qquad (11.16)$$

Substituting the values of y_0 and θ_0 from Eqs. (11.16) and (11.15) respectively, into Eq. (11.5), we get

$$\left.\begin{aligned} y_1 &= \omega^2(C_{11}y_1 + C_{12}y_2 + \cdots + C_{1n}y_n) \\ y_2 &= \omega^2(C_{21}y_1 + C_{22}y_2 + \cdots + C_{2n}y_n) \\ &\cdots\cdots\cdots\cdots\cdots\cdots \\ y_n &= \omega^2(C_{n1}y_1 + C_{n2}y_2 + \cdots + C_{nn}y_n) \end{aligned}\right\} \tag{11.17}$$

Equation (11.17) written in matrix form becomes

$$\{Y\} = \omega^2[C]\{Y\} \tag{11.18}$$

Equation (11.18) is a typical eigenvalue problem.

The flexibility coefficient may be computed for the non-uniform beam by incorporating the shear deformation. Each flexibility coefficient consists of contribution, from both flexure and shear.

Thus

$$a_{ij} = a'_{ij} + a''_{ij} \tag{11.19}$$

where a_{ij} is the flexibility coefficient shown in Fig. 11.6 (a), a'_{ij} and a''_{ij} are the contributions to the total flexibility coefficient from bending and shear, respectively.

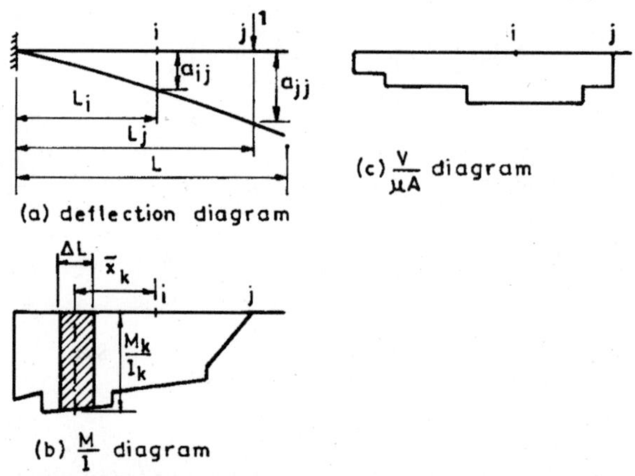

Fig. 11.6 $\frac{M}{I}$ and $\frac{V}{\mu A}$ diagrams

For the beam of Fig. 11.6, we have

$$a'_{ij} = \sum_{k=0}^{i} \frac{M_k}{I_k E}\bar{x}_k \Delta L \tag{11.20}$$

$$a''_{ij} = \sum_{k=0}^{i} \frac{V_k}{\mu A_k G}\Delta L \tag{11.21}$$

where A_k is the cross-sectional area of station k

μ is the shear coefficient.

and $\triangle L$ is the length between stations.

11.5 THE FLEXIBILITY MATRIX METHOD FOR THE ANALYSIS OF COUPLED HORIZONTAL AND TORSIONAL VIBRATION

When we consider the vertical vibration of the hull of the ship, the hull vibrates in its plane of symmetry. But same is not true in case of horizontal vibration of ship's hull. In this case, the shear centre of the cross-sectional area and the centroidal axis do not coincide. As a result of which, horizontal bending vibration is always accompanied with torsional vibration of ships. As such, the resulting vibration is always coupled [11.7].

We again consider Fig. 11.5, in which a ship girder having variable mass loading and sectional properties has been indicated. It should be borne in mind that the horizontal vibration of the girder is considered here, though we maintain the same axis system. y_0, θ_0 and α_0 are the deflection, bending rotation and angle of twist respectively, at station zero.

Let a_{ij} $(i, j = 1, 2, ..., n)$ be the influence coefficients related to bending and shearing deformation and b_{ij} $(i, j = 1, 2, ..., n)$ be the influence coefficients related to torsion. Let y_i be the total deflection at station i from the equilibrium position and x_i be the horizontal distance from zero station to station i. If $F_1, F_2, ..., F_n$ are the inertia forces acting at stations 1, 2, ..., n, then the horizontal translatory equations of motion for the vibrating ship can be written as

$$\left.\begin{aligned} y_1 - y_0 - x_1\theta_0 &= a_{11}F_1 + a_{12}F_2 + \cdots + a_{1n}F_n \\ y_2 - y_0 - x_2\theta_0 &= a_{21}F_1 + a_{22}F_2 + \cdots + a_{2n}F_n \\ &\cdots\cdots\cdots\cdots\cdots\cdots \\ y_n - y_0 - x_n\theta_n &= a_{n1}F_1 + a_{n2}F_2 + \cdots + a_{nn}F_n \end{aligned}\right\} \quad (11.22)$$

If α_i is the total angle of twist of station from the equilibrium position, β_i represents the distance between the centre of mass and the centre of twist at station i and $T_1, T_2, T_3, \cdots, T_n$, are the twisting moments acting at stations 1, 2, ..., n, then a set of equations similar to Eq. (11.22) can be written as

$$\left.\begin{aligned} \alpha_1 - \alpha_0 &= b_{11}T_1 + b_{12}T_2 + \cdots + b_{1n}T \\ \alpha_2 - \alpha_0 &= b_{21}T_1 + b_{22}T_2 + \cdots + b_{2n}T \\ &\cdots\cdots\cdots\cdots\cdots\cdots \\ \alpha_n - \alpha_0 &= b_{n1}T_1 + b_{n2}T_2 + \cdots + b_{nn}T \end{aligned}\right\} \quad (11.23)$$

The mass is distributed along the centroidal axis. The given mass is replaced by the same mass distributed along the shear centre axis and a torque of intensity $m_i s_i g$ acts at station i and so is the treatment at all other stations. There is, therefore, coupled bending and torsion about the shear centre axis.

The system is assumed to oscillate in simple harmonic motion about the equilibrium position with frequency ω. Then the inertia force F_i acting at station due to translation is (Fig. 11.7).

$$F_i = \omega^2 m_i (y_i + s_i \alpha_i) \quad (11.24)$$

Similarly

$$T_i = \omega^2 m_i s_i (y_i + s_i \alpha_i) + \gamma I_{pi} \alpha_i$$

or $$T_i = \omega^2 [m_i s_i y_i + \{m_i (s_i)^2 + \gamma I_{pi}\} \alpha_i]$$

or $$T_i = \omega^2 [m_i s_i y_i + \bar{I}_i \alpha_i] \tag{11.25}$$

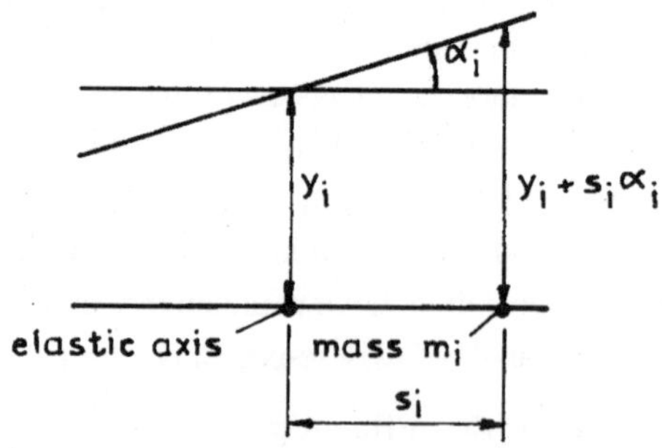

Fig. 11.7 A mass in coupled bending and torsion

where γ = mass density of the material.

and $\bar{I}_{pi}$ = polar moment of inertia for torsion at station i.

Substitution of the values of F_i and T_i from Eqs. (11.24) and (11.25) into Eqs. (11.22) and (11.23) result in the following two equations:

$$\left.\begin{aligned} y_1 - y_0 - x_1\theta_0 &= \omega^2[m_1 a_{11} y_1 + \cdots + m_n a_{1n} y_n \\ &\quad + m_1 s_1 a_{11} \alpha_1 + \cdots + m_n s_n a_{1n} \alpha_n] \\ &\cdots\cdots\cdots\cdots\cdots\cdots \\ y_n - y_0 - x_n\theta_0 &= \omega^2[m_1 a_{n1} y_1 + \cdots + m_n a_{nn} y_n \\ &\quad + m_1 s_1 a_{n1} \alpha_1 + \cdots + m_n s_n a_{nn} \alpha_n] \end{aligned}\right\} \tag{11.26}$$

and

$$\left.\begin{aligned} \alpha_1 - \alpha_0 &= \omega^2[m_1 s_1 b_{11} y_1 + \cdots + m_n s_n b_{1n} y_n \\ &\quad + \bar{I}_1 b_{11} \alpha_1 + \cdots + \bar{I}_n b_{1n} \alpha_n] \\ &\cdots\cdots\cdots\cdots\cdots\cdots \\ \alpha_n - \alpha_0 &= \omega^2[m_1 s_1 b_{n1} y_1 + \cdots + m_n s_n b_{nn} y_n \\ &\quad + \bar{I}_1 b_{n1} \alpha_1 + \cdots + \bar{I}_n b_{nn} \alpha_n] \end{aligned}\right\} \tag{11.27}$$

The following equations of equilibrium for the vibrating masses must be satisfied:

$$\sum F_i = 0, \quad \sum T_i = 0, \quad \sum (F_i x_i + I_0 \theta_0) = 0. \tag{11.28}$$

Making proper substitutions from Eqs (11.24) and (11.25) into Eq. (11.28), result in

$$\left.\begin{aligned}\sum_{i=0}^{n} m_i(y_i + s_i\alpha_i) = 0, \quad \sum_{i=0}^{n} (m_i s_i y_i + \bar{I}_i\alpha_i) = 0 \\ \sum_{i=0}^{n} m_i x_i(y_i + s_i\alpha_i) + I_0\theta_0 = 0\end{aligned}\right\} \tag{11.29}$$

First of Eq. (11.26) be multiplied by m_1, second by m_2 and so on. Similarly, first of Eq. (11.27) be multiplied by $m_1\ s_1$, the second by $m_2 s_2$ and so on. The results of these two equations when added, leads to

$$\sum_{i=0}^{n} m_i y_i + \sum_{i=0}^{n} m_i s_i \alpha_i - \sum_{i=0}^{n} m_i y_0 - \left(\sum_{i=0}^{n} m_i x_i\right)\theta_0 - \left(\sum_{i=0}^{n} m_i s_i\right)\alpha_0$$

$$= \omega^2[A_1 y_1 + \cdots + A_n y_n + B_1\alpha_1 + \cdots + B_n\alpha_n] \tag{11.30}$$

where $$A_n = \sum_{i=1}^{n} (m_i m_n a_{in} + m_i m_n s_i s_n b_{in}) \tag{11.31}$$

and $$B_n = \sum_{i=1}^{n} (m_i m_n s_n a_{in} + \bar{I}_n m_i s_i b_{in}) \tag{11.32}$$

Equation (11.30) is rewritten using the first relation of Eq. (11.29) as

$$-My_0 - P\theta_0 - Q\alpha_0 = \omega^2[A_1 y_1 + \cdots + A_n y_n + B_1\alpha_1 + \cdots + B_n\alpha_n] \tag{11.33}$$

where $$M = \sum m_i, \quad P = \sum m_i x_i \text{ and } Q = \sum m_i s_i$$

Similarly, the first of Eq. (11.26) is multiplied by $m_1 s_1$, second by $m_2 s_2$ and so on. In the same manner, Eq. (11.27) is multiplied by $\bar{I}_1$, second by $\bar{I}_2$ and so on. The results of these two equations when added, yields

$$\sum_{i=0}^{n} m_i s_i y_i + \sum_{i=0}^{n} \bar{I}_i\alpha_i - \left(\sum_{i=0}^{n} m_i s_i\right) y_0 - \left(\sum_{i=0}^{n} m_i s_i x_i\right)\theta_0$$

$$-\sum_{i=0}^{n} \bar{I}\alpha_0 = \omega^2[C_1 y_1 + \cdots + C_n y_n + D_1\alpha_1 + \cdots + D_n\alpha_n] \tag{11.34}$$

where $$C_n = \sum_{i=0}^{n} (m_i m_n s_i a_{in} + m_n s_n \bar{I}_i b_{in}) \tag{11.35}$$

and $$D_n = \sum_{i=1}^{n} (m_i m_n s_i s_n a_{in} + \bar{I}_i \bar{I}_n b_{in}) \tag{11.36}$$

Using the relations given in Eq. (11.29), Eq. (11.34) can be written as

$$-Qy_0 - R\theta_0 - U\alpha_0 = \omega^2[C_1 y_1 + \cdots + C_n y_n + D_1\alpha_1 + \cdots + D_n\alpha_n] \quad (11.37)$$

where $R = \sum_{i=1}^{n} m_i s_i x_i$ and $U = \sum_{i=1}^{n} \bar{I}_i$

Next, multiplying first of Eq. (11.26) by $m_1 x_1$, the second by $m_2 x_2$ and so on and similarly, multiplying first of Eq. (11.27) by $m_1 s_1 x_1$, the second by $m_2 s_2 x_2$ and so on, the steps similar to those are followed for obtaining Eqs. (11.37), and the following equation results

$$-Py_0 - V\theta_0 - R\alpha_0 = \omega^2[E_1 y_1 + \cdots + E_n y_n + G_1\alpha_1 + \cdots + G_n\alpha_n] \quad (11.38)$$

where
$$E_n = \sum_{i=1}^{n} (m_n m_i x_i a_{in} + m_i m_n s_i s_n x_n b_{in}) \quad (11.39)$$

$$G_n = \sum_{i=1}^{n} (m_i m_n x_i s_n a_{in} + m_i s_i x_i \bar{I}_n b_{in}) \quad (11.40)$$

$$V = \sum_{i=1}^{n} m_i x_i^2 \quad (11.41)$$

Solving Eqs. (11.33), (11.37) and (11.38), the values of y_0, θ_0 and α_0 can be obtained and they are expressed in terms $y_1, y_2, \cdots, y_n,\ \alpha_1, \alpha_2, \cdots \alpha_n$. Substitution of these values of y_0, θ_0 and α_0 into Eqs. (11.26) and (11.27) results in an equation of the following form:

$$\left.\begin{array}{l}
y_1 = \omega^2[k_{11}y_1 + \cdots + k_{1n}y_n + h_{11}\alpha_1 + \cdots + h_{1n}\alpha_n] \\
y_2 = \omega^2[k_{21}y_1 + \cdots + k_{2n}y_n + h_{21}\alpha_1 + \cdots + h_{2n}\alpha_n] \\
\cdots\cdots\cdots\cdots\cdots\cdots\cdots\cdots\cdots\cdots \\
y_n = \omega^2[k_{n1}y_1 + \cdots + k_{nn}y_n + h_{n1}\alpha_1 + \cdots + h_{nn}\alpha_n] \\
\\
\alpha_1 = \omega^2[p_{11}y_1 + \cdots + p_{1n}y_n + q_{11}\alpha_1 + \cdots + q_{1n}\alpha_n] \\
\alpha_2 = \omega^2[p_{21}y_1 + \cdots + p_{2n}y_n + q_{21}\alpha_1 + \cdots + q_{2n}\alpha_n] \\
\cdots\cdots\cdots\cdots\cdots\cdots\cdots\cdots\cdots\cdots \\
\alpha_n = \omega^2[p_{n1}y_1 + \cdots + p_{nn}y_n + q_{n1}\alpha_1 + \cdots + q_{nn}\alpha_n]
\end{array}\right\} \quad (11.42)$$

Equation (11.42) can be written in matrix form as

$$\{Y\} = \omega^2 [K]\{Y\} \quad (11.43)$$

Size of [*K*] is $2n \times 2n$. Equation (11.43) is a typical eigenvalue problem, which can be solved by using any suitable procedure. For the present case, [*K*] is an unsymmetric matrix.

11.6 NUMERICAL EXAMPLES

The flexibility matrix method presented in Sections 11.4 and 11.5 is somewhat mathematically involved. For ships or aircrafts, when a minimum of 20 divisions of the structure is required for a reliable result, the computations done in different stages are to be performed with the help of a computer by preparing the necessary software. Two results are presented in the following.

Example 11.1

The following data are given for an oil tanker. The ship is divided into 20 stations. The length between two consecutive stations is 11.17 m.

Determine the natural frequencies for the vertical vibration of the ship.

The problem has been solved both by the stiffness method and the flexibility method with same number of divisions of the ship. Results from both the methods are presented in the table. They are pretty close. It may be of some interest to note that incorporation of shear deformation results in a reduction of the natural frequency. This effect is more pronounced at higher modes.

Section number	Ship's weight (kN)	Entrained water mass (kN)	Moment of inertia (m^4)	Effective shear area (m^2)
1	4,310	880	47.3	0.249
2	10,850	6,960	145.0	0.765
3	12,340	19,740	200.0	1.054
4	11,090	34,590	200.0	1.054
5	9,860	48,410	182.8	0.962
6	60,190	59,210	192.5	1.013
7	57,020	64,900	192.5	1.013
8	42,960	66,770	192.5	1.013
9	42,960	66,770	192.5	1.013
10	43,040	66,770	192.5	1.013
11	35,110	66,770	192.5	1.013
12	27,310	66,770	192.5	1.013
13	30,500	65,910	192.5	1.013
14	33,460	64,890	192.5	1.013
15	37,730	61,770	192.5	1.013
16	40,250	55,540	192.5	1.013
17	40,020	43,700	172.3	0.908
18	37,810	26,310	145.2	0.765
19	16,540	7,340	109.0	0.574
20	2,610	1,110	72.8	0.383

Mode	Frequencies, cycles/min			
	Flexure only		Flexure plus shear	
	Stiffness method	Flexibility method	Stiffness method	Flexibility method
1	50.93	50.61	46.97	46.59
2	129.62	128.76	101.32	100.18
3	246.52	244.47	160.50	158.26
4	413.73	409.11	225.34	221.67
5	614.19	---	287.00	282.72

Example 11.2

Data of a ship divided into 20 divisions are given in the following table. Determine the natural frequencies corresponding to coupled horizontal and torsional vibrations.

x from aft end (m)	W (kN)	$AG \times 10^7$ (kN)	s (m)	$EI \times 10^9$ (kNm^2)	$GJ \times 10^9$ (kNm^2)	$I_p + ws^2 \times 10^4$ (kN)
0	5,300.0	2.9	−5.18	6.3	0.93	7.5
10	12,662.1	4.1	−2.58	15.7	3.53	17.6
20	14,500.0	5.7	−1.20	37.8	8.47	94.75
30	14,690.0	7.2	−1.20	65.0	12.60	120.75
40	13,205.3	8.2	−1.20	95.5	14.95	144.85
50	39,211.5	10.0	−1.20	120.0	17.30	265.13
60	40,655.0	11.3	−1.20	130.6	18.65	294.65
70	38,985.0	11.5	−1.20	130.6	19.30	288.39
50	18,348.0	11.5	−1.20	130.6	19.30	216.06
90	17,300.0	11.5	−1.20	130.6	19.30	207.55
100	15,265.0	11.5	−1.20	130.6	19.30	187.02
110	44,669.4	11.5	−1.20	130.6	19.30	338.80
120	24,382.6	11.5	−1.20	130.6	19.30	239.30
130	15,015.0	11.5	−1.20	130.6	19.30	200.00
140	14,705.0	11.5	−1.20	130.6	19.30	200.00
150	17,134.4	10.5	−1.20	129.5	19.30	202.90
160	15,956.0	9.2	−1.20	124.0	19.10	200.80
170	16,690.0	7.8	−1.20	104.0	17.45	179.50
180	41,081.4	6.3	−1.20	48.3	13.95	379.89
190	6,420.8	5.1	−1.20	23.1	10.57	339.20
208	6,868.0	3.2	−1.20	2.94	2.02	177.00

The problem has been solved with a computer program developed for this purpose. The results are presented in the following table:

Mode		Frequency (cycles/minute)
bending (node)	torsion (node)	
$\overline{2}$	1*	115.6
1	$\overline{1}$	165.02
$\overline{3}$	2	215.05
2	$\overline{2}$	319.078
$\overline{4}$	2	329.75
$\overline{5}$	3	412.92

*The bar indicates the deformation which is predominant in that mode. For example, the very first mode has 2-node bending and 1-node torsion and the mode is bending predominant.

References

11.1 O.C. Zienkiewicz and R.L Taylor, The Finite Element Method, Fourth Edition, McGraw Hill, 1989

11.2 M. Mukhopadhyay and S. Guha, A method for determining natural frequencies for vertical vibration of ships, Marine Engineering Division, Journal of the Institution of Engineers (India), May 1980, pp.36-39.

11.3 G. Anderson and K. Norrand, A method for the calculation of vertical vibration with several nodes and some other aspects of ship vibration, Transaction of the Royal Institution of Naval Architects, V. 111, 1969, pp. 367-383

11.4 D.K. Sunil and M. Mukhopadhyay, Finite element free vibration analysis of ship girder, Journal of Institution of Naval Architects, India, December 1993, pp. 83-92

11.5 M. Mukhopadhyay, Free vibration of free-free beam with rotary inertia effect - a flexibility matrix approach, Journal of Sound and Vibration, V.125, No.3, 1988, pp.565-569

11.6 F. H. Todd, Ship Hull Vibration, Edward Arnold Publishers Ltd., 1959

11.7 M. Mukhopadhyay, Determination of coupled horizontal and torsional vibration of ships, Proceedings of the Second International Symposium on Practical Design of Shipbuilding, Tokyo, October 16-21, 1983

11.8 T. Kumai, Added mass moment of inertia of ships, European Shipbuilding, No. 6, 1958, pp. 33-38.

EXERCISE 11

11.1 Write a computer program for determining vertical vibration characteristics of an aircraft by the stiffness method.

11.2 Write a computer program for determining the vertical vibration characteristics of a ship by the flexibility matrix method.

11.3 Write a computer program for determining the vibration characteristics of a ship in coupled horizontal and torsional vibration, by the flexibility matrix method.

11.4 Develop the necessary equations for determining natural frequencies of a ship or an aircraft, by the stiffness method.

11.5 Determine the natural frequencies and mode shapes for the ship, whose data are given in the following table for vertical vibration.

Station	Ship's weight (kN)	Virtual weight of water (kN)	Moment of inertia (m^4)	Effective shear area (m^2)
1	520	0	3.5	0.080
2	2880	250	9.6	0.194
3	2900	1030	14.3	0.238
4	2260	2450	15.6	0.238
5	2070	4910	18.6	0.254
6	3760	7420	21.4	0.258
7	4830	9660	24.2	0.252
8	5160	1,1120	29.5	0.250
9	5880	1,1740	29.5	0.280
10	5780	1,1860	29.5	0.275
11	6540	1,1640	29.5	0.298
12	4310	1,0890	32.5	0.241
13	2870	9300	28.0	0.241
14	2870	7230	25.1	0.256
15	2980	4710	23.2	0.250
16	2730	2340	20.3	0.263
17	2320	1060	17.5	0.260
18	2540	280	16.1	0.290
19	2190	60	12.9	0.226
20	770	0	5.1	0.123

Chapter 12

Finite Element Method in Vibration Analysis

12.1 INTRODUCTION TO THE FINITE ELEMENT METHOD

In the finite element method, the continuum is divided into a finite number of meshes by imaginary lines. For one-dimensional continuum, two adjacent elements placed side-by-side will meet at a point. Two dimensional continuum will have adjacent elements meeting at a common edge. Strictly speaking, the continuity requirement along the edge should be satisfied, but for the sake of simplicity, it is assumed that the elements are connected only at the nodal points and the continuum has to be continuous through those points. Total structure obtained as an assembly of elements is then analysed. In the finite element analysis, therefore, the continuum is divided into a finite number of elements, having finite dimensions and reducing the continuum having infinite degrees of freedom to finite degrees of unknowns. The term finite element was coined by Clough [12.1].

The finite element method has many advantages. The method can be efficiently applied to cater to the irregular geometry. Material anisotropy and nonhomogeneity can be dealt without much difficulty. The method takes into account any type of boundary condition in an efficient manner. Of the various approaches prevalent in the analysis by the finite element method, one based on the assumed displacement pattern within the element is widely used. It is perhaps the simplest among the finite element techniques and has been extensively used and applied successfully to a large variety of stress analysis problems [12.1-12.5].

12.2 TORSIONAL VIBRATION OF SHAFTS

A shaft element of Fig. 12.1 has two nodes at its two ends. The unknown displacements at each end are the angles of twist ϕ_1 and ϕ_2. The displacement function, which is the angle of twist in the present case is given by

$$f = \phi = \alpha_1 + \alpha_2 x \tag{12.1}$$

and

$$f = \phi = [1 \quad x]\begin{Bmatrix}\alpha_1 \\ \alpha_2\end{Bmatrix} \tag{12.2}$$

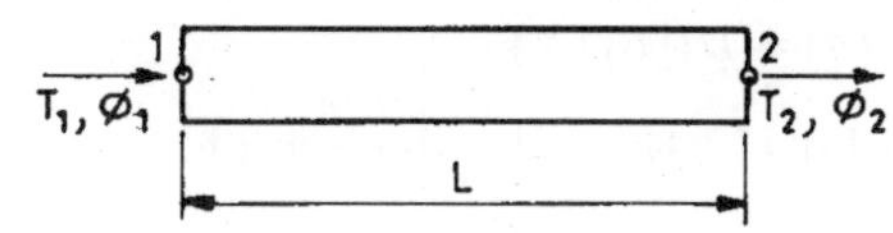

Fig. 12.1 Torsional element

Here, α_1 and α_2 are constants. Number of constants in a displacement function should be equal to the nodal displacements. The nodal values are

$$\begin{Bmatrix} \phi_1 \\ \phi_2 \end{Bmatrix} = \begin{bmatrix} 1 & 0 \\ 1 & L \end{bmatrix} \begin{Bmatrix} \alpha_1 \\ \alpha_2 \end{Bmatrix}$$

or
$$\begin{Bmatrix} \alpha_1 \\ \alpha_2 \end{Bmatrix} = \begin{bmatrix} 1 & 0 \\ 1 & L \end{bmatrix}^{-1} \begin{Bmatrix} \phi_1 \\ \phi_2 \end{Bmatrix} \tag{12.3}$$

Combining Eqs. (12.2) and (12.3), we get

$$f = \phi = [1 \quad x] \begin{bmatrix} 1 & 0 \\ 1 & L \end{bmatrix}^{-1} \begin{Bmatrix} \phi_1 \\ \phi_2 \end{Bmatrix}$$

or
$$f = \phi = [N]\{X\}_e \tag{12.4}$$

where
$$[N] = \left[\left(1 - \frac{x}{L}\right) \quad \frac{x}{L} \right] \tag{12.5}$$

and
$$\{X\}_e = \begin{Bmatrix} \phi_1 \\ \phi_2 \end{Bmatrix} \tag{12.6}$$

[*N*] is called the interpolation matrix, or the shape function matrix. The equivalent stress-strain relationship of the shaft is

$$T = GJ \frac{\partial \phi}{\partial x} \tag{12.7}$$

For generality, Eq. (12.7) is written as

$$\{\sigma\} = [D]\{\in\} \tag{12.8}$$

where
$$\{\sigma\} = T, \quad [D] = GJ \text{ and } \{\in\} = \frac{\partial \phi}{\partial x} \tag{12.9}$$

From Eq. (12.4), by differentiating both sides with respect to *x*, we can write

$$\{\in\} = \begin{Bmatrix} \dfrac{\partial \phi}{\partial x} \end{Bmatrix} = \begin{bmatrix} -\dfrac{1}{L} & \dfrac{1}{L} \end{bmatrix} \begin{Bmatrix} \phi_1 \\ \phi_2 \end{Bmatrix} \tag{12.10}$$

or
$$\{\in\} = [B]\{X\}_e \tag{12.11}$$

Combining Eqs. (12.8) and (12.11) leads to

$$\{\sigma\} = [D][B]\{X\}_e \tag{12.12}$$

Differentiating both sides of Eq. (12.4) twice with respect to *t*, we get

$$\{\ddot{\phi}\} = [N]\{\ddot{X}\}_e \tag{12.13}$$

Let us consider undamped free vibration case. Therefore, the inertia force $\bar{I}\ddot{\phi}$ is acting on the shaft, where $\bar{I}$ is the mass moment of inertia per unit length.

The potential energy of the shaft element is given by

$$\Phi = \frac{1}{2}\int_0^L \{\in\}^T \{\sigma\}\, dx - \int_0^L f^T (-\bar{I}\ddot{\phi})\, dx \qquad (12.14)$$

Combining Eqs (12.11), (12.12), (12.13) and (12.14), we get

$$\Phi = \frac{1}{2}\int_0^L \{X\}_e^T [B]^T [D][B]\, dx \{X\}_e + \int_0^L \{X\}_e^T [N]^T \bar{I}[N]\{\ddot{X}\}_e\, dx \qquad (12.15)$$

The principle of minimum potential energy requires

$$\frac{\partial \Phi}{\partial \{X\}_e} = \{0\} \qquad (12.16)$$

Therefore, performing partial differentiation of Eq. (12.15) and using the results of Eq. (12.16), we get

$$\frac{\partial \Phi}{\partial \{X\}_e} = \int_0^L [B]^T [D][B]\, dx \{X\}_e + \int_0^L [N]^T \bar{I}\; [N]\, dx \{\ddot{X}\}_e = \{0\} \qquad (12.17)$$

or

$$[K]_e \{X\}_e + [M]_e \{\ddot{X}\}_e = \{0\} \qquad (12.18)$$

where

$$[K]_e = \int_0^L [B]^T [D][B]\, dx \qquad (12.19)$$

$$[M]_e = \int_0^L [N]^T \bar{I}[N]\, dx \qquad (12.20)$$

$[K]_e$ is defined as the element stiffness matrix. After substituting the necessary values, $[K]_e$ for the shaft element is given by

$$[K]_e = \frac{GJ}{L}\begin{bmatrix} 1 & -1 \\ -1 & 1 \end{bmatrix} \qquad (12.21)$$

$[M]_e$ derived as above is known as the consistent mass matrix and for the shaft element, its value is given by

$$[M]_e = \frac{\bar{I}L}{6}\begin{bmatrix} 2 & 1 \\ 1 & 2 \end{bmatrix} \qquad (12.22)$$

The consistent mass matrix may be compared with lumped mass matrix derived earlier in Chapter 6. The consistent mass matrix possesses off diagonal elements and thus requires more storage space than the lumped mass matrix. For many categories of problems, consistent mass matrix gives a more accurate result with lesser number of elements than the lumped mass matrix formulation.

The principle of assembly of the elements of stiffness and mass matrices is same as that indicated in Appendix A.

When all the elements are assembled and the boundary conditions are incorporated, the final equations for free vibration become

$$[M]\{\ddot{X}\}+[K]\{X\}=\{0\} \tag{12.23}$$

which is identical to Eq. (6.5).

Assuming a solution

$$\{X\}=ae^{ipt}\{\phi\} \tag{12.24}$$

and following the steps similar to that given in Chapter 6, the final equation becomes

$$[K]^{-1}[M]\{\phi\}=\lambda\{I\}\{\phi\} \tag{12.25}$$

where $\lambda=\dfrac{1}{p^2}$

Equation (12.25) is a typical eigenvalue problem and its solution procedures have been discussed in Chapter 6.

Example 12.1

A uniform shaft having one end fixed and the other end free is divided into two equal elements (Fig. 12.2). Determine the natural frequency in free torsional vibration.

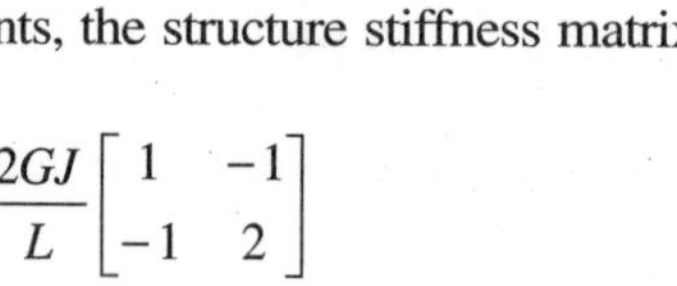

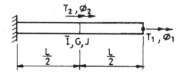

Fig. 12.2 Example 12.1

After assembling the two elements, the structure stiffness matrix is

$$[K]=\frac{2GJ}{L}\begin{bmatrix}1 & -1\\ -1 & 2\end{bmatrix}$$

and the total mass matrix is given by

$$[M]=\frac{\bar{I}L}{12}\begin{bmatrix}2 & 1\\ 1 & 4\end{bmatrix}$$

$$[K]^{-1}[M]=\frac{L}{2GJ}\begin{bmatrix}2 & 1\\ 1 & 1\end{bmatrix}\frac{\bar{I}L}{12}\begin{bmatrix}2 & 1\\ 1 & 4\end{bmatrix}=\frac{\bar{I}L^2}{24GJ}\begin{bmatrix}5 & 6\\ 3 & 5\end{bmatrix}$$

On solution of the eigenvalue problem, the natural frequency squared are

$$p^2 = 2.597 \frac{GJ}{\bar{I}L^2} \quad \text{and} \quad 31.687 \frac{GJ}{\bar{I}L^2}$$

If the system is subjected to forced vibration and damping is included in the system, Eq. (12.23) is modified as

$$[M]\{\ddot{X}\} + [C]\{\dot{X}\} + [K]\{X\} = \{P\} \tag{12.26}$$

If the system has viscous damping, then the damping matrix of an element will be

$$[C]_e = \int_0^L [N]^T \mu [N] \, dx \tag{12.27}$$

where μ, is some numerical value of viscous damping.

Equation (12.27) is identical to Eq. (7.20). As such when the structure stiffness, damping and mass matrices, as also the dynamic loading matrix have been formed, the procedure of analysis is same as that presented in Chapter 7. In the following, we limit ourselves to the derivation of element stiffness matrix and element consistent mass matrix, for different types of structural elements. The assembly of these elements can be done in the same manner as given in Appendix A and the necessary analysis can be performed as per Chapter 6 or 7, depending on the problem.

12.3 AXIAL VIBRATION OF RODS

A rod element of Fig. 12.3 has two nodes at two ends

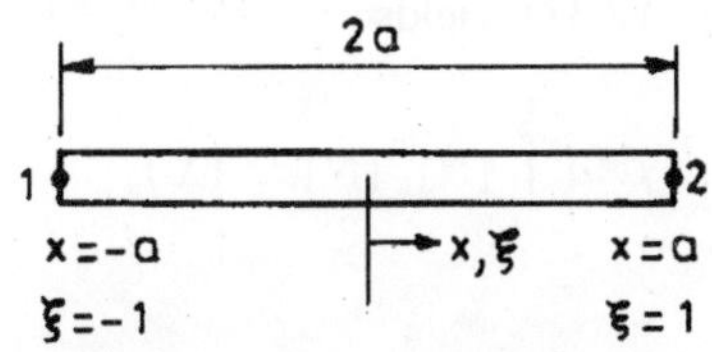

Fig 12.3 Axial element

The unknown displacements at each end are the axial displacements, X_1 and X_2. The displacement function in terms of non-dimensional coordinate ζ is given by

$$f = u = \alpha_1 + \alpha_2 \xi \tag{12.28}$$

where α_1 and α_2 are constants

or

$$f = u = [1 \quad \xi] \begin{Bmatrix} \alpha_1 \\ \alpha_2 \end{Bmatrix} \tag{12.29}$$

The nodal values are

$$\begin{Bmatrix} X_1 \\ X_2 \end{Bmatrix} = \begin{bmatrix} 1 & -1 \\ 1 & 1 \end{bmatrix} \begin{Bmatrix} \alpha_1 \\ \alpha_2 \end{Bmatrix}$$

or
$$\begin{Bmatrix} \alpha_1 \\ \alpha_2 \end{Bmatrix} = \begin{bmatrix} 1 & -1 \\ 1 & 1 \end{bmatrix}^{-1} \begin{Bmatrix} X_1 \\ X_2 \end{Bmatrix} \tag{12.30}$$

Combining Eq. (12.29) and (12.30), we get

$$f = u = [1 \quad \xi] \begin{bmatrix} 1 & -1 \\ 1 & 1 \end{bmatrix}^{-1} \begin{Bmatrix} X_1 \\ X_2 \end{Bmatrix}$$

or
$$f = u = [N]\{X\}_e \tag{12.31}$$

where
$$[N] = \frac{1}{2}[(1-\xi) \;\; (1+\xi)] \tag{12.32}$$

For the axial element shown in Fig. 12.3, the energy expressions are

$$T_e = \frac{1}{2}\int_{-a}^{a} \rho A \, \dot{u}^2 \, dx \tag{12.33}$$

$$U_e = \frac{1}{2}\int_{-a}^{a} EA\left(\frac{\partial u}{\partial x}\right)^2 dx \tag{12.34}$$

ρ is the mass density of the rod element, E is the modulus of elasticity and A is the cross-sectional area.

Substituting Eq. (12.32) into Eq. (12.33), yields

$$T_e = \frac{1}{2}\{\dot{X}\}_e^T \rho A a \int_{-1}^{1} [N]^T [N] d\xi \, \{\dot{X}\}_e \tag{12.35}$$

or
$$T_e = \frac{1}{2}\{\dot{X}\}_e^T [M]_e \{\dot{X}\}_e \tag{12.36}$$

where

$$[M]_e = \rho A a \int_{-1}^{1} [N]^T [N] \, d\xi \tag{12.37}$$

Combining Eq. (12.32) and Eq. (12.37), yields the explicit value of the consistent mass matrix $[M]_e$ which is

$$[M]_e = \frac{\rho A a}{3}\begin{bmatrix} 2 & 1 \\ 1 & 2 \end{bmatrix} \tag{12.38}$$

Now, let us look into strain energy expression of Eq. (12.34). Combining Eq. (12.34) and Eq. (12.31), yields

$$U_e = \frac{1}{2}\int_{-1}^{1} EA \frac{1}{a^2}\left(\frac{\partial u}{\partial \xi}\right)^2 a\, d\xi$$

$$= \frac{1}{2}\{X\}_e^T [K]_e \{X\}_e \qquad (12.39)$$

where

$$[K]_e = \frac{EA}{a}\int_{-1}^{1} [N']^T [N']\, d\xi \text{ and } [N'] = [\frac{\partial u}{\partial \xi}]$$

$$[K]_e = \frac{EA}{2a}\begin{bmatrix} 1 & -1 \\ -1 & 1 \end{bmatrix} \qquad (12.40)$$

Example 12.2

For the rod shown in the Fig. 12.4, determine the lowest frequency by the finite element method by considering it as an element.

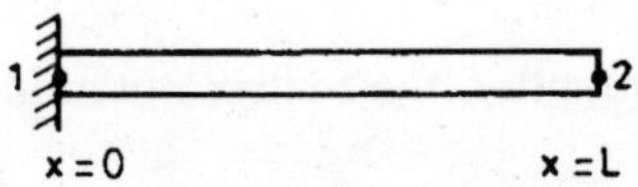

Fig. 12.4 A rod

The boundary condition indicates that the displacement at the left hand support is zero. Nodes 1 and 2 have been indicated in Fig. 12.4 at two ends.

Based on Eq. (12.38) and (12.40), the eigenvalue problem is

$$\frac{\rho A a}{3}\begin{bmatrix} 2 & 1 \\ 1 & 2 \end{bmatrix}\begin{Bmatrix} X_1 \\ X_2 \end{Bmatrix} = \lambda \frac{EA}{2a}\begin{bmatrix} 1 & -1 \\ -1 & 1 \end{bmatrix}\begin{Bmatrix} X_1 \\ X_2 \end{Bmatrix}$$

Now, $X_1 = 0$, therefore

$$\frac{\rho A a}{3} . 2X_2 = \lambda \frac{EA}{2a} X_2$$

or,

$$\frac{1}{\lambda} = p^2 = \frac{EA}{2a^2} \cdot \frac{3}{2\rho A}$$

or,

$$p_1^2 = 3\frac{E}{4\rho a^2}$$

or,

$$p_1 = 1.732\left(\frac{E}{\rho L^2}\right)^{1/2}$$

12.4 FLEXURAL VIBRATION OF BEAMS

Consider a uniform beam element of length L, cross-sectional area A and mass density ρ. Modulus of elasticity of the material of the beam is E and I the second moment of the area. The unknown displacements of this element are the deflection and the rotation at two ends - they total four in all. The displacement function is represented by the deflection equation. The slopes can be obtained by differentiating the deflection. The displacement function is given by

$$\{f\} = w = \alpha_1 + \alpha_2 x + \alpha_3 x^2 + \alpha_4 x^3 \tag{12.41}$$

where α_1 and α_2 are constants

or
$$\{f\} = [1 \quad x \quad x^2 \quad x^3] \begin{Bmatrix} \alpha_1 \\ \alpha_2 \\ \alpha_3 \\ \alpha_4 \end{Bmatrix} \tag{12.42}$$

or
$$\{f\} = [C]\{\alpha\} \tag{12.43}$$

Though $\{f\}$ indicates a single quantity, the matrix notation of $\{f\}$ is given for the sake of generality.

Based on Eq. (12.41) and its derivative, the nodal displacements can be written as [Fig. 12.5].

$$\begin{Bmatrix} X_1 \\ X_2 \\ X_3 \\ X_4 \end{Bmatrix} = \begin{bmatrix} 1 & 0 & 0 & 0 \\ 0 & 1 & 0 & 0 \\ 1 & L & L^2 & L^3 \\ 0 & 1 & 2L & 3L^2 \end{bmatrix} \begin{Bmatrix} \alpha_1 \\ \alpha_2 \\ \alpha_3 \\ \alpha_4 \end{Bmatrix} \tag{12.44}$$

or
$$\{X\}_e = [A]\{\alpha\} \tag{12.45}$$

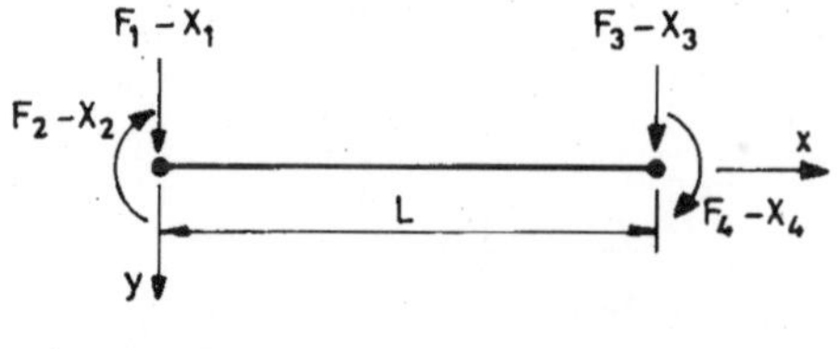

Fig. 12.5 Beam element

Combining Eqs. (12.43) and (12.45), we get

$$\{f\} = w = [N]\{X\}_e \tag{12.46}$$

where
$$[N] = [C][A]^{-1} \tag{12.47}$$

$$[N] = \left[\left(1 - \frac{3x^2}{L^2} + \frac{2x^3}{L^3}\right) \left(x - \frac{2x^2}{L} + \frac{x^3}{L^2}\right) \left(\frac{3x^2}{L^2} - \frac{2x^3}{L^3}\right) \left(-\frac{x^2}{L} + \frac{x^3}{L^2}\right)\right] \tag{12.48}$$

The stress-strain relation of the beam element is nothing but the bending moment-curvature relationship.

$$M = -EI\frac{d^2w}{dx^2} \tag{12.49}$$

Equation (12.49) written in matrix form for generality

$$\{\sigma\} = [D]\{\in\} \tag{12.50}$$

where $$\{\sigma\} = M, \quad [D] = EI \text{ and } \{\in\} = -\frac{d^2w}{dx^2} \tag{12.51}$$

Differentiating both sides of Eq. (12.48) twice with respect to x, we get

$$\{\in\} = \left[\left(\frac{6}{L^2} - \frac{12x}{L^3}\right) \left(\frac{4}{L} - \frac{6x}{L^2}\right) \left(-\frac{6}{L^2} + \frac{12x}{L^3}\right) \left(\frac{2}{L} - \frac{6x}{L^2}\right)\right]\{X\}_e \tag{12.52}$$

or $$\{\in\} = [B]\{X\}_e \tag{12.53}$$

Combining Eqs. (12.50) and (12.53), we get

$$\{\sigma\} = [D][B]\{X\}_e \tag{12.54}$$

Based on the minimum potential energy principle shown in the previous section, the stiffness matrix of the element is given by

$$[K]_e = \int_0^L [B]^T [D][B]\,dx \tag{12.55}$$

or $$[K]_e = \frac{EI}{L^3}\begin{bmatrix} 12 & & & \textit{Symmetrical} \\ 6L & 4L^2 & & \\ -12 & -6L & 12 & \\ 6L & 2L^2 & -6L & 4L^2 \end{bmatrix} \tag{12.56}$$

The consistent mass matrix is given by

$$[M]_e = A\int_0^L [N]^T \rho[N]\,dx \tag{12.57}$$

or $$[M]_e = \frac{\rho AL}{420}\begin{bmatrix} 156 & & & \textit{Symmetrical} \\ 22L & 4L^2 & & \\ 54 & 13L & 156 & \\ -13L & -3L^2 & -22L & 4L^2 \end{bmatrix} \tag{12.58}$$

For flexural vibrations of bar elements, it has been found that the consistent mass matrix formulation improves the results significantly. However, there is not much to gain by consistent mass matrix formulation, for torsional or longitudinal vibration problems.

Example 12.3

For a simply supported uniform beam, give a one-element solution for the lowest natural frequency

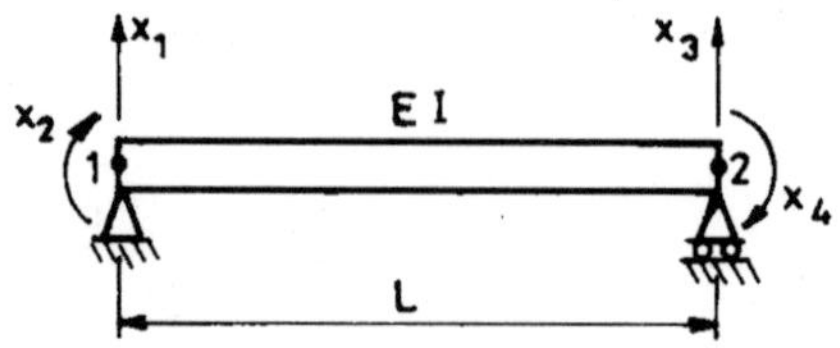

Fig. 12.6 Example 12.3

Two nodes have been placed at the two ends of the beam. Substituting Eqs. (12.56) and (12.58) into Eq. (6.20) results in

$$p^2\frac{\rho AL}{420}\begin{bmatrix}156 & 22L & 54 & -13L\\ 22L & 4L^2 & 13L & -3L^2\\ 54 & 13L & 156 & -22L\\ -13L & -3L^2 & -22L & 4L^2\end{bmatrix}\begin{Bmatrix}X_1\\X_2\\X_3\\X_4\end{Bmatrix}$$

$$=\frac{EI}{L^3}\begin{bmatrix}12 & 6L & -12 & 6L\\ 6L & 4L^2 & -6L & 2L^2\\ -12 & -6L & 12 & -6L\\ 6L & 2L^2 & -6L & 4L^2\end{bmatrix}\begin{Bmatrix}X_1\\X_2\\X_3\\X_4\end{Bmatrix} \quad \text{(a)}$$

Boundary conditions are $X_1 = X_3 = 0$, Therefore, Eq. (a) reduces to

$$p^2\frac{\rho AL}{420}\begin{bmatrix}4L^2 & -3L^2\\ -3L^2 & 4L^2\end{bmatrix}\begin{Bmatrix}X_2\\X_4\end{Bmatrix}=\frac{EI}{L^3}\begin{bmatrix}4L^2 & 2L^2\\ 2L^2 & 4L^2\end{bmatrix}\begin{Bmatrix}X_2\\X_4\end{Bmatrix} \quad \text{(b)}$$

Let $\dfrac{p^2\rho AL^4}{420EI}=\lambda$, then

$$\left(\lambda\begin{bmatrix}4 & -3\\ -3 & 4\end{bmatrix}-\begin{bmatrix}4 & 2\\ 2 & 4\end{bmatrix}\right)\begin{Bmatrix}X_2\\X_4\end{Bmatrix}=\begin{Bmatrix}0\\0\end{Bmatrix}$$

or,

$$\begin{bmatrix}(4\lambda-4) & (-3\lambda-2)\\ (-3\lambda-2) & (4\lambda-4)\end{bmatrix}\begin{Bmatrix}X_2\\X_4\end{Bmatrix}=0 \quad \text{(c)}$$

A non-trivial solution of Eq. (c) is

$$\begin{vmatrix} (4\lambda - 4) & -(3\lambda + 2) \\ -(3\lambda + 2) & (4\lambda - 4) \end{vmatrix} = 0$$

or,

$$7\lambda^2 - 44\lambda + 12 = 0 \quad \text{(d)}$$

Roots of Eq. (d) are

$$\lambda_{1,2} = 0.286,\ 6.00$$

The lowest natural frequency of the system is

$$p_1 = (420 \times 0.286)^{1/2} \left(\frac{EI}{\rho A L^4} \right)^{1/2} = 10.94 \left(\frac{EI}{\rho A L^4} \right)^{1/2}$$

The correct value of the coefficient is 9.86.
Thus, the one element solution gives reasonable accurate value.

12.5 VIBRATION OF TIMOSHENKO BEAMS

It has been pointed out earlier that the classical Bernoulli-Euler theory predicts the frequencies of flexural vibration of the lower modes of the slender beam, with adequate precision. But with increasingly large depth of the beam and with increased modes of vibration, the effect of transverse shear deformation and rotary inertia becomes increasingly important, which are not accounted for in the classical theory. The inclusion of these effects makes the analysis much more involved.

A large variety of Timoshenko beam finite elements have been proposed. The element is termed as simple when each node of the straight beam element has two degrees of freedom, and if additional degrees of freedom is incorporated at the nodes, it is termed as complex. It has been observed that the simple Timoshenko beam element is aptly adequate for general purpose use. We present the derivation of one such element.

The Timoshenko beam element shown in Fig. 12.7 has two nodes at its two ends. At each end, the degrees of freedom are the total deflection w and the bending rotation θ.

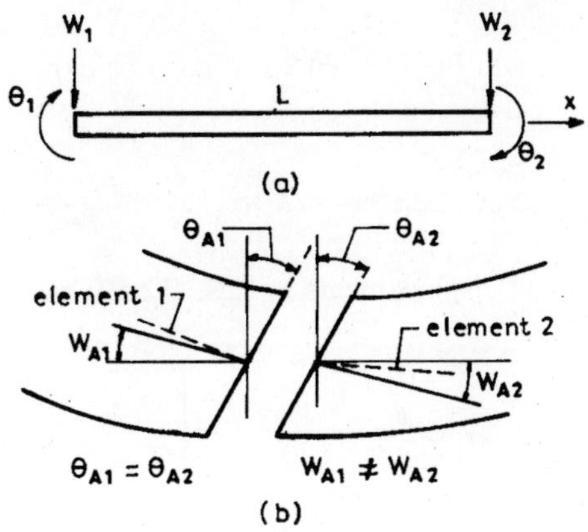

Fig. 12.7 Timoshenko beam element

The deflection function is given by

$$w = \alpha_0 + \alpha_1 x + \alpha_2 x^2 + \alpha_3 x^3 \tag{12.59}$$

The relation between transverse shear strain γ, w' and θ is

$$w' = \theta + \gamma \tag{12.60}$$

Transverse shear strain is assumed to be independent of x, that is

$$\gamma = \beta_0 \tag{12.61}$$

The bending moment-curvature relationship is

$$M = -EI\frac{\partial \theta}{\partial x} \tag{12.62}$$

The shear force V is related to transverse shear strain g by

$$V = \mu AG\gamma \tag{12.63}$$

The shear force and bending moment is related as follows:

$$\frac{\partial M}{\partial x} = V \tag{12.64}$$

Equation (12.64) means that the rotary inertia term is ignored in the moment equilibrium equation within the element, but the effect of rotary inertia will nevertheless be included in lumped form at the nodes.

Using Eqs. (12.59) to (12.63) in Eq. (12.64) gives

$$\beta_0 = -\alpha_3 \frac{\phi}{2} L^2 \tag{12.65}$$

where
$$\phi = \frac{12EI}{\mu AGL^2} \tag{12.66}$$

L is the length of the beam element.

Using Eqs. (12.59) to (12.61) and Eq. (12.65) allow θ to be expressed as a polynomial in x as follows:

$$\theta = \alpha_1 + 2\alpha_2 x + \left(3x^2 - \frac{\phi}{2} L^2\right)\alpha_3 \tag{12.67}$$

Substituting the values of nodal displacements in Eqs. (12.59) and (12.67), we get

$$\left.\begin{aligned} w_1 &= \alpha_1 \\ \theta_1 &= \alpha_1 - \frac{\phi}{2} L^2 \alpha_3 \\ w_2 &= \alpha_1 + \alpha_2 L + \alpha_3 L^2 + \alpha_4 L^3 \\ \theta_2 &= \alpha_1 + 2\alpha_2 L + 3L^2\alpha_3 - \frac{\phi}{2} L^2 \alpha_3 \end{aligned}\right\} \tag{12.68}$$

Equation (12.68) written in matrix form yields

$$\begin{Bmatrix} w_1 \\ \theta_1 \\ w_2 \\ \theta_2 \end{Bmatrix} = \begin{bmatrix} 1 & 0 & 0 & 0 \\ 1 & 0 & -\frac{\phi}{2}L^2 & 0 \\ 1 & L & L^2 & L^3 \\ 1 & 2L & \left(3-\frac{\phi}{2}\right)L^2 & 0 \end{bmatrix} \begin{Bmatrix} \alpha_1 \\ \alpha_2 \\ \alpha_3 \\ \alpha_4 \end{Bmatrix} \tag{12.69}$$

or
$$\{X\}_e = [A]\{\alpha\} \tag{12.70}$$

or
$$\{\alpha\} = [A]^{-1}\{X\}_e \tag{12.71}$$

Substituting values of $\{\alpha\}$ from Eq. (12.71) into Eq. (12.59), we get

$$w = [1 \quad x \quad x^2 \quad x^3][A]^{-1}\{X\}_e \tag{12.72}$$

or
$$w = \frac{1}{1+\phi}\Big[(1-3\xi^2+2\xi^3)+(1-\xi)\phi\,(\xi-2\xi^2+\xi^3)$$

$$+\frac{1}{2}(\xi-\xi^2)\phi L(3\xi-2\xi^2+\xi\phi)\left\{-\xi^2+\xi^3-\frac{1}{2}(\xi-\xi^3)\phi\, L\right\}\Big] \tag{12.73}$$

where $\xi = \dfrac{x}{L}$.

The expression for strain energy is

$$U = \frac{EI}{2}\int_0^L \left(\frac{\partial\theta}{\partial x}\right)^2 dx + \frac{\mu AG}{2}\int_0^L \gamma^2\, dx \tag{12.74}$$

Following steps similar to that mentioned in Art. 12.2, the stiffness matrix is given by

$$[K]_e = \frac{EI}{1+\phi}\begin{bmatrix} \frac{12}{L^3} & & & \textit{Symmetrical} \\ \frac{6}{L^2} & \frac{4+\phi}{L} & & \\ -\frac{12}{L^3} & -\frac{6}{L^2} & \frac{12}{L^3} & \\ \frac{6}{L^2} & \frac{2-\phi}{K} & -\frac{6}{L^2} & \frac{4+\phi}{L} \end{bmatrix} \tag{12.75}$$

The kinetic energy of the Timoshenko beam is

$$T = \frac{1}{2}\int_0^L \rho A\left(\frac{\partial w}{\partial t}\right)^2 dx + \frac{1}{2}\int_0^L \rho I\left(\frac{\partial \theta}{\partial t}\right)^2 dx \qquad (12.76)$$

Following steps similar to that presented earlier, the mass matrix is given by Eq. (12.77).

The first term of Eq. (12.77) represents the translational mass inertia, while second term represents the rotary inertia of the beam.

$$
[M]_e = \frac{\rho AL}{420}\begin{bmatrix}
\frac{13}{35}+\frac{7}{10}\phi+\frac{1}{3}\phi^2 & & & \\
\left(\frac{11}{210}+\frac{11}{120}\phi+\frac{1}{24}\phi^2\right)L & \left(\frac{1}{105}+\frac{\phi}{60}+\frac{\phi^2}{120}\right)L^2 & & \textit{Symmetrical} \\
\left(\frac{9}{70}+\frac{3\phi}{10}+\frac{1}{6}\phi^2\right) & \left(\frac{13}{420}+\frac{3}{40}\phi+\frac{1}{24}\phi^2\right)L & \left(\frac{13}{35}+\frac{7}{10}\phi+\frac{\phi^3}{3}\right) & \\
-\left(\frac{13}{420}+\frac{3}{40}\phi+\frac{1}{24}\phi^2\right)L & -\left(\frac{1}{140}+\frac{\phi}{60}+\frac{\phi^2}{120}\right)L^2 & -\left(\frac{11}{210}+\frac{11}{120}\phi+\frac{1}{24}\phi^2\right)L & \left(\frac{1}{105}+\frac{\phi}{60}+\frac{\phi^2}{120}\right)L^2
\end{bmatrix}
$$

$$
+\frac{\rho AL}{(1+\phi)^2}\left(\frac{r}{L}\right)^2\begin{bmatrix}
\frac{6}{5} & & & \\
\left(\frac{1}{10}-\frac{\phi}{2}\right)L & \left(\frac{2}{15}+\frac{\phi}{6}+\frac{\phi^2}{3}\right)L^2 & & \textit{Symmetrical} \\
-\frac{6}{5} & \left(-\frac{1}{10}+\frac{\phi}{2}\right)L & \frac{6}{5} & \\
\left(\frac{1}{10}-\frac{\phi}{2}\right)L & \left(-\frac{1}{30}-\frac{\phi}{6}+\frac{\phi^2}{6}\right)L^2 & \left(-\frac{1}{10}+\frac{\phi}{2}\right)L & \left(\frac{2}{15}+\frac{\phi}{6}+\frac{\phi^2}{3}\right)L^2
\end{bmatrix} \tag{12.77}
$$

where r = radius of gyration = $\sqrt{\frac{I}{A}}$

12.6 INPLANE VIBRATION OF PLATES

12.6.1 Linear Triangular Element

The vibration of the plate in its own plane is considered here. The simplest element available in the plane elasticity problem is the three-noded triangular element. This element is also referred to as constant strain triangle (CST), as the strain at any point within the element remains constant.

A triangular element of arbitrary shape with nodes at the corners denoted as 1, 2, and 3 is shown in Fig. 12.8. The coordinates of 1, 2 and 3 are (x_1, y_1), (x_2, y_2) and (x_3, y_3). The element has a constant thickness equal to *t*. The displacements parallel to the axis system at points 1, 2 and 3 are (u_1, v_1), (u_2, v_2) and (u_3, v_3). The displacement vector is given by

$$\{X\}_e^T = \{u_1 \quad v_1 \quad u_2 \quad v_2 \quad u_3 \quad v_3\} \tag{12.78}$$

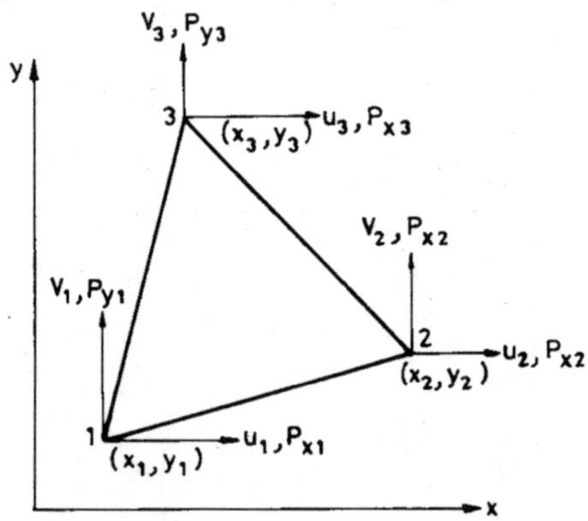

Fig. 12.8 Linear triangular element

The corresponding force vector is given by

$$\{P\}_e^T = \{P_{x1} \quad P_{y1} \quad P_{x2} \quad P_{y2} \quad P_{x3} \quad P_{y3}\} \tag{12.79}$$

The displacement function is represented by two linear polynomials

$$\left.\begin{aligned} u &= \alpha_1 + \alpha_2 x + \alpha_3 y \\ v &= \alpha_4 + \alpha_5 x + \alpha_6 y \end{aligned}\right\} \tag{12.80}$$

where α_1 to α_6 are constants and *u* and *v*, are the displacements in the *x* and *y* directions. The displacement function is written as

$$\{f\} = \begin{Bmatrix} u \\ v \end{Bmatrix} = \begin{bmatrix} 1 & x & y & 0 & 0 & 0 \\ 0 & 0 & 0 & 1 & x & y \end{bmatrix} \begin{Bmatrix} \alpha_1 \\ \alpha_2 \\ \vdots \\ \alpha_6 \end{Bmatrix} \tag{12.81}$$

or
$$\{f\}=[C]\{\alpha\} \tag{12.82}$$

The nodal displacements can be written in terms of nodal coordinates as follows:

$$\begin{Bmatrix} u_1 \\ v_1 \\ u_2 \\ v_2 \\ u_3 \\ v_3 \end{Bmatrix} = \begin{bmatrix} 1 & x_1 & y_1 & 0 & 0 & 0 \\ 0 & 0 & 0 & 1 & x_1 & y_1 \\ 1 & x_2 & y_2 & 0 & 0 & 0 \\ 0 & 0 & 0 & 1 & x_2 & y_2 \\ 1 & x_3 & y_3 & 0 & 0 & 0 \\ 0 & 0 & 0 & 1 & x_3 & y_3 \end{bmatrix} \begin{Bmatrix} \alpha_1 \\ \alpha_2 \\ \alpha_3 \\ \alpha_4 \\ \alpha_5 \\ \alpha_6 \end{Bmatrix} \tag{12.83}$$

or
$$\{X\}_e=[A]\{\alpha\} \tag{12.84}$$

Combining Eqs. (12.82) and (12.84), we get

$$\{f\}=[N]\{X\}_e \tag{12.85}$$

where
$$[N]=[C][A]^{-1} \tag{12.86}$$

Strain components for the plane stress problem is given by

$$\{\in\} = \begin{Bmatrix} \in_x \\ \in_y \\ \gamma_{xy} \end{Bmatrix} = \begin{Bmatrix} \dfrac{\partial u}{\partial x} \\ \dfrac{\partial v}{\partial y} \\ \dfrac{\partial u}{\partial y}+\dfrac{\partial v}{\partial x} \end{Bmatrix} \tag{12.87}$$

where $\in_x, \in_y$ are the axial strains in the x and y directions, γ_{xy} is the shearing strain.

Combining Eqs. (12.85) and (12.87), we get

$$\{\in\} = \begin{bmatrix} \dfrac{\partial N_1}{\partial x} & 0 & \dfrac{\partial N_2}{\partial x} & 0 & \dfrac{\partial N_3}{\partial x} & 0 \\ 0 & \dfrac{\partial N_1}{\partial y} & 0 & \dfrac{\partial N_2}{\partial y} & 0 & \dfrac{\partial N_3}{\partial y} \\ \dfrac{\partial N_1}{\partial y} & \dfrac{\partial N_1}{\partial x} & \dfrac{\partial N_2}{\partial y} & \dfrac{\partial N_2}{\partial x} & \dfrac{\partial N_3}{\partial y} & \dfrac{\partial N_3}{\partial x} \end{bmatrix} \{X\}_e \tag{12.88}$$

or
$$\{\in\}=[B]\{X\}_e \tag{12.89}$$

N_1, N_2 and N_3 are linear functions of x and y. As such the elements of $[B]$ matrix are constants, which indicate that the strain is constant within the element.

The stress-strain relationship for an isotropic plate is given by

$$\begin{Bmatrix} \sigma_x \\ \sigma_y \\ \tau_{xy} \end{Bmatrix} = \frac{E}{1-\nu^2} \begin{bmatrix} 1 & \nu & 0 \\ \nu & 1 & 0 \\ 0 & 0 & \frac{1-\nu}{2} \end{bmatrix} \begin{Bmatrix} \in_x \\ \in_y \\ \gamma_{xy} \end{Bmatrix} \tag{12.90}$$

where σ_x and σ_y are the normal stresses and τ_{xy} is the shearing stress, E is the Young's modulus of elasticity and ν is the Poisson's ratio.

Equation (12.90) can be written in compact form as

$$\{\sigma\} = [D]\{\in\} \tag{12.91}$$

Combining Eqs. (12.89) and (12.91), we get

$$\{\sigma\} = [D][B]\{X\}_e \tag{12.92}$$

Total strain energy of the plate is given by

$$U = \frac{1}{2}\int_v \{\in\}^T \{\sigma\}\, d_{\text{vol}} \tag{12.93}$$

where d_{vol} is the elementary volume of the element.

Combining Eqs. (12.89), (12.92) and (12.93), we get

$$U = \frac{1}{2}\int_{\text{vol}} \{X\}_e^T [B]^T [D][B]\{X\}_e\, d_{\text{vol}} \tag{12.94}$$

As $[B]$, $[D]$ matrices are independent of x and y, therefore, Eq. (12.94) becomes

$$U = \frac{1}{2}\{X\}_e^T [B]^T [D][B]\{X\}_e\, \Delta \cdot t \tag{12.95}$$

where $\triangle$ is the area of the triangular element and t is its thickness.

Now applying Castigliano's theorem [see Ref. (6.1)]

$$\frac{\partial U}{\partial \{X\}_e} = [B]^T [D][B]\{X\}_e\, \Delta \cdot t \tag{12.96}$$

or

$$\{P\}_e = [K]_e \{X\}_e \tag{12.97}$$

where $\{P\}_e$ are the nodal forces and the element stiffness matrix is given by

$$[K]_e = [B]^T [D][B]\,\Delta \cdot t \tag{12.98}$$

In general, expression for consistent mass matrix is

$$[M]_e = \int [N]^T \rho\, [N]\, d_{\text{vol}} \tag{12.99}$$

where ρ is the mass density of the plate. For a plate of constant thickness t subjected to inplane forces, Eq. (12.99) becomes

$$[M]_e = \rho t \int [N]^T [I][N]\, dA \tag{12.100}$$

where [I] is identity matrix and dA is the elementary cross-sectional area. Substituting [N] from Eq. (12.86) into Eq. (12.100), we get

$$[M]_e = \rho \frac{t\Delta}{12} \begin{bmatrix} 2 & & & & & \\ 0 & 2 & & & Symmertical & \\ 1 & 0 & 2 & & & \\ 0 & 1 & 0 & 2 & & \\ 1 & 0 & 1 & 0 & 2 & \\ 0 & 1 & 0 & 1 & 0 & 2 \end{bmatrix} \tag{12.101}$$

12.6.2 **Linear Rectangular Element**

For structures having regular shaped boundaries, rectangular elements can be effectively used for the analysis. A four-noded rectangular element having sides a and b is shown in Fig. 12.9. The thickness of the element is t. The nodes are numbered such that the coordinates of nodes 1, 2, 3 and 4 are (0, 0), (0, b), (a, 0) and (a, b) respectively. The displacements at node 1 parallel to the axis system are u_1, v_1 and so for other nodes.

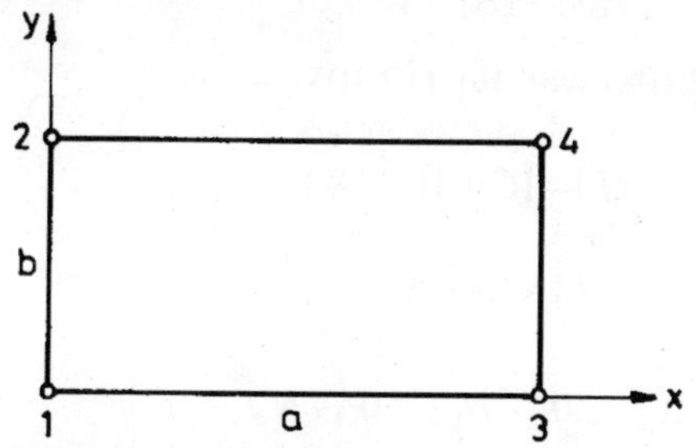

Fig. 12.9 A 4-noded rectangular element

The displacement vector is given by

$$\{X\}_e^T = \{u_1 \quad v_1 \quad u_2 \quad v_2 \quad u_3 \quad v_3 \quad u_4 \quad v_4\} \tag{12.102}$$

The displacement functions are given by

$$\left.\begin{aligned} u &= \alpha_1 + \alpha_2 x + \alpha_3 y + \alpha_4 xy \\ v &= \alpha_5 + \alpha_6 x + \alpha_7 y + \alpha_8 xy \end{aligned}\right\} \tag{12.103}$$

or, $$\{f\} = \begin{Bmatrix} u \\ v \end{Bmatrix} = \begin{bmatrix} 1 & x & y & xy & 0 & 0 & 0 & 0 \\ 0 & 0 & 0 & 0 & 1 & x & y & xy \end{bmatrix} \begin{Bmatrix} \alpha_1 \\ \alpha_2 \\ \vdots \\ \alpha_7 \\ \alpha_8 \end{Bmatrix} \tag{12.104}$$

or, $$\{f\}=[C]\{\alpha\} \tag{12.105}$$

Nodal displacements can be expressed in terms of nodal coordinates with the help of Eq. (12.104).

$$\begin{Bmatrix} u_1 \\ v_1 \\ u_2 \\ v_2 \\ u_3 \\ v_3 \\ u_4 \\ v_4 \end{Bmatrix} = \begin{bmatrix} 1 & 0 & 0 & 0 & 0 & 0 & 0 & 0 \\ 0 & 0 & 0 & 0 & 1 & 0 & 0 & 0 \\ 1 & 0 & b & 0 & 0 & 0 & 0 & 0 \\ 0 & 0 & 0 & 0 & 1 & 0 & b & 0 \\ 1 & a & 0 & 0 & 0 & 0 & 0 & 0 \\ 0 & 0 & 0 & 0 & 1 & a & 0 & 0 \\ 1 & a & b & ab & 0 & 0 & 0 & 0 \\ 0 & 0 & 0 & 0 & 1 & a & b & ab \end{bmatrix} \begin{Bmatrix} \alpha_1 \\ \alpha_2 \\ \alpha_3 \\ \alpha_4 \\ \alpha_5 \\ \alpha_6 \\ \alpha_7 \\ \alpha_8 \end{Bmatrix} \tag{12.106}$$

or, $$\{X\}_e=[A]\{\alpha\} \tag{12.107}$$

From Eq. (12.107), one can write

$$\{\alpha\}=[A]^{-1}\{X\}_e \tag{12.108}$$

Replacing $\{\alpha\}$ from Eq. (12.108) with Eq. (12.105), we get

$$\{f\}=[C][A]^{-1}\{X\}_e$$

or, $$\{f\}=[N]\{X\}_e \tag{12.109}$$

where $$[N]=\begin{bmatrix} N_1 & 0 & N_2 & 0 & N_3 & 0 & N_4 & 0 \\ 0 & N_1 & 0 & N_2 & 0 & N_3 & 0 & N_4 \end{bmatrix} \tag{12.110}$$

$$\left.\begin{aligned} N_1 &= \left(1-\frac{x}{a}\right)\left(1-\frac{y}{b}\right) \\ N_2 &= \left(1-\frac{x}{a}\right)\frac{y}{b} \\ N_3 &= \frac{x}{a}\left(1-\frac{y}{b}\right) \\ N_4 &= \frac{xy}{ab} \end{aligned}\right\} \tag{12.111}$$

Strain components for a plane problem is given by

$$\{\in\} = \begin{Bmatrix} \in_x \\ \in_y \\ \gamma_{xy} \end{Bmatrix} = \begin{bmatrix} \dfrac{\partial}{\partial x} & 0 \\ 0 & \dfrac{\partial}{\partial y} \\ \dfrac{\partial}{\partial y} & \dfrac{\partial}{\partial x} \end{bmatrix} = \begin{Bmatrix} u \\ v \end{Bmatrix} \tag{12.112}$$

Combining Eqs. (12.109) and (12.112), yields

$$\{\in\} = [B]\{X\}_e \tag{12.113}$$

The relationship of stress with strain is given by

$$\{\sigma\} = [D]\{\in\} = [D][B]\{X\}_e \tag{12.114}$$

$[D]$ is same as Eq. (12.90) for an isotropic plate.

As already shown, the element stiffness matrix is given by

$$[K]_e = t\int_A [B]^T[D][B]\,dx\,dy \tag{12.115}$$

$[K]_e$ in explicit form for an isotropic plate is given by Eq. (12.116).

The consistent mass matrix is given by Eq. (12.99).

Substituting Eqs. (12.110) in Eq. (12.99), yields

$$[M]_e = \frac{mab}{36}\begin{bmatrix} 4 & 0 & 2 & 0 & 1 & 0 & 2 & 0 \\ 0 & 4 & 0 & 2 & 0 & 1 & 0 & 2 \\ 2 & 0 & 4 & 0 & 2 & 0 & 1 & 0 \\ 0 & 2 & 0 & 4 & 0 & 2 & 0 & 1 \\ 1 & 0 & 2 & 0 & 4 & 0 & 2 & 0 \\ 0 & 1 & 0 & 2 & 0 & 4 & 0 & 2 \\ 2 & 0 & 1 & 0 & 2 & 0 & 4 & 0 \\ 0 & 2 & 0 & 1 & 0 & 2 & 0 & 4 \end{bmatrix} \tag{12.117}$$

12.7 FLEXURAL VIBRATION OF PLATES

Let us consider a rectangular plate bending element. The plate element dimensions, the node numbering and the coordinate system have been shown in Fig. 12.10. Each node has three degrees of freedom-one vertical displacement and two rotations along x and y-axes. The displacements at node 1 therefore, are w_1, θ_{x1} and θ_{y1}. The complete displacement vector for this element is

$$\{X\}_e^T = \{w_1\ \ \theta_{x1}\ \ \theta_{y1}\ \ w_2\ \ \theta_{x2}\ \ \theta_{y2}\ \ w_3\ \ \theta_{x3}\ \ \theta_{y3}\ \ w_4\ \ \theta_{x4}\ \ \theta_{y4}\} \tag{12.118}$$

The displacement function is given by

$$[K]_e = \frac{Dt}{12}\begin{bmatrix}
4p^{-1}+2(1-\nu) & & & & & & & \\
3(1+\nu) & 4p+2(1-\nu)^{-1}p & & & \textit{Symmetrical} & p=\frac{a}{b} & & \\
2p^{-1}+4p & -\frac{3\nu+3}{2} & 4p^{-1}+2(1-\nu)\,p & & & & & \\
-\frac{9\nu-3}{2} & -4p+(1-\nu)\,p^{-1} & -\frac{3\nu+3}{2} & 4p+2(1-\nu)\,p^{-1} & & & & \\
-4p^{-1}+(1-\nu)\,p & -\frac{9\nu-3}{2} & -2p^{-1}+2(1-\nu)\,p & \frac{3\nu+3}{2} & 4p^{-1}+2(1-\nu)\,p & & & \\
\frac{9\nu-3}{2} & 2p-2(1-\nu)\,p^{-1} & \frac{3\nu+3}{2} & -2p-(1-\nu)\,p^{-1} & -\frac{3\nu+3}{2} & 4p+2(1-\nu)p^{-1} & & \\
-2p^{-1}+p(1-\nu) & -\frac{3\nu+3}{2} & -4p^{-1}+(1-\nu)\,p & \frac{9\nu-3}{2} & 2p^{-1}-2(1-\nu)\,p & \frac{3\nu+3}{2} & 4p^{-1}+2(1-\nu)\,p & \\
-\frac{3\nu+3}{2} & -2p^{-1}-(1-\nu)\,p^{-1} & -\frac{9\nu-3}{2} & 2p-2(1-\nu)\,p^{-1} & \frac{9\nu-3}{2} & -4p+(1-\nu)\,p^{-1} & \frac{3\nu+3}{2} & 4p+2(1-\nu)^{-1}\,p
\end{bmatrix} \qquad (12.116)$$

$$\{f\} = \{w\} = \alpha_1 + \alpha_2 x + \alpha_3 y + \alpha_4 x^2 + \alpha_5 xy + \alpha_6 y^2 + \alpha_7 x^3$$
$$+ \alpha_8 x^2 y + \alpha_9 xy^2 + \alpha_{10} y^3 + \alpha_{11} x^3 y + \alpha_{12} xy^3 \qquad (12.119)$$

or $$f = [1 \;\; x \;\; y \;\; x^2 \;\; xy \;\; y^2 \;\; x^3 \;\; x^2 y \;\; xy^2 \;\; y^3 \;\; x^3 y \;\; xy^3] \begin{Bmatrix} \alpha_1 \\ \alpha_2 \\ \vdots \\ \alpha_{12} \end{Bmatrix} \qquad (12.120)$$

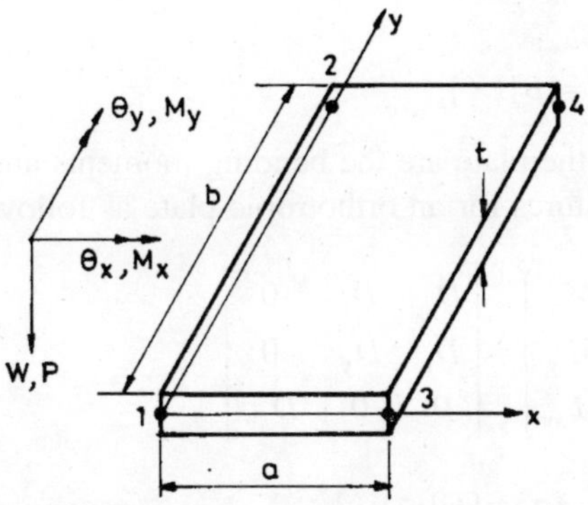

Fig. 12.10 A rectangular plate in bending

or $$\{f\} = [C]\{\alpha\} \qquad (12.121)$$

The rotations are given by

$$\theta_x = -\frac{\partial w}{\partial y} = -(\alpha_3 + \alpha_5 x + 2\alpha_6 y + \alpha_8 x^2 + 2\alpha_9 xy$$
$$+ 3\alpha_{10} y^2 + \alpha_{11} x^3 + 3\alpha_{12} xy^2)$$

$$\theta_y = \frac{\partial w}{\partial x} = (\alpha_2 + 2\alpha_4 x + \alpha_5 y + 3\alpha_7 x^2 + 2\alpha_8 xy + \alpha_9 y^2$$
$$+ 3\alpha_{11} x^2 y + \alpha_{12} y^3) \qquad (12.122)$$

Using Eqs. (12.119) and (12.122) and substituting the nodal displacements with proper coordinates, we can write in matrix form

$$\{X\}_e = [A]\{\alpha\} \qquad (12.123)$$

where $[A]$ matrix contains elements which are functions of nodal coordinates. Combining Eqs. (12.121) and (12.123), we get

$$f = w = [N]\{X\}_e \qquad (12.124)$$

where $$[N] = [C][A]^{-1} \qquad (12.125)$$

The 'strains' for the plate bending element are curvatures.

The strain matrix is given by

$$\{\in\} = \begin{Bmatrix} -\dfrac{\partial^2 w}{\partial x^2} \\ -\dfrac{\partial^2 w}{d y^2} \\ 2\dfrac{\partial^2 w}{\partial x\, \partial y} \end{Bmatrix} \tag{12.126}$$

By performing the necessary differentiation in Eq. (12.124), the resulting equation can be written as

$$\{\in\} = [B]\{X\}_e \tag{12.127}$$

The 'stress' components of the plate are the bending moments and the twisting moment, and they are related to the curvatures for an orthotropic plate as follows:

$$\{\sigma\} = \begin{Bmatrix} M_x \\ M_y \\ M_{xy} \end{Bmatrix} = \begin{bmatrix} D_x & D_1 & 0 \\ D_1 & D_y & 0 \\ 0 & 0 & D_{xy} \end{bmatrix} \tag{12.128}$$

or

$$\{\sigma\} = [D](\in\} \tag{12.129}$$

D_x, D_y, D_1 and D_{xy} relate to the geometrical and material properties of the plate.

Combining Eqs. (12.127) and (12.128), we get

$$\{\sigma\} = [D][B]\{X\}_e \tag{12.130}$$

Following steps similar to that given in the previous sections of this chapter, the stiffness matrix for the rectangular plate element is given by

$$[K]_e = \int_0^a \int_0^b [B]^T [D][B]\, dx\, dy \tag{12.131}$$

The consistent mass matrix for the rectangular plate bending element is given by

$$[M]_e = \rho t \int_A [N]^T [I][N]\, dA \tag{12.132}$$

Substituting [*N*] from Eq. (12.125) into (12.132), we get

$$[M]_e = \rho t \int_0^a \int_0^b ([A]^{-1})^T [C]^T [C][A]^{-1}\, dA \tag{12.133}$$

Equation (12.133) is explicitly evaluated and it is given below

$[M]_e =$

$$\eta \times \begin{bmatrix} 3454 & & & & & & & & & & & & \eta = \dfrac{\rho tab}{6300} \\ -461 & 80 \\ -461 & -63 & 80 \\ 1226 & -274 & 199 & 3454 \\ 274 & -60 & 42 & 461 & 80 \\ 199 & -42 & 40 & 461 & 63 & 80 \\ 1226 & -199 & 274 & 394 & 116 & 116 & 3454 \\ -199 & 40 & -42 & -116 & -30 & -28 & -461 & 80 \\ -274 & 42 & -60 & -116 & -28 & -30 & -461 & 63 & 80 \\ 394 & -116 & 116 & 1226 & 199 & 274 & 1226 & -274 & -199 & 3454 \\ 116 & -30 & 28 & 199 & 40 & 42 & 274 & -60 & -42 & 461 & 80 \\ -116 & 28 & -30 & -274 & -42 & -60 & -199 & 42 & 40 & -461 & -63 & 80 \end{bmatrix}$$

(12.134)

12.8 FLEXURAL VIBRATIONS OF PLATES USING ISOPARAMETRIC ELEMENTS

Isoparametric bending elements are very versatile elements for the analysis of plate flexure. There are two major advantages of using isoparametric element. They are capable of taking care of the boundary curvature. The element can incorporate shear deformation in an elegant manner. The element is named isoparametric, as both the geometry and the displacement are denoted by the same parameter.

In the theory of thin plates it has been assumed that normal to the middle surface before deformation remains normal after deformation of the plate. When shearing deformation is considered, this assumption does not remain valid and the assumption requires modification. The derivation of the element is based on Mindlin's theory which states that normal to the middle plane of the plate before bending remain straight, but not necessarily normal to the middle surface after bending. Therefore, the rotations θ_x and θ_y of the plate are given by [Fig. 12.11]

$$\begin{Bmatrix} \theta_x \\ \theta_y \end{Bmatrix} = \begin{Bmatrix} \dfrac{\partial w}{\partial x} + \phi_x \\ \dfrac{\partial w}{\partial y} + \phi_y \end{Bmatrix} \qquad (12.135)$$

where ϕ_x and ϕ_y are average shear rotations over the thickness of the plate.

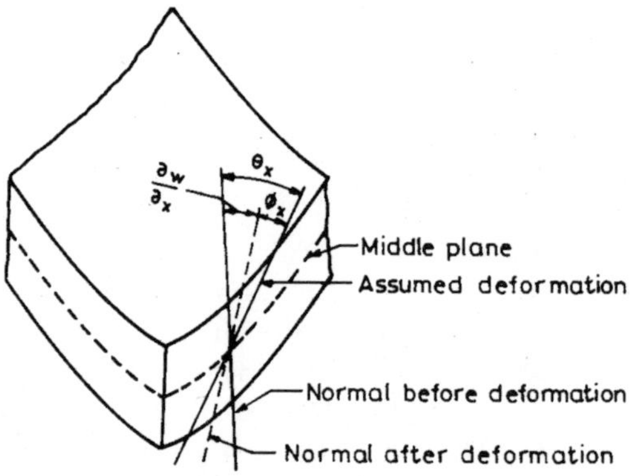

Fig. 12.11 Deformation of plate in *XZ* plane

The displacement of the plate may be prescribed by

$$\{f\} = \begin{Bmatrix} w \\ \theta_x \\ \theta_y \end{Bmatrix} \tag{12.136}$$

A general formulation is presented so that the derivation is valid for all types of isoparametric elements - linear, quadratic, cubic etc. Only the appropriate shape function is to be substituted. The coordinates within an element are

$$\begin{Bmatrix} x \\ y \end{Bmatrix} = \sum_{r=1}^{n} \begin{bmatrix} 1 & 0 \\ 0 & 1 \end{bmatrix} N_r \begin{Bmatrix} x_r \\ y_r \end{Bmatrix} \tag{12.137}$$

x_r , y_r are nodal coordinates and N_r is the shape function.

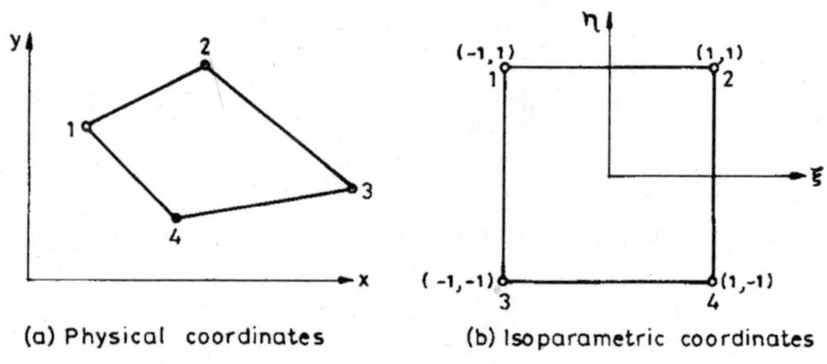

Fig. 12.12 Isoparametric linear element

The shape functions for the linear element of Fig. 12.12 are given by

$$N_r = \frac{1}{4}(1+\xi\xi_r)(1+\eta\eta_r) \tag{12.138}$$

The shape functions for the quadratic element of Fig. 12.13 are given below.

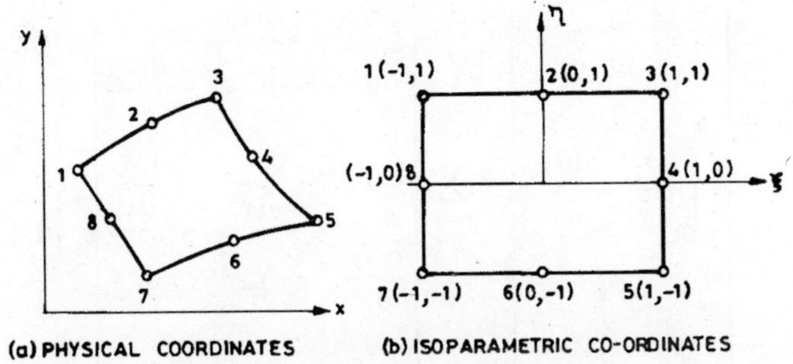

Fig. 12.13 Transformation from physical to isoparametric coordinates for quadratic elements

For the corner nodes,

$$N_r = \frac{1}{4}(1+\xi\xi_r)(1+\eta\eta_r)(\xi\xi_r+\eta\eta_r-1) \tag{12.139}$$

For midside nodes along ξ-axis,

$$N_r = (1-\eta^2)(1+\xi\xi_r)/2 \tag{12.140}$$

For midside nodes along η-axis,

$$N_r = (1-\xi^2)(1+\eta\eta_r)/2 \tag{12.141}$$

Similarly, the displacement function is given by

$$\begin{Bmatrix} w \\ \theta_x \\ \theta_y \end{Bmatrix} = \sum_{r=1}^{n} N_r \begin{bmatrix} 1 & 0 & 0 \\ 0 & 1 & 0 \\ 0 & 0 & 1 \end{bmatrix} \begin{Bmatrix} w_r \\ \theta_{xr} \\ \theta_{yr} \end{Bmatrix} \tag{12.142}$$

or,

$$\{f\} = [N]\{X\}_e \tag{12.143}$$

The generalised strains are

$$\{\in\}^T = \left\{ -\frac{\partial \theta_x}{\partial x} \quad -\frac{\partial \theta_y}{\partial y} \quad -\left(\frac{\partial \theta_x}{\partial y} + \frac{\partial \theta_y}{\partial x}\right) \quad -\phi_x \quad -\phi_y \right\} \tag{12.144}$$

Therefore, performing necessary differentiation with respect to different quantities given by Eq. (12.142), yields

$$\left\{\begin{matrix} -\dfrac{\partial \theta_x}{\partial x} \\ -\dfrac{\partial \theta_y}{\partial y} \\ -\left(\dfrac{\partial \theta_x}{\partial y}+\dfrac{\partial \theta_y}{\partial x}\right) \\ -\phi_x \\ -\phi_y \end{matrix}\right\} = \sum_{r=1}^{n} \begin{bmatrix} 0 & -\dfrac{\partial N_r}{\partial x} & 0 \\ 0 & 0 & -\dfrac{\partial N_r}{\partial y} \\ 0 & -\dfrac{\partial N_r}{\partial y} & -\dfrac{\partial N_r}{\partial x} \\ \dfrac{\partial N_r}{\partial x} & -N_r & 0 \\ \dfrac{\partial N_r}{\partial y} & 0 & -N_r \end{bmatrix} \begin{Bmatrix} w_r \\ \theta_{xr} \\ \theta_{yr} \end{Bmatrix} \tag{12.145}$$

while deriving Eq. (12.144), use has been made of Eq. (12.135).

Eq. (12.144) can be written in compact form as given below

$$\{\in\} = \sum_{r=1}^{n} [B]_r \{X_r\}_e = [B]\{X\}_e \tag{12.146}$$

The quantities $\partial N_r / \partial x$ and $\partial N_r / \partial y$ are to be computed to evaluate [*B*] matrix.

$\dfrac{\partial N_r}{\partial \xi}$ and $\dfrac{\partial N_r}{\partial \eta}$ are related to $\dfrac{\partial N_r}{\partial x}$ and $\dfrac{\partial N_r}{\partial y}$ as follows:

$$\begin{Bmatrix} \dfrac{\partial N_r}{\partial \xi} \\ \dfrac{\partial N_r}{\partial \eta} \end{Bmatrix} = \begin{bmatrix} \dfrac{\partial x}{\partial \xi} & \dfrac{\partial y}{\partial \xi} \\ \dfrac{\partial x}{\partial \eta} & \dfrac{\partial y}{\partial \eta} \end{bmatrix} \begin{Bmatrix} \dfrac{\partial N_r}{\partial x} \\ \dfrac{\partial N_r}{\partial y} \end{Bmatrix} \tag{12.147}$$

Eq. (12.147) can now be written as

$$\begin{Bmatrix} \dfrac{\partial N_r}{\partial x} \\ \dfrac{\partial N_r}{\partial y} \end{Bmatrix} = [J]^{-1} \begin{Bmatrix} \dfrac{\partial N_r}{\partial \xi} \\ \dfrac{\partial N_r}{\partial \eta} \end{Bmatrix} \tag{12.148}$$

where

$$[J]^{-1} = \begin{bmatrix} \dfrac{\partial \xi}{\partial x} & \dfrac{\partial \eta}{\partial x} \\ \dfrac{\partial \xi}{\partial y} & \dfrac{\partial \eta}{\partial y} \end{bmatrix} \tag{12.149}$$

The elements of [J] are calculated by

$$\left.\begin{aligned}\frac{\partial x}{\partial \xi} &= \sum_{r=1}^{n} \frac{\partial N_r}{\partial \xi} x_r \\ \frac{\partial y}{\partial \xi} &= \sum_{r=1}^{n} \frac{\partial N_r}{\partial \xi} y_r \\ \frac{\partial x}{\partial \eta} &= \sum_{r=1}^{n} \frac{\partial N_r}{\partial \eta} x_r \\ \frac{\partial y}{\partial \eta} &= \sum_{r=1}^{n} \frac{\partial N_r}{\partial \eta} y_r\end{aligned}\right\} \tag{12.150}$$

The stress components to be considered in this is

$$\{\sigma\}^T = \{M_x \quad M_y \quad M_{xy} \quad V_x \quad V_y\} \tag{12.151}$$

The stress-strain relationship for an isotropic plate is

$$\begin{Bmatrix} M_x \\ M_y \\ M_{xy} \\ V_x \\ V_y \end{Bmatrix} = \begin{bmatrix} D & \nu D & 0 & 0 & 0 \\ \nu D & D & 0 & 0 & 0 \\ 0 & 0 & \frac{1-\nu}{2} D & 0 & 0 \\ 0 & 0 & 0 & S_x & 0 \\ 0 & 0 & 0 & 0 & S_y \end{bmatrix} \begin{Bmatrix} -\frac{\partial \theta_x}{\partial x} \\ -\frac{\partial \theta_y}{\partial x} \\ -\left(\frac{\partial \theta_x}{\partial y} + \frac{\partial \theta}{\partial x}\right) \\ -\phi_x \\ -\phi_y \end{Bmatrix} \tag{12.152}$$

where S_x and S_y are shear rigidities. For an isotropic plate, they are

$$S_x = S_y = \frac{Gt}{1.2} \tag{12.153}$$

In compact form, Eq. (12.152) becomes

$$\{\sigma\} = [D]\{\in\} \tag{12.154}$$

The element stiffness matrix can be derived as

$$[K]_e = \int_{-1}^{1} \int_{-1}^{1} [B]^T [D][B] |J| \, d\xi \, d\eta \tag{12.155}$$

It is extremely tedious to evaluate explicitly $[K]_e$ in Eq. (12.155). The integration is carried out numerically by employing Gaussian quadrature. For the isoparametric plate bending element, it has been found that the best results of the 'integration can be obtained by reduced integration technique'.

12.8.1 Reduced Integration Technique

The Gaussian integration of n points gives exact values of integral of polynomials of degree $(2n - 1)$ in x. In order to evaluate the final values of the element stiffness matrix, elements of $[B]^T\,[D][B]$ are to be studied in details. The shear strain components are functions of N_r and its derivations, the flexural strain components are functions of derivatives of N_r only. Therefore, for all the cases of linear, quadratic and other isoparametric elements, shear components are one order higher than the flexural strain components which introduces an inconsistency in the analysis, that is responsible for the poor behaviour of these elements in shear - a phenomena known as shear locking. We limit our discussion to the isoparametric quadratic element only.

For an isoparametric quadratic element, the shape functions are quadratic in ζ and η and their derivations are linear. It can be seen in $[B]$ matrix of Eq. (12.145) that flexural strain terms are derivatives of shape function and vary linearly whereas shear strain terms are functions of both the shape functions and its derivatives and they have a quadratic variation. Thus, the shear strain terms have a higher order variation which introduces inconsistency. Thus, for the exact evaluation of flexural components in Eq. (12.155), a 2×2 Gaussian quadrature rule is to be employed, whereas exact evaluation of the shear contribution is to be done by 3×3 Gaussian quadrature. Thus results of shear forces obtained by 3×3 Gaussian integration for the thin plate have been found to be erratic. On the other hand, it has been found that a reduced integration of 2×2 (which is also economic in computation) gives vastly superior results.

It is preferable to have all terms of moments and shears having the same bilinear variation. Fig. 12.14 shown a least square line fit to a parabola and it has been shown that they intersect at Gauss point values. Thus if a 2×2 Gaussian integration is performed, the quadratic function is sampled at points such that the results will be identical to that obtained by use of 'smoothed' (linear) shape functions. So both the functions are reduced to the same order by the least square approximation. Thus, the quadratic function is replaced by a linear function.

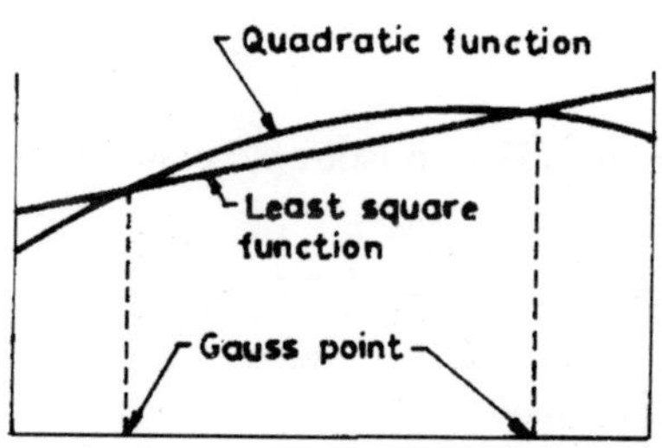

$\xi = -1 \quad \xi = \frac{1}{\sqrt{3}} \qquad\qquad \xi = \frac{1}{\sqrt{3}} \quad \xi = +1$

Fig. 12.14 A least square straight line fitting the quadratic function

12.8.2 Consistent mass matrix

At any point within the element

$$\begin{Bmatrix} \ddot{w} \\ \ddot{\theta}_x \\ \ddot{\theta}_y \end{Bmatrix} = \sum_{r=1}^{n} \begin{bmatrix} N_r & 0 & 0 \\ 0 & N_r & 0 \\ 0 & 0 & N_r \end{bmatrix} \begin{Bmatrix} \ddot{w}_r \\ \ddot{\theta}_{xr} \\ \ddot{\theta}_{yr} \end{Bmatrix} \tag{12.156}$$

The acceleration field of the plate is given by

$$\{f\} = \begin{Bmatrix} \ddot{U} \\ \ddot{V} \\ \ddot{W} \end{Bmatrix} = \begin{Bmatrix} -z\ddot{\theta}_x \\ -z\ddot{\theta}_y \\ \ddot{w} \end{Bmatrix} \tag{12.157}$$

Combining Eqs. (12.156) and (12.157), yields

$$\{f\} = [G][N] \sum_{r=1}^{n} \begin{Bmatrix} \ddot{w}_r \\ \ddot{\theta}_{xr} \\ \ddot{\theta}_{yr} \end{Bmatrix} \tag{12.158}$$

where

$$[G] = \begin{bmatrix} 0 & -z & 0 \\ 0 & 0 & -z \\ 1 & 0 & 0 \end{bmatrix} \tag{12.159}$$

$[N]$ has already been defined by Eq. (12.143).

The inertia forces are expressed as

$$\{F_e\}_I = \begin{bmatrix} \rho & 0 & 0 \\ 0 & \rho & 0 \\ 0 & 0 & \rho \end{bmatrix} \{f\} = [\rho]\{f\} \tag{12.160}$$

where ρ is the mass density.

The virtual work expression is

$$\delta W = \delta \iiint \{F_e\}^T \{f\}\, dx\, dy\, dz$$

$$= \delta \iiint \{\ddot{X}\}_e^T [N]^T [G]^T [\rho][G][N]\{\ddot{X}\}_e\, dx\, dy\, dz \tag{12.161}$$

Integration with respect to z-direction has been performed separately

$$[m] = \int_{-t/2}^{t/2} [G]^T [\rho][G]\, dz \tag{12.162}$$

or,
$$[m]=\begin{bmatrix} \rho t & 0 & 0 \\ 0 & \dfrac{\rho t^3}{12} & 0 \\ 0 & 0 & \dfrac{\rho t^3}{12} \end{bmatrix} \tag{12.163}$$

Substituting the above expression in Eq. [12.161],

$$\delta W = \iint \delta \{\ddot{X}\}_e^T [N]^T [m][N]\{\ddot{X}\}_e \, dx\, dy \tag{12.164}$$

For virtual displacements at nodes, the work done by the nodal forces

$$\delta W = \delta \{\ddot{X}\}_e^T \{P\} \tag{12.165}$$

Equating δW of Eq. (12.164) to that of Eq. (12.165), yields

$$\{P\} = \iint [N]^T [m][N]\, dx\, dy$$

or,
$$[M]_e \{\ddot{X}\}_e = \{P\}_e \tag{12.166}$$

where $[M]_e$ is the consistent mass matrix and is given by

$$[M]_e = \iint [N]^T [M][N]\, dx\, dy$$

or,
$$[M]_e = \int_{-1}^{1}\int_{-1}^{1} [N]^T [m][N]|J|\, dx\, dy \tag{12.167}$$

12.9 PERIODIC STRUCTURES

The type of structure which contains a series of identical structural elements connected in an identical manner to one another is referred to as periodic structure. This suggests that in this type of construction, there is a regularly repeating section. There are many examples of such structures, mostly arising in civil, aerospace and ocean engineering and naval architecture. A tall building has uniform and identical stories, a ship or an aircraft consists of a uniform shell reinforced with periodically arranged stiffeners. A jack-up rig or a space station is constructed of truss-like components, where truss structures are essentially assemblage of identical bays and thus are spatially periodic. Other applications include rails, shafts with bearings and discs and bridges which consist of identical subsystems connected in identical ways. Other types of periodic structures are those which possess material periodicity, such as composite laminated structures which is beyond the scope of this present section.

Considerable work has been carried out on the analysis of periodic structures [12.7]. A good summary of the earlier work may be found in Brillounin's classic work [12.8]. Early work concerning with engineering application began with simple periodic beams [12.9]. Periodic structures are subjected to dynamic load of high intensity. As such their vibration analysis is of great importance. Cremer and Leilich, from their investigation of flexural motion of periodic beam structures concluded that waves can propagate in some frequency bands and not in others, that is, only waves of certain frequencies pass through the beam. These frequencies are called pass-bands and the beam behaves as a pass-band filter. The wave propagation in periodic beams is discussed in details by Mead [12.10]. Mead and his co-research workers have worked extensively on the problem and have taken the subject to a mature stage [12.11 - 12.18].

Taking advantage of the spatially periodic property of the structures enables carrying out the vibration analysis in a more convenient manner. This aids in catering the whole problem with a much less computational effort. The computation is now based on typical periodic element with boundary conditions at its extreme ends. For example, for a structure having '*N*' elements and each element has '*i*' degrees of freedom, a routine analysis will pertain to the solution of '$N \times i$' equations. Treating the structure as a periodic structure, a maximum of '*i*' equations have to be solved.

For the response analysis of the structure it becomes almost unnecessary to determine principal modes at all, as the vibrating characteristics of the periodic structure is best understood in terms of propagating and non propagating free wave motion. The natural free waves assist in understanding the vibration caused by moving loads or pressure fields. The wave motion approach forms an important topic for understanding the subject.

12.9.1 Different Types of Periodic Structures

A periodic structure consisting of a line of periodic elements joined together end-to-end is referred to as one-dimensional periodic structure. Let us take the example of the rail of a railway track supported over equidistant sleeper in an identical manner. Each element is connected to its neighbour through two coordinates- the transverse deflection and the bending rotation at the support. This is called a bi-coupled system. If out of the two, one of the deformation is eliminated, such as the vertical deflection by providing a vertical rigid support, then the system becomes mono-coupled. Similarly, multi-coupled system will have higher coupling coordinates between adjacent elements.

A periodic structure consisting of plate type periodic elements joined together end-to-end and side-by-side is called a two-dimensional periodic structure. This will result in a multi-coupled periodic system. If a periodic plate structure is of finite width and have equally spaced unidirectional stiffener, the problem as such is multi-coupled, but can be reduced to bi-coupled periodic system if the plate edges perpendicular to stiffeners are simply supported.

A periodic structure consisting of 'layers' of two-dimensional structures coupled on one-on-top of another is known as three-dimensional period structure. Typical examples are a jack-up-rig used for oil drilling in the ocean bed and multi-storey building in modular construction.

12.9.2 Free harmonic wave motion through a mono-coupled periodic beam

A beam extending to infinity and having equally spaced simple supports is considered. One of the bays is subjected to transverse harmonic force. The variation of the response of the forcing point with the forcing frequency is given by Fig.12.15. The resonant peaks shown in the figure, occur at natural frequencies of a single beam element and with its ends either simple supported or fully fixed.

At low frequencies, the displacement response of an undamped beam varies as shown in

Fig. 12.16. Each bay vibrates in an identical mode, but with reduced amplitude and in counterphase with its neighbours. The amplitude at any point of one bay is $e^{-\mu}$ times the corresponding point of the previous bay. μ is a function of frequency only and not on the pair of points chosen. It is a distinct measure of the rate of decay of the particular type of motion being generated and is known as *attenuation constant* of this periodic system.

With further increase of the frequency, the amplitude of the point of excitation increases resulting in a drop in the decay rate. At the frequency λ_1, the decay rate is zero. The vibration of each bay will be in counterphase with its neighbour having the same amplitude as its neighbour.

As the frequency is further increased, each bay vibrates with a finite amplitude and in the same mode as its neighbour, but a phase difference $\in$ (between 0 and $\pm\ \pi$) exists between motions at all pairs of corresponding points in adjacent bays. In this case, the response at a point in one unloaded bay is then $e^{-i\in}$ times the response at the corresponding point in the preceding unloaded bay. $\in$ is called the phase constant.

Within a single bay, phase difference exists between displacements of different points. As such the complex flexural motion cannot be represented by a single wave. When $\in = \pi / 2$, the total motion may appear at a particular instant shown in Fig. 12.16c. Later, the pattern will have moved over one or more bays.

At the frequency λ_2, both $\in$ and μ are zero. As such adjacent bays vibrate indentically in phase in the mode as shown in Fig. 12.16d.

Fig. 12.17 shows the variation of μ and ε over an extended range for a beam on periodic simple supports. Figure indicates that when μ is non-zero, $\in$ is constant and this results in the decay of all wave motion. The frequency region are known as attenuation zones. On the other hand when μ is zero and/or constant, $\in$ varies. Energy is then propagated and the frequency regions are known as *propagation zones*. It can be seen from Fig. 12.17 that the wave motion which involves the propagation of energy is only possible in particular frequency bands. When limits outside this bands are considered, it results in a decay of the wave motions when they tend to spread out. The beam acts as a pass-band filter and this characterizes a periodic structure. These bands are very important from the study of vibration characteristics of the periodic structure. Within the bands are expected to result in the greatest vibration response. Natural frequencies of many finite periodic structures live within these bands [12.10, 12.16]. As such it is evident that the phase constant with the band significantly influences the vibration response and natural frequencies.

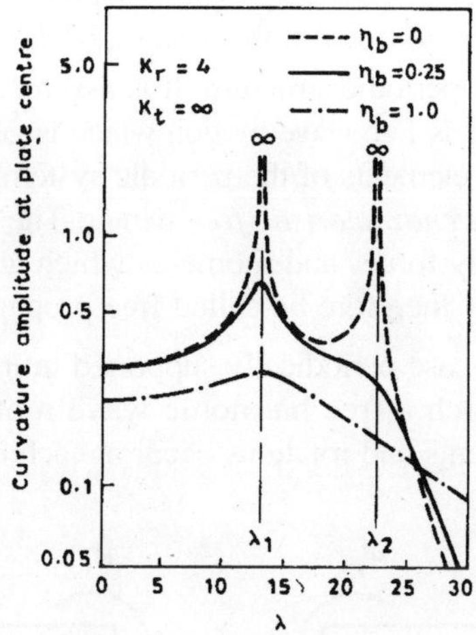

Fig. 12.15 Variation with frequency and plate damping of the curvature amplitude at the centre of the loaded bay of an infinite stiffened plate single point load

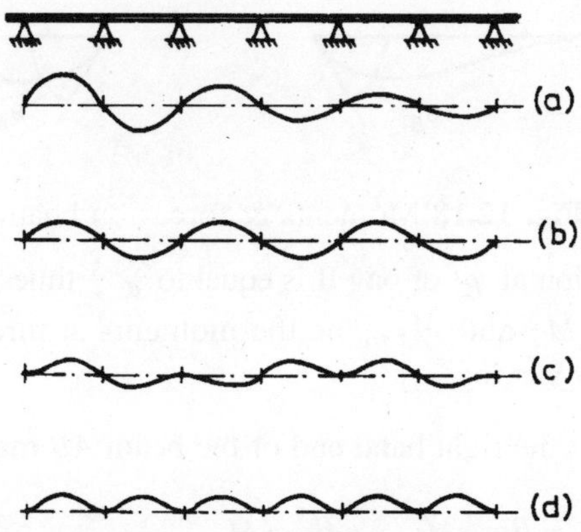

Fig. 12.16 Modes of displacement of a periodic beam at different frequencies

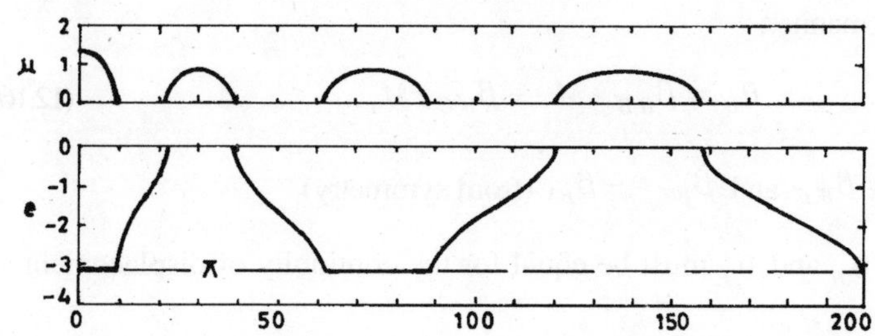

Fig. 12.17 The real and imaginary parts of propagation constant for a beam on periodic simple supports.

The complex quantity $\mu + i\in$ is termed as the complex propagation constant and will be denoted by δ which has nothing to do with the loading of the loaded bay, but is dependent on the single element of the entire periodic structure. It is assumed that wave motion is free in all bays except the loaded bay. This is free wave motion which is generated at a particular frequency which is a characteristic of the elements of the periodic system and not of the loaded bay. Due to this the waves are termed as *characteristic free waves*. The free transmission of the motion from one bay to the next is only forces and moments which are internal to the periodic beam. The propagation constants may therefore be called free propagation constants.

Let us now investigate the case periodically supported infinite beam. For this, two adjacent bays *AB* and *BC* through which a free harmonic wave motion is passing is taken up for consideration. Harmonic moments and rotations occur at each end of each bay (Fig. 12.18).

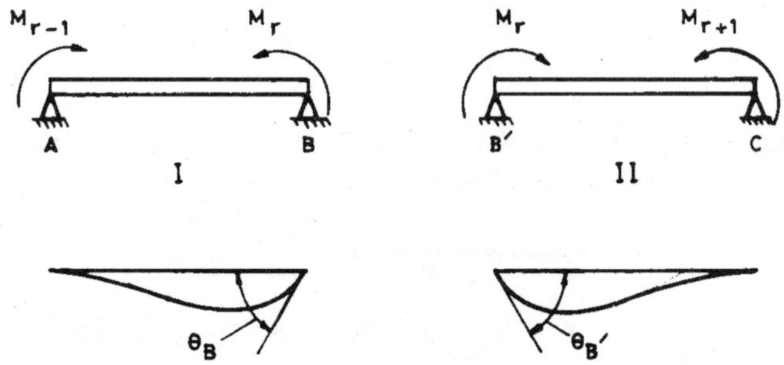

Fig. 12.18 Moments & forces on beam elements

Moment and rotation at B' of bay II is equal to $e^{-\delta}$ times the moment and rotation at end of bay I. Let M_{r-1}, M_r and M_{r+1} be the moments at three supports- they are complex quantities harmonically varying.

The rotation θ_B at the right hand end of the beam *AB* may be expressed in the form.

$$\theta_B = \beta_{BA} \cdot M_{r-1} + \beta_{BB} \cdot M_r \qquad (12.168)$$

where β_{BA} is the harmonic rotation at *A* due to a unit harmonic moment at *A*. β_{BB} is the rotation at *B* due to a unit moment at *B*.

In the same manner,

$$\theta_{B'} = \beta_{B'B'} \cdot M_r + \beta_{B'C} \cdot M_{r+1} \qquad (12.169)$$

Now, $\beta_{BB} = \beta_{B'B'}$ and $\beta_{BC} = -\beta_{BA}$ (from symmetry)

Slopes θ_B and θ_E must be equal for the continuity of displacement and slope across the support.

Hence equating θ_B and θ_E from Eqs. (12.168) and (12.169), yields

$$M_{r-1} + 2\frac{\beta_{BB}}{\beta_{BA}} + M_{r+1} = 0 \qquad (12.170)$$

An identical equation applies to moments at each support between free bays of the beam. Now,

$$\left.\begin{aligned} M_{r+1} &= e^{\mu} M_r \\ M_r &= e^{\mu} M_{r-1} \end{aligned}\right\} \tag{12.171}$$

Substituting Eq. (12.171) into Eq. (12.169), yields

$$e^{\mu} + e^{-\mu} = -2\frac{\beta_{BB}}{\beta_{BA}}$$

or,

$$\cosh\mu = -\frac{\beta_{BB}}{\beta_{BA}} \tag{12.172}$$

Eq. (12.172) gives the equation for propagation constant μ. The right hand side of the equation is strongly frequency dependent and will be complex if the damping exists in the beam.

The receptance functions β_{BB} and β_{BA} are transcendental functions of frequency, the mass of the beam, the stiffness and damping Eq. (12.172) can be expressed as

$$\cosh\mu = -\frac{F_5}{F_8} \tag{12.173}$$

where

$$F_5 = \cos\lambda L \sinh\lambda L - \sin\lambda L \cosh\lambda L \tag{12.174}$$

$$F_8 = \sin\lambda L - \sinh\lambda L$$

$\lambda^4 = \dfrac{mp^2}{EI}$, m is the mass of the beam per unit length, EI is the flexural rigidity of the beam and p is the frequency.

The expression $w(x)$, the transverse displacement of any bay is given by

$$w(x) = A\cosh\lambda x + B\sinh\lambda x + C\cos\lambda x + D\sin\lambda x \tag{12.175}$$

The boundary conditions are

$$w(0) = w(L) = 0 \tag{12.176}$$

where L is the length of the bay.

Furthermore, slopes at $x = 0$ and $x = L$ are related by

$$w'(L) = e^{\mu} w'(0) \tag{12.177}$$

Substituting the boundary conditions given by Eqs. (12.176) and (12.177) into Eq. (12.175) result in

$$w(x) = A\left\{\cosh\lambda x - \cos\lambda x + \sinh\lambda x\left(\frac{F_1 - F_2 - e^{\mu}F_{10}}{F_5 - e^{\mu}F_7}\right)\right.$$

$$+ \sin \lambda x \left(\frac{2F_{10}}{F_7 + F_8} - \frac{(F_7 - F_8)(F_1 - F_2 - e^{\mu} F_{10})}{(F_7 + F_8)(F_5 - e^{\mu} F_7)} \right) \Bigg\} \quad (12.178)$$

where

$$F_1 = \sin \lambda L \sinh \lambda L, \quad F_2 = \cos \lambda L \cosh \lambda L, \quad F_5 = \cos \lambda L \sinh \lambda L - \cosh \lambda L \sin \lambda L$$

$$F_7 = \sin \lambda L + \sinh \lambda L, \quad F_8 = \sin \lambda L - \sinh \lambda L, \quad F_{10} = \cos \lambda L - \cosh \lambda L \quad (12.179)$$

It now can be very clearly seen that for a given value of λL, that is, for a given beam and frequency, the mode of displacement is uniquely determined by the propagation constant μ. Since there is one value of m for each frequency, there is only one mode of free wave displacement for each frequency [12.10].

12.9.3 Finite Element Analysis of Periodic Structures

Consider an infinite one-dimensional periodic structure as shown in Fig. 12.19. Each periodic component $AA'B'B$, $BB'C'C$ etc. can be described by an identical arrangement of elements. If one of the components, e, is represented by a finite element model, then when it vibrates harmonically with angular frequency p, the equation of motion is

$$([K^e]-p^2[M^e])\{\delta^e\}=\{F^e\} \quad (12.180)$$

where $\{\delta^e\}$ is the nodal degrees of freedom and $\{F^e\}$ the equivalent nodal forces.

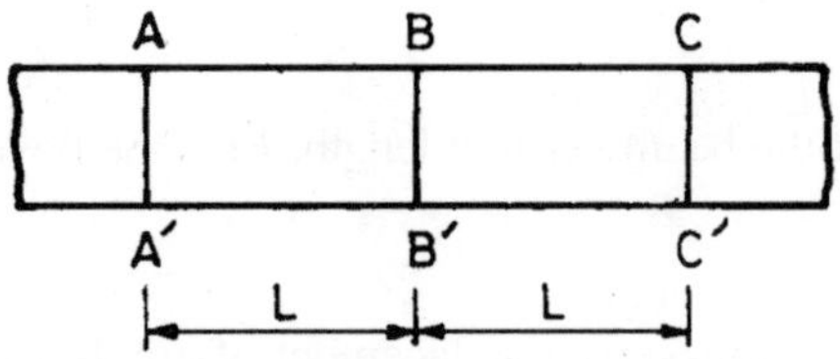

Fig. 12.19 Periodic beam

The displacement vector $\{\delta^e\}$ is split into three parts- $\{\delta_L^e\}$, the degrees of freedom corresponding to the nodes at the left hand boundary, $\{\delta_R^e\}$, the degrees of freedom corresponding to the nodes at the right hand boundary and $\{\delta_I^e\}$ are the displacements corresponding to the remaining degrees of freedom. Thus

$$\{\delta^e\} = \begin{Bmatrix} \{\delta_L^e\} \\ \{\delta_I^e\} \\ \{\delta_R^e\} \end{Bmatrix} \quad (12.181)$$

Now,

$$\{\delta_R^e\} = \left\{\delta_L^{e+1}\right\} = e^{\mu} \{\delta_L^e\} \quad (12.182)$$

Forces $\{F^e\}$ can be partitioned in a similar manner

$$\{F^e\} = \begin{Bmatrix} \{F_L^e\} \\ \{F_I^e\} \\ \{F_R^e\} \end{Bmatrix} \tag{12.183}$$

Since a free wave is propagating, $\{F_I^e\}$ in fact will be zero. The forces at the boundaries are responsible for transmission of wave motion from one component to the next. Therefore $\{F_L^e\}$ and $\{F_R^e\}$ are not zero.

Equilibrium conditions at the boundary BB', yield

$$\{F_R^e\} + \{F_L^{e+1}\} = \{0\} \tag{12.184}$$

Therefore $$\{F_R^e\} = -\{F_L^{e+1}\} = -e^{\mu}\{F_L^e\} \tag{12.185}$$

The relationships of Eqs. (12.182) and (12.185) give rise to the following transformations.

$$\{\delta^e\} = \begin{Bmatrix} \{\delta_L^e\} \\ \{\delta_I^e\} \\ \{\delta_R^e\} \end{Bmatrix} = \begin{bmatrix} [I] & [0] \\ [0] & [I] \\ e^{\mu}[I] & [0] \end{bmatrix} \begin{Bmatrix} \{\delta_L^e\} \\ \{\delta_I^e\} \end{Bmatrix} = [W]\{\bar{\delta}^e\} \tag{12.186}$$

and

$$\{F^e\} = \begin{Bmatrix} \{F_L^e\} \\ \{F_I^e\} \\ \{F_R^e\} \end{Bmatrix} = \begin{bmatrix} [I] & [0] \\ [0] & [I] \\ e^{\mu}[I] & [0] \end{bmatrix} \begin{Bmatrix} \{F_L^e\} \\ \{F_I^e\} \end{Bmatrix} \tag{12.187}$$

when the transformation equations (12.186) and (12.187) are applied to Eq. (12.182), the unknown boundary forces can be eliminated by premultiplying by the matrix $[W]^H$. H is used to denote taking the complex conjugate of the matrix and then transposing it. Therefore,

$$[W]^H = \begin{bmatrix} [I] & [0] & e^{\mu}[I] \\ 0 & [I] & [0] \end{bmatrix} \tag{12.188}$$

Taking $\{F_I^e\} = \{0\}$ results in the equation

$$[[K(\mu)] - \omega^2[M(\mu)]] \begin{Bmatrix} \{\delta_L^e\} \\ \{\delta_I^e\} \end{Bmatrix} = \{0\} \tag{12.189}$$

where

$$[K(\mu)] = [W]^H [K^e][W] \tag{12.190}$$

and

$$[M(\mu)] = [W]^H [M^e][W] \tag{12.191}$$

Both $[K(\mu)]$ and $[M(\mu)]$ are Hermitian matrices, since

$$([W]^H [K^e][W])^H = [W]^H [K^e][W] \tag{12.192}$$

and

$$([W]^H [M^e][W])^H = [W]^H [M^e][W] \tag{12.193}$$

(Note, $[A]$ is Hermitian if $[A]^H = [A]$)

Relationship given by Eq. (12.189) can be derived in another way.

Eq. (12.180) can be written as

$$\left(\begin{bmatrix} [K^e_{LL}] & [K^e_{LI}] & [K^e_{LR}] \\ [K^e_{IL}] & [K^e_{II}] & [K^e_{IR}] \\ [K^e_{RL}] & [K^e_{RI}] & [K^e_{RR}] \end{bmatrix} - p^2 \begin{bmatrix} [M^e_{LL}] & [M^e_{LI}] & [M^e_{LR}] \\ [M^e_{IL}] & [M^e_{II}] & [M^e_{IR}] \\ [M^e_{RL}] & [M^e_{RI}] & [M^I_{RR}] \end{bmatrix}\right) \begin{Bmatrix} \{\delta^e_L\} \\ \{\delta^e_I\} \\ \{\delta^e_R\} \end{Bmatrix} = \begin{Bmatrix} \{F^e_L\} \\ \{0\} \\ \{F^e_R\} \end{Bmatrix} \tag{12.194}$$

In Eq. (12.194), $[K^e]$ and $[M^e]$ matrices are partitioned corresponding to the degrees of freedom of three types of modes - those on the left hand boundary, all interior nodes and those on the right hand boundary.

Three matrices equations in terms of partitioned matrices can be written as

$$\begin{aligned}[K^e_{LL}]\{\delta^e_L\} + [K^e_{LI}]\{\delta^e_I\} + [K^e_{LR}]e^{\mu}\{\delta^e_L\} - p^2([M^e_{LL}]\{\delta^e_L\} \\ + [M^e_{LI}]\{\delta^e_I\} + [M^e_{LR}]e^{\mu}\{\delta^e_L\}) = \{F^e_L\}\end{aligned} \tag{12.195}$$

$$\begin{aligned}([K^e_{IL}]\{\delta^e_L\} + [K^e_{II}]\{\delta^e_I\}) + [K^e_{IR}]e^{\mu}\{\delta^e_L\} - p^2([M^e_{IL}]\{\delta^e_L\} \\ + [M^e_{II}]\{\delta^e_I\} + [M^e_{IR}]e^{\mu}\{\delta^e_L\}) = \{0\}\end{aligned} \tag{12.196}$$

$$\begin{aligned}([K^e_{RL}]\{\delta^e_L\} + [K^e_{RI}]\{\delta^e_I\} + [K^e_{RR}]e^{\mu}\{\delta^e_L\}) - p^2([M^e_{RL}]\{\delta^e_L\} \\ + [M^e_{RI}\}\{\delta^e_I\} + [M^e_{RR}]e^{\mu}\{\delta^e_L\}) = -e^{\mu}\{F^e_L\}\end{aligned} \tag{12.197}$$

Eliminating $\{F^e_L\}$ from Eq. (12.195) and (12.197), we get

$$\begin{aligned}&([K^e_{LL}] + [K^e_{RR}] + [K^e_{IR}]e^{\mu} + e^{-\mu}[K^e_{RI}])\{\delta^e_L\} + ([K^e_{LI}] + e^{-\mu}[K^e_{RI}]) \\ &\{\delta^e_I\} + ([M^e_{LL}] + [M^e_{RR}] + e^{\mu}[M^e_{LR}] + e^{-\mu}[M^e_{RI}])\{\delta^e_L\} \\ &+ ([M^e_{LI}] + e^{-\mu}[M_{RI}])\{\delta^e_I\} = \{0\}\end{aligned} \tag{12.198}$$

Writing Eqs. (12.196) and (12.198) in partitioned forms, we get

$$\left[\begin{array}{c|c} [K_{LL}^e]+[K_{RR}^e]+e^{\mu}[K_{LR}^e]+e^{-\mu}[K_{RL}^e] & [K_{LI}^e]+e^{-\mu}[K_{RI}^e] \\ \hline [K_{IL}^e]+e^{\mu}[K_{IR}^e] & [K_{II}^e] \end{array}\right]\begin{Bmatrix}\{\delta_L^e\}\\ \{\delta_R^e\}\end{Bmatrix}$$

$$-\,p^2\left[\begin{array}{c|c} [M_{LL}^e]+[M_{RR}^e]+e^{\mu}[M_{LR}^e]+e^{-\mu}[M_{RL}^e] & [M_{LI}^e]+e^{-\mu}[M_{RI}^e] \\ \hline [M_{IL}^e]+e^{\mu}[M_{IR}^e] & [M_{II}^e] \end{array}\right]$$

$$\begin{Bmatrix}\{\delta_L^e\}\\ \hline \{\delta_I^e\}\end{Bmatrix}=\{0\} \qquad (12.199)$$

In abbreviation form, Eq. (6.239) is same as Eq. (12.189), that is

$$[K(\mu)-p^2[M(\mu)]\begin{Bmatrix}\{\delta_L^e\}\\ \{\delta_I^e\}\end{Bmatrix}=\{0\} \qquad (12.200)$$

The matrices $[K(\mu)]$ and $[M(\mu)]$ are real functions of complex variable μ. Eq. (12.200) represents a eigenvalue problem. For a given value of μ, there will be a discrete range of frequencies for which a wave motion with particular propagation constant is possible.

If w is assumed to have some values, then Eq. (12.200) may be organised as a eigenvalue problem for μ For diagonal mass matrix, this type of arrangement is possible.

But when the finite element method is applied to this type of problem, $[K(\mu)]$ and $[M(\mu)]$ can be of high order and $[M(\mu)]$ may be banded and not diagonal. In such cases however, the solution of the eigenvalue problem given by Eq. (12.200) by a general, complex eigenvalue routine can be extremely tedious.

Sengupta [12.16] has shown that the frequencies of the finite structure can be determined for the purely imaginary part of the propagation constant. Thus, in assessing the modal response of the structure, the variation of the imaginary part of the propagation constant is a more important quantity. If m is purely imaginary, matrices in Eqs. (12.199) and (12.200) are Hermitian. Wilkinson [12.22] has shown that the complex eigenvalue problem can be reduced to a real symmetric matrix. Let

$$[K(\mu)]=[K]^R+i[K]^I \qquad (12.201)$$

$$[M(\mu)]=[M]^R+i[M]^I \qquad (12.202)$$

$$\begin{Bmatrix}\{\delta_L^e\}\\ \{\delta_I^e\}\end{Bmatrix}=\{\delta\}^R+i\{\delta\}^I \qquad (12.203)$$

Superscript R indicates the real part and superscript I, the imaginary part.

Eq. (12.200) can be written as

$$[([K]^R+i[K]^I)-p^2([M]^R+i[M]^I)](\{\delta\}^R+i\{\delta\}^I)=\{0\} \qquad (12.204)$$

By separating the real and imaginary parts and combining the resulting two equations, gives

$$\left(\begin{bmatrix}[K]^R & -[K]^I \\ [K]^I & [K]^R\end{bmatrix} - p^2\begin{bmatrix}[M]^R & -[M]^I \\ [M]^I & [M]^R\end{bmatrix}\right)\begin{Bmatrix}\{\delta\}^R \\ \{\delta\}^I\end{Bmatrix} = \{0\} \qquad (12.205)$$

As the matrices $[K(\mu)]$ and $[M(\mu)]$ are Hermitian, $[K]^I = -([K]^I)^T$ and $[M]^I = -([M]^I)^T$.

Thus, Eq. (12.205) represents a real symmetric eigenvalue problem. For any given value of μ which is real, Eq. (12.205) can be solved. The finite element method can now be applied with ease. All the existing finite element subroutines can be used without affecting any change and other standard approaches related to assembly and procedure for satisfaction of boundary conditions can be applied.

A discrete set of frequencies will occur in equal pairs corresponding to each value of μ in Eq. (12.205). Corresponding to each frequency, there will be a eigenvector. This eigenvector defines the wave motion in the periodic section at that frequency. The motion in an adjacent section will differ from that in the previous by e^{μ}.

Example

A uniform beam periodically spaced on simple supports is considered. The finite element technique using a beam element with two degrees of freedom - a deflection and a rotation is applied for the analysis. One span of the beam is considered as a repeating section. There is only one unconstrained degree of freedom in this case. Each basic component is divided into four elements. The frequencies found by the analysis is plotted as a function of the propagation constant. The finite element results are compared with exact solution [12.17]. From the result given in Table 12.1, it is evident that the finite element method gives very accurate values for the imaginary part of the propagation constant.

Table 12.1 [N]

Comparison of limiting frequencies for an infinite periodic beam

		Exact frequency	Finite element solution
First band	Lower bound	2.4674	2.4680
	Upper bound	5.5947	5.6007
Second band	Lower bound	9.8696	9.9086
	Upper bound	15.4056	15.5608
Third band	Lower bound	22.2096	22.6124
	Upper bound	30.2257	30.6714

References

12.1 M. Mukhopadhyay, Structures: Matrix and Finite Element, 3rd Edition, A. A. Balkema, The Netherlands, 1993.

12.2 M. Petyt, Introduction to Finite Element Vibration Analysis, Cambridge University Press, U.K., 1990.

12.3 W. Weaver, Jr. and P.R. Johnston, Structural Dynamics by Finite Elements, Prentice-Hall Inc., New Jersey, 1987.

12.4 R.D. Cook, D.S. Malkus and M.E. Plesha, Concepts and Applications of Finite Element Analysis, John Wiley & Sons, USA, 1989.

12.5 O.C. Zienkiewicz and R.L. Taylor, The Finite Element Method, 4th Edition, McGraw-Hill, USA, 1989.

12.6 E. Hinton, A. Razzaque, O.C. Zienkiewicz and J.O. Davies, A simple finite element solution for plate of homogenous, sandwich and cellular construction, Proceedings of the Institution of Civil Engineers (London), Part 2, V.59, March 1975, pp. 43-65.

12.7 Dong Li and Haym Benaroya, Dynamics of Periodic and Near-periodic Structures, Applied Mechanics Review, V. 45, No. 11, Nov. 1992, pp. 447-459.

12.8 L. Brillouin, Wave Propagation in Periodic Structures, First Edition, Dover Publications Inc., New York, 1946.

12.9 L. Cremer and H.O. Leilich, Zür Theorie der Biegekettenleiter, Archiv der Electrichen Übertiagung, V.7, 1953, pp. 263.

12.10 D.J. Mead, Free wave propagation in periodically supported infinite beams, Journal of Sound and Vibration, V. 11, 1970, pp. 181-197.

12.11 D.J. Mead, Vibration response and wave propagation in periodic structures, Transactions of American Society of Mechanical Engineers, Journal of Engineering Industry, V. 93, 1971, pp. 783-792.

12.12 D.J. Mead, A general theory of harmonic wave propagation in periodic structures with multiple coupling, Journal of Sound and Vibration, V. 17, 1973, pp. 235-260.

12.13 D.J. Mead, Wave propagation and natural modes in periodic system, I: Mono-coupled system. Journal of Sound and Vibration, V. 40, 1975, pp. 1-18.

12.14 D.J. Mead, Wave propagation and natural modes in periodic system. I: Multi-coupled systems, with and without damping, Journal of Sound and Vibration, V.40, 1975, pp. 19.

12.15 D.J. Mead and S. Parthan, Free wave propagation in two-dimensional periodic plates, Journal of Sound and Vibration, V.64, No. 3, 1979, pp. 325-348.

12.16 G. Sengupta, Natural flexural waves and the normal modes of periodically supported beams and plates, Journal of Sound and Vibration, V. 13, 1970, pp. 89-101.

12.17 R.M. Orris and M. Petyt, A finite element study of harmonic and propagation in periodic structures, Journal and Sound and Vibration, V. 33, No. 2, 1974.

12.18 S. Mukherjee and S. Parthan, Normal modes and acoustic response of finite one and two-dimensional multibay periodic panels, Journal of Sound and Vibration, V. 197, No. 5, 1996, pp. 527-545.

12.19 R.G. White and J.G. Walker (Editors). Noise and Vibration, Ellis Harwood Limited Publishers, U.K., 1982.

12.20 R.E.D. Bishop and D.C. Johnson, The Mechanics of Vibration, Cambridge University Press, 1966.

12.21 G.L. Rogers, Dynamics of Framed Structures, John Wiley, 1959.

12.22 J.H. Wilkinson, The Algebraic Eigenvalue Problem, The Clarendon Press, Oxford, U.K., 1965.

12.23 D.A. Wells, Theory and Problems of Lagrangian Dynamics, Schaum Outline Series, McGraw Hill, New York, 1967.

EXERCISE 12

12.1 Consider a bar element subjected to axial vibration. The length of the element is L, cross-sectional area A, the density of the material ρ and Young's modulus of the material of the bar E. Derive the stiffness matrix and consistent mass matrix of the bar element.

12.2 Consider a uniform bar of length L. The bar is fixed at one end and free at the other end. The bar is divided into two elements. The element has two nodes, one at each end. Determine the natural frequency of the free vibration of the bar.

12.3 Consider a shaft element subjected to torsion. The element has three nodes, including one midside node. Find the consistent mass matrix and the stiffness matrix for the element.

12.4 Consider a beam clamped at both ends. Treating the beam as uniform and dividing it into two elements, determine the fundamental frequency of the beam.

12.5 Consider a rectangular plate element subjected to inplane vibrations. Derive the stiffness matrix and the consistent mass matrix for the plate element.

12.6 For the rhombic element shown in the figure of Prob. 12.5. Determine the stiffness matrix and consistent mass matrix for inplane vibrations.

12.7 For the rhombic element shown in the figure, determine the consistent mass matrix and the stiffness matrix for the plate bending problem.

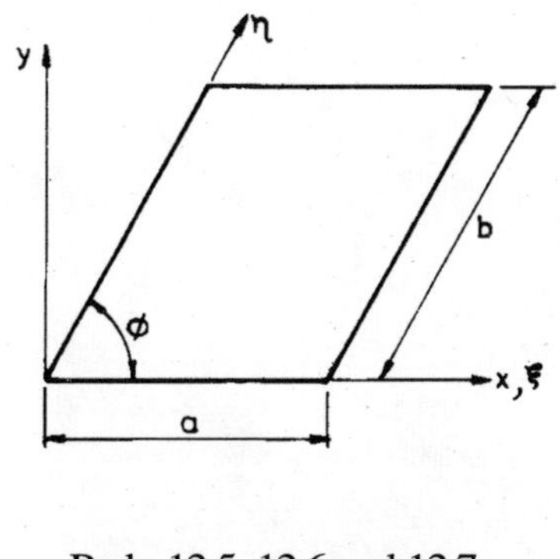

Prob. 12.5, 12.6 and 12.7

12.8 Consider a triangular-plate bending element, having the following displacement function:

$$w = \alpha_1 + \alpha_2 x + \alpha_3 y + \alpha_4 x^2 + \alpha_5 xy + \alpha_6 y^2 + \alpha_7 x$$

$$+ \alpha_8 (x^2 y + xy^2) + \alpha_9 y^3$$

Derive the stiffness matrix and the consistent mass matrix for the above element.

12.9 Determine the consistent mass and the stiffness matrix for the nonuniform rod shown in the figure in longitudinal vibration.

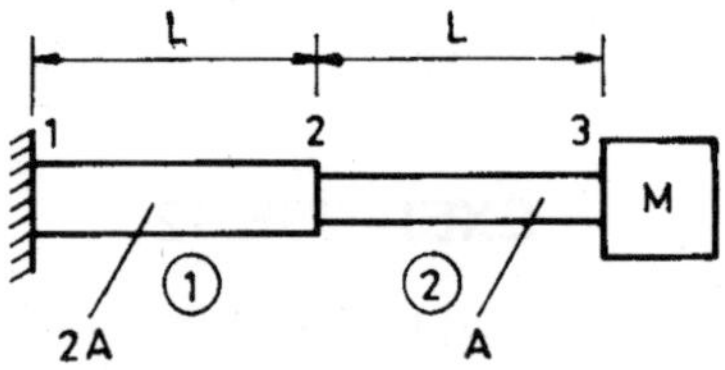

Fig. Prob. 12.9

12.10 A Uniform cantilever beam consists of two equal elements of length $L/2$. Form the stiffness matrix and the mass matrix of the total structure.

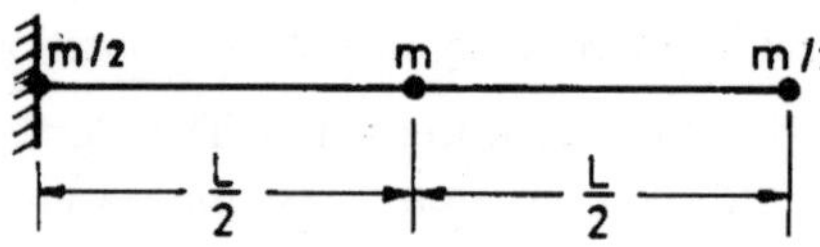

Fig. Prob. 12.10

12.11 Determine the two natural frequencies in torsional vibration of a uniform shaft, fixed at one end and free at the other end using two elements of equal length.

12.12 Explain the reasons why the elements in the first, third and fifth rows and columns of the consistent mass matrix of the constant strain triangle is equal to the mass of the element.

12.13 Show that rigid body motion do not cause any strain in the linear quadratic element.

12.14 How will you take care of the shear locking effect of the isoparametric linear element?

Chapter 13

Finite Difference Method for the Vibration Analysis of Beams and Plates

13.1 INTRODUCTION TO THE FINITE DIFFERENCE METHOD

The finite difference technique is another versatile numerical method for the solution of vibration problems. The method has been explained in Chapter 4 with respect to the time function. It is applied here with respect to spatial variables. The differential equation is the starting point of the method. The continuum is divided in the form of a mesh and the unknowns in the problem are those at the nodes. The derivatives of the equation are expressed in the finite difference form. The differential equation, split in this discrete form is applied at each node. This results in a set of simultaneous equations.

In the finite element method, the final differential equation is surpassed, whereas it is this differential equation, which is the starting point in the finite difference method. The approximation involved in the finite element method is physical in nature, as the actual continuum is replaced by finite elements. The element formulation is mathematically exact. The finite difference technique involves the exact representation of the continuum, in terms of the differential equation and on this actual physical problem, the mathematical model is approximated. [13.1-13.6]

13.2 CENTRAL DIFFERENCE METHOD

A variable w, which is a function of x is shown in Fig. 13.1. The function is divided into equally spaced intervals of h. Let i be any station. Stations to its right are indicated as $i+1$, $i+2$, ... and the corresponding values of the function are $w_{i+1}, w_{i+2}, \cdots$. Stations to the left of i are designated as $i-1$, $i-2$.... and the corresponding values of the function are $w_{i-1}, w_{i-2}, \cdots,$.

The first derivative is given by

$$\left(\frac{dw}{dx}\right)_i = \frac{1}{2h}(w_{i+1} - w_{i-1}) \tag{13.1}$$

Equation (13.1) suggests that the slope of the curve between stations $i + 1$ and i – 1 has been considered to be constant.

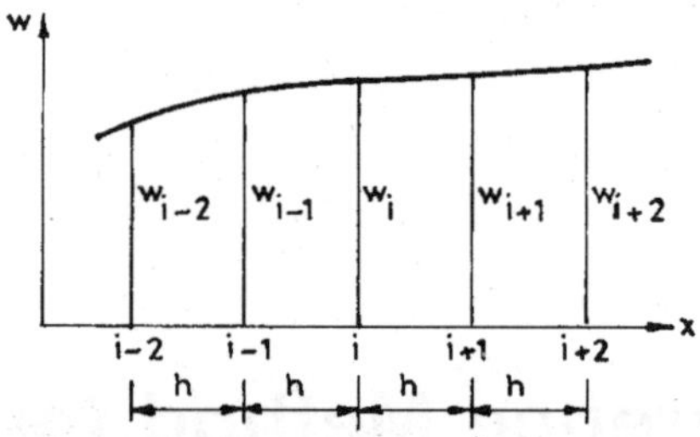

Fig. 13.1 Representation of a curve

The second derivative can be approximated by

$$\left(\frac{d^2w}{dx^2}\right)_i = \frac{1}{h}\left[\left(\frac{dw}{dx}\right)_{i+1/2} - \left(\frac{dw}{dx}\right)_{i-1/2}\right] \tag{13.2}$$

It can be shown that

$$\left(\frac{dw}{dx}\right)_{i+1/2} = \frac{1}{h}(w_{i+1} - w_i)$$

and

$$\left(\frac{dw}{dx}\right)_{i-1/2} = \frac{1}{h}(w_i - w_{i-1}) \tag{13.3}$$

Combining Eqs. (13.2) and (13.3), we get

$$\left(\frac{d^2w}{dx^2}\right)_i = \frac{1}{h^2}(w_{i+1} - 2w_i + w_{i-1}) \tag{13.4}$$

Likewise, the third and fourth derivatives can be shown to be

$$\left(\frac{d^3w}{dx^3}\right)_i = \frac{1}{2h^3}(w_{i+2} - 2w_{i+1} + 2w_{i-1} - w_{i-2}) \tag{13.5}$$

and $$\left(\frac{d^4w}{dx^4}\right)_i = \frac{1}{h^4}(w_{i+2} - 4w_{i+1} + 6w_i - 4w_{i-1} + w_{i-2}) \tag{13.6}$$

The central finite difference schemes have been adopted in the above derivations. It may be noted that the nodes are located symmetrically about the point, at which the derivatives are of interest to us. Though there are two other schemes, viz. the backward differences and the forward differences, it is the central difference schemes that are adopted for the solution of vibration problems, vis-à-vis the structural mechanics problems. The above four derivatives are shown schematically in Fig. 13.2.

13.3 FREE VIBRATION OF BEAMS

The differential equation for the free vibration of a beam is given by

$$\frac{d^2}{dx^2}\left(EI\,\frac{d^2 w}{dx^2} \right) = p^2 mw \tag{13.7}$$

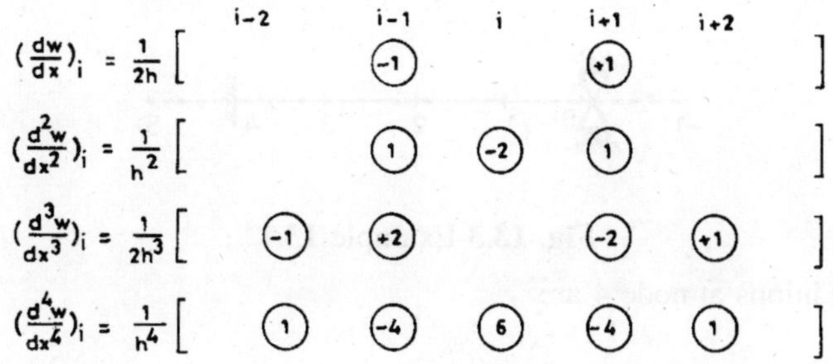

Fig. 13.2 Schematic representation of derivatives

where m is the mass per unit length.

For a uniform beam, Eq. (13.7) becomes

$$EI\,\frac{d^4 w}{dx^4} = p^2 mw \tag{13.8}$$

When expressed in finite difference form, Eq. (13.8) becomes

$$\frac{EI}{h^4}(w_{i+2} - 4w_{i+1} + 6w_i - 4w_{i-1} + w_{i-2}) = p^2 mw_i \tag{13.9}$$

Introducing $\lambda = mp^2h^4 / EI$, Eq. (13.9) becomes

$$(w_{i+2} - 4w_{i+1} + 6w_i - 4w_{i-1} + w_{i-2}) = \lambda w_i \tag{13.10}$$

In order to study the free vibration of a beam, it is divided into a number of segments. The finite difference Eq. (13.10) is applied at each node and a set of equations is thus generated. The boundary conditions are then applied. The resulting equations are in the form of eigenvalue problem, which on solution yields the natural frequencies and the mode shapes. The method is demonstrated with the help of the following example.

Example 13.1

Determine the fundamental frequency of the beam of Fig. 13.3, by dividing it into four equal parts. Draw the first mode shape.

The beam has been divided into four equal parts and the node numbers have been indicated in Fig. 13.3. The deflections at nodes 1, 2 and 3 are w_1, w_2 and w_3. Two ends are denoted as 0 and 4. The imaginary node beyond 0 has been shown as − 1 and that beyond 4 as 5. The boundary conditions at node 0 are

$$w_0 = 0 \quad \text{and} \quad \left(EI \frac{d^2 w}{dx^2} \right)_0 = 0 \tag{a}$$

Writing in the finite difference form, the second of Eq. (a) results in

$$w_1 + w_{-1} = 0 \tag{b}$$

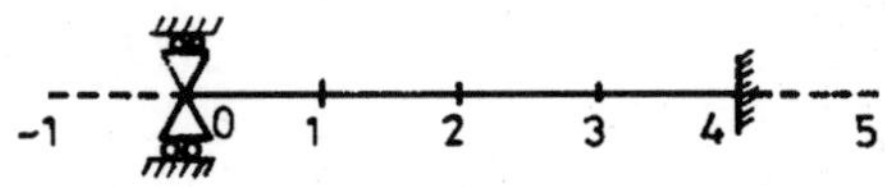

Fig. 13.3 Example 13.1

The boundary conditions at node 4 are

$$w_4 = 0 \quad \text{and} \quad \left(\frac{dw}{dx} \right)_4 = 0 \tag{c}$$

Writing second of Eq. (c) in finite difference form gives

$$w_3 + w_5 = 0 \tag{d}$$

Equations (b) and (d) relate the external nodes to the internal nodes. Eq. (13.10) is applied to the nodes 1, 2 and 3 respectively, and the appropriate boundary conditions are incorporated. The resulting equations are

$$\begin{bmatrix} 5 & -4 & 1 \\ -4 & 6 & -4 \\ 1 & -4 & 7 \end{bmatrix} \begin{Bmatrix} w_1 \\ w_2 \\ w_3 \end{Bmatrix} = \lambda \begin{bmatrix} 1 & 0 & 0 \\ 0 & 1 & 0 \\ 0 & 0 & 1 \end{bmatrix} \begin{Bmatrix} w_1 \\ w_2 \\ w_3 \end{Bmatrix} \tag{e}$$

Equation (e) is an eigenvalue problem. For the non-trivial solution of Eq. (e)

$$\begin{vmatrix} 5-\lambda & -4 & 1 \\ -4 & 6-\lambda & -4 \\ 1 & -4 & 7-\lambda \end{vmatrix} = 0 \tag{f}$$

The determinant when expanded gives

$$(5-\lambda)\,[(6-\lambda)\,(7-\lambda) - 16] + 4[-4(7-\lambda) + 4] + [16 - 6 + \lambda] = 0 \tag{g}$$

Equation (g) is a cubic equation, whose lowest root is

$$\lambda_1 = 0.72$$

Therefore

$$\frac{p^2 h^4 m}{EI} = 0.72$$

or

$$p_1 = 13.6\sqrt{\frac{EI}{mL^4}} \tag{h}$$

It is interesting to note that the exact value of

$$p_1 = 15.9\sqrt{\frac{EI}{mL^4}}$$

Assuming $w_1 = 1$ and substituting the value of $\lambda = 0.72$ in the second and third of Eq. (e), we get

$$\left.\begin{aligned} 5.26w_2 - 4w_3 &= 4 \\ -4w_2 + 6.28w_3 &= -1 \end{aligned}\right\} \tag{i}$$

On solution, $w_2 = 1.231$ and $w_3 = 0.625$.

The first mode shape has been shown in Fig. 13.4.

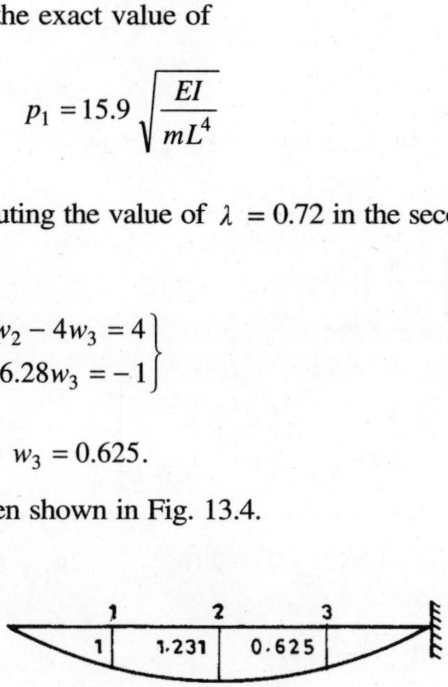

Fig. 13.4 1st mode shape

13.4 FREE VIBRATION OF RECTANGULAR PLATES

The free vibration equation of the plate is given by

$$\frac{\partial^4 w}{\partial x^4} + 2\frac{\partial^4 w}{\partial x^2\,\partial y^2} + \frac{\partial^4 w}{\partial y^4} = \frac{mp^2}{D}w \tag{13.11}$$

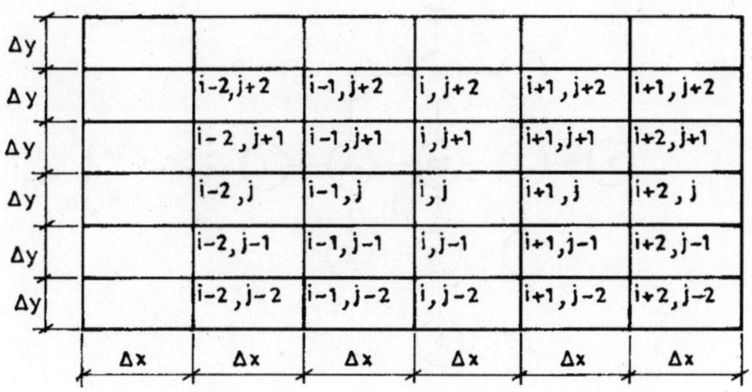

Fig. 13.5 Finite difference representation of a plate node

The plate is divided into equally spaced square mesh, having $\Delta x = \Delta y = h$ (Fig. 13.5). The finite difference expressions are written with respect to the node (i, j). 'i's are numbered in the x-direction and 'j's are in y-direction. Therefore,

$$\left(\frac{\partial^4 w}{\partial x^4}\right)_{i,j} = \frac{1}{h^4}[w_{i+2,j} - 4w_{i+1,j} + 6w_{i,j} - 4w_{i-1,j} + w_{i-2,j}] \qquad (13.12)$$

$$\left(\frac{\partial^4 w}{\partial y^4}\right)_{i,j} = \frac{1}{h^4}[w_{i,j+2} - 4w_{i,j+1} + 6w_{i,j} - 4w_{i,j-1} + w_{i,j-2}] \qquad (13.13)$$

$$\left(\frac{\partial^4 w}{\partial x^2\, \partial y^2}\right)_{i,j} = \left[\frac{\partial^2}{\partial y^2}\left(\frac{\partial^2 w}{\partial x^2}\right)\right]_{i,j}$$

$$= \frac{1}{h^2}\left[\left(\frac{\partial^2 w}{\partial x^2}\right)_{i,j+1} - 2\left(\frac{\partial^2 w}{\partial x^2}\right)_{i,j} + \left(\frac{\partial^2 w}{\partial x^2}\right)_{i,j-1}\right]$$

$$= \frac{1}{h^4}[(w_{i+1,j+1} - 2w_{i,j+1} + w_{i-1,j+1}) - 2(w_{i+1,j} - 2w_{i,j} + w_{i-1,j})$$

$$+ (w_{i+1,j-1} - 2w_{i,j-1} + w_{i-1,j-1})]$$

or $$\left(\frac{\partial^4 w}{\partial x^2\, \partial y^2}\right)_{i,j} = \frac{1}{h^4}[w_{i+1,j+1} - 2w_{i,j+1} + w_{i-1,j+1}$$

$$- 2w_{i+1,j} + 4w_{i,j} - 2w_{i-1,j} + w_{i+1,j-1} - 2w_{i,j-1} + w_{i-1,j-1}] \qquad (13.14)$$

By substituting the finite difference expressions of the different terms of Eq.(13.11), the final equation for a node (i, j) is shown in Fig. 13.6 and is denoted as Eq. (13.15).

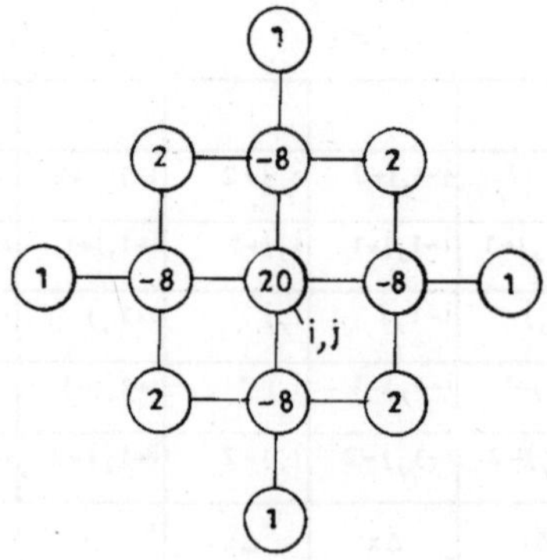

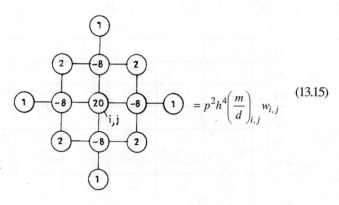

$$= p^2 h^4 \left(\frac{m}{d}\right)_{i,j} w_{i,j} \qquad (13.15)$$

Fig. 13.6 Schematic diagram of finite difference equation

13.5 SEMI-ANALYTIC FINITE DIFFERENCE METHOD FOR FREE VIBRATION ANALYSIS OF RECTANGULAR PLATES

A semi-analytic method has been developed, which has been successfully applied to the analysis of isotropic and orthotropic plates having various boundary conditions [13.7-13.12]. In this method, a displacement function satisfying the boundary conditions along two opposite edges in one direction is assumed. The displacement function is then substituted into the differential equation of the free vibration of the plate. By certain transformations, the resulting equation is reduced to an ordinary differential equation. This equation is expressed in finite difference form. The solution of the eigenvalue problem yields the natural frequencies of the free vibration of the plate. The method is presented in the following for the free vibration of rectangular plates.

The differential equation of an orthotropic rectangular plate in free vibration is given by

$$D_x \frac{\partial^4 w}{\partial x^4} + 2H \frac{\partial^4 w}{\partial x^2 \partial y^2} + D_y \frac{\partial^4 w}{\partial y^4} + \overline{m} \frac{\partial^2 w}{\partial t^2} = 0 \qquad (13.16)$$

where D_x, H and D_y are dependent on the geometric and material properties of the plate and $\overline{m}$ is the mass per unit area. w, the transverse displacement at any point of the plate is a function of x, y and time t. For harmonic vibration

$$w(x, y, t) = W(x, y) \sin pt \qquad (13.17)$$

where $W(x, y)$ is the shape function describing the modes of vibration, and p is the natural frequency of the plate.
Substitution of w from Eq. (13.17) into Eq. (13.16), yields

$$D_x \frac{\partial^4 W}{\partial x^4} + 2H \frac{\partial^4 W}{\partial x^2 \partial y^2} + D_y \frac{\partial^4 W}{\partial y^4} = \overline{m} p^2 W \qquad (13.18)$$

The transverse deflection W can be assumed to be of the form

$$W(x, y) = \sum_{m=1}^{r} W_m(x) Y_m(y) \qquad (13.19)$$

where Y_m is the characteristic function or the basic function along y-axis, satisfying the boundary conditions along y-axis for harmonic number m (Fig. 13.7) and r is the number of terms considered. Except for the edges which are free, the basic function Y_m is same as those presented in Arts. 8.6 and 8.7. W_m is the displacement function for the harmonic number m.

Substituting W from Eq. (13.19) into Eq. (13.18), we get

$$\sum_{m=1}^{r} \left(D_x Y_m \frac{d^4 W_m}{dx^4} + 2H Y_m'' \frac{d^2 W_m}{dx^2} + D_y Y_m^{iv} W_m \right) = p^2 \sum_{m=1}^{r} \overline{m} W_m Y_m \qquad (13.20)$$

Here $\qquad Y_m'' = d^2 Y_m / dy^2 \quad$ and $\quad Y_m^{iv} = d^4 Y_m / dy^4$

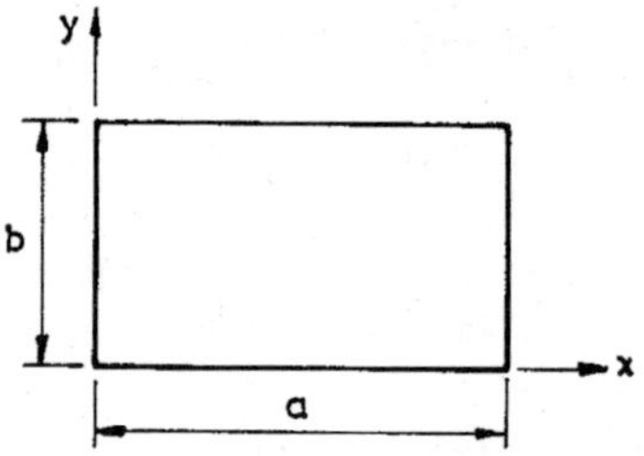

Fig. 13.7 A plate

Multiplying both sides of Eq.(13.20) by Y_n and integrating with respect to y from 0 to b, we get

$$\sum_{m=1}^{r}\left(D_x a_{mn}\frac{d^4 W_m}{dx^4} + 2H c_{mn}\frac{d^2 W_m}{dx^2} + D_y e_{mn} W_m\right) = p^2 \sum_{m=1}^{r} \overline{m}\, a_{mn} W_m$$

$$a_{mn} = \int_0^b Y_m Y_n\, dy, \quad c_{mn} = \int_0^b Y_m'' Y_n\, dy, \quad e_{mn} = \int_0^b Y_m^{iv} Y_n\, dy \qquad (13.21)$$

where

$$a_{mn} = \int_0^b Y_m Y_n\, dy, \quad c_{mn} = \int_0^b Y_m'' Y_n\, dy, \quad e_{mn} = \int_0^b Y_m^{iv} Y_n\, dy \qquad (13.22)$$

We know from the properties of the basic function given by Chapter 8, that the first and the third integrals of Eq. (13.22) are zero when in $m \neq n$. Taking advantage of this relationship, Eq. (13.21) can be simplified as

$$D_x a_{nn}\frac{d^4 W_n}{dx^4} + 2H \sum_{m=1}^{r} c_{mn}\frac{d^2 W_m}{dx^2} + D_y e_{nn} W_n = p^2 \overline{m}\, a_{nn} W_n \qquad (13.23)$$

In the x-direction, the plate is divided into a discrete number of nodal lines. Equation (13.23) when expressed in finite difference form at node i, becomes

$$D_x a_{nn} W_{n_{i-2}} - 4 a_{nn} D_x W_{n_{i-1}} + (6 D_x a_{nn} + D_y e_{nn} h^4) W_{n_i}$$

$$- 4 a_{nn} D_x W_{n_{i+1}} + D_x a_{nn} W_{n_{i+2}} + 2Hh^2 \sum_{m=1}^{r} c_{mn} W_{m_{i-1}}$$

$$-4 Hh^2 \sum_{m=1}^{r} c_{mn} W_{m_i} + 2 Hh^2 \sum_{m=1}^{r} c_{mn} W_{m_{i+1}} = p^2 \overline{m}\, a_{nn} W_{n_i} h^4 \quad (13.24)$$

where h is the width of each division. By writing the equations at the s discrete nodal lines, by which the plate is divided and incorporating the appropriate boundary conditions in the x-direction, the following simultaneous equations in matrix form are obtained

$$\begin{bmatrix} [K]_{11} & [K]_{12} & \cdots & [K]_{1n} \\ [K]_{21} & [K]_{22} & \cdots & [K]_{2n} \\ \cdots & \cdots & \cdots & \cdots \\ [K]_{n1} & [K]_{n2} & \cdots & [K]_{nn} \\ \cdots & \cdots & \cdots & \cdots \\ [K]_{r1} & [K]_{r2} & & [K]_{rn} \end{bmatrix} \begin{Bmatrix} \{W\}_1 \\ \{W\}_2 \\ \vdots \\ \{W\}_n \\ \vdots \\ \{W\}_r \end{Bmatrix}$$

$$= p^2 \begin{bmatrix} [M]_1 & & 0 & & \\ & [M]_2 & & & \\ & & \ddots & & \\ 0 & & [M]_n & & \\ & & & \ddots & \\ & & & & [M]_r \end{bmatrix} \begin{Bmatrix} \{W\}_1 \\ \{W\}_2 \\ \vdots \\ \{W\}_n \\ \vdots \\ \{W\}_r \end{Bmatrix} \qquad (13.25)$$

Each submatrix will have a size s by s. In compact form Eq. (13.25) becomes

$$[K]\{W\} = p^2 [M]\{W\} \qquad (13.26)$$

or

$$\lambda [K]\{W\} = [M]\{W\} \qquad (13.27)$$

Equation (13.26) is a typical eigenvalue problem. It can be solved by the methods described in Chapter 6.

Example 13.2

Determine the natural frequencies for the plate of Fig. 13.8.

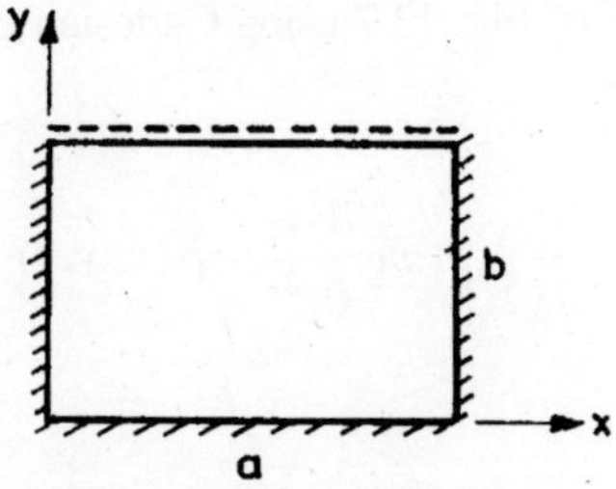

Fig. 13.8 Example 13.2

The plate has three edges clamped and one edge simply supported. The basic function for one edge clamped and the other edge simply supported is assumed. The basic function for these end conditions given by Eq. (8.75) is reproduced

$$Y_m = \sin\frac{\mu_m y}{b} - \alpha_m \sinh\frac{\mu_m y}{b}$$

where

$$\mu_m = \frac{4m+1}{4}\pi$$

$$\alpha_m = \frac{\sin \mu_m}{\sinh \mu_m}$$

A suitable computer program has been made and results for plates having three aspect ratios are presented in Table 13. 1.

Table 13.1 Frequency parameter $\lambda = pa^2 \sqrt{\frac{\rho t}{D}}$

a/b					
0.4 20 divisions $r = 2$		1.0 10 divisions $r = 3$		2.5 10 divisions $r = 5$	
SFDM	Ref.*	SFDM	Ref.*	SFDM	Ref.*
23.333	23.648	31.532	31.829	105.40	107.07
26.970	27.817	61.594	63.347	131.63	139.66
34.427	35.446	71.102	71.084	193.36	194.41
44.498	46.702	100.11	100.83	264.57	270.48
58.109	61.554	109.70	116.40	310.34	322.55
63.146	63.100	130.56	130.37	334.06	353.43

SFDM = Semi-analytic finite difference method.
Divisions refer to that along the x-direction
* Ref. 13.12

13.6 SEMI-ANALYTIC FINITE DIFFERENCE METHOD FOR FORCED VIBRATION ANALYSIS OF PLATES

The method presented in the previous section can be easily extended to the forced vibration analysis of plates.

For rectangular isotropic plate of Fig. 13.7 using Cartesian coordinate, the differential equation of forced vibration is given by

$$D\left(\frac{\partial^4 w}{\partial x^4} + 2\frac{\partial^4 w}{\partial x^2\, \partial y^2} + \frac{\partial^2 w}{\partial y^4}\right) + \overline{m}\,\frac{\partial^2 w}{\partial t^2} = p(x, y, t) \qquad (13.28)$$

where

$D = \dfrac{Et^3}{12\,(1 - v^2)}$ = flexural rigidity of the plate,
t = thickness of the plate,
ν = Poisson's ratio
$\overline{m}$ = mass per unit area
w = transverse displacement at any point of the plate, and
p = forcing function.

Let us assume the forcing function to be harmonic, such that

$$p = p_0 \sin \omega t \qquad (13.29)$$

For steady-state vibration

$$w(x, y, t) = W(x, y) \sin \omega t \tag{13.30}$$

where $W(x, y)$ is the shape function describing the modes of vibration.
Substituting W from Eq. (13.30) into Eq. (13.28), we get

$$D\left(\frac{\partial^4 W}{\partial x^4} + 2\frac{\partial^4 W}{\partial x^2\,\partial y^2} + \frac{\partial^4 W}{\partial y^4}\right) - \overline{m}\omega^2 W = p_0 \tag{13.31}$$

The transverse deflection W has been assumed as

$$W = \sum_{m=1}^{r} W_m Y_m \tag{13.32}$$

where, as explained in the previous section W_m is the displacement coefficient for harmonic number m and Y_m is the basic function.

The amplitude p_0 of the exciting force is resolved into basic function series in the y-direction

$$p_0 = \sum_{m=1}^{r} p_m Y_m \tag{13.33}$$

For orthogonality properties of the basic functions, it can be shown that

$$p_m = \frac{\int_0^b p_0 Y_m\, dy}{\int_0^b Y_m^2\, dy} \tag{13.34}$$

Combining Eqs. (13.31) to (13.34), we get

$$\sum_{m=1}^{r}\left[D\left(Y_m \frac{d^4 W_m}{dx^4} + 2Y_m'' \frac{d^2 W_m}{dx^2} + Y_m^{iv} W_m\right) - \overline{m}\omega^2 Y_m W_m\right] = \sum_{m=1}^{r} p_m Y_m \tag{13.35}$$

Multiplying both sides of Eq. (13.35) by Y_n, then integrating with respect to y from 0 to b and finally using the orthogonality relationship we get

$$\left(\int_0^b Y_m Y_n\, dy = 0 \quad \text{and} \quad \int_0^b Y_m^{iv} Y_n\, dy = 0 \quad \text{for} \quad m \neq n\right)$$

$$Da_{nn}\frac{d^4 W_n}{dx^4} + 2D\sum_{m=1}^{r} c_{mn}\frac{d^2 W_m}{dx^2} + D\,e_{nn} W_n - \overline{m}\omega^2 a_{nn} W_n = p_n a_{nn} \tag{13.36}$$

$$\text{where} \quad a_{nn} = \int_0^b Y_n^2\, dy, \quad c_{mn} = \int_0^b Y_m''\, Y_n\, dy, \quad e_{nn} = \int_0^b Y_n^{iv}\, Y_n\, dy \tag{13.37}$$

In the x-direction, the plate is divided into a discrete number of nodal lines. Eq. (13.37), when expressed in the finite difference form at node i, becomes

$$D a_{nn}\left(W_{n_{i-2}} - 4W_{n_{i-1}} + 6W_{n_i} - 4W_{n_{i+1}} + W_{n_{i+2}}\right) + (6a_{nn} D + e_{nn} D h^4$$

$$- \bar{m}\omega^2 a_{nn} h^4) W_{n_i} + 2Hh^2 \sum_{m=1}^{r} c_{mn} W_{m_{i-1}} - 4Hh^2 \sum_{m=1}^{r} c_{nn} W_{m_i}$$

$$+ 2Hh^2 \sum_{m=1}^{r} c_{mn} W_{m_{i-1}} = p_n a_{nn} h^4 \tag{13.38}$$

where h is the width of each division.

By writing the equations at s discrete nodal lines, the final equation in compact form becomes

$$[K]\{W\} = \{P\} \tag{13.39}$$

The size of $[K]$ is $(r \times s) \times (r \times s)$. Final values of w can be obtained by substituting into Eq. (13.30) the values of W_m, for different harmonic numbers as obtained from the solution of Eq. (13.39).

The bending moments can similarly be expressed in the finite difference form, in terms of the basic function. M_x and M_y given by Eqs (8.172) and (8.173) at any point j along the y-axis on the ith nodal line are given by

$$(M_x)_{i,j} = \sum_{m=1}^{r} - D\left[1/h^2\left(W_{m_{i-1}} - 2W_{m_i} + W_{m_{i+1}}\right)(Y_m)_j + \nu\,(Y_m'')_{j W_{m_i}}\right] \tag{13.40}$$

$$(M_y)_{i,j} = \sum_{m=1}^{r} - D\left[\nu / h^2\left(W_{m_{i-1}} - 2W_{m_i} + W_{m_{i+1}}\right)(Y_m)_j + (Y_m'')_j\, W_{m_i}\right] \tag{13.41}$$

13.7 SPLINE FINITE STRIP METHOD OF ANALYSIS OF PLATE VIBRATION

For structures having regular boundaries, the vibration analysis by the finite element method may entail a substantial computer time. The finite strip method proposed by Cheung [13.13] is an elegant method to take care of this problem so as to obtain an economic solution. The method is semianalytic in nature similar to what has been presented earlier in this chapter on semianalytic finite difference method. The commonality of both these methods lies in the choice of the displacement function satisfying boundary conditions along two opposite edges in one direction thus reducing a two-dimensional plate problem to one dimension. In the semianalytic finite difference method, the other direction is discretized by the finite difference equations. But in the semianalytic finite method, it is the finite element method which is applied to the other direction. It has been extensively used for various types of regular shaped plates.

But the semianalytic finite strip method suffers from a number of drawbacks. It fails for plates having mixed boundary conditions, internal cutouts, for continuous span, having discrete supports at strip ends. As the method uses characteristic beam function in one direction which is continuously differentiable, it imposes limitations on the plate problems involving abrupt changes of properties, concentrated and plate loads, internal supports etc. To overcome the difficulties mentioned above and to retain the advantage of the finite strip method, the spline finite strip method is developed [13.14 - 13.17].

13.7.1 The spline Functions

Spline is originally the name of a small flexible wooden strip employed by draughtsman as a tool for drawing a continuous smooth curve segment by segment. It has now been used as a mathematical tool for the solution of engineering problems.

A variety of spline functions are available. The spline function adopted here to represent displacement is the B_3 spline of equal section length.

$$y = \sum_{r=1}^{m+1} \alpha_i \phi_i \tag{13.42}$$

in which each local B_3 spline ϕ_i has non-zero values over four successive sections with section-knot $x = xi$ as the centre, is defined as

$$\phi_i = 1/6h^3 \begin{Bmatrix} 0, & x < x_{i-2} \\ (x-x_{i-2})^3, & x_{i-2} \le x \le x_{i-1} \\ h^3 + 3h^2(x-x_{i-1}) + 3h(x-x_{i-1})^2 - 3(x-x_{i-1})^3, & x_{i-1} \le x \le x_i \\ h^3 + 3h^2(x_{i+1}-x) + 3h(x_{i+1}-x)^2 - 3(x_{i+1}-x)^3, & x_i \le x \le x_{i+1} \\ (x_{i+2}-x)^3, & x_{i+1} \le x \le x_{i+2} \\ 0, & x_{i+2} < x \end{Bmatrix}$$

(13.43

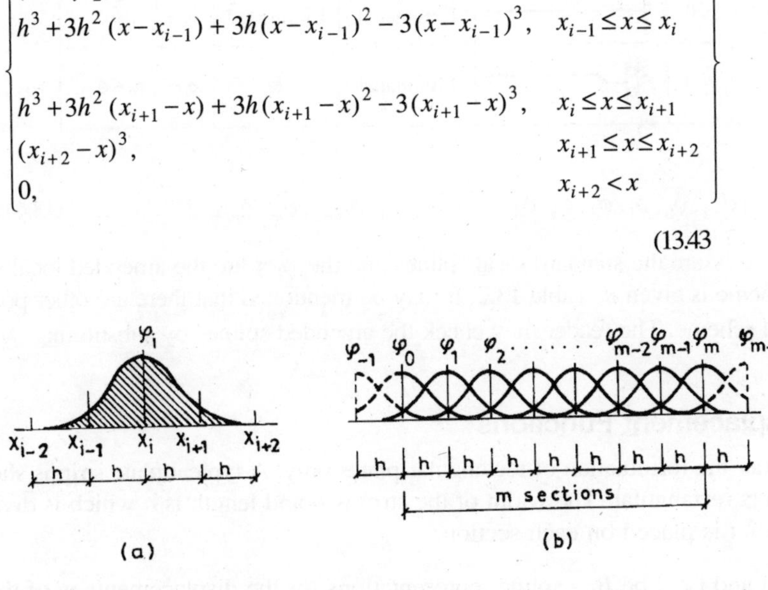

Fig. 13.9 (a) Typical B_3 - spline; (b) Basis of B_3 - spline expression

The use of B_3 - splines offers certain distinct advantages when compared with the conventional finite element method and semi-analytical finite strip method.

(1) It is computationlly more efficient than the finite element method.

When using B_3 spline displacement functions, continuity is ensured up to second order (C_2 - continuity). However, to achieve the same continuity conditions for the conventional finite element, it is necessary to have three times as many unknowns at the nodes.

(2) It is more flexible than the semi-analytical finite strip method in the boundary conditions treatment.

Only several local spline around the boundary point need to be amended to fit any specified boundary conditions.

The B_3 - spline represents and can be easily managed to adapt to various prescribed boundary condition. Due to the localization of the B_3 - splines, only three boundary local splines have to be amended, i.e.,

Table 13.2

Amended Scheme for Boundary Local Spline

Boundary condition			Amended local spline		
			$\bar{\phi}_{-1}$	$\bar{\phi}_0$	$\bar{\phi}_1$
Free end	$y(x_0)\neq 0$ $y'(x_0)\neq 0$		ϕ_{-1}	$\phi_0 - 4\phi_{-1}$	$\phi_1 - \frac{1}{2}\phi_0 + \phi_{-1}$
Simply supported end	$y(x_0)=0$ $y'(x_0)\neq 0$		Eliminated	$\phi_0 - 4\phi_{-1}$	$\phi_1 - \frac{1}{2}\phi_0 + \phi_{-1}$
Clamped end	$y(x_0)=0$	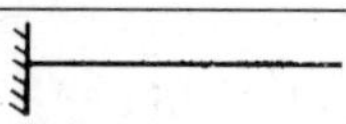	Eliminated	Eliminated	$\phi_1 - \frac{1}{2}\phi_0 + \phi_{-1}$
Sliding clamped end	$y(x_0)\neq 0$	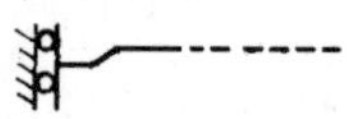	Eliminated	ϕ_0	$\phi_1 - \frac{1}{2}\phi_0 + \phi_{-1}$

$$[\Phi] = [\bar{\phi}_{-1}, \bar{\phi}_0, \bar{\phi}_1, \phi_2, \phi_3, \phi_4, \cdots\cdots, \phi_{m-3}, \bar{\phi}_{m-1}, \bar{\phi}_m, \bar{\phi}_{m+1}] \tag{13.44}$$

in which the ϕ_i's are the standard local splines and the $\bar{\phi}_i$'s are the amended local splines. The amended scheme is given in Table 13.2. It may be mentioned that there are other possibilities of the amended scheme. The reader may check the amended splines by substituting ϕ_i's from Eq. (13.43).

13.7.2 Displacement Functions

Let us consider the formulation of rectangular plates only. A typical plate strip is shown in Fig. 13.10 which is rectangular. The width of the strip is b and length is a which is divided into m sections, a knot is placed on each section.

Let $[\phi_1]$ and $[\phi_3]$ be B_3 - spline representations for the displacements w of the nodal lines i and j respectively (m-sections) and $[\phi_2]$ and $[\phi_4]$ be B_3-spline representations for the rotations θ of nodal lines i and j, respectively (m-sections), $\{w_i\}$ and $\{w_j\}$ be the displacement parameter vectors

$[w_{i_{-1}}, w_{i_0}, w_{i_1}, w_{i_2}, \cdots\cdots, w_{i_{m-1}}, w_{i_m}, w_{i_{m+1}}]^T$ & $[w_{j_{-1}}, w_{j_0}, w_{j_1}, w_{j_2}, \cdots\cdots,$

$w_{j_{m-1}}, w_{j_m}, w_{j_{m+1}}]^T$

corresponding to ϕ_1 and ϕ_3, respectively, and $\{\theta_i\}$ and $\{\theta_j\}$ be the rotation parameter vectors $[\theta_{i_{-1}}, \theta_{i_0}, \theta_{i_1}, \theta_{i_2}, ^T \cdots, \theta_{i_{m-1}}, \theta_{i_m}, \theta_{i_{m+1}}]$ and $[\theta_{j_{-1}}, \theta_{j_0}, \theta_{j_1}, \theta_{j_2}, \cdots\cdots,$

$\theta_{j_{m-1}}, \theta_{j_m}, \theta_{j_{m+1}}]^T$ corresponding to ϕ_2 and ϕ_4 respectively. Then, with a cubic polynomial interpolation in the x-direction, the displacement function for the strip will be

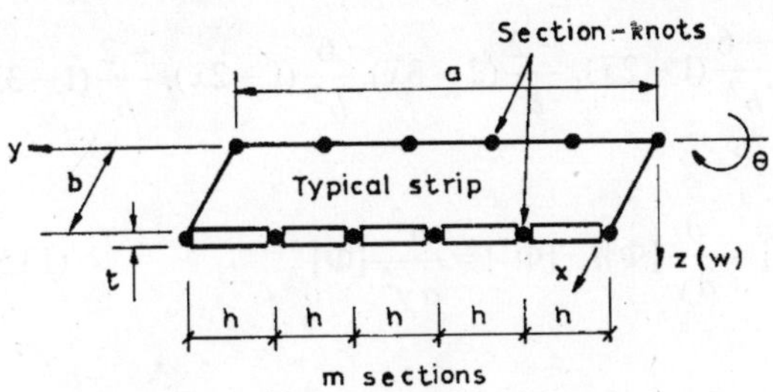

Fig 13.10 Typical strip

$$\{f\} = w = [N_1, N_2, N_3, N_4]\begin{bmatrix} [\phi_1] & & & 0 \\ & [\phi_2] & & \\ & & [\phi_3] & \\ 0 & & & [\phi_4] \end{bmatrix}\begin{Bmatrix} \{w_i\} \\ \{\theta_i\} \\ \{w_j\} \\ \{\theta_j\} \end{Bmatrix} \tag{13.45}$$

where

$$N_1 = 1 - 3\bar{x}^2 + 2\bar{x}^3,\ N_2 = x(1 - 2\bar{x} + \bar{x}^2)\ N_3 = (3\bar{x}^2 - 2\bar{x}^3),\ N_4 = x(\bar{x}^2 - \bar{x}) \tag{13.46}$$

and $\bar{x} = x/b$. In concise matrix form,

$$\{f\} = [N][\Phi]\{X\}_e \tag{13.47}$$

13.7.3 Strain-Displacement Relationship

The strain-displacement relationship is

$$\{\varepsilon\} = \begin{Bmatrix} -\dfrac{\partial^2 w}{\partial x^2} \\ -\dfrac{\partial^2 w}{\partial y^2} \\ 2\dfrac{\partial^2 w}{\partial x\, \partial y} \end{Bmatrix} = \begin{bmatrix} -[N''] & 0 & 0 \\ 0 & -[N] & 0 \\ 0 & 0 & 2[N'] \end{bmatrix} \begin{bmatrix} [\Phi] \\ [\Phi''] \\ [\Phi'] \end{bmatrix} \{\delta\}_e \qquad (13.48)$$

$$[N'] = \frac{\partial}{\partial x}[N] = \left[\frac{1}{b}(-6\bar{x} + 6\bar{x}^2),\ (1 - 4\bar{x} + 3\bar{x}^2),\ \frac{1}{b}(6\bar{x} - 6\bar{x}^2),\ (3\bar{x}^2 - 2\bar{x})\right] \qquad (13.49)$$

$$[N''] = \frac{\partial^2}{\partial x^2}[N] = \left[\frac{-6}{b^2}(1 - 2\bar{x}),\ \frac{-2}{b}(2 - 3\bar{x}),\ \frac{6}{b^2}(1 - 2\bar{x}),\ \frac{-2}{b}(1 - 3\bar{x})\right] \qquad (13.50)$$

$$[\Phi'] = \frac{\partial}{\partial y}[\Phi], \quad [\Phi''] = \frac{\partial^2}{\partial y^2}[\Phi] \qquad (13.51, 52)$$

In matrix form

$$\{\varepsilon\} = [B]\{\delta\}_e \qquad (13.53)$$

13.7.4 Stiffness Matrix

The stiffness matrix [K] for an orthotropic plate strip may be worked out by hand and expressed in an explicit form as follows,

$$[K] = \int [B]^T [D][B]\, d\,(\text{area}) = \int [[\Phi]^T, [\Phi'']^T, [\Phi']^T]$$

$$\begin{bmatrix} -[N'']^T & 0 & 0 \\ 0 & -[N]^T & 0 \\ 0 & 0 & 2[N']^T \end{bmatrix} \begin{bmatrix} D_x & D_1 & 0 \\ D_1 & D_y & 0 \\ 0 & 0 & D_{xy} \end{bmatrix}$$

$$\begin{bmatrix} -[N'']^T & 0 & 0 \\ 0 & -[N]^T & 0 \\ 0 & 0 & 2[N']^T \end{bmatrix} \begin{bmatrix} [\Phi] \\ [\Phi''] \\ [\Phi'] \end{bmatrix} d\,(\text{area})$$

$$= \int_0^a [[\Phi]^T,\ [\Phi'']^T,\ [\Phi']^T] \begin{bmatrix} [C_1] & [C_3] & 0 \\ [C_2] & [C_4] & 0 \\ 0 & 0 & [C_5] \end{bmatrix} \begin{bmatrix} [\Phi] \\ [\Phi''] \\ [\Phi'] \end{bmatrix} dy$$

$$= \int_0^a \{[\Phi]^T [C_1][\Phi] + [\Phi]^T [C_3][\Phi''] + [\Phi'']^T [C_2][\Phi]$$

$$+ [\Phi'']^T [C_4][\Phi''] + [\Phi']^T [C_5][\Phi']\}\, dy \tag{13.54}$$

$$[C_1] = \int_0^b [N'']^T D_x [N'']\, dx, \quad [C_2] = \int_0^b [N'']^T D_1 [N]\, dx,$$

$$[C_3] = \int_0^b [N]^T D_1 [N'']\, dx, \quad [C_4] = \int_0^b [N]^T D_y [N]\, dx,$$

$$[C_5] = \int_0^b [N']^T D_{xy} [N']\, dx \tag{13.55}$$

D_{xy}, D_y etc., being the usual orthotropic plate constants.

The integrations in Eq. (13.55) can be worked out explicitly since only simple polynomials are involved while the integrations for the coupling matrices $\int_0^a \phi_i^{(k)} \phi_j^{(l)}\, dy$ in Eq. (13.54) may be reduced to summation operations of the integrals $\int \phi_i^{(k)} \phi_j^{(l)}\, dy$ over a typical segment. Such segment integrals have also been tabulated explicitly.

Table 13.3
Values of cross-sectional integrations of two standard
B_3-splines, $\int \phi_{(1)} \phi_{(2)}\, dy$

$\phi_{(1)}$ S_{ij} $\phi_{(2)}$	Section L_1	Section L_2	Section L_3	Section L_4	
Section L_4	1	60	129	20	$\times \frac{1}{140} \times \frac{h}{36}$
Section L_3	60	933	1188	129	
Section L_2	129	1188	933	60	
Section L_1	20	129	60	1	

Table 13.4
Values of cross-segmental integrations of standard B_3-spline and its first derivative $\int \phi_{(1)} \phi'_{(2)} \, dy$

$\phi_{(1)}$ S_{ij} $\phi_{(2)}$	Section L_1	Section L_2	Section L_3	Section L_4	
Section L_4	– 1	– 38	– 71	– 10	$\times \frac{1}{20} \times \frac{1}{36}$
Section L_3	– 18	– 183	– 150	– 9	
Section L_2	9	150	183	18	
Section L_1	10	71	38	1	

Table 13.5
Values of cross-segmental integrations of the first derivatives of standard B_3-splines, $\int \phi'_{(1)} \phi'_{(2)} \, dy$

$\phi'_{(1)}$ S_{ij} $\phi'_{(2)}$	Section L_1	Section L_2	Section L_3	Section L_4	
Section L_4	– 3	– 36	21	18	$\times \frac{1}{10} \times \frac{1}{36h}$
Section L_3	– 36	– 87	102	21	
Section L_2	21	102	– 87	– 36	
Section L_1	18	21	– 36	– 3	

Table 13.6

Values of cross-segmental integrations of standard B_3-spline and its second $\int \phi'_{(1)} \phi''_{(2)} \, dy$

$\phi_{(1)}$ S_{ij} $\phi''_{(2)}$	Section L_1	Section L_2	Section L_3	Section L_4	
Section L_4	3	66	99	12	
Section L_3	6	− 33	− 132	− 21	$\times \frac{1}{10} \times \frac{1}{36h}$
Section L_2	− 21	− 132	− 33	6	
Section L_1	12	99	66	3	

Table 13.7

Values of cross - segmental integrations of the second derivatives of standard B_3-splines, $\int \phi''_{(1)} \phi''_{(2)} \, dy$

$\phi''_{(1)}$ S_{ij} $\phi''_{(2)}$	Section L_1	Section L_2	Section L_3	Section L_4	
Section L_4	6	0	− 18	12	
Section L_3	0	− 18	36	− 18	$\times \frac{1}{36h^3}$
Section L_2	− 18	36	− 18	0	
Section L_1	12	− 18	0	6	

They are given in Tables 13.3 - 13.7

The explicit value of stiffness matrix is given in Table 13.8 [13.15]

Table 13.8

Bending stiffness matrix of rectangular strip

$[K] = \dfrac{1}{420b^3}$

$5040 \; D_x \; I1_{11}$ $-504 \; b^2 \; D_1 \; I2_{11}$ $-504 \; b^2 \; D_1 \; I3_{11}$ $+156 \; b^4 \; D_y \; I4_{11}$ $+2016 \; b^2 \; D_{xy} \; I4_{11}$	$2520 \; b \; D_x \; I1_{12}$ $-462 \; b^3 \; D_1 \; I2_{12}$ $-42 \; b^3 \; D_1 \; I3_{12}$ $+22 \; b^5 \; D_y \; I4_{12}$ $+168 \; b^3 \; D_{xy} \; I5_{12}$	$-5040 \; D_x \; I1_{13}$ $+504 \; b^2 \; D_1 \; I2_{13}$ $+504 \; b^2 \; D_1 \; I3_{13}$ $+54 \; b^4 \; D_y \; I4_{13}$ $-2016 \; b^2 \; D_{xy} \; I4_{13}$	$2520 \; b \; D_x \; I1_{14}$ $-42 \; b^3 \; D_1 \; I2_{14}$ $-42 \; b^3 \; D_1 \; I3_{14}$ $-13 \; b^5 \; D_y \; I4_{14}$ $+168 \; b^3 \; D_{xy} \; I5_{14}$
	$1680 \; b^2 \; D_x \; I1_{22}$ $-56 \; b^4 \; D_1 \; I2_{22}$ $-56 \; b^4 \; D_1 \; I3_{22}$ $+4 \; b^6 \; D_y \; I4_{22}$ $+224 \; b^4 \; D_{xy} \; I5_{22}$	$-2520 \; b \; D_x \; I1_{23}$ $+42 \; b^3 \; D_1 \; I2_{23}$ $+42 \; b^3 \; D_1 \; I3_{23}$ $+13 \; b^5 \; D_y \; I4_{23}$ $168 \; b^3 \; D_{xy} \; I5_{23}$	$840 \; b^2 \; D_x \; I1_{24}$ $+14 \; b^4 \; D_1 \; I2_{24}$ $+14 \; b^4 \; D_1 \; I3_{24}$ $-3 \; b^6 \; D_y \; I4_{24}$ $-56 \; b^4 \; D_{xy} \; I5_{24}$
		$5040 \; D_x \; I1_{33}$ $-504 \; b^2 \; D_1 \; I2_{33}$ $-504 \; b^2 \; D_1 \; I3_{33}$ $+156 \; b^4 \; D_y \; I4_{33}$ $+2016 \; b^2 \; D_{xy} \; I4_{33}$	$-2520 \; b \; D_x \; I1_{34}$ $+462 \; b^3 \; D_1 \; I2_{34}$ $+42 \; b^3 \; D_1 \; I3_{34}$ $-22 \; b^5 \; D_y \; I4_{34}$ $-168 \; b^3 \; D_{xy} \; I5_{34}$
			$1680 \; b^2 \; D_x \; I1_{44}$ $-56 \; b^4 \; D_1 \; I2_{44}$ $-56 \; b^4 \; D_1 \; I3_{44}$ $+4 \; b^6 \; D_y \; I4_{44}$ $+224 \; b^4 \; D_{xy} \; I5_{44}$

$I_{1_j} = \int_0^a t\,\phi_i^T \phi_j \,dy, \quad I_{2_j} = \int_0^a t\,\phi_i''^T \phi_j \,dy, \quad I_{3_j} = \int_0^a t\,\phi_i^T \phi_j'' \,dy, \quad I_{4_j} = \int_0^a t\,\phi_i''^T \phi_j'' \,dy, \quad I_{5_j} = \int_0^a t\,\phi_i'^T \phi_j'^T \,dy$

13.7.5 **Consistent Mass Matrix of the Plate Strip**

A consistent mass matrix is formulated on the basis of lateral displacement w. The displacement is given by Eq. (13.45)

$$\{f\} = [N][\Phi]\{X\}_e = [N_{mw}]\{X\}_e \tag{13.56}$$

$[N_{mw}]$ is defined by Eq. (13.45) to (13.47).

The acceleration components of this point is given by

$$\{\ddot{f}\} = \{\ddot{w}\} = [N_{mw}]\{\ddot{X}\}_e \tag{13.57}$$

The components of the inertia force of small elements of volume dV at that point is given by

$$\{f_I\} = \rho dV\{\ddot{w}\} = \rho dV[N_{mw}]\{\ddot{X}\}_e \tag{13.58}$$

where ρ is the mass density of the plate material.

Applying virtual work principles, we get

$$\{X\}_e^T\{F_I\} = \int_v (dF)^T\{F_I\} \tag{13.59}$$

where $\{F_I\}$ is the nodal inertia force parameters.

Combining Eqs. (13.56), (13.58) and (13.59), yields

$$\{X\}_e^T\{F_I\} = \int_v \{dX_e\}^T[N_{mw}]^T \rho dV[N_{mw}]\{\ddot{X}\}_e \tag{13.60}$$

Eq. (13.60) can be written as

$$\{F_I\} = \rho\int_v [N_{mw}]^T[N_{mw}]\,dV\,\{\ddot{X}\}_e = [M]_e\{\ddot{X}\}_e \tag{13.61}$$

where

$$[M]_e = \rho\int [N_{mw}]^T[N_{mw}]\,dV \tag{13.62}$$

$[M]_e$ for a plate strip of uniform thickness t is given below when rotational displacements are included

$$[M]_e = \frac{\rho t^3}{12}\iint\left([N_{mw}]_{,x}^T[N_{mw}]_{,x} + [N_{mw}]_{,y}^T[N_{mw}]_{,y}\right)dx\,dy$$
$$+\ \rho t\iint [N_{mw}]^T[N_{mw}]\,dx\,dy \tag{13.63}$$

References

13.1 P.C. Wang, Numerical Methods in Structural Mechanics, John Wiley & Sons, 1980.

13.2 F. Schied, Numerical Analysis, Schaum Series, McGraw-Hill, 1968.

13.3 E.V. Krishnamoorthy and S.K. Sen, Computer Based Numerical Algorithms, 2nd Edition, Affiliated East-West Press, 1998.

13.4 R. Ali, Finite difference methods in vibration analysis, Shock and Vibration Digest, V.15, No. 1, 1983, pp 3-7.

13.5 F.B. Hilderbrand, Introduction to Numerical Analysis, McGraw-Hill, 1976.

13.6 M.G. Salvadori and M.L. Baron, Numerical Methods in Engineering, Prentice-Hall, New Jersey, 1952.

13.7 M. Mukhopadhyay, A semianalytic solution for free vibration of rectangular plates, Journal of Sound and Vibration, V.60, 1978, pp. 71-85.

13.8 M. Mukhopadhyay, Free vibration of rectangular plates with edges having different degrees of rotational restraint, Journal of Sound and Vibration, V.73, 1979, pp. 459-458.

13.9 M. Mukhopadhyay, Vibrational analysis of elastically restrained rectangular plates with concentrated mass, Journal of Sound and Vibration, V.113, 1987, pp. 547-558.

13.10 M. Mukhopadhyay, A semianalytic solution for free vibration of annular sector plates, Journal of Sound and Vibration, V.63, 1979, pp. 87-95.

13.11 M. Mukhopadhyay, Free vibration of annular sector plates with edges having different degrees of rotational restraints, Journal of Sound and Vibration, V.73, 1981, pp. 275-279.

13.12 A.W. Leissa, The free vibration of rectangular plates, Journal of Sound and Vibration, V.31, 1973, pp. 257-273.

13.13 Y.K. Cheung, Finite Strip Method in Structural Analysis, Pergamon Press, 1976.

13.14 Y.K. Cheung, S.C. Fan and C.Q. Wu, Spline finite strip in structural analysis, Proceedings of the International Conference on Finite Element, Shanghai, China, 1982, pp. 704-709.

13.15 S.C. Fan and Y.K. Cheung, Flexural vibrations of rectangular plates with complex support conditions, Journal of Sound and Vibration, V.93 (1), 1984, pp. 81-94.

13.16 W.Y. Li, Y.K. Cheung and L.G. Tham, Spline finite strip analysis of general plates, Journal of Engineering Mechanics Division, ASCE, V.112, 1986, pp. 43-54.

13.17 Abdul Hamid Sheikh, Linear and Nonlinear Analysis of Stiffened Plate Structures under Static and Dynamic Loading by the Spline Finite Strip Method, Ph. D. Thesis, Indian Institute of Technology, Kharagpur, 1995.

EXERCISE 13

13.1 Using finite difference method, determine the fundamental frequency of the uniform beam shown in the figure, by dividing the beam into four equal parts.

13.2 A simply supported square plate with mesh division is shown in the figure. Determine the fundamental frequency. Take advantage of the symmetry about both axes.

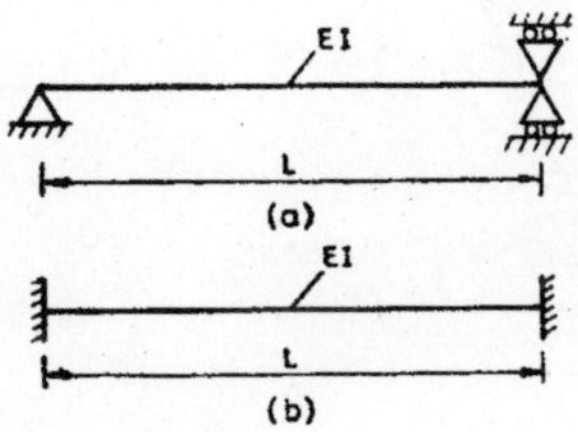

Prob. 13.1

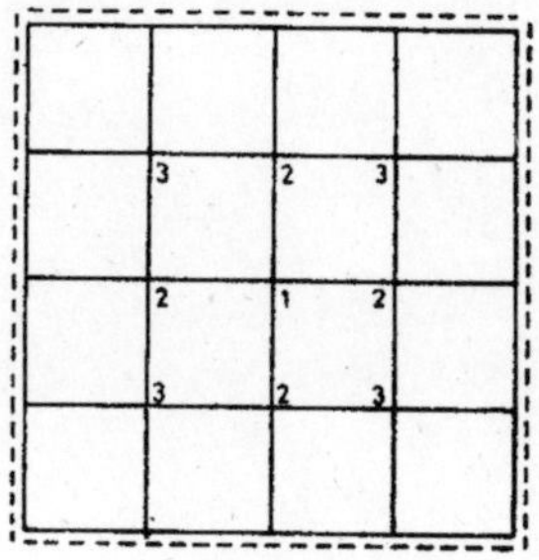

Prob. 13.2

13.3 Write a computer program for free vibration analysis of plates using semi-analytic finite difference method.

13.4 Write a computer program for steady-state vibration analysis of plates in harmonic excitation, using semi-analytical finite difference method.

13.5 Derive the equations related to the forced vibration analysis of plates by the semianalytic finite difference method, when the forcing function is not harmonic.

Chapter 14

Nonlinear Vibration

14.1 INTRODUCTION

In all the preceding chapters, we have discussed different linear systems. All the differential equations that have been formed and dealt with so far, whether a single equation in the case of SDF system or a set of coupled equations in the case of MDF system, are linear. But there are certain vibration phenomena, which cannot be predicted by the linear theory. The vibration of a string, the belt friction drive, the shimming of automobile wheels and the inclusion of nonlinear damping are some of the examples of nonlinear vibration. [14.1-14.5]

The principle of superposition does not hold good for nonlinear systems. If the excitation to a system is trebled, it does not mean that the response will also be trebled. Thus the Duhamel's integral cannot be applied to determine the response of these SDF systems. Another characteristic of nonlinear system is that, it can have more than one equilibrium position, whereas a linear system can have only one equilibrium position.

Nonlinear problems are classified into two categories: geometric nonlinearity and material nonlinearity. So far we have dealt with small strains, Certain assumptions in the derivation have been introduced to make the problem linear. But accurate representation of the displacement may involve geometric nonlinearity. For instance, membrane stresses in the neutral surface which are neglected in the beam analysis, may reduce displacements significantly, though the value of the displacement may remain quite small. These types of problems are also referred to as large deflection problems, and this will result in nonlinear equations. There is another class of problem, where the stress-strain relationship of the material of the structure is not linear. For a material like steel, the stress-strain relation is linear up to yield point, but beyond it, the curve deviates from linearity. So, for a structure made of steel, if an analysis is to be carried out beyond yield point, it will involve the nonlinear material behaviour. For certain other materials, the stress varies in a curvilinear manner with the strain.

Nonlinear vibration problems can be solved by a number of methods:

(1) The Method of Direct Integration.

(2) Graphical methods such as the Phase Plane Method, Pell's Method, Isoclinic Method, etc.

(3) Analytical Methods, such as the Method of Iteration, the Method of Perturbation, the Method of Fourier series, the Method of Harmonic Balance, The Variation of Parameter Method, etc.

(4) Numerical Methods such as Step-by-Step Integration, Wilson q Method, Runge-Kutta Method, Newmark's Method etc.

In the following, we deal with some of the above methods.

14.2 PERTURBATION METHOD

When the parameter associated with the nonlinear term of a differential equation is small, the perturbation method can be successfully applied. The proposed solution in this method is expressed in terms of a power series, which consists of a generating solution and added corrective terms.

Example 14.1

A taut spring with a centrally attached mass m is shown in Fig. 14.1. The spring has an initial tension T_0. The tension in the spring is assumed to be linear with its elongation. Determine the equation of motion of the spring. Also determine by the perturbation method the response of the system in free vibration.

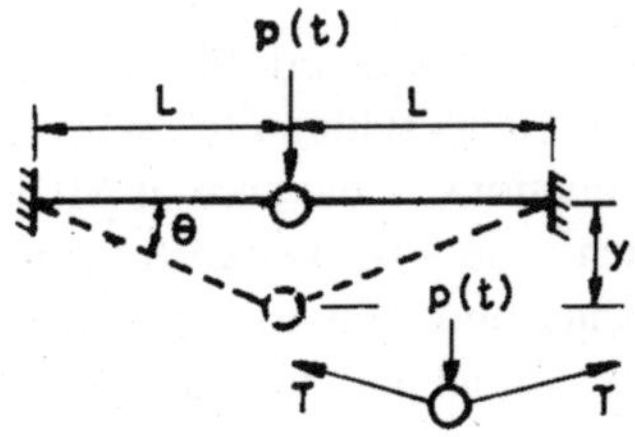

Fig. 14.1 Example 14.1

The tension in the string is

$$T = T_0 + \left(\frac{AE}{L}\right)\delta \tag{a}$$

where AE is the axial rigidity of the string and δ is its elongation.

From Fig 14.1

$$L + \delta = \sqrt{L^2 + y^2}$$

or

$$\delta = \sqrt{L^2 + y^2} - L \tag{b}$$

From the freebody diagram in Fig. 14.1, the equation of motion is found to be

$$\Sigma F_y = m\ddot{y}$$

or

$$p(t) - 2T\sin\theta = m\ddot{y} \tag{c}$$

where

$$\sin\theta = \frac{y}{\sqrt{L^2 + y^2}} \tag{d}$$

Combining Eqs. (a) to (d), we get

$$m\ddot{y} + 2\left[T_0 + \left(\frac{AE}{L}\right)\left\{\sqrt{L^2 + y^2} - L\right\}\right]\left[\frac{y}{\sqrt{L^2 + y^2}}\right] = p(t) \tag{e}$$

For small displacements $y << L$, Eq. (e) can be approximated as

$$m\ddot{y} + 2T_0\left(\frac{y}{L}\right) = p(t) \tag{f}$$

In Eq. (b), expanding the right hand side, we get

$$\delta = L\left[1 + \frac{1}{2}\left(\frac{y}{L}\right)^2\right] - L = \left(\frac{1}{2L}\right)y^2 \tag{g}$$

In Eq. (g), the expansion is preserved up to the second term on the right hand side. The second term is the square of the displacement and its value is considered to be significant

$$\sin\theta = \frac{y}{L} \tag{h}$$

Substituting Eqs. (g) and (h) into Eq. (c) we get

$$m\ddot{y} + \left(\frac{2T_0}{L}\right)y + \left(\frac{AE}{L^3}\right)y^3 = p(t) \tag{i}$$

Equation (i) is a nonlinear equation.

Equation (i) is written for free vibration as

$$\ddot{y} + p_0^2 y + hy^3 = 0 \tag{j}$$

where $p_0 = \sqrt{2T_0 / Lm}$ and $h = \dfrac{AE}{mL^3}$.

Let the solution be assumed as

$$y(t) = y_0(t) + hy_1(t) + h^2 y_2(t) + \cdots \tag{k}$$

where $y_0(t)$ is the generating solution, h is the perturbation parameter and $y_1(t)$, $y_2(t)$, etc., are added corrective terms.

The frequency is perturbed as follows:

$$p^2 = p_0^2 + h\alpha_1 + h^2\alpha_2 + \cdots$$

$$p_0^2 = p^2 - h\alpha_1 - h^2\alpha_2 - \cdots \tag{l}$$

Substituting Eqs. (k) and (l) into Eq. (j), we get

$$(\ddot{y}_0 + h\ddot{y}_1 + h^2\ddot{y}_2 + \cdots) + (p^2 - h\alpha_1 - h^2\alpha_2 - \cdots)\times$$

$$(y_0 + hy_1 + h^2 y_2 + \cdots) + h(y_0 + hy_1 + h^2 y_2 + \cdots)^3 = 0 \qquad \text{(m)}$$

The coefficients of various powers of h are equated to zero which results in the following system of equations.

$$\ddot{y}_0 + p^2 y_0 = 0 \qquad \text{(n)}$$

$$\begin{aligned} \ddot{y}_1 + p^2 y_1 &= \alpha_1 y_0 - y_0^3 \\ &\vdots \end{aligned} \qquad \text{(o)}$$

The generating solution of Eq. (n) is

$$y_0 = A \cos pt + B \sin pt \qquad \text{(p)}$$

where A and B are constants dependent on the initial conditions.

Let at $t = 0$, $\quad y_0 = 0$, $\quad$ and $\quad \dot{y}_0 = A$

Then

$$y_0 = A / p \sin pt \qquad \text{(q)}$$

We substitute the value of y_0 from Eq. (q) into Eq. (o), which gives

$$\ddot{y}_1 + p^2 y_1 = \alpha_1 \frac{A}{p} \sin pt - \frac{A^3}{p^3} \sin^3 pt$$

$$= \alpha_1 \frac{A}{p} \sin pt - \frac{A^3}{p^3} \left[\frac{3}{4} \sin pt - \frac{1}{4} \sin 3pt \right]$$

or

$$\ddot{y}_1 + p^2 y_1 = \left(\alpha_1 \frac{A}{p} - \frac{3}{4} \frac{A^3}{p^3} \right) \sin pt + \frac{1}{4} \frac{A^3}{p^3} \sin 3pt \qquad \text{(r)}$$

The total solution of Eq. (r) is

$$y_1 = C_1 \cos pt + C_2 \sin pt + \frac{pt}{2} \left(\alpha_1 \frac{A}{p} - \frac{3}{4} \frac{A^3}{p^3} \right) \cos pt$$

$$+ \frac{A^3}{32 p^5} \sin 3pt \qquad \text{(s)}$$

The new solution must satisfy the initial conditions of y_0, i.e. at $t = 0$, $y_1 = 0$ and $\dot{y}_1 = A$. Therefore

$$C_1 = 0 \quad \text{and} \quad C_2 = -\frac{3}{32} \frac{A^3}{p^5}$$

On substitution of the above C_1 and C_2 values in Eq. (s), we get

$$y = \left(\alpha_1 \frac{A}{p} - \frac{3}{4}\frac{A^3}{p^3}\right)\frac{pt}{2}\cos pt - (3\sin pt - \sin 3pt)\frac{A^3}{32p^5} = 0 \qquad \text{(t)}$$

The first term of Eq. (t) having $t \cos pt$ corresponds to resonance condition, which prohibits the initial stipulation that the motion is periodic. As such, we impose the condition that

$$\alpha_1 \frac{A}{p} - \frac{3}{4}\frac{A^3}{p^3} = 0 \qquad \text{(u)}$$

which gives

$$\alpha_1 = \frac{3}{4}\frac{A^2}{p^2} \qquad \text{(v)}$$

The solution of y_1 is

$$y_1 = -\frac{A^3}{32p^5}(3\sin pt - \sin 3pt) \qquad \text{(w)}$$

The first order solution of y is, therefore

$$y(t) = y_0(t) + hy_1(t)$$

or

$$y(t) = \frac{A}{p}\sin pt - \frac{A^3 h}{32p^5}(3\sin pt - \sin 3pt) \qquad \text{(x)}$$

$$p^2 = p_0^2 + \frac{3}{4}h\frac{A^2}{p^2} \qquad \text{(y)}$$

or

$$p_2 = \frac{3}{8}hA^2 + p_0\sqrt{1 + \left(\frac{3hA^2}{2p_0}\right)} \qquad \text{(z)}$$

The solution is found to be periodic and the fundamental frequency of p increases with the amplitude. The effect of the nonlinear spring is to distort the oscillation by introducing higher harmonic terms in the solution.

Higher order corrections may be similarly performed.

14.3 STEP-BY-STEP INTEGRATION

The numerical methods presented in Chapter 4 can be applied to the solution of nonlinear problems. The freebody diagram and the force-deformation curve for the spring is shown in Fig. 14.2. F_S and F_D represent the spring force and the damping force respectively. $F(t)$ is the exciting force to the system. At time instant t_i, the equation of motion can be written as

$$m\ddot{x}_i + c_i\dot{x}_i + k_i x_i = F_i \qquad (14.1)$$

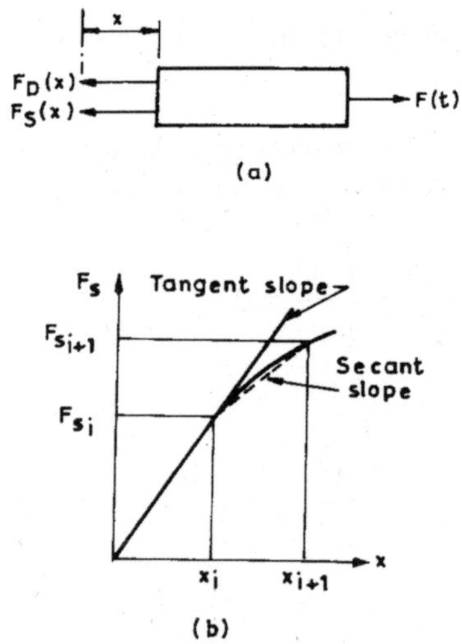

Fig. 14.2 Nonlinear System

At time t_{i+1}, the equation of motion is

$$m\ddot{x}_{i+1} + c_{i+1}\,\dot{x}_{i+1} + k_{i+1}\,x_{i+1} = F_{i+1} \tag{14.2}$$

Subtracting Eq. (14.2) from Eq. (14.1), we get

$$m\Delta\ddot{x}_i + c_i\Delta\dot{x}_i + k_i\Delta x_i = \Delta F_i \tag{14.3}$$

Step-by-step integration is to be carried out on the basis of Eq. (14.3).

Because both c_i and k_i may be nonlinear, the values of k_i and c_i are to be evaluated at each time step. k_i is the tangent slope at x_i as shown in Fig. 14.2(b). Similarly, c_i is the slope of the corresponding damping force curve.

An example is presented below, which applies linear acceleration method of Art. 4.2.2 for calculating the dynamic response of a nonlinear SDF system.

Example 14.2

A SDF system of Fig. 14.3 has an elastic, perfectly plastic force-deformation behaviour. The system is subjected to a triangular excitation. The system is initially at rest. Using the linear acceleration method, compute the response.

$$m = 0.3,\ k = 40 \text{ and } F_{sy} = 5. \text{ Take } \Delta t = 0.1$$

The equation of motion for this problem is

$$\ddot{x}_i = \frac{F_i}{m} - \frac{F_{si}}{m} \tag{a}$$

and $\quad x_1 = 0$ and $\dot{x}_1 = 0$ (given).

Assume $\ddot{x}_1 = \ddot{x}_2 \doteq 18$

Substituting these values of Eqs. (4.12) and (4.13), we get

$$\dot{x}_2 = 0 + \frac{1}{2} \times 18 \times 0.1 = 0.9$$

and $$x_2 = 0 + 0 + \frac{1}{6} \times (0.1)^2 \times [2 \times 18 + 18] = 0.09$$

Substituting the above values in Eq. (a), we get

$$\ddot{x}_2 = \frac{1.5}{0.3} - \frac{40 \times 0.09}{0.3} = -7$$

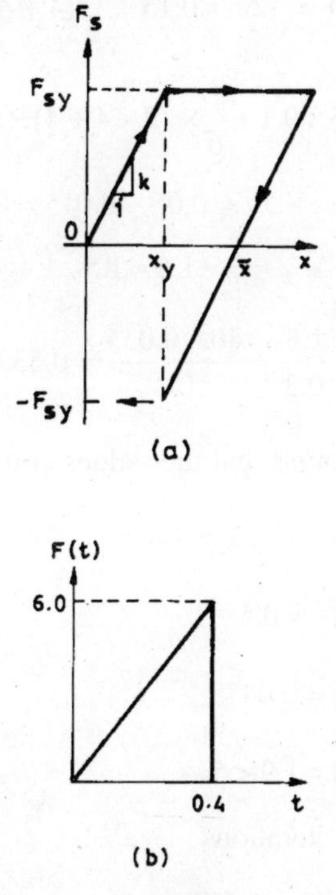

Fig. 14.3 Example 14.2

The assumed value $\ddot{x}_2$ of is not correct. Assume a new value $\ddot{x}_3 = 3$.

Equations (4.12) and (4.13) give

$$\dot{x}_2 = \frac{1}{2} \times 3 \times 0.1 = 0.15$$

and $$x_2 = \frac{1}{6} \times (0.1)^2 \times [2 \times 3 + 3] = 0.015$$

Equation (a) now gives

$$\ddot{x}_2 = \frac{1.5}{0.3} - \frac{40 \times 0.015}{0.3} = 3$$

Therefore, calculation may be started for the next time step at $t = 0.2$s.

Here, we introduce the incremental concept and use Eq. (14.3) as

$$\Delta \ddot{x}_i = \frac{\Delta F_i}{m} - k_i \Delta x_i \tag{b}$$

Let us assume $\Delta \ddot{x}_2 = 1$

$$\ddot{x}_3 = \ddot{x}_2 + \Delta \ddot{x}_2 = 3 + 1 = 4$$

Therefore $$\dot{x}_3 = \dot{x}_2 + \frac{1}{2}(\ddot{x}_2 + \ddot{x}_3)\,\Delta t = 0.15 + \frac{1}{2}(4 + 4) \times 0.1 = 0.55$$

$$x_3 = 0.015 + 0.15 \times 0.1 + \frac{1}{6} \times (2 \times 4 + 4) \times (0.1)^2 = 0.05$$

$$\Delta x_2 = x_3 - x_2 = 0.05 - 0.015 = 0.035$$

$$\Delta F_2 = 3 - 1.5 = 1.5$$

and $$\Delta \ddot{x}_2 = \frac{1.5}{0.3} - \frac{40 \times 0.035}{0.3} = 0.333$$

After a few iterations, it will be found that the values converge at $\Delta \ddot{x}_2 = 0.6$.

$$\ddot{x}_3 = \ddot{x}_2 + \Delta \ddot{x}_2 = 3.6$$

$$\dot{x}_3 = 0.5$$

$$x_3 = 0.0475$$

The spring force $F_3 = 0.0475 \times 40 = 1.9 < 5$.

For the next time step, after a few iterations

$$\ddot{x}_4 = 0.24, \quad \dot{x}_4 = 0.692 \quad \text{and} \quad x_4 = 0.110$$

The spring force $F_{s4} = 0.110 \times 40 = 4.4 < 5$.

Let us assume $\ddot{x}_5 = -1$, which means

$$\Delta \ddot{x}_4 = 0.24 - 1 = -0.76,$$

$$\dot{x}_5 = 0.692 + \frac{1}{2}(0.24 - 1)(0.1) = 0.654,$$

$$x_5 = 0.110 + 0.692 \times 0.1 + \frac{1}{6} \times (2 \times 0.24 - 1)\,(0.1)^2 = 0.178, \text{ and}$$

$$\Delta x_4 = 0.178 - 0.110 = 0.068$$

The spring force $= 40 \times 0.178 = 7.12 > 5$

The spring at yield point = 5, therefore

$$\Delta \ddot{x}_4 = 5 - 5 = 0$$

We assume $\ddot{x}_5 = 0.24$, so that $\Delta \ddot{x}_4 = 0$

$$\dot{x}_5 = 0.692 + \frac{1}{2}(0.24 + 0.24) \times 0.1 = 0.716 \text{ and}$$

$$x_5 = 0.110 + 0.692 \times 0.1 + \frac{1}{6}(2 \times 0.24 + 0.24)\,(0.1)^2 = 0.1804$$

As the maximum $F_s = 5$ corresponding to F_{sy}

$$\Delta \ddot{x}_4 = 5 - 5 = 0$$

So the assumed value of $\ddot{x}_5$ is correct.

The above steps can be repeated for other time instants. The pattern of the response curve expected is shown in Fig. 14.4.

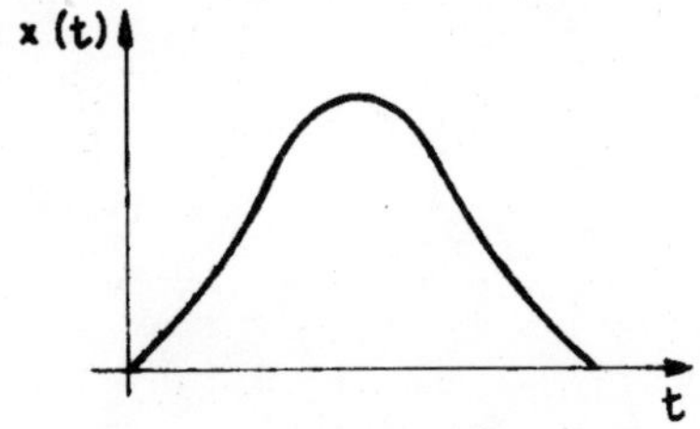

Fig. 14.4 Response curve

The method can be easily extended to MDF systems. Nonlinear vibration has been discussed very briefly in this chapter. The main purpose is nothing beyond mere introduction of the topic. A number of textbooks are available on the subject, which may be consulted for further details [14.1 - 14.5].

References

14.1 W.J. Cunnigham, Introduction to Nonlinear Analysis, McGraw-Hill, New York, 1958.

14.2 F. Dinca and C. Theodosin, Nonlinear and Random Vibration, Academic Press, 1973.

14.3 N. Minorsky, Nonlinear Oscillations, D. Van Nostrand, New Jersey, 1962.

14.4 J.J. Stoker, Nonlinear Vibration, Interscience Publishers Inc., New York, 1950.

14.5 N.V. Butenin, Elements of the Theory of Nonlinear Oscillations, Blaisdell Publishing Co., New York, 1965.

Chapter 15

Random Vibrations

15.1 INTRODUCTION

During the last three decades, a great deal of activity has taken place in studying the loads acting on the structure in a realistic manner and the necessary response thereof. The analytical methods during this period have made tremendous strides and a need has been felt to account for the load in its truer perspective. The dynamic loads which have been considered in all the previous chapters, have fixed or definite values of amplitude, frequency, period and phase. But for many cases, such as wave forces on offshore structures, earthquake effects on buildings, bridges and dams, air pressure on aeroplanes and vibration of ships in rough seas, there is an uncertainty involved with the exactness of the loading parameters. These uncertainties are related to the random time functions.

The input to the problem, that is the loading, is rather complex and extremely irregular in many cases. Exact mathematical representation of them is practically impossible. The only rational way to deal with such cases is to treat them as random process, so that the loading can be treated in a probabilistic sense. The input being random, the response of the structure will also be random. [15.1 - 15.7]

15.2 RANDOM PROCESS

Based on the behaviour of the parameters, a process may be categorised into the following two types: deterministic process and random process.

If an experiment is conducted a number of times under identical conditions and the records obtained are always alike, then the process is termed as deterministic.

If an experiment is conducted a number of times, maintaining the variables identified by the experimenter same in all cases, but the output or records continually differ from each other, then the process is said to be random. The degree of randomness in a process depends on, (1) the understanding of the variable parameters associated with the experiment and (2) the ability to control them.

The wave elevation at a given location is intended to be measured as a function of time. The wave elevation at certain wind speed, duration, etc., obtained at different times is shown in Fig. 15.1. The striking feature of these records is that, they are not identical. Thus, based on macroscopic parameters such as wind direction, speed, duration, etc., the wave elevation at a given location cannot exactly be predicted. The wave generation is therefore treated as a random process. Each of the wave elevation record is called a sample function. Total number of records together is referred to as an ensemble. To form a rational basis of the understanding of the random process, concepts of probability is used.

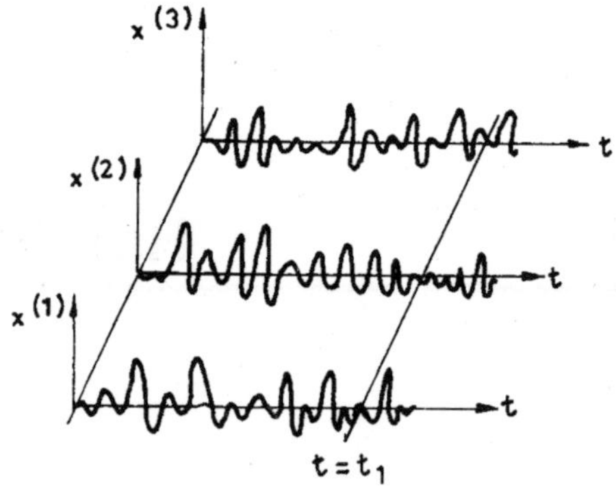

Fig. 15.1 Variation of wave elevation with time

15.3 PROBABILITY DISTRIBUTIONS

We consider infinitely many sample functions, each of which consists of a continuous function of *t*. At a particular value $t = t_i$ the value of *X*(*t*) from each of the three samples $x^{(1)}(t)$, $x^{(2)}(t)$ and $x^{(3)}(t)$ be $x^{(1)}(t_1)$, $x^{(2)}(t_1)$ and $x^{(3)}(t_1)$ [Fig. 15.1]. In the same way, for a large number of samples, we obtain the value of the variable X(t) at $t = t_i$.

Probability can be defined as the fraction of the favourable events, out of all possible events. If an experiment is repeated a great number of times under essentially the same conditions, then the ratio of the number of times the event happens, to the total number of experiments is called the probability of happening of the event.

Probability may be discrete or continuous. We shall deal only with the continuous variables. For example, the wave height is a continuous variable, because it can take any value in the given interval $a \leq X \leq b$ of the arithmetic continuum. The probability distributions related to these variables are known as continuous probability distributions.

The probability that a random variable *X* lies between *x* and *x* + *dx* is $p(x)\,dx$ [Fig. 15.21], where *p*(*x*) is known as the probability density function.

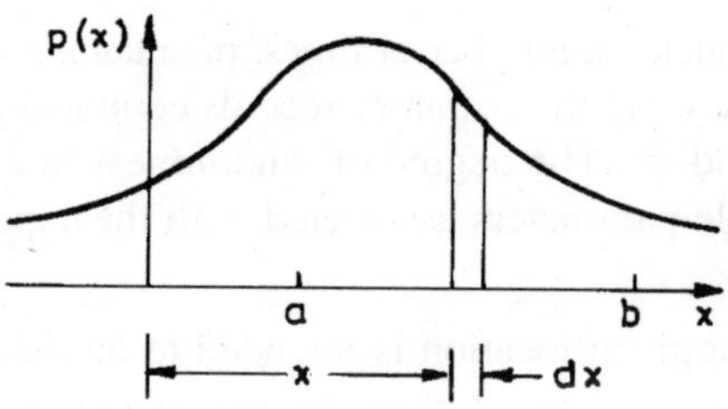

Fig. 15.2 Probability distribution

Mathematically

$$P[x < X < x + dx] = p(x)\,dx \tag{15.1}$$

The probability that a sample lies between a and b is just the sum of the probabilities, that a sample lies in each of the individual dx interval

$$P[a < X < b] = \int_a^b p(x)\,dx \tag{15.2}$$

The probability that the random variable X lies between $-\infty$ and $+\infty$ is one, that is

$$\int_{-\infty}^{\infty} p(x)\,dx = 1 \tag{15.3}$$

If the random variable is equal to some preselected value, say $x = a$, then the probability is zero

$$P[x = a] = 0 \tag{15.4}$$

Example 15.1

Draw the probability distribution diagram for the wave elevation data given below.

Probability density function for each range of the wave elevation is first calculated. The respective values are shown in the same table.

Wave elevation (X) m	No. of occurrences i.e., frequencies	Probability density of occurrences
3.5 – 4.5	18	0.096
2.5 – 3.5	20	0.106
1.5 – 2.5	22	0.117
0.5 –1.5	24	0.128
– 0.5 – 0.5	30	0.160
– 1.5 – (– 0.5)	28	0.148
– 2.5 – (– 1.5)	24	0.128
– 3.5 – (– 2.5)	22	0.117
	188	1.000

Total number of occurrences = 188.

No. of occurrences in the range (3.5- 4.5) m = 18 m.

Therefore probability density function in that range = $\dfrac{18}{188} = 0.096$

The calculation is repeated in each range of wave elevation. Based on these values, a histogram is drawn first and a mean curve through it indicates the probability distribution diagram as shown in Fig. 15.3.

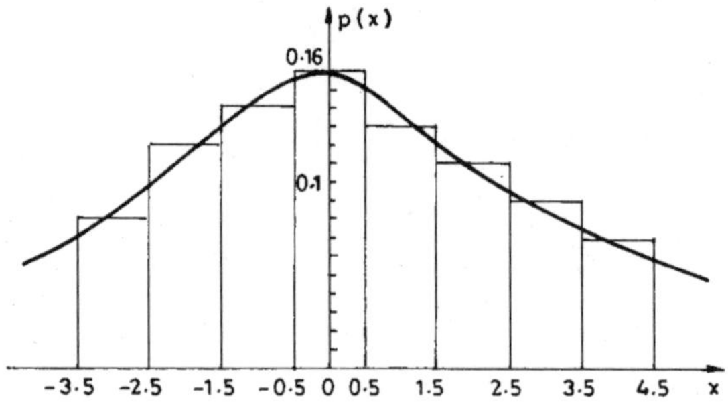

Fig. 15.3 Probability distribution diagram

15.3.1 Second-order Probability Distribution

At two time instants t_1 and t_2, the random variable be denoted as $X(t_1)$ and $X(t_2)$ or simply X_1 and X_2. The probability distribution of X_1 and X_2 is called the joint probability density function and is given by a surface $p(x_1, x_2)$ [Fig. 15.4].

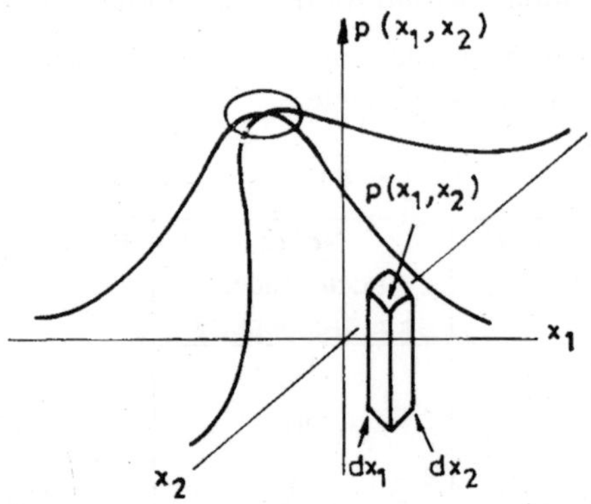

Fig. 15.4 Joint probability distribution

Joint probability density function $p(x_1, x_2)$ is defined, so that the element volume $p(x_1, x_2)$ $dx_1\, dx_2$ represents the probability that a pair of sampled values will be within the region $x_1 < X_1 < x_1 + dx_1$ and $x_2 < X_2 < x_2 + dx_2$. Mathematically

$$P[x_1 < X_1 < x_1 + dx_1 \quad \text{and} \quad x_2 < X_2 < x_2 + dx_2] = p(x_1, x_2)\, dx_1\, dx_2 \qquad (15.5)$$

Extending the one variable concept to two, the above definition leads to the following equation

$$P[a < X_1 < b \quad \text{and} \quad c < X_2 < d] = \int_a^b \int_c^d p(x_1, x_2)\, dx_1\, dx_2 \qquad (15.6)$$

and
$$\int_{-\infty}^{\infty}\int_{-\infty}^{\infty} p(x_1, x_2)\, dx_1\, dx_2 = 1.0 \tag{15.7}$$

First-order probability information can be obtained from the second-order probability, that is, the joint probability distribution. The marginal probability density function $p(x_1)$ is defined as $p(x_1)\,dx_1$ such that the chances that a sampled value X_1 will be in the range $x_1 < X_1 < x_1 + dx_1$, regardless of the value of X_2 sampled.

$$P[x_1 < X < x_1 + dx_1 \quad \text{regardless of} \quad X_2] = p(x_1)\, dx_1$$

$$= \int_{-\infty}^{\infty} p(x_1, x_2)\, dx_1\, dx_2$$

$$p(x_1) = \int_{-\infty}^{\infty} p(x_1, x_2)\, dx_2 \tag{15.8}$$

or Likewise
$$p(x_2) = \int_{-\infty}^{\infty} p(x_1, x_2)\, dx_1 \tag{15.9}$$

The derivation of Eqs. (15.8) and (15.9) is left as an exercise to the reader.

Example 15.2

The wave height and wave period are taken as random variables. Number of occurrences in the different ranges are given in the following table:

Wave period (s)	Wave height (m)			Σ
	3 - 4	4 - 5	5 - 6	
5 - 10	6	8	7	21
10 - 15	3	2	9	14
15 - 20	6	6	8	20
Σ	15	16	24	55

Determine

(a) joint probability density function that the wave period will be in the range 10 - 15 s and wave height 4-5 m,

(b) marginal probability density function that the wave period will be 10-15 s, irrespective of any value of wave height.

The values in all the rows and columns in the table are added up. Total number of occurrences = 55.

(a) Number of occurrences when wave height is 4-5 m and wave period is 10-15 s is = 2.

Joint probability density function of wave height 4-5 m and wave period 10-15s $= \frac{2}{55}$ $= 0.0364$.

(b) Marginal probability density function of the wave period 10-15 s, irrespective of any value of wave height $= \frac{14}{55} = 0.255$.

15.4 ENSEMBLE AVERAGES, MEAN AND AUTOCORRELATION

In this section, we deal with some important averages across the ensemble.

Let $f(x)$ be a function of x, whose first-order probability density function is $p(x)$. The expected value of $f(x)$, which is considered as a continuous variable is

$$E[f(X)] = \int_{-\infty}^{\infty} f(x)\, p(x)\, dx \tag{15.10}$$

This ensemble average is called the mathematical expectation of $f(x)$ and the operator E is used to denote this kind of average.

When $f(x) = X$, then

$$E(X) = \int_{-\infty}^{\infty} xp(x)\, dx = \text{mean of ensemble average}$$

$$= \text{expected value of } X. \tag{15.11}$$

When $f(x) = x^2$, then Eq. (15.10) becomes

$$E[X^2] = \int_{-\infty}^{\infty} x^2 p(x)\, dx \tag{15.12}$$

= mean square of the random variable x.

The expected value of X and its probability density function is shown in Fig. 15.5.

The variance σ^2 is defined as the ensemble average of the square of the deviation from the mean

$$\sigma^2 = E[X - (E[X])^2] = \int_{-\infty}^{\infty} (x - E[X])^2\, p(x)\, dx \tag{15.13}$$

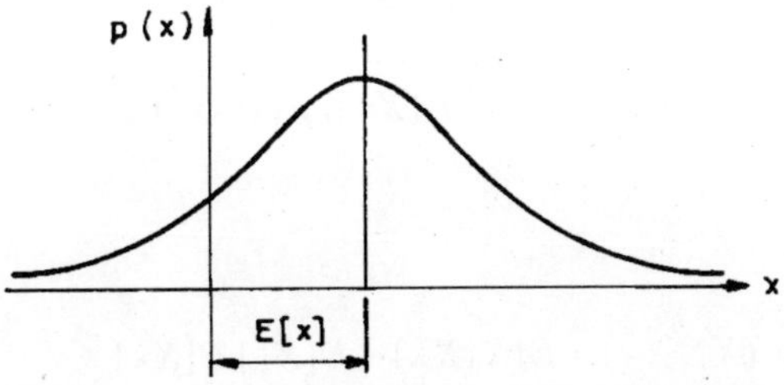

Fig. 15.5 Probability distribution and expected value

Expanding right hand side of Eq. (15.13), we have

$$\sigma^2 = \int_{-\infty}^{\infty} x^2 p(x)\, dx - \int_{-\infty}^{\infty} 2xE[X]\, p(x)\, dx + \int_{-\infty}^{\infty} (E[X])^2 p(x)\, dx$$

$$= E[X^2] - 2(E[X])^2 + (E[X])^2$$

or

$$\sigma^2 = E[X^2] - (E[X])^2 \tag{15.14}$$

Standard deviation of the random variable X is given by

$$\sigma = \sqrt{\sigma^2} = \sqrt{\text{variance of } X} \tag{15.15}$$

It indicates the spread of the random variable about the mean.

From single random variable, we pass on to two random variables. The measure of simultaneous variation of X_1 and X_2 is to be obtained. If $f(x_1)$ and $g(x_2)$ are functions of x_1 and x_2 respectively, then the mathematical expectation of their product is

$$E[f(x_1)\, g(x_2)] = \int_{-\infty}^{\infty}\int_{-\infty}^{\infty} f(x_1)\, g(x_2)\, p(x_1, x_2)\, dx_1\, dx_2 \tag{15.16}$$

when $f(x_1) = X_1$ and $g(x_2) = X_2$

then

$$E[X_1 X_2] = \int_{-\infty}^{\infty}\int_{-\infty}^{\infty} x_1 x_2\, p(x_1, x_2)\, dx_1\, dx_2 \tag{15.17}$$

= average across the ensemble of all products X_1X_2

$E[X_1X_2]$ is called the autocorrelation function. As X_1 and X_2 represent the product of the values of the same sample function, the prefix auto is used.

A Covariance is obtained by averaging the deviations from the mean at two time instants

$$\text{Cov}(X_1, X_2) = E[\{X_1 - E(X_1)\}\{X_2 - E(X_2)\}]$$

$$= \int_{-\infty}^{\infty} \int_{-\infty}^{\infty} (X_1 - E[X_1])(X_2 - E[X_2])\, p(x_1, x_2)\, dx_1\, dx_2 \qquad (15.18)$$

It can be shown that

$$\text{Cov}(X_1, X_2) = E[X_1 X_2] - E[X_1]\,E[X_2] \qquad (15.19)$$

If X_1 and X_2 have zero means, the covariance is identical to the autocorrelation.

From the above statistical averages, we obtain a gross description of the probability distribution. Just as in rigid body dynamics, the total mass, the mass moment of inertia, the product moment of inertia and centroid give only gross information, which do not enable us to make more detailed study about certain quantities, such as the internal force; similarly, mean, variance and so on, only provide a somewhat limited information. In order to make a more refined study of the random processes, detailed information about probability distributions are required.

15.5 STATIONARY PROCESS, ERGODIC PROCESS AND TEMPORAL AVERAGES

We have so far considered the properties of random process at a fixed time instant or at two fixed time instants. The variation of these properties, when time instants are assumed to vary is discussed next.

15.5.1 Stationary Process

A random process is said to be stationary, if its probability distributions are invariant under a shift of time scale, that is, independent of the origin. This suggests that the probability density at any time instant is valid for later times as well. This is somewhat analogous to the assumptions made in the case of steady state vibration, where the resulting vibration is independent of the initial conditions.

Wave and wind generation are random processes, which can be treated as stationary processes without introducing much error in the analysis. Earthquake ground motions however cannot be treated as stationary processes, as the probability distributions vary with time. For a process to be strictly stationary, it can have no beginning and no end and each sample should extend from $-\infty$ to $+\infty$.

For stationary processes, the first-order probability distribution $p(x)$ is independent of time. As such $E[X]$ and σ^2 are also independent of time.

In case of second-order probability density function, $p(x_1, x_2)$ is a function of time lag between t_1 and t_2 and not a function of t_1 and t_2 individually. Time lag $\tau = t_2 - t_1$. Then the autocorrelation function is

$$E[X_1 X_2] = E[X(t)\, X(t+\tau)] = R(\tau) \qquad (15.20)$$

Since by definition of stationary process, the probability density function $p(t, t+\tau)$ is independent of t, the autocorrelation function is related to time lag only. $R(\tau)$ will be used to denote autocorrelation function of a stationary random process.

Some of the properties of $R(\tau)$ is as follows:

(1) When $\tau = 0$, $R(0) = E[X^2] =$ ensemble mean square of the process. Further, for this process if $E[X] = 0$, that is, the process has a zero mean, then the mean square is identical to the variance, that is, $R(0) = \sigma^2$.

(2) Autocorrelation function is always an even function of t, that is, $R(\tau) = R(-\tau)$.

(3) If $X(t) = X(t+\tau)$, then $R(\tau) = E(X^2)$. For this case, autocorrelation function is independent of τ.

(4) If $X(t)$ is independent of $X(t+\tau)$, then $R(\tau) = E[X(t)\, E[X(t+\tau)]$. In this case the autocorrelation function becomes the square of the mean. When the mean is zero, $R(\tau) = 0$.

15.5.2 Temporal Statistics and Ergodic Hypothesis

So far, the discussion has been limited to the consideration of ensemble averages. For obtaining necessary statistics, a large number of sample functions are required for such cases. However, for certain stationary processes, it has been found that a single but long sample function of time can provide the necessary statistical information, when averaging is done with respect to time along the sample. Such an average is called the temporal average.

Temporal mean and temporal mean square is defined as

$$< X(t) > = \lim_{T \to \infty} \frac{1}{T} \int_{-T/2}^{T/2} x(t)\, dt \tag{15.21}$$

and

$$< X(t)^2 > = \lim_{T \to \infty} \frac{1}{T} \int_{-T/2}^{T/2} x^2(t)\, dt \tag{15.22}$$

Temporal autocorrelation is defined as

$$\phi(\tau) = < X(t)\, X(t+\tau) > = \lim_{T \to \infty} \frac{1}{T} \int_{-T/2}^{T/2} x(t)\, x(t+\tau)\, dt \tag{15.23}$$

When $\tau = 0$, $\phi(0) = < X^2(t) > =$ temporal mean square.

If it is assumed that a single sample function is typical of all other sample functions, the stationary random process is referred to as ergodic process. Therefore, ergodic process is one for which ensemble averages are equal to temporal averages

$$< X(t) > = E[X(t)] \tag{15.24}$$

$$\phi(\tau) = R(\tau) \tag{15.25}$$

The symbol $< >$ is used to indicate temporal averages.

When samples obtained at a particular location is spatially valid at other locations, the process is said to be a homogeneous process.

A random process may be stationary without being ergodic. Let us consider wave data at a particular location of the sea. The data will indicate stationary process. But the statistical properties at the different locations of the ocean will be different, so that one sample function may not be typical at all places in the ocean. Therefore, the wave data are stationary without being ergodic.

When we calculate the mean of the product of the variable obtained from two different random processes, the term cross-correlation is used.

Example 15.3

Determine the ensemble mean, ensemble mean square, temporal mean, temporal mean square, the temporal autocorrelation and the ensemble autocorrelation of the following function

$$X(t) = A \sin(\omega t + \alpha)$$

where A and ω have fixed values, but the phase angle ω is a random variable. The probability density function of $p(\alpha)$ is shown in Fig. 15.6.

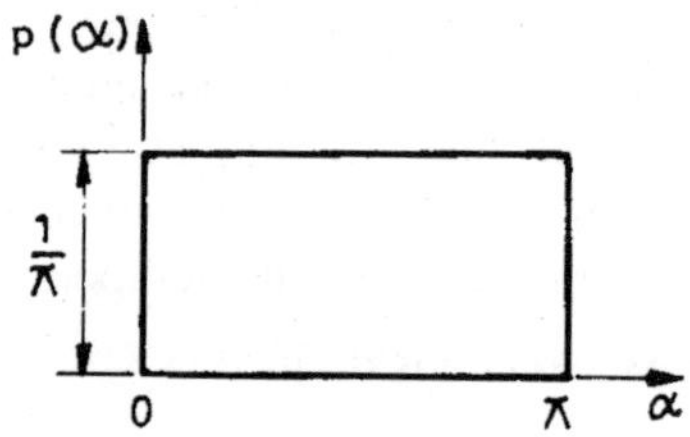

Fig. 15.6 Example 15.3

The ensemble mean is

$$E[X(t)] = A\int_0^{\pi} \sin(\omega t + \alpha)\, p(\alpha)\, d\alpha$$

$$= \frac{A}{\pi}\int_0^{\pi} \sin(\omega t + \alpha)\, d\alpha = \frac{2A}{\pi}\cos\omega t$$

The ensemble mean square is

$$E[X^2(t)] = A^2\int_0^{\pi} \sin^2(\omega t + \alpha)\, p(\alpha)\, d\alpha$$

$$= \frac{A^2}{\pi}\int_0^{\pi} \frac{1 - \cos 2(\omega t + \alpha)}{2}\, d\alpha = \frac{A^2}{2}$$

The temporal mean is

$$< X(t) > = \lim_{T\to\infty} \frac{1}{T}\int_0^{T} A \sin(\omega t + \alpha)\, dt = 0$$

The temporal mean square is

$$< X^2(t) > = \lim_{T \to \infty} \frac{1}{T} \int_0^T A^2 \sin^2(\omega t + \alpha)\, dt = \frac{A^2}{2}$$

The temporal autocorrelation is

$$\phi(\tau) = \lim_{T \to \infty} \frac{1}{T} \int_0^T A \sin(\omega t + \alpha)\, A \sin(\omega t + \alpha + \tau)\, dt$$

$$= \frac{A^2}{2} \cos \omega \tau$$

The ensemble autocorrelation is

$$R(\tau) = \int_0^{\pi} A \sin(\omega t + \alpha)\, A \sin(\omega t + \omega\tau + \alpha) \frac{1}{\pi}\, d\alpha$$

$$= \frac{2A}{\pi} \cos \omega t \cos \omega \tau$$

Comparing ensemble and temporal averages of the problem, it is clear that the random process described is not stationary.

15.6 POWER SPECTRAL DENSITY

For a stationary ergodic process, the temporal mean square is

$$< F^2(t) > = \lim_{T \to \infty} \frac{1}{T} \int_{-T/2}^{T/2} F(t)\, F(t + \tau)\, dt \qquad (15.26)$$

Writing $\omega = 2\pi f$, Eqs. (3.124) and (3.125) are written as

$$F(t) = \int_{-\infty}^{\infty} F(t)\, e^{i2\pi ft}\, df \qquad (15.27)$$

and

$$F(f) = \int_{-\infty}^{\infty} F(t)\, e^{-i2\pi ft}\, dt \qquad (15.28)$$

Substituting one $F(t)$ of the integral of Eq. (15.26) by Eq. (15.27), we get

$$< F^2(t) > = \lim_{T \to \infty} \frac{1}{T} \int_{-T/2}^{T/2} F(t) \int_{-\infty}^{\infty} F(f)\, e^{i2\pi ft}\, df\, dt$$

and
$$= \lim_{T \to \infty} \int_{-\infty}^{\infty} F(f) \frac{1}{T} \left[\int_{-T/2}^{T/2} F(t)\, e^{i2\pi ft}\, dt \right] df \tag{15.29}$$

A comparison of the integral within the parenthesis of Eqs. (15.29) to (15.28) reveals that the integral is the complex conjugate of $F(f)$. We shall indicate it as $F^*(f)$.

Therefore, Eq. (15.29) becomes

$$< F^2(t) > = \lim_{T \to \infty} \int_{-\infty}^{\infty} \frac{1}{T} F(f)\, F^*(f)\, df$$

$$= \lim_{T \to \infty} \int_{-\infty}^{\infty} \frac{1}{T} |F(f)|^2\, df$$

or
$$< F^2(t) > = \int_{-\infty}^{\infty} S(f)\, df \tag{15.30}$$

where
$$S(f) = \frac{1}{T} |F(f)|^2 \tag{15.31}$$

$S(f)$ is called the power spectral density and can be obtained from the square of the Fourier transform of the random process. If the spectral density can be obtained for the whole range of frequencies, the mean square of the process can be determined from Eq. (15.31).

15.7 RELATIONSHIP BETWEEN AUTOCORRELATION FUNCTION AND POWER SPECTRAL DENSITY

For a stationary ergodic process

$$\phi(\tau) = \lim_{T \to \infty} \frac{1}{T} \int_{-T/2}^{T/2} F(t)\, F(t+\tau)\, dt \tag{15.32}$$

From Eq. (15.27) we can write

$$F(t+\tau) = \int_{-\infty}^{\infty} F(f)\, e^{i2\pi f\,(t+\tau)}\, df \tag{15.33}$$

and
$$F(t) = \int_{-\infty}^{\infty} F(t+\tau)\, e^{-i2\pi f\,(t+\tau)}\, dt \tag{15.34}$$

Combining Eqs. (15.32) and (15.34), we get

$$\phi(\tau) = \lim_{T \to \infty} \frac{1}{T} \int_{-T/2}^{T/2} F(t) \int_{-\infty}^{\infty} F(f)\, e^{i2\pi f(t+\tau)}\, df$$

or $$\phi(\tau) = \lim_{T \to \infty} \frac{1}{T} \int_{-\infty}^{\infty} F(f) \left[\int_{-T/2}^{T/2} F(t)\, e^{i2\pi ft}\, dt \right] e^{i2\pi f\tau}\, df \qquad (15.35)$$

Combining Eqs. (15.28) and (15.35), we get

$$\phi(\tau) = \lim_{T \to \infty} \int_{-\infty}^{\infty} \frac{1}{T} F(f)\, F^*(f)\, e^{i2\pi f\tau}\, dt$$

or $$\phi(\tau) = \lim_{T \to \infty} \int_{-\infty}^{\infty} \frac{1}{T} |F(f)|^2\, e^{i2\pi f\tau}\, df \qquad (15.36)$$

Combining Eqs. (15.31) and (15.36), we get

$$\phi(\tau) = \int_{-\infty}^{\infty} S(f)\, e^{i2\pi f\tau}\, df \qquad (15.37)$$

This implies that the autocorrelation function $\phi(\tau)$ is the Fourier transform of the spectral density *S(f)* of the process. Therefore

$$S(f) = \int_{-\infty}^{\infty} \phi(\tau)\, e^{-i2\pi f\tau}\, d\tau \qquad (15.38)$$

The following properties of the power spectral density may be noted:
From Eq. (15.37), when $\tau = 0$

(1) $$\phi(0) = \int_{-\infty}^{\infty} S(f)\, df = \langle F(t)^2 \rangle \qquad (15.39)$$

(2) *S(f)* is real and even function of *f*, because $\phi(\tau)$ is an even and real function of τ.

(3) *S(f)* should always be a positive quantity.

Example 15.4

Determine the mean square and the autocorrelation function for the stationary random process *X(t)*, whose spectral density is shown in Fig. 15.7.

The mean square of the process is

$$E[X^2(t)] = \int_{-\infty}^{\infty} S(f)\, df$$

$$= 2S_0(f_2 - f_1)$$

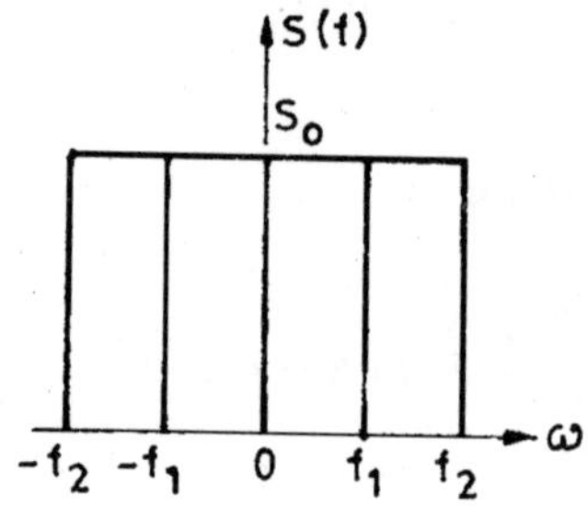

Fig. 15.7 Example 15.4

The autocorrelation function is

$$\phi(\tau) = \int_{-\infty}^{\infty} S(f)\, e^{i2\pi f\tau}\, df$$

$$= \int_{-\infty}^{\infty} S(f) \cos 2\pi\tau\, df$$

since $S(f)$ is an even function of f. So

$$\phi(\tau) = 2\int_{f_1}^{f_2} S_0 \cos 2\pi f\tau\, df$$

$$= \frac{2S_0}{2\pi\tau} \sin 2\pi f\tau \Big|_{f_1}^{f_2}$$

$$= \frac{S_0}{\pi\tau} (\sin 2\pi f_2\tau - \sin 2\pi f_1\tau)$$

$$= \frac{2S_0}{\pi\tau} \cos[\pi(f_1 + f_2)\tau] \sin[\pi(f_2 - f_1)\tau]$$

A plot of $\phi(\tau)$ with τ is shown in Fig. 15.8.

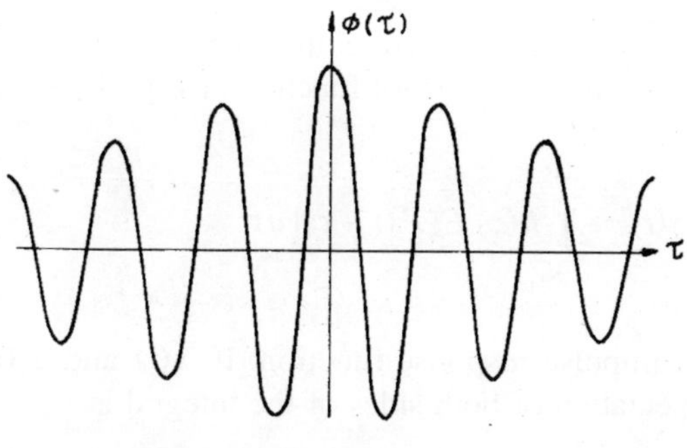

Fig. 15.8 Variation of $\phi(\tau)$ with τ

Example 15.5

A random process is given by the following equation

$$x(t) = A \sin(2\pi f_0 t + \phi)$$

where the amplitude A and frequency f_0 are deterministic and the phase angle ϕ is random. Determine the autocorrelation function and based on this relation determine the spectral density of the process.

The autocorrelation of the same function has been calculated in Example 15.3, and the value is

$$\phi(\tau) = \frac{A^2}{2} \cos 2\pi f_0 \tau$$

The spectral density in terms of autocorrelation is

$$S(f) = \int_{-\infty}^{\infty} \phi(\tau) e^{-i2\pi f\tau} d\tau$$

$$= \int_{-\infty}^{\infty} \frac{A^2}{2} \cos 2\pi f_0 \tau \, e^{-i2\pi fr} d\tau$$

$$= \frac{A^2}{2}\left[\frac{\sin 2\pi(f - f_0)\tau}{2\pi(f - f_0)}\right]$$

15.8 RANDOM RESPONSE OF SDF SYSTEMS

It has been pointed out in Chapter 3 that there are two approaches of studying the response of a system: they are, time domain analysis and frequency domain analysis. These concepts are extended here for a SDF system, whose input excitation is random.

15.8.1 Time Domain Analysis

Indicating $x(t)$ as the response of the structure, the Duhamel's integral given by Eq. (3.91) can be written in terms of unit impulse response function of Eq. (3.132), as

$$x(t) = \int_{-\infty}^{\infty} h(\tau)\, E[F(t-\tau)]\, d\tau \tag{15.40}$$

where $h(\tau)$ is the unit impulse response function. If $x(t)$ and $F(t)$ are stationary random processes, then the expectation of both sides of the integral is

$$E[x(t)] = \int_{-\infty}^{\infty} h(\tau) E[F(t-\tau)]\, d\tau \tag{15.41}$$

If m_x and m_F denote the mean values of $x(t)$ and $F(t)$, then Eq. (15.41) can be written as

$$m_x = m_F \int_{-\infty}^{\infty} h(\tau)\, d\tau \tag{15.42}$$

Equation (15.42) gives the relationship between the mean values of the input and output of the process.

If we consider the autocorrelation $R_{FF}(\tau)$ for the excitation and $R_{xx}(\tau)$ for the response, then

$$R_{xx}(\tau) = E[x(t)\, x(t+\tau)] = \int_{-\infty}^{\infty}\int_{-\infty}^{\infty} h(\theta_1)\, h(\theta_2)\, E[F(t-\theta_1)$$

$$F(t+\tau-\theta_2)]\, d\theta_1\, d\theta_2$$

$$= \int_{-\infty}^{\infty}\int_{-\infty}^{\infty} h(\theta_1)\, h(\theta_2)\, R_{FF}(\tau+\theta_1-\theta_2)\, d\theta_1\, d\theta_2 \tag{15.43}$$

15.8.2 Frequency Domain Analysis

The frequency domain analysis is much more popular than time domain analysis. If $x(t)$ is the response of the system due to an excitation $F(t)$, then from Eq. (3.129), we know

$$X(\omega) = H(\omega)\, F(\omega) \tag{15.44}$$

where $X(\omega)$ and $F(\omega)$ are Fourier transforms of $x(t)$ and $F(t)$ and $H(\omega)$ is the complex frequency response function. For a single degree freedom system having damping and stiffness, we know from Eq. (3.111) that

$$H(\omega) = \frac{1}{-m\omega^2 + ic\omega + k} = \frac{1/k}{\left[1 - \left(\frac{\omega}{p}\right)^2\right] + i2\zeta\left(\frac{\omega}{p}\right)}$$

Taking complex conjugate of all the quantities in Eq. (15.44), we get

$$X^*(\omega) = H^*(\omega)\, F^*(\omega) \tag{15.45}$$

Multiplying Eq. (15.44) by Eq. (15.45), we obtain

$$X(\omega)\, X^*(\omega) = H(\omega)\, H^*(\omega)\, F(\omega)\, F^*(\omega)$$

or

$$X(\omega)\, X^*(\omega) = |H(\omega)|^2\, F(\omega)\, F^*(\omega) \tag{15.46}$$

Dividing both sides of Eq. (15.46) by T, and taking limits as $T \to \infty$, we get

$$\lim_{T\to\infty} \frac{1}{T} |X(\omega)|^2 = |H(\omega)|^2 \lim_{T\to\infty} \frac{1}{T} |F(\omega)|^2 \tag{15.47}$$

From the definition of power spectral density as given by Eq. (15.31), we can write

$$S_{xx}(\omega) = \lim_{T\to\infty} \frac{1}{T} |X(\omega)|^2 \tag{15.48}$$

and

$$S_{FF}(\omega) = \lim_{T\to\infty} \frac{1}{T} |F(\omega)|^2 \tag{15.49}$$

$S_{xx}(\omega)$ and $S_{FF}(\omega)$ are the spectral densities of the response and the excitation. Combining Eqs. (15.47) to (15.49), we get

$$S_{xx}(\omega) = |H(\omega)|^2\, S_{FF}(\omega) \tag{15.50}$$

Equation (15.50) gives the relationship between the spectral densities of excitation and response. In the remaining part of the chapter, we shall analyse the different systems by frequency domain approach.

Example 15.6

Determine the response spectral density and mean square of response of a SDF system having mass m, damping constant c and spring stiffness k due to a white noise of magnitude S_0.

If the spectral density of a random process remains constant at all ranges of the frequency, then it is referred to as white noise (Fig. 15.9). From Eq. (3.136) we get

$$|H(f)|^2 = \frac{1}{(k - 4\pi^2 m f^2)^2 + 4\pi^2 f^2 c^2}$$

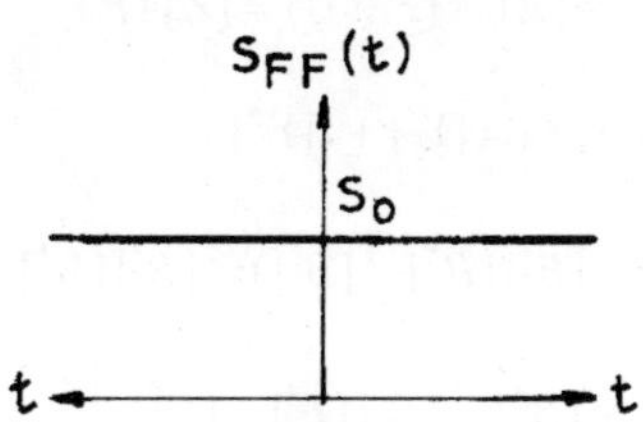

Fig. 15.9 White noise

From Eq. (15.50), we get

$$S_{xx}(f) = \frac{1}{(k - 4\pi^2 mf^2)^2 + 4\pi^2 f^2 c^2}, \quad S_{FF}(f) = \frac{S_0}{(k - 4\pi^2 mf^2)^2 + 4\pi^2 f^2 c^2}$$

The mean square of the response is given by

$$< x^2 > = \int_{-\infty}^{\infty} S_{xx}(f)\, df = 2\int_{0}^{\infty} S_{xx}(f)\, df$$

$$= 2\int_{0}^{\infty} \frac{S_0}{(k - 4\pi^2 mf^2)^2 + 4\pi^2 f^2 c^2}\, df$$

15.9 RANDOM RESPONSE OF MDF SYSTEMS

Complex structures have multiple degrees of freedom. The random response analysis of these structures can be carried out in the frequency domain. There are two approaches to the problem.

15.9.1 Complex Matrix Inversion Method

The equation of motion for a system having n degrees of freedom is given by Eq. (6.4), which is reproduced with usual notations

$$[M]\{\ddot{x}\} + [C]\{\dot{x}\} + [K]\{x\} = \{F(t)\} \tag{6.4}$$

Let us assume $\{x(t)\}$ and $\{F(t)\}$ as stationary random processes, with zero mean value. Taking Fourier transform of Eq. (6.4), we get

$$([K] + i\omega[C] - \omega^2[M])\{X(\omega)\} = \{F(\omega)\} \tag{15.51}$$

where ω is the frequency of vibration. Equation (15.51) is written as

$$[A]\{X\} = \{F\} \tag{15.52}$$

where $[A]$ is a complex matrix of order $n \times n$.

$$[A] = [B'] + i[B'']$$

where $[B']$ and $[B'']$ are real square matrices of the same order as that of $[A]$.

Solution of Eq. (15.51) is

$$\{X(\omega)\} = [A]^{-1}\{F(\omega)\} = [Z]\{F\} \tag{15.53}$$

and

$$[Z] = [Y'] + i[Y''] \tag{15.54}$$

where

$$[Y'] = ([B''] + [B'][B'']^{-1}[B'])^{-1}[B'][B'']^{-1} \tag{15.55a}$$

$$[Y''] = [B'']^{-1}[B'][Y'] - [B'']^{-1} \tag{15.55b}$$

It is assumed in Eqs. (15.55a) and (15.55b) that $[B'']$ is a non-singular matrix.

Writing the ith equation of Eq. (15.53)

$$X_i = \sum_{j=1}^{n} Z_{if} F_j^* \tag{15.56}$$

Based on the relation of Eq. (15.56), we can write the complex conjugate of X_m as

$$X_m^* = \sum_{l=1}^{n} Z_{ml}^* F_l^* \tag{15.57}$$

Forming the product of Eqs. (15.56) and (15.57) and dividing both sides by T and taking limits as $T \to \infty$, we get

$$\lim_{T\to\infty} \frac{1}{T} X_i X_m^* = \lim_{T\to\infty} \frac{1}{T} \sum_{j=1}^{n} \sum_{l=1}^{n} Z_{ij} Z_{ml}^* F_j F_l^* \tag{15.58}$$

Equation (15.58) can be interpreted in terms of cross-spectral density of the responses and the generalised forces, which is written as

$$S_{x_i x_m} = \sum_{j=1}^{n} \sum_{l=1}^{n} Z_{ij} Z_{ml}^* S_{F_j F_l} \tag{15.59}$$

$S_{x_i x_m}$ can be separated into real and imaginary parts as follows:

$S'_{x_i x_m}$ = Real part of $S_{x_i x_m}$

$$= \sum_{j=1}^{n} \sum_{l=1}^{n} (Y'_{ij} Y'_{ml} + Y''_{ij} Y''_{ml}) S'_{F_j F_l} - (Y''_{ij} Y'_{ml} - Y'_{ij} Y''_{ml}) S''_{F_j F_l} \tag{15.60a}$$

$S''_{x_i x_m}$ = imaginary part of $S_{x_i} S_{x_m}$

$$= \sum_{j=1}^{n} \sum_{l=1}^{n} (Y'_{ij} Y'_{ml} + Y''_{ij} Y''_{ml}) S''_{F_j F_l} - (Y''_{ij} Y'_{ml} - Y'_{ij} Y''_{ml}) S'_{F_j F_l} \tag{15.60b}$$

Likewise, taking the combination of i's and m's, we can form the cross-spectral density matrix $[S_{xx}]$. In order to determine the mean square value of a displacement $< x_i^2 >$, we pick up the diagonal element of $[S_{xx}]$ and integrate over the entire frequency range.

The frequencies are discretised in this method and the procedure mentioned requires the inversion of matrices at each frequency step. As such, the method demands a high computer time.

15.9.2 Normal Mode Method

We have seen in Chapters 6 and 7 that the displacement in a multiple degrees of freedom system, can be expressed in terms of the normal coordinates.

$$\{x\} = [\Phi]\{\xi\} \tag{15.61}$$

where $[\phi]$ is the modal matrix and $[\xi]$ is the normal coordinate.

Multiplying both sides of Eq. (6.4) by $[\phi]^T$ and using Eq. (15.61), we obtain

$$[\Phi]^T[M][\Phi]\{\ddot{\xi}\} + [\Phi]^T[C][\Phi]\{\dot{\xi}\} + [\Phi]^T[K][\Phi]\{\xi\} = [\Phi]^T\{F(t)\} \tag{15.62}$$

Equation (15.62) is written as follows:

$$[\overline{M}]\{\ddot{\xi}\} + [\overline{C}]\{\dot{\xi}\} + [\overline{K}]\{\xi\} = [\Phi]^T\{F(t)\} \tag{15.63}$$

It is assumed that $[\overline{M}]$, $[\overline{C}]$ and $[\overline{K}]$ are diagonal matrices. This aspect has been discussed in Chapters 6 and 7.

The ith equation from the set given by Eq. (15.63) will have the form

$$\overline{M}_i\ddot{\xi}_i + \overline{C}_i\dot{\xi}_i + \overline{K}_i\xi_i = \phi_i^{(1)}F_1 + \phi_i^{(2)}F_2 + \cdots\cdots + \phi_i^{(n)}F_n \tag{15.64}$$

The complex frequency response function is given by

$$H_j(f) = \frac{1}{\overline{K}_j - \overline{M}_j\omega^2 + i\overline{C}_j\omega^2} \tag{15.65}$$

The direct and cross-spectral densities of the generalised forces are

$$S_{f_i f_j}(f) = \sum_{r=1}^{n}\sum_{s=1}^{n} \phi_i^{(r)}\phi_j^{(s)} S_{F_r F_s}(f) \tag{15.66}$$

It may be noted from Eq. (15.66) that the direct spectral densities are real and the cross-spectral densities are complex.

The spectral densities of the generalised modal coordinates are given by an equation similar to Eq. (15.59)

$$S_{\xi_i\xi_j}(f) = H_i H_j^* S_{f_i f_j}(f) \tag{15.67}$$

Combining Eqs. (15.66) and (15.67), we get

$$S_{\xi_i\xi_j}(f) = \sum_{r=1}^{n}\sum_{s=1}^{n} \phi_i^{(r)}\phi_i^{(s)} H_i H_j^* S_{F_r F_s}(f) \tag{15.68}$$

The spectral densities in terms of the original coordinate system can be related to modal spectral densities as follows:

$$S_{x_i x_j}(f) = \sum_{r=1}^{n}\sum_{s=1}^{n} \phi_i^{(r)}\phi_i^{(s)} S_{\xi_r\xi_s}(f) \tag{15.69}$$

Sometimes it is convenient to note the following relationship

$$S_{\xi_r\xi_s}(f) = S^*_{\xi_r\xi_s}(f) \quad \text{for} \quad r \neq s \tag{15.70}$$

Using the relation given by Eq. (15.70), Eq. (15.69) is rewritten for direct spectral densities as

$$S_{x_i x_j}(f) = \sum_{r=1}^{n} (\phi_i^{(r)})^2 \, S_{\xi_i\xi_j}(f) + 2\sum_{r=1}^{n}\sum_{s=1}^{n} \phi_i^{(r)}\phi_i^{(s)} \, \text{Re}\{S_{\xi_r\xi_s}(f)\} \tag{15.71}$$

where Re{ } denotes the real part of the function. Further, for most practical problems, second term on the right hand side of Eq. (15.71) may be neglected. Substituting $S_{\xi_i\xi_j}(f)$ from Eq. (15.68) into Eq. (15.69), will give the element $S_{x_i x_j}(f)$ of the spectral density matrix. Computing all such elements, the spectral density matrix can be written as

$$[S_{xx}(f)] = [\Phi][H(f)][\Phi]^T \, [S_{FF}(f)][\Phi][H(f)][\Phi]^T \tag{15.72}$$

Once the spectral densities are obtained, the mean square value of the response can be derived by following the same procedure as mentioned earlier.

Example 15.7

A two-degree of freedom system model is shown in Fig. 15.10. The natural frequencies and mode shapes for the system based on the coordinate system of the figure are

$$\{p^2\} = \begin{Bmatrix} 1 \\ 3 \end{Bmatrix} \frac{k}{m} \quad \text{and} \quad [\Phi] = \begin{bmatrix} 1 & 1 \\ 1 & -1 \end{bmatrix}$$

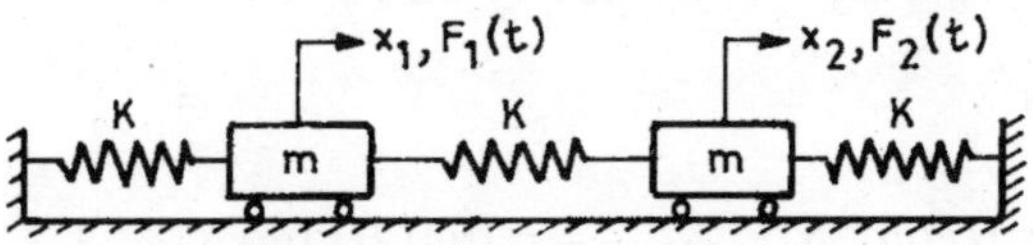

Fig. 15.10 Example 15.7

Determine the power spectral density function $S_{x_1x_1}$ corresponding to random displacement x_1, when the excitation is a single stationary random process $F_1(t)$ with a white noise S_0. Also determine the stationary mean square value of $E[X_1^2]$.

The mass matrix is

$$[M] = \begin{bmatrix} m & 0 \\ 0 & m \end{bmatrix}$$

$$[\Phi]^T[M][\Phi] = \begin{bmatrix} 1 & 1 \\ 1 & -1 \end{bmatrix}\begin{bmatrix} m & 0 \\ 0 & m \end{bmatrix}\begin{bmatrix} 1 & 1 \\ 1 & -1 \end{bmatrix} = m\begin{bmatrix} 2 & 0 \\ 0 & 2 \end{bmatrix}$$

$$[\Phi]^T[K][\Phi] = [\Phi]^T[M][\Phi][p^2] = m\begin{bmatrix} 2 & 0 \\ 0 & 2 \end{bmatrix}\frac{k}{m}\begin{bmatrix} 1 & 0 \\ 0 & 3 \end{bmatrix} = k\begin{bmatrix} 2 & 0 \\ 0 & 6 \end{bmatrix}$$

Assume that $F_1(t)$ and $F_2(t)$ are acting on the masses. Then from Eq. (7.4), we get

$$2m\begin{bmatrix}1 & 0\\ 0 & 1\end{bmatrix}\begin{Bmatrix}\ddot{\xi}_1\\ \ddot{\xi}_2\end{Bmatrix} + 2k\begin{bmatrix}1 & 0\\ 0 & 3\end{bmatrix}\begin{Bmatrix}\xi_1\\ \xi_2\end{Bmatrix} = \begin{Bmatrix}F_1(t)\\ F_2(t)\end{Bmatrix} \quad \text{(a)}$$

Two uncoupled equations are

$$2m\ddot{\xi}_1 + 2k\xi_1 = F_1(t) \quad \text{(b)}$$

$$2m\ddot{\xi}_2 + 6k\xi_2 = F_2(f) \quad \text{(c)}$$

$$H_1(\omega) = \frac{1}{2k - 2m\omega^2} = \frac{1}{2m\left(\dfrac{k}{m} - \omega^2\right)} = \frac{1}{2m(p_1^2 - \omega^2)} \quad \text{(d)}$$

$$H_2(\omega) = \frac{1}{6k - 2m\omega^2} = \frac{1}{2m\left(\dfrac{3k}{m} - \omega^2\right)} = \frac{1}{2m(p_2^2 - \omega^2)} \quad \text{(e)}$$

p_1 and p_2 are the two natural frequencies of the two-degree of freedom system

$$S_{F_1F_1} = S_0 \quad \text{and} \quad S_{F_1F_2} = S_{F_2F_1} = S_{F_2F_2} = 0$$

Therefore, from Eq. (15.66)

$$S_{f_1f_1} = S_0 \quad \text{and} \quad S_{f_2f_1} = S_{f_1f_2} = S_{f_2f_2} = 0$$

From Eq. (15.68), we get

$$S_{\xi_1\xi_1} = H_1(\omega)^2 S_0, \quad S_{\xi_1\xi_2} = H_1(\omega)\, H_2^*(\omega)\, S_0$$

and $$S_{\xi_2\xi_1} = H_1^*(\omega)\, H_2(\omega)\, S_0, \quad S_{\xi_2\xi_1} = H_2(\omega)^2 S_0$$

From Eq. (15.69), we get

$$S_{x_1x_1}(\omega) = |H_1(\omega)|^2 S_0 + H_1(\omega)\, H_2^*(\omega)\, S_0 + H_1^*(\omega)\, H_2(\omega)\, S_0 + |H_2(\omega)|^2 S_0 \quad \text{(f)}$$

Substituting all the necessary values of $H_1(\omega)$ etc., in Eq. (*f*), we get

$$S_{x_1x_1}(\omega) = \frac{S_0}{4m^2}\left[\frac{1}{(p_1^2 - \omega^2)^2} + \frac{2}{(p_1^2 - \omega^2)(p_2^2 - \omega^2)} + \frac{1}{(p_2^2 - \omega^2)^2}\right] \quad \text{(g)}$$

Therefore, the mean square value of x_1 is

$$E[X_1^2] = \int_{-\infty}^{\infty} S_{x_1x_1}(\omega)\, d\omega = \infty$$

15.10 RESPONSE OF FLEXURAL BEAMS UNDER RANDOM LOADING

The relative displacement of a uniform flexural beam to ground motion excitation is given by [Eq. (9.57)] as

$$Z(x,t) = -\sum_{r=1}^{\infty} Y_r(x) \frac{\int_0^L Y_r\,dx}{\int_0^L Y_r^2\,dx} \int_{-\infty}^{t} \ddot{y}_s(t-\theta)\,h_j(\theta)\,d\theta \tag{15.73}$$

where $h_j(\theta)$ is given by Eq. (3.131) and $\ddot{y}_g(t)$ is the ground acceleration. Further, it may be noted that as the process is random stationary, the shift of origin has been made in the integral of Eq. (15.73).

The autocorrelation function at any location x of $[R_{zz}(x,\tau)]$ is given by

$$R_{zz}(x,\tau) = E[z(x,t)\,z(x,t+\tau)] \tag{15.74}$$

The ground acceleration $\ddot{y}_g$ is assumed to be random stationary. Using Eqs. (15.73) and (15.74), we get

$$R_{zz}(x,\tau) = \sum_{j=1}^{\infty}\sum_{k=1}^{\infty} Y_j(x)\,Y_k(x) \frac{\int_0^L Y_j(x_1)\,dx_1 \int_0^L Y_k(x_2)\,dx_2}{\int_0^L Y_j^2(x_1)\,dx_1 \int_0^L Y_k^2(x_2)\,dx_2}$$

$$\times \int_{-\infty}^{\infty}\int_{-\infty}^{\infty} R_g(\tau-\theta_2+\theta_1)\,h_j(\theta_1)\,h_k(\theta_2)\,d\theta_1\,d\theta_2 \tag{15.75}$$

where R_g = autocorrelation of the ground acceleration

$$= E[\ddot{y}_g(t-\theta_1)\,\ddot{y}_g(t+\tau-\theta_2)]$$

In the frequency domain analysis, the spectral density function for relative displacement $z(x_1,t)$ is defined by [Eq. (15.38)]

$$S_{zz}(x,\omega) = \frac{1}{2\pi}\int_{-\infty}^{\infty} R_{zz}(x,\tau)\,e^{-i\omega\tau}\,d\tau \tag{15.76}$$

Substituting $R_{zz}(x,\tau)$ from Eq. (15.75) into Eq. (15.76,), we get

$$S_{zz}(x,\omega) = S_g(\omega)\sum_{j=1}^{\infty}\sum_{k=1}^{\infty} a_{jk}\,Y_j(x)\,Y_k(x)\,H_j(\omega)\,H_k(-\omega) \tag{15.77}$$

where

$S_g(\omega)$ = spectral density of ground acceleration $\ddot{y}_g(t)$

$$a_{jk} = \frac{\int_0^L Y_j(x_1)\,dx_1 \int_0^L Y_k(x_2)\,dx_2}{\int_0^L Y_j^2(x_1)\,dx_1 \int_0^L Y_k^2(x_2)\,dx_2}$$

$Y_j(x)$ = jth normal mode, and

$H_j(\omega)$ = frequency response function of the jth normal coordinate

$$= \frac{1}{m(p_j^2 + i2\zeta\omega p_j - \omega^2)}$$

where m is the mass per unit length.

15.11 FINITE ELEMENT RANDOM RESPONSE OF PLATES

We take a rectangular plate and divide it into a number of finite elements. In Art. 12.6, the element stiffness and the element mass matrices are derived. All the element matrices are assembled to form the overall matrices of the structure. Appropriate boundary conditions are then incorporated. Overall loading matrix is calculated by assembling the element loading matrices. The equation of motion of the plate after including the damping then becomes

$$[M]\{\ddot{x}\} + [C]\{\dot{x}\} + [K]\{x\} = \{F(t)\} \tag{15.78}$$

The treatment for finite element random analysis of plates is similar to that of multiple degrees of freedom system.

Let us express the displacements in terms of normal coordinates

$$\{x\} = [\Phi]\{\xi] \tag{15.79}$$

Substituting $\{x\}$ from Eq. (15.79) into Eq. (15.78) and premultiplying both sides by $[\phi]^T$, we get

$$[\Phi]^T[M][\Phi]\{\ddot{\xi}\} + [\Phi]^T[C][\Phi]\{\dot{\xi}] + [\Phi]^T[K][\Phi]\{\xi] = [\Phi]^T\{F(t)\} \tag{15.80}$$

Using the orthogonality relationship, the values of $[\phi]$ are adjusted such that (see Art .6.6)

$$[\Phi]^T[M][\Phi] = [\lambda] \tag{15.81}$$

and

$$[\Phi]^T[K][\Phi] = [I] = \text{identity matrix} \tag{15.82}$$

where $[\lambda]$ is a diagonal matrix consisting of the eigenvalues of the system.

Further, it is assumed that

$$[\Phi][C][\Phi] = [D] = \text{a diagonal matrix.}$$

The ith equation of motion may be written as

$$\lambda_i \ddot{\xi}_i + D_i \dot{\xi}_i + \xi_i = \{\phi_i\}^T \{F(t)\} \tag{15.83}$$

Taking the Fourier transform of the unit impulse applied to all degrees of freedom in succession, the following result is obtained

$$\bar{\xi}_j(\omega) = \frac{1}{-\omega^2 \lambda_j + i\omega D_j + 1} \{\phi_j\}^T \tag{15.84}$$

or

$$\bar{\xi}_j(\omega) = H_j(\omega)\{\phi_j\}^T \tag{15.85}$$

$\bar{\xi}_j(\omega)$ is the Fourier transform of $\xi_j(t)$, $H_j(\omega)$ is the complex frequency response function for the jth mode. Taking Fourier transform of Eq. (15.79) and using Eq. (15.85), we get

$$\{X_j(\omega)\} = \{\phi_j\} H_j(\omega) \{\phi_j\}^T \tag{15.86}$$

The procedure above for jth mode is applied to all degrees of freedom in succession and the resulting equations are arranged in matrix form as

$$[X(\omega)] = [\Phi] \lceil H(\omega) \rfloor [\Phi]^T \tag{15.87}$$

The matrix $[X(\omega)]$ is a diagonal matrix. $[X(\omega)]$ in Eq. (15.87) is defined as the complex admittance matrix for the whole system and is given a new notation

$$[H(\omega)] = [\Phi][H(\omega)][\Phi]^T \tag{15.88}$$

15.11.1 Cross-spectral Density Matrix for Generalised Forces

We consider transverse loading on the plate only. The overall generalised nodal force vectors at an instant t for a distributed pressure is expressed as

$$\{F(t)\} = \sum_{i=1}^{NE} \iint [N]_i \{p(x, y, t)\}\, dx\, dy \tag{15.89}$$

where NE indicates the number of elements. The correlation matrix for the generalised nodal forces between two instants [Fig. 15.11]

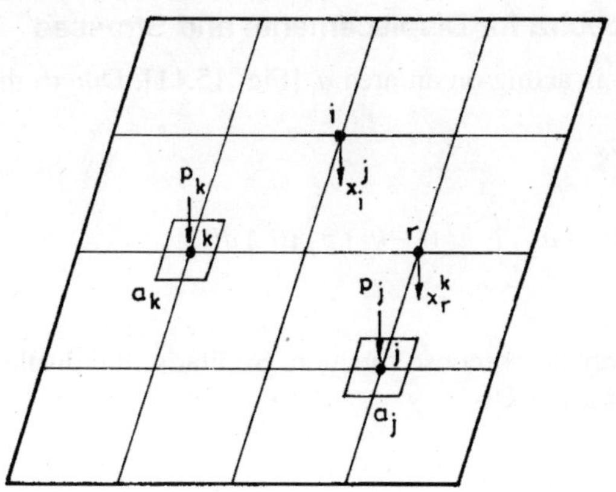

Fig. 15.11 A plate

$$[R_{FF}(t_1, t_2)] = E[\{F(t_1)\}\{F(t_1)\}^T] \tag{15.90}$$

Using the relation in eq. (15.89), Eq. (15.90) becomes

$$[R_{FF}(t_1, t_2)] = \sum_{i=1}^{NE} \iint [N]_i \sum_{j=1}^{NE} \iint [N]_j \, E[p_i\{x_1, y_1, t_1\}$$

$$\times p_j\{x_2, y_2, t_2\}] \, dx\, dy\, dx\, dy \tag{15.91}$$

Since the normal pressure spectra is assumed to be stationary, the correlation function $[R_{FF}(t_1, t_2)]$ is only a function of the time shift τ or $(t_2 - t_1)$.

The cross-spectral density matrix for the generalised forces can be formulated by taking Fourier transform of Eq. (15.91) as

$$S_{FF}(\omega) = \sum_{i=1}^{NE} \iint [N]_i \sum_{j=1}^{NE} \iint [N]_j [S_{pp}(r_1, r_2, \omega)] \, dx\, dy\, dx\, dy \tag{15.92}$$

where $[S_{pp}(r_1, r_2, \omega)]$ is the cross-spectral density matrix of the normal pressure and r_1 and r_2 are arbitrary points. Equation (15.92) reveals that the integration over the whole domain is to be performed twice for the calculation of the generalised nodal force spectra, from the generalised normal pressure spectra at each step of frequency. As such, this integration is tedious and a few simplifications have been made, to reduce the complexity of the problem as well as the computation time. It is assumed that the same cross - spectral density of the pressure nets over the whole contributing area-which modifies Eq. (15.92) as

$$[S_{FF}(\omega)] = \sum_{i=1}^{NE} \iint [N]_i \, dx\, dy \, [S_{pp}(\omega)] \sum_{j=1}^{NE} \iint [N]_j \, dx\, dy$$

or

$$[S_{FF}(\omega)] = \lceil A \rfloor [S_{pp}(\omega)] \lceil A \rfloor \tag{15.93}$$

where $\lceil A \rfloor$ is the diagonal matrix of the area associated with each node.

15.11.2 Response Spectra for Displacements and Stresses

At node j, a pressure p_j is acting on an area a_j [Fig. 15.11]. Due to this, the response at point i of the plate is

$$x_i^j(t) = a_j \int_{-\infty}^{\infty} h_i^j(t - \theta_1)\, p_j(\theta_1)\, d\theta_1 \tag{15.94}$$

where $h(t)$ is the unit impulse response function. Similarly, the displacement at point r due to pressure load applied at point k is

$$x_r^k(t) = a_k \int_{-\infty}^{\infty} h_r^k(t - \theta_2)\, p_k(\theta_2)\, d\theta_2 \tag{15.95}$$

The loads and the responses are stationary random processes. The correlation between the two displacements is

$$R_{x_i^j x_r^k}(\tau) = \lim_{T \to \infty} \frac{1}{T} \int_{-\infty}^{\infty} x_i^j(t)\, x_r^k(t+\tau)\, dt \qquad (15.96)$$

where τ is the time lag.

Combining Eqs. (15.96), (15.94) and (15.95), we get

$$R_{x_i^j x_r^k}(\tau) = a_j a_k \int_{-\infty}^{\infty} \int_{-\infty}^{\infty} R_{p_j p_k}(\tau - \theta_2 + \theta_1)\, h_i^j(\theta_1)\, h_r^k(\theta_2)\, d\theta_1\, d\theta_2 \qquad (15.97)$$

where $R_{p_j p_k}$ is the cross-correlation of the pressure at points j and k. θ_1 and θ_2 are the time transformations.

Taking Fourier transform of Eq. (15.97), we get

$$S_{x_i^j x_r^k}(\omega) = a_j a_k S_{p_j p_k}(\omega)\, H_i^{j*}(\omega)\, H_r^k(\omega) \qquad (15.98)$$

where $S_{p_j p_k}$ is the cross-spectral density of the pressure at points j and k.

The generalised displacement at point i is that resulting from each load applied at different load points, whose number is n, and is given by

$$x_i(t) = \sum_{j=1}^{n} q_i^j(t) \qquad (15.99)$$

Cross-spectral density of two displacements, when loads are applied at all points is

$$R_{x_i x_r}(\tau) = \sum_{j=1}^{n} \sum_{k=1}^{n} R_{x_i^j x_r^k}(\tau) \qquad (15.100)$$

Cross-spectral density of displacements at i and r is

$$S_{x_i x_r}(\omega) = \sum_{j=1}^{n} \sum_{k=1}^{n} S_{x_i^j x_r^k} \qquad (15.101)$$

Combining Eqs (15.98) and (15.101), we get

$$S_{x_i x_r}(\omega) = \sum_{j=1}^{n} \sum_{k=1}^{n} a_j a_k S_{p_j p_k} H_i^{j*}(\omega)\, H_r^k(\omega) \qquad (15.102)$$

Equation (15.102) when written for all cross-spectral densities, becomes

$$[S_{xx}(\omega)] = [H^*(\omega)] \lceil A \rfloor [S_{pp}(\omega)] \lceil A \rfloor [H(\omega)]^T \qquad (15.103)$$

Combining Eqs (15.93) and (15.103), we get

$$[S_{xx}(\omega)] = [(\omega)]\lceil S_{FF}(\omega)\rfloor[H(\omega)]^T \tag{15.104}$$

$[S_{xx}(\omega)]$ is the cross-spectral density matrix for the generalised displacements.

Substituting $[H(\omega)]$ from Eq. (15.88) into Eq. (15.104), we get

$$[S_{xx}(\omega)] = [\Phi][H^*(\omega)][\Phi]^T[S_{FF}(\omega)][\Phi][H(\omega)][\Phi]^T \tag{15.105}$$

The relationship between the nodal displacements of an element and stresses at any point r_1 within the element, given in Eq. (12.101), is reproduced after expressing it as a function of time

$$\sigma(r_1, t) = [D]_{r_1}[B]_{r_2}\{X\}_e \tag{15.106}$$

The cross-correlation matrix for stationary response for stresses between an arbitrary pair of points r_1 and r_2 is

$$[R_{\sigma_1\sigma_2}(r_1, r_2, \omega)] = [D]_{r_1}[B]_{r_1}[R_{xx}(\omega)][B]_{r_2}^T[D]_{r_2}^T \tag{15.107}$$

Taking the Fourier transform of Eq. (15.107), the cross-spectral density matrix of the generalised stresses is given by

$$[S_{\sigma_1\sigma_2}(r_1, r_2, \omega] = [D]_{r_1}[B]_{r_2}[S_{xx}(\omega)][B]_{r_2}^T[D]_{r_2}^T \tag{15.108}$$

Hence, once the cross-spectral densities for the displacements for the nodes of the element are known, the cross-spectral densities of stresses at any point within the element can be determined.

References

15.1 J.G. Bendat and A.G. Piersol, Measurement and Analysis of Random Data, John Wiley and Sons, New York, 1966.

15.2 S.H. Crandall, Random Vibration, The Technology Press of MIT, Massachusetts, Cambridge, 1962.

15.3 F. Dinca and C. Theodosin, Nonlinear and Random Vibrations, Academic Press, 1973.

15.4 Y.K. Lin, Probabilistic Theory of Structural Dynamics, McGraw-Hill, New York, 1967.

15.5 N.H. Newman and E. Rosenblueth, Fundamentals of Earthquake Engineering, Prentice Hall, New Jersey, 1973.

15.6 R.W. Clough and J. Penzien, Dynamics of Structures, 2nd Edition, McGraw-Hill, New York, 1993.

15.7 A.K. Chopra, Dynamics of Structures, Prentice Hall, New Jersey, 1995.

EXERCISE 15

15.1 Compute the autocorrelation of the function shown in the figure.

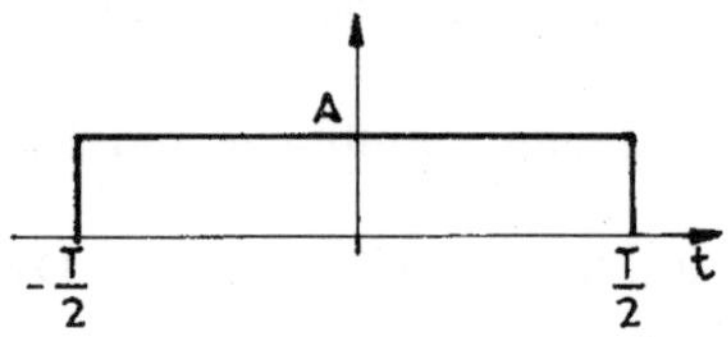

Prob. 15.1

15.2 A random process has a constant spectral density

$$S(f) = 0.5 \text{ mm}^2/\text{cps}$$

within the range 50-1500 cps. The values of the spectral density beyond this range are zeroes. The mean value of the process is 250 mm. Determine its root mean square value and the standard deviation.

15.3 Using the result of autocorrelation of Prob. 15.1, determine the spectral density.

15.4 A simple spring-mass system having damping ratio $\zeta = 0.10$, is excited by a force $F(t) = F_0 \cos \omega t + F_0 \cos 2\omega t + F_0 \cos 3\omega t$. Determine the mean square response of the system.

15.5 A cosine wave is given by

$$x = a_0 + b_0 \cos \omega t$$

Determine the expected values $E[X]$ and $E[X^2]$.

15.6 Determine the autocorrelation function for the wave shown in the figure.

15.7 An idealised water tower having top mass 10,000 kg is subjected to wind loads acting on the mass as shown. The tower has a natural frequency 1 Hz. Determine

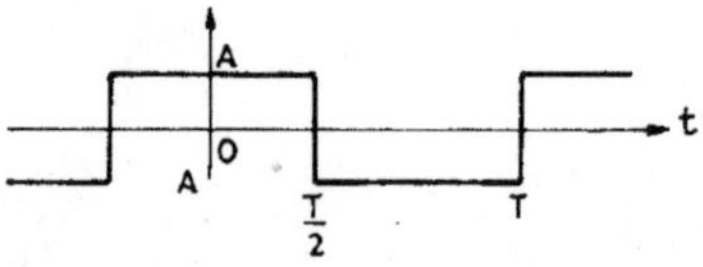

Prob. 15.6

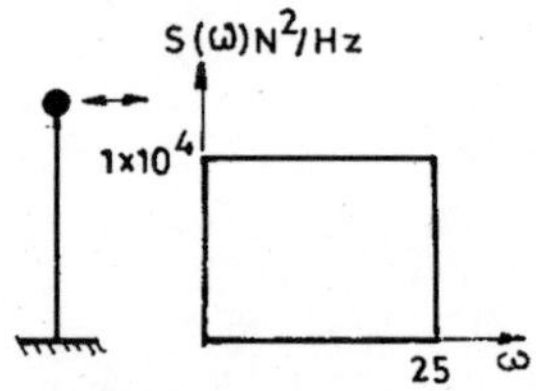

Prob. 15.7

the mean square response and the probability of the tower exceeding a vibration amplitude of 0.01 m. Assume $\zeta = 0.02$.

15.8 Following the steps given in Art. 15.10, determine the random response of a shear beam due to a distributed random load.

15.9 For a random process having a constant spectral density diagram of magnitude S_0 over a band of frequency between ω_1 and ω_2 and zero outside the band, determine the autocorrelation function.

15.10 Consider a random process

$$y(x,t) = A\sin(kx - \omega t)$$

in which k and ω are deterministic, but the amplitude A is random. The probability density of A is shown in the figure. Determine $E[Y]$, $E[Y^2]$ and $R_y(\tau)$.

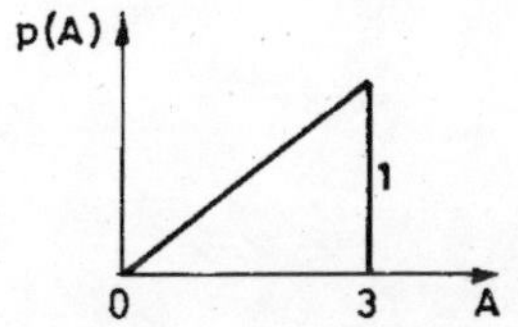

Prob. 15.10

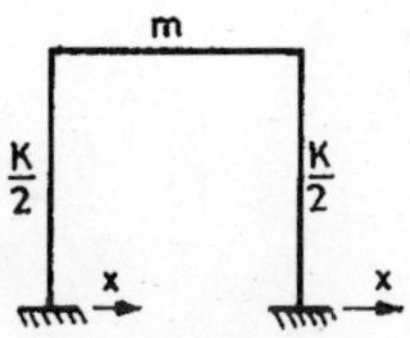

Prob. 15.11

15.11 For the frame shown in the figure, determine the mean square $E[Z^2]$ and the autocorrelation function $R_z(\tau)$ of the relative displacement z, when the ground acceleration $\ddot{x}_g(t)$ is idealised as white noise, with autocorrelation $R_{\ddot{x}\ddot{x}}(\tau) = 2\pi S_0 \delta(\tau)$ where $\delta(\tau)$ is the Dirac-delta function.

15.12 The equation of motion of a SDF system is as follows:

$$\ddot{x} + 2\zeta p\dot{x} + p^2 x = \frac{F(t)}{m}$$

$F(t)$ is a stationary random process with zero mean and a white spectrum S_0. Determine $E[X^2]$ when $\xi = 0$ and 0.1.

15.13 The mass and stiffness matrices of a two-degree of freedom system are as follows:

$$[M] = \begin{bmatrix} 10 & 0 \\ 0 & 20 \end{bmatrix} \quad \text{and} \quad [K] = 5000\begin{bmatrix} 1 & -1 \\ -1 & 2 \end{bmatrix}$$

The free vibration analysis of the system gave the following frequencies and mode shapes:

$$\{p\} = \begin{Bmatrix} 12.1 \\ 29.2 \end{Bmatrix} \quad \text{and} \quad [\Phi] = \begin{bmatrix} 1 & 1 \\ 0.707 & -0.707 \end{bmatrix}$$

Assume $\zeta = 0.05$ and the exscitation force vector as

$$\{F(t)\} = \begin{Bmatrix} 1 \\ 0 \end{Bmatrix} F_e(t)$$

where $F_0(t)$ is a stationary random process with zero mean and white spectral density S_0.

Find the response spectral density and *m* the mean square of the displacement of the first mass.

15.14 Determine the random response at the centre of a uniform flexural beam simply supported at both ends, when a random concentrated load acts at the midspan. The load is considered to be random stationary, with zero mean and with white spectral density $S_0(\omega) = S_0$.

Chapter 16

Computer Programs in Vibration Analysis

16.1 INTRODUCTION

Based on the chapters dealt with, it is revealed that the linear vibration analysis of a structure will involve one or more of the following types: free vibration analysis, forced vibration analysis and random vibration analysis. The free vibration analysis is essentially an eigenvalue problem. If the problem is to be solved on the basis of matrix operations, then the mass matrix and the stiffness matrix of the structure are to be formed. It may be achieved by using the direct stiffness method. Once they are formed, the natural frequencies and the mode shapes may be determined by using any one of the well-known techniques discussed in Chapter 6. In fact, the computer program for the solution of the eigenvalue problem is readily available in text-books [16.1,16.2] and journals [16.3 - 16.5].

16.2 COMPUTER PROGRAM FOR FORCED VIBRATION ANALYSIS

The forced vibration analysis of structures can be carried out in two ways: the direct integration method and the mode superposition method. As in free vibration analysis, the mass matrix and the stiffness matrix of the structure are to be formed. Numerical methods such as Newmark's b method presented in Chapter 7 may be taken resort to for the determination of the response. In order to obtain the response by the mode superposition method, the first step lies in the free vibration analysis of the structure. The natural frequencies and mode shapes thus determined become the input to the problem. In this approach, coupled equations are reduced to n equations of single degree of freedom system. The individual equation of normal mode can be solved by using numerical techniques involving time integration. The outline of the program on forced vibration by mode superposition method, after the free vibration analysis is carried out and the results stored in a low speed magnetic tape, is given in Table 16.1.

16.3 COMPUTER PROGRAM FOR RANDOM VIBRATION ANALYSIS

The computer program for determining response under stochastic loading is much more involved than the deterministic approach. It has been shown in the previous chapter, that the mode superposition method can be applied to the solution of the random vibration problems. The analysis in this case has to be carried out in the frequency domain. This system admittance matrix and the cross-spectral density matrix of the generalised forces are populated square matrices of the order of total number of equations of the problem, If they are stored in the computer memory, they would occupy a large space. Hence, storing these two matrices in a disk is desirable. The outline of the computer program for the random vibration analysis is given in Table 16.2.

Table 16.1 Outline of Computer Program for Program RESPONSE

```
            FORCED VIBRATION
   Input    from low speed memory
            Natural frequencies and mode shapes
            Mass matrix
            Geometric and material properties
   Input    from user
            Damping coefficient
            Time step, duration
            Pressure function
            Initial displacements and velocities
            Displacement and stress output points
   Call     ORTHO (* to orthonormalise the eigenvectors [Φ]^T [M] [Φ]*)
   Call     NEWMARK (*to determine the constants of Newmark's scheme of integration*)
10 Call     FORCE (*to find out nodal forces from pressure function*)
   Call     TRANS (*to calculate [Φ]^T {F (t)}*)
   Cal I    INTIG (*to evaluate displacements, velocities and accelerations at
            the next time step*)
   Write    (displacement, velocity and acceleration)
   If       (stress is required)
Then Call   STRESS (*to generate element stress from nodal displacement*)
            Write (stresses)
   End      If
NSTEP : =   NSTEP + 1
   If       (End of time function is reached) RETURN
            Go to 10
            End
```

The computer program for the free vibration of the framed structure is given in the next section.

Table 16.2 Outline of Computer Program for Random Vibration Analysis

```
Program     RANDOM
Input       from low speed memory
            Natural frequencies and mode shapes
            Mass matrix
            Geometric and material properties
Input       from user
            Damping coefficient
            Starting frequency, terminating frequency, frequency step
            Pressure cross-spectral density function
            No. of blocks for which force spectra is to be calculated and
            block frequency
            Displacements and stress output points
Call        ORTHO (* to orthonormalise the eigenvectors)
```

$$[\Phi]^T [M] [\Phi] = ([I]*0$$

```
Do        i = 1, no. of blocks
          ω = ω block (i)
Call      FORCE (* to form nodal force spectral density matrix*)
          ω = ω (start)
          Do j = 1, nstep (i)
          Call Bw (* to form [H (ω)]*)
          Do K = 1, no. of equations
          Call MULT (* to Multiply [Φ]^T [H (ω)] [Φ]^T for one row*)
          Write results on file
End Do
          Call DISPEC (* to calculate the displacement spectra at desired points
from the inputs of file *)
          Call STRSPC (* to calculate stress spectra *)
          ω : = ω + step
End       Do
End       Do
          Return
          End
```

16.4 COMPUTER PROGRAM FOR FREE VIBRATION ANALYSIS OF FRAMED STRUCTURES

A computer program in FORTRAN is presented in Table 16.3 for the free vibration analysis of space frame structures, having six degrees of freedom per node - three transactions and three rotations. The notations and different steps in the computer program have been explained with the help of comment cards. Table 16.3 gives the main program along with a number of subroutines. GSTIF and GMASS are the overall mass matrix and overall stiffness matrix of the framed structure. ESTIF of the main program is the element stiffness matrix in local coordinates. The elements of ESTIF matrix are generated with the help of the subroutines ELST. They are then pre and post - multiplied by the transpose of the transformation matrix TRA and the transformation matrix respectively, to yield the elements of ESTG matrix. It may be noted that ESTG matrix in the main program is the element stiffness matrix in global coordinates bearing local numbering. The elements of the transformation matrix are generated in subroutine TRAN. ESTG matrix for the individual element is assembled into GSTIF matrix.

It may be noted that while ESTG matrix is a square matrix, GSTIF is a matrix of single array. The purpose of subroutine CRUNCH and SEARCH is to reduce the size of GSTIF matrix, eliminating the zero elements lying in the extreme ends of the band in rows. There is yet another purpose of CRUNCH . It labels the diagonal elements of the overall stiffness matrix. As GSTIF is arranged in a single array, this label is indicated by KIND in the main program, which is referred to as pointer vector.

The boundary restraints have been taken into account by inserting a very high number in the diagonal element of the stiffness matrix corresponding to that restraint. The eigenvalue problem is solved on the basis of total degrees of freedom.

As mentioned in Chapter 6, there are a large number of methods available for the solution of eigenvalue problem. As only the first few natural frequencies are of interest, it is desirable to apply a technique which will serve this purpose. The natural frequencies and mode shapes are determined with the help of subroutine R8USIV. It is based on the simultaneous iteration method of Corr and Jennings [6.12]. The listing of the subroutine is given in the computer program of Table 16.3. All the output statements are in the main program.

Table16.3 Computer Program for the Free Vibration Analysis of Space Frame

```
*        PROGRAM FOR FREE VIBRATION ANALYSIS OF SPACE FRAMES

         parameter (ielt=50,inod=50,ntdf=inod*6,inb=50)
         parameter (ibh=25,itdf=ntdf*ibh)

         common /a/ ntv,nbh,gstif(itdf),gmass(itdf),gload(ntdf),kind(ntdf),
        >mind(ntdf)
         common /b/ nod(ielt,2),xg(inod),yg(inod),zg(inod)
         common /c/ area(ielt),ixx(ielt),iyy(ielt),izz(ielt)

         dimension estif(12,12),tra(12,12),emas(12,12),estg(12,12)
         dimension nodlod(inod),load(inod,6),nbd(inb,7)
         dimension freq(20)

         dimension title(15)

         character ans*2

         real ixx,iyy,izz,len,load,length

         open(1,file='frame.in')
         open(2,file='frame.out')

150      format(15a4)
151      format(/,'Youngs Modulus.....= ',g12.4,/'Density.....= ',g12.4)
152      format(/'No. of elements.....= ',i5,/'No. of nodes.....= ',i5)
153      format(/'Cross-Sectional Properties',/'Element Area
        >Ixx Iyy Izz',/(i5,4g12.4))
154      format(/'Nodal Connectivity',/'Element Nodal
        >Connectivity',/(i5,5x,2i8))
155      format(/'Nodal Coordinates',/'Node X-Cor Y-Cor
        >Z-Cor',/(i5,3g15.6))
156      format(/'No. of boundary points......= ',i5)
157      format(/'Bounary Codes',/'Node Boundary Codes',/(i3,7x,6i2))
158      format(/'No. of loads acting on the structure......= ',i5)
159      format(/'Nodes on which load is acting and its magnitude',/
        >(i3,6g10.4))
160      format(/'Total d.o.f in the structure.....= ',i5,/'Half Bandwidth
        >.....= ',i5)
161      format(/'Element Stiffness Matrix',/(6g12.4))
162      format(/'Element Mass Matrix',/(6g12.4))
163      format(/'Overall Stiffness Matrix',/(6g12.4))
164      format(/'Overall Load Matrix',/(6g12.4))
165      format(/'Overall Mass Matrix',/(6g12.4))
166      format(/'Nodal Displacements',/(i4,6g12.4))
167      format(/'Natural Frequencies',/(i3,5x,g12.4))
168      format(a2)

         print *,'Do you want to carry Static Analysis? If yes enter : y
        >or Y.'
         print *,'Else Free Vibration Analysis is carried out'
         read 168,ans

         read (1,150)title
```

```
      write(2,150)title

*     Specify the material properties
      read(1,*)emod,dens
      write(2,151)emod,dens

*     Specify the no. of elements and nodes in the structure
      read(1,*)nelm,nnod
      write(2,152)nelm,nnod

*     Specify the properties of the elements
      read(1,*)(area(iel),ixx(iel),iyy(iel),izz(iel),iel=1,nelm)

      rite(2,153)(iel,area(iel),ixx(iel),iyy(iel),izz(iel),iel=1,nelm)

*     Specify the nodal connectivity of the elements
      read(1,*)((nod(iel,ind),ind=1,2),iel=1,nelm)
      write(2,154)(iel,(nod(iel,ind),ind=1,2),iel=1,nelm)

*     Specify the nodal coordinates
      read(1,*)(xg(ind),yg(ind),zg(ind),ind=1,nnod)
      write(2,155)(ind,xg(ind),yg(ind),zg(ind),ind=1,nnod)

*     Specift the no. of bounary nodes
      read(1,*)nbp
      write(2,156)nbp

*     Specify the boundary codes
      if(nbp.ne.0)then
      print *,'Specify boundary codes'
      read(1,*)((nbd(ibp,ind),ind=1,7),ibp=1,nbp)

      write(2,157)((nbd(ibp,ind),ind=1,7),ibp=1,nbp)
      endif

      if(ans.eq.'y'.or.ans.eq.'Y')then
*     Specify the loads acting on the members
      read(1,*)nload
      read(1,*)(nodlod(ild),(load(ild,ind),ind=1,6),ild=1,nload)

      write(2,158)nload
      write(2,159)(nodlod(ild),(load(ild,ind),ind=1,6),ild=1,nload)
      endif

      ntv=nnod*6

*     Half bandwidth

      nbh=0
      do iel=1,nelm
      do in =1,2
      do jn =1,2
      nbl = abs(nod(iel,in)-nod(iel,jn))
      if(nbl.gt.nbh)nbh=nbl
      enddo
      enddo
      enddo
      nbh=(nbh+1)*6

      write(2,160)ntv,nbh

*     Calculation of stiffness and mass matrices of the elements

      do 101 iel=1,nelm
```

```
      x1=xg(nod(iel,1))
      x2=xg(nod(iel,2))

      y1=yg(nod(iel,1))
      y2=yg(nod(iel,2))

      z1=zg(nod(iel,1))
      z2=zg(nod(iel,2))

      len = sqrt((x1-x2)*(x1-x2)+(y1-y2)*(y1-y2)+(z1-z2)*(z1-z2))

      call elst(iel,emod,len,estif)
      call tran(iel,x1,x2,y1,y2,z1,z2,len,tra)
      call ttkt(iel,estif,tra,estg)

      if(ans.eq.'y'.or.ans.eq.'Y') go to 103
      call elmas(iel,dens,len,emas)

      do irow=1,12
      estg(irow,irow) = estg(irow,irow)+alpha*emas(irow,irow)
      enddo

103   if(iel.eq.1)then
      write(2,161) (estg(itg,itg),itg=1,12)
      endif

*     Assembly of Element Stiffness Matrix

      call assem(1,iel,estg)

      if(iel.eq.1)then
      write(2,162) (emas(itg,itg),itg=1,12)
      endif

      if(ans.eq.'y'.or.ans.eq.'Y') go to 101
*     Assembly of Element Mass Matrix
      call assem(2,iel,emas)

101   continue

*     Crunching of Overall Stiiffness and Mass matrices

      call crunch(1)

      write(2,163) (gstif(kind(itv)),itv=1,ntv)

      if(ans.eq.'y'.or.ans.eq.'Y') then
      do iload=1,nload
      ind=(nodlod(iload)-1)*6
      do jnd=1,6
      gload(ind+jnd)=gload(ind+jnd)+load(iload,jnd)
      enddo
      enddo

      write(2,164) (gload(itv),itv=1,ntv)

      else
      call crunch(2)

      write(2,165) (gmass(mind(itv)),itv=1,ntv)
      endif

*     Setting Boundary Conditions
```

```
        do 102 i=1,nbp
        ii=nbd(i,1)
        do 102 j=1,6
        if(nbd(i,j+1).eq.0)go to 102
        iii=(ii-1)*6+j
        gstif(kind(iii))=gstif(kind(iii))*1.0e20
102     continue
        if(ans.eq.'y'.or.ans.eq.'Y') then
*       Calculation of Nodal Displacements

        call decom(ier)
        call for
        call bac

        write (2,166)  (ind,(gload((ind-1)*6+jnd),jnd=1,6),ind=1,nnod)

        else
*       Calculation of Natural Frequencies

        print *,'Specify the number of frequencies to be calculated'
        read(*,*)nrqd

        call r8usiv(nrqd,freq,alpha)

*       write(2,167)(irqd,freq(irqd),irqd=1,nrqd)
*       print 167,(irqd,freq(irqd),irqd=1,nrqd)

        endif

        stop
        end

*       Subroutine to calculate the stiffness matrix of the element


        subroutine elst(iel,emod,len,estif)

        parameter (ielt=50,inod=50,ntdf=inod*6,inb=50)

        common /c/ area(ielt),ixx(ielt),iyy(ielt),izz(ielt)

        dimension estif(12,12)

        real len,l,ix,iy,iz,ixx,iyy,izz

        e    =    emod
        l    =    len
        a    =    area(iel)
        g    =    emod/(2.*(1+0.3))
        ix   =    ixx(iel)
        iy   =    iyy(iel)
        iz   =    izz(iel)
*       g    =    80.0E6

*       Geneartion of upper triangle of the stiffness matrix

        estif(1,1)     = e*a/l
        estif(1,7)     =-e*a/l

        estif(2,2)     = 12.0*e*iz/l**3
        estif(2,6)     = 6.0*e*iz/l**2
        estif(2,8)     =-12.0*e*iz/l**3
        estif(2,12)    = 6.0*e*iz/l**2
```

```
      estif(3,3)     = 12.0*e*iy/l**3
      estif(3,5)     =-6.0*e*iy/l**2
      estif(3,9)     =-12.0*e*iy/l**3
      estif(3,11) =-6.0*e*iy/l**2

      estif(4,4)     = g*ix/l
      estif(4,10) =-g*ix/l

      estif(5,5)     = 4.0*e*iy/l
      estif(5,9)     = 6.0*e*iy/l**2
      estif(5,11)    = 2.0*e*iy/l

      estif(6,6)     = 4.0*e*iz/l
      estif(6,8)     =-6.0*e*iz/l**2
      estif(6,12)    = 2.0*e*iz/l

      estif(7,7)     = e*a/l

      estif(8,8)     = 12.0*e*iz/l**3
      estif(8,12)    =-6.0*e*iz/l**2

      estif(9,9)     = 12.0*e*iy/l**3
      estif(9,11)    = 6.0*e*iy/l**2

      estif(10,10)   = g*ix/l

      estif(11,11)   = 4.0*e*iy/l

      estif(12,12)   = 4.0*e*iz/l

*     Geneartion of lower triangle of the stiffness matrix

      do irow=1,12
      do icol=1,12
      estif(icol,irow)=estif(irow,icol)
      enddo
      enddo

      return
      end

*     Subroutine to calculate the transformation matrix of the element

      subroutine tran(iel,x1,x2,y1,y2,z1,z2,len,tra)

      dimension tra(12,12)

      real len

      do irow=1,12
      do icol=1,12
      tra(irow,icol)=0.0
      enddo
      enddo

      cx  = (x2-x1)/len
      cy  = (y2-y1)/len
      cz  = (z2-z1)/len
      cxz = sqrt(cx*cx+cz*cz)

      alpha = 0.0
      pi    = atan(1.)*4.
      alpha = alpha*pi/180.0
```

```
      cosa = cos(alpha)
      sina = sin(alpha)

      if(x1.eq.x2.and.z1.eq.z2.and.y1.ne.y2) then

      tra(1,2) = cy
      tra(2,1) =-cy*cosa
      tra(2,3) = sina
      tra(3,1) = cy*sina
      tra(3,3) = cosa

      else
      tra(1,1) = cx
      tra(1,2) = cy
      tra(1,3) = cz

      tra(2,1) = (-cx*cy*cosa-cz*sina)/cxz
      tra(2,2) = cxz*cosa
      tra(2,3) = (-cy*cz*cosa+cx*sina)/cxz

      tra(3,1) = (cx*cy*sina-cz*cosa)/cxz
      tra(3,2) =-cxz*sina
      tra(3,3) = cy*cz*sina+cx*cosa/cxz

      endif

      do iro=1,3
      do jro=1,3
      do jco=1,3
      ir=iro*3+jro
      ic=iro*3+jco
      tra(ir,ic) = tra(jro,jco)
      enddo
      enddo
      enddo

      return
      end

*     Subroutine to calculate the global stiffness matrix of the element

      subroutine ttkt(iel,estif,tra,estg)

      dimension estif(12,12),estg(12,12),tra(12,12),est(12,12)

      do iro=1,12
      do ico=1,12
      est(iro,ico) = 0.0
      do jro=1,12
      est(iro,ico)=est(iro,ico)+tra(jro,iro)*estif(jro,ico)
      enddo
      enddo
      enddo

      do iro=1,12
      do ico=1,12
      estg(iro,ico) = 0.0
      do jro=1,12
      estg(iro,ico)=estg(iro,ico)+est(iro,jro)*tra(jro,ico)
      enddo
      enddo
      enddo
```

```
      return
      end

*     Subroutine to calculate the global mass matrix of the element

      subroutine elmas(iel,dens,len,emas)

      parameter (ielt=50,inod=50,ntdf=inod*6,inb=50)

      common /c/ area(ielt),ixx(ielt),iyy(ielt),izz(ielt)

      dimension emas(12,12)

      real len,ixx,iyy,izz,mass

      mass = dens*area(iel)*len/2.

      do iro=1,12
      do ico=1,12
      emas(iro,ico) = 0.0
      enddo
      enddo

      emas(1,1) = mass
      emas(2,2) = mass
      emas(3,3) = mass

      emas(7,7) = mass
      emas(8,8) = mass
      emas(9,9) = mass

      return
      end

*     Subroutine for assembling stiffness coefficients

      subroutine assem(ind,ielm,estif)

      parameter (ielt=50,inod=50,ntdf=inod*6,inb=50)
      parameter (ibh=25,itdf=ntdf*ibh)

      common /a/ ntv,nbh,gstif(itdf),gmass(itdf),gload(ntdf),kind(ntdf),
     >mind(ntdf)
      common /b/ nod(ielt,2),xg(inod),yg(inod),zg(inod)

      dimension estif(12,12)

      ndf =6
      nnpe=2

      do 60 i=1,nnpe
      ig=(nod(ielm,i)-1)*ndf
      do 60 j=1,ndf
      mmi=ig+j
      mmm=mmi-1
      imi=(i-1)*ndf+j
      do 60 k=1,nnpe
      jg=(nod(ielm,k)-1)*ndf
      do 60 l = 1,ndf
      nnj=jg+l
      jnj=(k-1)*ndf+l
      if(nnj.gt.mmi) go to 55
      if (mmi.le.nbh) then
      kjj=mmm*(mmm+1)/2+nnj
```

```
      kkj=kjj
      else
      kkj=nbh*(nbh+1)/2
      lrow=(mmm-nbh+1)*nbh+nnj-mmi
      kkj=kkj+lrow
      endif
      if(ind.eq.1) gstif(kkj)=gstif(kkj)+estif(imi,jnj)
      if(ind.eq.2) gmass(kkj)=gmass(kkj)+estif(imi,jnj)
55    continue
60    continue
      return
      end


*     Subroutine crunch

      subroutine crunch(ind)

      parameter (ielt=50,inod=50,ntdf=inod*6,inb=50)
      parameter (ibh=25,itdf=ntdf*ibh)

      common /a/ ntv,nbh,gstif(itdf),gmass(itdf),gload(ntdf),kind(ntdf),
     >mind(ntdf)
      common /b/ nod(ielt,2),xg(inod),yg(inod),zg(inod)

      dimension ak(200)

      ist=1
      iend=0
      do 10 jj=1,nbh
      i=(jj*(jj-1))/2
      do 2 ii=1,jj
      if(ind.eq.1) ak(ii)=gstif(i+ii)
      if(ind.eq.2) ak(ii)=gmass(i+ii)
  2   continue
      call search(ak,jj,ist,iend,ind)
      if(ind.eq.1) kind(jj)=iend
      if(ind.eq.2) mind(jj)=iend
 10   continue
      do 20 jj=nbh+1,ntv
      i=i+nbh
      do 4 ii=1,nbh
      if (ind.eq.1) ak(ii)=gstif(i+ii)
      if (ind.eq.2) ak(ii)=gmass(i+ii)
  4   continue
      call search(ak,nbh,ist,iend,ind)
      if (ind.eq.1) kind(jj)=iend
      if (ind.eq.2) mind(jj)=iend
 20   continue
      return
      end
*     Subroutine search

      subroutine search(ak,nb,ist,iend,ind)

      parameter (ielt=50,inod=50,ntdf=inod*6,inb=50)
      parameter (ibh=25,itdf=ntdf*ibh)

      common /a/ ntv,nbh,gstif(itdf),gmass(itdf),gload(ntdf),kind(ntdf),
     >mind(ntdf)

      dimension ak(200)

      do 20 ii=1,nb
```

```
      if(ak(ii))10,20,10
10    ibl=ii
      go to 30
20    continue
      ibl=nb
30    continue
      icount=nb-ibl+1
      iend=iend+icount
      do 40 ii=ist,iend
      jj=ii-ist+ibl
      if (ind.eq.1) gstif(ii)=ak(jj)
      if (ind.eq.2) gmass(ii)=ak(jj)
40    continue
      ist=ist+icount
      return
      end


*     Subroutine Bac

      subroutine bac
**    Solves lt*v=u by Back Substitution **

      parameter (ielt=50,inod=50,ntdf=inod*6,inb=50)
      parameter (ibh=25,itdf=ntdf*ibh)
      common /a/ npt,nbh,l(itdf),g(itdf),u(ntdf),ld(ntdf),gd(ntdf)
      real l
      integer q

      do 20 i=1,npt
20    u(i)=u(i)/l(ld(i))
      do 22 it=2,npt
      i=npt+2-it
      i1=i-1
      q=ld(i)-i
      m1=1+ld(i-1)-q
      if(m1-i)10,22,10
10    continue
      do 21 j=m1,i1
      u(j)=u(j)-l(j+q)*u(i)/l(ld(j))
21    continue
22    continue
      return
      end

*     Subroutine For
      subroutine for
**    Solves lv=u by Forward Substitution **

      parameter (ielt=50,inod=50,ntdf=inod*6,inb=50)
      parameter (ibh=25,itdf=ntdf*ibh)

      common /a/ npt,nbh,l(itdf),g(itdf),u(ntdf),ld(ntdf),gd(ntdf)
      real l
      integer q

      u(1)=u(1)/l(1)
      do 13 i=2,npt
      q=ld(i)-i
      m1=1+ld(i-1)-q
      el=0.0
      if(m1-i)10,13,10
10    continue
      i1=i-1
```

```
        do 14 j=m1,i1
14      el=el+l(j+q)*u(j)
13      u(i)=(u(i)-el)/l(ld(i))
        return
        end

*       Subroutine Decom

        subroutine decom(ier)
**      Decomposes a Symmetric matrix into Cholesky Factors **

        parameter (ielt=50,inod=50,ntdf=inod*6,inb=50)
        parameter (ibh=25,itdf=ntdf*ibh)

        common /a/ npt,nbh,l(itdf),g(itdf),u(ntdf),ld(ntdf),gd(ntdf)
        real l
        integer q

        n=npt
        xcm=1.0
        l(1)=sqrt(l(1))
        do 1 i=2,n
        q=ld(i)-i
        m1=1+ld(i-1)-q
        do 2 j=m1,i
        nn=0
        el=l(j+q)
        if(j-1)10,2,10
10      continue
        kk=ld(j)-j
        nn=1+ld(j-1)-kk
        nn=max0(nn,m1)
        if(nn-j)15,2,15
15      continue
        j1=j-1
        do 3 k=nn,j1
3       el=el-l(q+k)*l(kk+k)
2       l(j+q)=el/l(ld(j))
        if(el .le. 0)go to 5
        xc=1./l(i+q)
        if(xc-xcm)1,20,20
20      continue
        xcm=xc
        jcm=i
1       l(ld(i))=sqrt(el)
        icm=int(alog10(xcm))+1
        ier=2
        write(*,*) ' n = ',n
        go to 6
5       ier=3
        n=-n
        write(*,*) ' n = ',n
        write(*,*)'error*** Stiffness Matrix Not Positive Definite'
6       return
        end

C       SUBROUTINE TO CALCULATE EIGEN VECTORS
        SUBROUTINE R8USIV(NRQD,BD,ALPHA)
        DIMENSION U(300,20),V(300,20),W(300,20),BD(20),HZ(20),ERR(20)
        COMMON/A/NNF,NBH,L(7500),G(7500),GL(300),LD(300),GD(300)
        INTEGER GD
        REAL L,HZ
        DATA NOI,TOLVEC,GES,IRENT,LET,LOCK,INT/80,1.E-06,0.,0,0,1,1/
        N=NNF
```

```
      ND=N
      NA=GD(N)
      RQD=NRQD
      M=NRQD*2.
      CALL R8URED(NK,N,IER)
      IF(N .LT. 0)RETURN
C     IF(IRENT .EQ. 2)CALL R8UPRE(ND,W,U,NK,M,N,1,IRENT)
      CALL R8URAN(ND,U,N,M,IRENT,GES)
      CALL R8UORT(ND,U,N,M,1)
 50   CALL R8UBAC(ND,U,V,NK,N,M,LOCK)
      CALL R8UPRE1(ND,V,W,NA,N,M,LOCK,0)
      CALL R8UFOR(ND,W,V,NK,N,M,LOCK)
      CALL R8UDEC(ND,U,V,W,BD,N,M,LOCK,TOLVEC)
      CALL R8URAN(ND,W,N,M,IRENT,GES)
      CALL R8UORT(ND,W,N,M,LOCK)
      CALL R8UERR(ND,U,W,BD,ERR,N,M,NRQD,LOCK,TOLVEC,LET)
      DO 20 I=1,N
      DO 20 J=LOCK,M
 20   U(I,J)=W(I,J)
      IF(LET-1)15,10,15
 15   CONTINUE
      IF(NOI-INT)11,16,16
 16   CONTINUE
      INT=INT+1
      GO TO 50
 10   IER=0
      GO TO 120
 11   IER=1
 120  CALL R8UBAC(ND,U,W,NK,N,M,1)
      DO 121 J=1,M
      EL=0.0
      IF(BD(J) .LE. 0.0) GO TO 140
      IF(J .LE. NRQD) BD(J)=1./SQRT(BD(J))
 140  CONTINUE
      DO 122 I=1,N
      IF(ABS(W(I,J))-EL)130,130,125
 125  CONTINUE
      EL=ABS(W(I,J))
 130  CONTINUE
 122  CONTINUE
      IF (EL .EQ. 0.) GO TO 150
      EL=1./EL
      DO 123 I=1,N
 123  W(I,J)=W(I,J)*EL
 121  CONTINUE
 150  CONTINUE

      PI=ATAN(1.)*4.
      DO 135 I=1,NRQD
      BD(I) = ABS(BD(I)*BD(I)-ALPHA)
      BD(I) = SQRT(BD(I))
      HZ(I) = BD(I)/2./PI
      HZ(I) = SQRT(BD(I))
 135  CONTINUE

      WRITE(2,216)
 216  FORMAT(/12X,'***** FREE VIBRATION RESULTS *****')
      WRITE(2,211)INT,IER
 211  FORMAT(/'ITEARATION = ',I5,4X,'ERROR CODE = ',I5)
      WRITE(2,212)(I,BD(I),HZ(I),I=1,NRQD)
      PRINT 212,(I,BD(I),HZ(I),I=1,NRQD)
 212  FORMAT(/'EIGENVALUES ARE'/'Mode in rad/sec
     >in Hz'/(I3,5x,F15.8,3x,F15.8))
      WRITE(2,213)
```

```
213     FORMAT(/'MODE SHAPES')
        DO 210 I=1,NRQD
210     WRITE(2,214)I,BD(I),(LL1,(W((LL1-1)*6+J,I),J=1,6),LL1=1,NNF/6)
214     FORMAT(/'MODE SHAPE = ',I3,10X,'NATURAL FREQUENCY = ',F12.6,
       >/(I3,6G12.4))
*       WRITE(2,215)(I,ERR(I),I=1,NRQD)
215     FORMAT(/'ERROR :',/(I3,5X,G15.6))
        RETURN
        END
***************************SUBROUTINE R8URED*****************************
        SUBROUTINE R8URED(NK,N,IER)
**      DECOMPOSES A SYMMETRIC MATRIX INTO CHOLESKY FACTORS **
        COMMON/A/NNF,NBH,L(7500),G(7500),GL(300),LD(300),GD(300)
        INTEGER Q,GD
        REAL L
        XCM=1.0
        L(1)=SQRT(L(1))
        DO 1 I=2,N
        Q=LD(I)-I
        M1=1+LD(I-1)-Q
        DO 2 J=M1,I
        NN=0
        EL=L(J+Q)
        IF(J-1)10,2,10
10      CONTINUE
        KK=LD(J)-J
        NN=1+LD(J-1)-KK
        NN=MAX0(NN,M1)
        IF(NN-J)15,2,15
15      CONTINUE
        J1=J-1
        DO 3 K=NN,J1
3       EL=EL-L(Q+K)*L(KK+K)
2       L(J+Q)=EL/L(LD(J))
        IF(EL .LE. 0)GO TO 5
        XC=1./L(I+Q)
        IF(XC-XCM)1,20,20
20      CONTINUE
        XCM=XC
        JCM=I
1       L(LD(I))=SQRT(EL)
        ICM=INT(ALOG10(XCM))+1
        IER=2
        GO TO 6
5       IER=3
        N=-N
        WRITE(2,*)'ERROR*** STIFFNESS MATRIX NOT POSITIVE DEFINITE'
6       RETURN
        END
***************************SUBROUTINE R8UPRE1****************************
        SUBROUTINE R8UPRE1(ND,U,V,NA,N,M,LOCK,IRENT)
**      PERFORMS PREMATMULTIPLICATION V=AU **
        DIMENSION U(300,20),V(300,20)
        COMMON/A/NNF,NBH,G(7500),L(7500),GL(300),GD(300),LD(300)
        INTEGER Q,GD
        REAL L,LK
        DO 1 I=1,N
        LK=L(LD(I))
        DO 1 K=LOCK,M
1       V(I,K)=U(I,K)*LK
        DO 8 I=2,N
        Q=LD(I)-I
        M1=1+LD(I-1)-Q
        DO 8 K=LOCK,M
```

```
        IF(M1-I)10,8,10
 10     CONTINUE
        IF(IRENT-2)15,6,15
 15     CONTINUE
        I1=I-1
        DO 5 J=M1,I1
 5      V(I,K)=V(I,K)+L(J+Q)*U(J,K)
 6      CONTINUE
        DO 4 J=M1,I1
        V(J,K)=V(J,K)+L(J+Q)*U(I,K)
 4      CONTINUE
 8      CONTINUE
        RETURN
        END
*************************SUBROUTINE R8URAN******************************
        SUBROUTINE R8URAN(ND,U,N,M,IRENT,GES)
**      INPUTS RANDOM NUMBERS INTO U **
        DIMENSION U(300,20)
        K=1
        IF(IRENT .NE. 0)K=M
        IRENT=1
        DO 1 I=1,N
        DO 1 J=K,M
        IF(GES .EQ. 0.0)GES=0.31415926
        Y=GES*GES
 2      Y=Y*10.0
        IF(Y-1.0)2,5,5
 5      CONTINUE
        NZ=Y
        Y=Y-NZ
        U(I,J)=Y
 1      GES=Y
        RETURN
        END
*************************SUBROUTINE R8UORT******************************
        SUBROUTINE R8UORT(ND,W,N,M,LOCK)
**      ORTHONORMALISES VECTORS IN W BY SCHMIDT PROCESS **
        DIMENSION W(300,20)
        DO 1 I=LOCK,M
        DO 1 J=1,I
        EL=0.0
        DO 4 K=1,N
 4      EL=EL+W(K,J)*W(K,I)
        IF(I-J)10,5,10
 10     CONTINUE
        DO 6 K=1,N
 6      W(K,I)=W(K,I)-EL*W(K,J)
        GO TO 1
 5      D=1.0/SQRT(EL)
        DO 3 K=1,N
        W(K,I)=D*W(K,I)
 3      CONTINUE
 1      CONTINUE
        RETURN
        END
*************************SUBROUTINE R8UBAC******************************
        SUBROUTINE R8UBAC(ND,U,V,NK,N,M,LOCK)
**      SOLVES LT*V=U BY BACK SUBSTITUTION **
        DIMENSION U(300,20),V(300,20)
        COMMON/A/NNF,NBH,L(7500),G(7500),GL(300),LD(300),GD(300)
        INTEGER Q,GD
        REAL L
        DO 20 K=LOCK,M
        DO 20 I=1,N
```

```
 20       V(I,K)=U(I,K)/L(LD(I))
          DO 22 IT=2,N
          I=N+2-IT
          I1=I-1
          Q=LD(I)-I
          M1=1+LD(I-1)-Q
          IF(M1-I)10,22,10
 10       CONTINUE
          DO 21 J=M1,I1
          DO 21 K=LOCK,M
          V(J,K)=V(J,K)-L(J+Q)*V(I,K)/L(LD(J))
 21       CONTINUE
 22       CONTINUE
          RETURN
          END
***************************SUBROUTINE R8UFOR*******************************
          SUBROUTINE R8UFOR(ND,U,V,NK,N,M,LOCK)
**        SOLVES LV=U BY FORWARD SUBSTITUTION **
          DIMENSION U(300,20),V(300,20)
          COMMON/A/NNF,NBH,L(7500),G(7500),GL(300),LD(300),GD(300)
          INTEGER GD,Q
          REAL L
          DO 12 K=LOCK,M
 12       V(1,K)=U(1,K)/L(1)
          DO 13 I=2,N
          Q=LD(I)-I
          M1=1+LD(I-1)-Q
          DO 13 K=LOCK,M
          EL=0.0
          IF(M1-I)10,13,10
 10       CONTINUE
          I1=I-1
          DO 14 J=M1,I1
 14       EL=EL+L(J+Q)*V(J,K)
 13       V(I,K)=(U(I,K)-EL)/L(LD(I))
          RETURN
          END
***************************SUBROUTINE R8UDEC*******************************
          SUBROUTINE R8UDEC(ND,U,V,W,BD,N,M,LOCK,TOLVEC)
**        SORTS VECTORS IN U,V IN ASCENDING ORDER OF EIGENVALUES **
          DIMENSION U(300,20),V(300,20),W(300,20),BD(20)
 C        THIS SECTION CALCULATES THE EIGENVALUES
          DO 1 J=LOCK,M
          EL=0.0
          DO 2 I=1,N
   2      EL=EL+U(I,J)*V(I,J)
   1      BD(J)=EL
 C        THIS SECTION TESTS THAT THE EIGEN VALUES ARE IN
 C        DECENDING ORDER AND IF NOT SORTS THEM
          LOC1=LOCK+1
          DO 3 J=LOC1,M
          I=J
   4      K=I-1
          PQ=ABS(BD(K))-ABS(BD(I))
          IF(PQ)20,20,3
  20      CONTINUE
          EL=BD(I)
          BD(I)=BD(K)
          BD(K)=EL
          DO 5 II=1,N
          EL=U(II,I)
          EL1=V(II,I)
          U(II,I)=U(II,K)
          V(II,I)=V(II,K)
```

```
      U(II,K)=EL
      V(II,K)=EL1
  5   CONTINUE
      I=I-1
      KKL=I-1-LOCK
      IF(KKL)3,3,4
  3   CONTINUE
C     THIS SECTION CALCULATES THE INTERACTION BETWEEN
C     THE ITERATION VECTORS AND DECOUPLE THEM
 39   DO 6 I=1,N
      DO 6 J=LOCK,M
  6   W(I,J)=V(I,J)
      DO 8 I=LOCK,M
      I1=I-1
      IF(I-1)25,8,25
 25   CONTINUE
      DO 7 J=LOCK,I1
      EL=0.0
      DO 9 K=1,N
  9   EL=EL+U(K,I)*V(K,J)
      EL=-2.*EL
      Q=BD(I)-BD(J)
      IF(ABS(Q/BD(J))-TOLVEC)30,30,10
 30   CONTINUE
      IF(ABS(EL/BD(J))-TOLVEC)40,40,10
 40   CONTINUE
      EL=0.0
      GO TO 7
 10   ELL=SQRT(Q*Q+4.*EL*EL)
      IF(Q .LT. 0.0)ELL=-ELL
      EL=EL/(Q+ELL)
      DO 13 K=1,N
      W(K,I)=W(K,I)-EL*V(K,J)
 13   W(K,J)=W(K,J)+EL*V(K,I)
  7   CONTINUE
  8   CONTINUE
      RETURN
      END
***************************SUBROUTINE R8UERR*******************************
      SUBROUTINE R8UERR(ND,U,W,BD,ERR,N,M,NRQD,LOCK,TOLVEC,LET)
**    ESTIMATES ERRORS IN COMPUTING EIGENVECTORS **
      DIMENSION U(300,20),W(300,20),BD(20),ERR(20)
      AN=N
      DO 2 J=LOCK,M
      ER=0.0
      IF(BD(J))6,6,50
 50   CONTINUE
      DO 1 I=1,N
      EL=U(I,J)-W(I,J)
  1   ER=ER+EL*EL
      GO TO 2
  6   DO 8 I=1,N
      EL=U(I,J)+W(I,J)
  8   ER=ER+EL*EL
  2   ERR(J)=SQRT(ER/AN)
  5   IF(TOLVEC-ERR(LOCK))4,3,3
  3   ERR(LOCK)=-ERR(LOCK)
      LOCK=LOCK+1
      IF(NRQD-LOCK)10,5,5
 10   CONTINUE
      LET=1
  4   RETURN
      END
```

```
*        emod   = Young's modulus of elasticity
*        dens   = density of the material
*        nelm   = n0. of elements
*        node   = no. of nodes in the structure
*        area   = cross sectional area of the member
*        ixx, lyy
         izz    = second moment
*        nod    = nodal connectivity
*        nbp    = no. of boundary points
*        nbd    = boundary codes at each boundary point
*        ntv    = half-band width
*        estif = elements element stiffness matrix
*        emas   = element mass matrix
*        gstif =  global stiffness matrix
*        nrgd   = no. of frequencies to be calculated
*        frcq   = natural frequency
```

16.4.1 A plane frame problem

A portal frame structure shown in Fig. 16.1 has been analysed by the computer program given in Table 16.3.

As such in order to use the computer program for the analysis of plane frame, constraints have been put in the boundary conditions along the out of plane direction. The node numbers and the member numbers have been indicated in Fig. 16.1. The displacement labels are related to the nodes. Total degrees of freedom in the computer program are the total number of nodes in the structure multiplied by the degrees of freedom per node. It can be seen in Fig. 16.1 that each member of the frame has been divided into four elements. The input data of the problem is given in Table 16.4. The output file has been presented in Table 16.5.

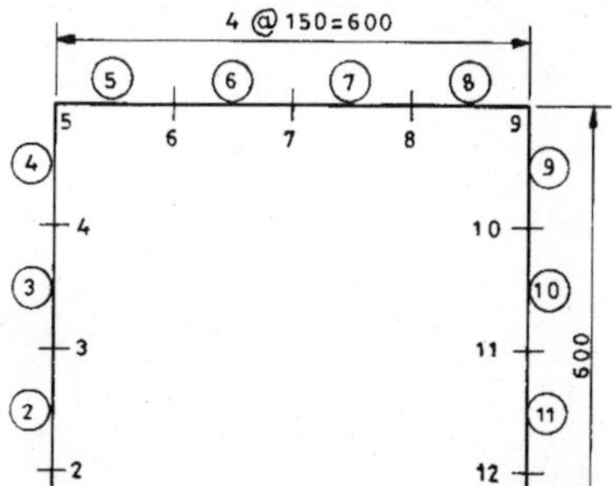

Fig. 16.1 A plane frame

It may be mentioned that Table 16.3 contains the computer program of the space frame.

Table 16.4 Input Data for the Portal Frame of Fig. 16.1

```
FREE VIBRATION ANALYSIS OF A PORTAL FRAME
30.0E6 0.00073381
12 13
29.39 1.0E8 1.0E8 3021.0
29.39 1.0E8 1.0E8 3021.0
```

```
29.39  1.0E8  1.0E8  3021.0
29.39  1.0E8  1.0E8  3021.0
29.39  1.0E8  1.0E8  3021.0
29.39  1.0E8  1.0E8  3021.0
29.39  1.0E8  1.0E8  3021.0
29.39  1.0E8  1.0E8  3021.0
29.39  1.0E8  1.0E8  3021.0
29.39  1.0E8  1.0E8  3021.0
29.39  1.0E8  1.0E8  3021.0
29.39  1.0E8  1.0E8  3021.0
1  2
2  3
3  4
4  5
5  6
6  7
7  8
8  9
9  10
10  11
11  12
12  13
0.0      0.0      0.0
0.0      150.0    0.0
0.0      300.0    0.0
0.0      450.0    0.0
0.0      600.0    0.0
150.0    600.0    0.0
300.0    600.0    0.0
450.0    600.0    0.0
600.0    600.0    0.0
600.0    450.0    0.0
600.0    300.0    0.0
600.0    150.0    0.0
600.0    0.0      0.0
13
1   1  1  1  1  1  1
2   0  0  1  1  1  0
3   0  0  1  1  1  0
4   0  0  1  1  1  0
5   0  0  1  1  1  0
6   0  0  1  1  1  0
7   0  0  1  1  1  0
8   0  0  1  1  1  0
9   0  0  1  1  1  0
10  0  0  1  1  1  0
11  0  0  1  1  1  0
12  0  0  1  1  1  0
13  1  1  1  1  1  1
```

Table 16.5 Output of the Portal Frame of Fig. 16.1

```
FREE  VIBRATION  ANALYSIS  OF  A  PORTAL  FRAME
```

```
Youngs Modulus.....=  0.3000E+08
Density.....=  0.7338E-03

No. of elements.....=  12
No. of nodes.....=  13
Cross-Sectional Properties
Element   Area         Ixx            Iyy          Izz
  1       29.39     0.1000E+09     0.1000E+09     3021.
  2       29.39     0.1000E+09     0.1000E+09     3021.
  3       29.39     0.1000E+09     0.1000E+09     3021.
  4       29.39     0.1000E+09     0.1000E+09     3021.
  5       29.39     0.1000E+09     0.1000E+09     3021.
  6       29.39     0.1000E+09     0.1000E+09     3021.
  7       29.39     0.1000E+09     0.1000E+09     3021.
  8       29.39     0.1000E+09     0.1000E+09     3021.
  9       29.39     0.1000E+09     0.1000E+09     3021.
 10       29.39     0.1000E+09     0.1000E+09     3021.
 11       29.39     0.1000E+09     0.1000E+09     3021.
 12       29.39     0.1000E+09     0.1000E+09     3021.

Nodal Connectivity
Element      Nodal       Connectivity
   1           1             2
   2           2             3
   3           3             4
   4           4             5
   5           5             6
   6           6             7
   7           7             8
   8           8             9
   9           9            10
  10          10            11
  11          11            12
  12          12            13

Nodal Coordinates
Node      X-Cor            Y-Cor            Z-Cor
  1    0.000000E+00    0.000000E+00    0.000000E+00
  2    0.000000E+00      150.000       0.000000E+00
  3    0.000000E+00      300.000       0.000000E+00
  4    0.000000E+00      450.000       0.000000E+00
  5    0.000000E+00      600.000       0.000000E+00
  6      150.000         600.000       0.000000E+00
  7      300.000         600.000       0.000000E+00
  8      450.000         600.000       0.000000E+00
  9      600.000         600.000       0.000000E+00
 10      600.000         450.000       0.000000E+00
 11      600.000         300.000       0.000000E+00
 12      600.000         150.000       0.000000E+00
 13      600.000       0.000000E+00    0.000000E+00

No. of boundary points......=  13

Bounary Codes
Node       Boundary Codes
 1          1  1  1  1  1  1
```

```
     2        0  0  1  1  1  0
     3        0  0  1  1  1  0
     4        0  0  1  1  1  0
     5        0  0  1  1  1  0
     6        0  0  1  1  1  0
     7        0  0  1  1  1  0
     8        0  0  1  1  1  0
     9        0  0  1  1  1  0
    10        0  0  1  1  1  0
    11        0  0  1  1  1  0
    12        0  0  1  1  1  0
    13        1  1  1  1  1  1

Total  d.o.f  in  the  structure.....=  78
Half  Bandwidth.....=  12

Element  Stiffness  Matrix
0.3222E+060.5878E+070.1067E+110.8000E+140.7692E+13 0.2417E+10
0.3222E+060.5878E+070.1067E+110.8000E+140.7692E+13 0.2417E+10

Element  Mass  Matrix
1.618     1.618    1.618     0.0000E+00    0.0000E+00    0.0000E+00
1.618     1.618    1.618     0.0000E+00    0.0000E+00    0.0000E+00

Overall  Stiffness  Matrix
0.3222E+06 0.5878E+070.1067E+110.8000E+14 0.7692E+130.2417E+10
0.6445E+06 0.1176E+080.2133E+110.1600E+15 0.1538E+140.4834E+10
0.6445E+06 0.1176E+080.2133E+110.1600E+15 0.1538E+140.4834E+10
0.6445E+06 0.1176E+080.2133E+110.1600E+15 0.1538E+140.4834E+10
0.6200E+07 0.6200E+070.2133E+110.8769E+14 0.8769E+140.4834E+10
0.1176E+08 0.6445E+060.2133E+110.1538E+14 0.1600E+150.4834E+10
0.1176E+08 0.6445E+060.2133E+110.1538E+14 0.1600E+150.4834E+10
0.1176E+08 0.6445E+060.2133E+110.1538E+14 0.1600E+150.4834E+10
0.6200E+07 0.6200E+070.2133E+110.8769E+14 0.8769E+140.4834E+10
0.6445E+06 0.1176E+080.2133E+110.1600E+15 0.1538E+140.4834E+10
0.6445E+06 0.1176E+080.2133E+110.1600E+15 0.1538E+140.4834E+10
0.6445E+06 0.1176E+080.2133E+110.1600E+15 0.1538E+140.4834E+10
0.3222E+06 0.5878E+070.1067E+110.8000E+14 0.7692E+13   0.2417E+10

Overall  Mass  Matrix
1.618     1.618    1.618   0.0000E+00    0.0000E+00    0.0000E+00
3.235     3.235    3.235   0.0000E+00    0.0000E+00    0.0000E+00
3.235     3.235    3.235   0.0000E+00    0.0000E+00    0.0000E+00
3.235     3.235    3.235   0.0000E+00    0.0000E+00    0.0000E+00
3.235     3.235    3.235   0.0000E+00    0.0000E+00    0.0000E+00
3.235     3.235    3.235   0.0000E+00    0.0000E+00    0.0000E+00
3.235     3.235    3.235   0.0000E+00    0.0000E+00    0.0000E+00
3.235     3.235    3.235   0.0000E+00    0.0000E+00    0.0000E+00
3.235     3.235    3.235   0.0000E+00    0.0000E+00    0.0000E+00
3.235     3.235    3.235   0.0000E+00    0.0000E+00    0.0000E+00
3.235     3.235    3.235   0.0000E+00    0.0000E+00    0.0000E+00
3.235     3.235    3.235   0.0000E+00    0.0000E+00    0.0000E+00
1.618     1.618    1.618   0.0000E+00    0.0000E+00    0.0000E+00

                   *****  FREE  VIBRATION  RESULTS  *****

   ITEARATION  =  7  ERROR  CODE  =  0
```

```
EIGENVALUES ARE
Mode        in  rad/sec          in  Hz
  1        18.16512299        4.26205635
  2        71.57073212        8.45994854
  3       115.78981018       10.76056767
```

```
MODE  SHAPES

MODE SHAPE = 1                   NATURAL FREQUENCY = 18.165123
1   -0.1276E-21   -0.4709E-23   0.0000E+00   0.0000E+00   0.0000E+00    0.5374E-23
2   -0.1357       -0.4709E-03   0.0000E+00   0.0000E+00   0.0000E+00    0.1639E-02
3   -0.4417       -0.9417E-03   0.0000E+00   0.0000E+00   0.0000E+00    0.2276E-02
4   -0.7723       -0.1412E-02   0.0000E+00   0.0000E+00   0.0000E+00    0.1987E-02
5   -0.9996       -0.1883E-02   0.0000E+00   0.0000E+00   0.0000E+00    0.9333E-03
6   -0.9999        0.5135E-01   0.0000E+00   0.0000E+00   0.0000E+00   -0.1092E-03
7   -1.000        -0.8320E-07   0.0000E+00   0.0000E+00   0.0000E+00   -0.4589E-03
8   -0.9999       -0.5135E-01   0.0000E+00   0.0000E+00   0.0000E+00   -0.1092E-03
9   -0.9996        0.1883E-02   0.0000E+00   0.0000E+00   0.0000E+00    0.9333E-03
10  -0.7723        0.1412E-02   0.0000E+00   0.0000E+00   0.0000E+00    0.1987E-02
11  -0.4417        0.9417E-03   0.0000E+00   0.0000E+00   0.0000E+00    0.2276E-02
12  -0.1357        0.4709E-03   0.0000E+00   0.0000E+00   0.0000E+00    0.1639E-02
13  -0.1276E-21    0.4709E-23   0.0000E+00   0.0000E+00   0.0000E+00    0.5374E-23

MODE SHAPE = 2                   NATURAL FREQUENCY = 71.570732
1   -0.4535E-21   -0.3435E-22   0.0000E+00   0.0000E+00   0.0000E+00    0.9954E-23
2   -0.2079       -0.3435E-02   0.0000E+00   0.0000E+00   0.0000E+00    0.2168E-02
3   -0.4903       -0.6861E-02   0.0000E+00   0.0000E+00   0.0000E+00    0.1135E-02
4   -0.4603       -0.1027E-01   0.0000E+00   0.0000E+00   0.0000E+00   -0.1661E-02
5   -0.1571E-02   -0.1364E-01   0.0000E+00   0.0000E+00   0.0000E+00   -0.4266E-02
6   -0.7866E-03   -0.6844       0.0000E+00   0.0000E+00   0.0000E+00   -0.3865E-02
7    0.1831E-06   -1.000        0.0000E+00   0.0000E+00   0.0000E+00   -0.4025E-09
8    0.7869E-03   -0.6844       0.0000E+00   0.0000E+00   0.0000E+00    0.3865E-02
9    0.1571E-02   -0.1364E-01   0.0000E+00   0.0000E+00   0.0000E+00    0.4266E-02
10   0.4603       -0.1027E-01   0.0000E+00   0.0000E+00   0.0000E+00    0.1661E-02
11   0.4903       -0.6861E-02   0.0000E+00   0.0000E+00   0.0000E+00   -0.1135E-02
12   0.2079       -0.3435E-02   0.0000E+00   0.0000E+00   0.0000E+00   -0.2168E-02
13   0.4535E-21   -0.3435E-22   0.0000E+00   0.0000E+00   0.0000E+00   -0.9954E-23

MODE SHAPE = 3                   NATURAL FREQUENCY = 115.789810
1   -0.1549E-20    0.9854E-23   0.0000E+00   0.0000E+00   0.0000E+00    0.2781E-22
2   -0.5244        0.9854E-03   0.0000E+00   0.0000E+00   0.0000E+00    0.4928E-02
3   -1.000         0.1964E-02   0.0000E+00   0.0000E+00   0.0000E+00    0.2897E-03
4   -0.5435        0.2927E-02   0.0000E+00   0.0000E+00   0.0000E+00   -0.5706E-02
5    0.3642        0.3869E-02   0.0000E+00   0.0000E+00   0.0000E+00   -0.4750E-02
6    0.3683       -0.2998       0.0000E+00   0.0000E+00   0.0000E+00    0.4783E-03
7    0.3697       -0.1090E-05   0.0000E+00   0.0000E+00   0.0000E+00    0.2759E-02
8    0.3683        0.2998       0.0000E+00   0.0000E+00   0.0000E+00    0.4784E-03
9    0.3642       -0.3869E-02   0.0000E+00   0.0000E+00   0.0000E+00   -0.4750E-02
10  -0.5435       -0.2927E-02   0.0000E+00   0.0000E+00   0.0000E+00   -0.5706E-02
11  -1.000        -0.1964E-02   0.0000E+00   0.0000E+00   0.0000E+00    0.2898E-03
12  -0.5244       -0.9855E-03   0.0000E+00   0.0000E+00   0.0000E+00    0.4928E-02
13  -0.1549E-20   -0.9855E-23   0.0000E+00   0.0000E+00   0.0000E+00    0.2780E-22
```

16.5 COMPUTER PROGRAM FOR THE FREE VIBRATION ANALYSIS OF SHIPS BY FLEXIBILITY MATRIX METHOD

A computer program in FORTRAN is presented in Table 16.6 for the free vibration analysis of ships by the flexibility matrix method which has been discussed in section 11.4. The input data has been explained with the help of comment cards. There are two subroutines placed after the main program. The subroutine INCOF generates the influence coefficients at different stations. It takes into account both the bending and the shear deformation. The eigenvalue solver is subroutine GENSS. It deals with solid matrix. GENSS is based on QR transformation. As such it can solve eigenvalue problems dealing with unsymmetric matrices. It does not give results unless all eigenvalues are calculated. As the size of the matrices are rather small for this problem, no attempt has been made to make arrangements for the economic storage of matrices.

Input data is presented for the oil tanker given in Example 11.1. Table 16.7. Data included all the particulars of 20 solutions at which the ship is divided. Output given in Table 16.8 contains only the natural frequencies and mode shapes for first four modes.

Table 16.6 Computer program for Free Vibration Analysis of Ships by the flexibility Matrix Method.

```
C         DETERMINATION OF NATURAL FREQUENCIES AND MODE SHAPES OF SHIPS
          DIMENSION A(50,50),AM(30),AA(50),BB(50),X(50),T(50,50)
          DIMENSION AI(50),XX(50),AS(50)
          DIMENSION Y(50),YN(50),YON(50),YNN(50),A1(50),B1(50)
          DIMENSION XON(50),ADDM(50),YS(50),YT(50),YZ(12),CRLC(50)
          DIMENSION ADDC(50)
          DIMENSION AMM(50),DIF(50,50),ASEC(50),AC(50,50)
          COMMON A,T
          OPEN(10,FILE='sin.dat')
          OPEN(12,FILE='sout')
C         MATRIX ANALYSIS OF FREE VIBRATION OF SHIPS
C         ALL UNITS ARE IN METERS AND TONNES
          E=2.076*(10.0**7.0)
C         MOD IS THE NUMBER OF MODES FOR WHICH FREQUENCIES ARE TO BE
          STUDIED
          READ(10,*)MOD,N,D
          G=9.81
          READ(10,*)(YZ(I),I=1,MOD)
C         D=DISTANCE BETWEEN THE STATIONS
C         N=TOTAL NUMBER OF STATIONS
          MODE=1
          L=N-1
          READ(10,*)(AI(I),I=1,N)
C         AI IS THE MOMENT OF INERTIA AT N STATIONS
          DO 305 I=1,N
          AI(I)=AI(I)*E
          BX=I-1
          XX(I)=D*BX
    305   CONTINUE
C         READ(10,*)(XX(I),I=1,N)
          DO 2 I=2,N
          CALL INCOF(AI,XX,N,I,D,AS)
          LL=I-1
          DO 2 J=1,L
          AC(J,LL)=AS(J)
    2     CONTINUE
          READ(10,*)(AMM(I),I=1,N)
C         AMM=WEIGHT AT N STATIONS
          READ(10,*)(ADDM(I),I=1,N)
          READ(10,*)(ASEC(I),I=1,N)
          DK=0.0
          I=1
```

```
  902     J=I+1
          DIF(I,I)=DK+(2.6*D)/(E*ASEC(J))
C         DIF(I,I)=DK+(2.6*D)/(ASEC(J))
          DK=DIF(I,I)
          DO 901 K=J,L
          DIF(I,K)=DIF(I,I)
  901     DIF(K,I)=DIF(I,I)
          IF(J-L)903,904,904
  903     I=I+1
          GO TO 902
  904     DIF(J,J)=DK+(2.6*D)/(E*ASEC(J+1))
C 904     DIF(J,J)=DK+(2.6*D)/(ASEC(J+1))
          DO 421 I=1,L
          DO 421 J=1,L
          421 AC(I,J)=AC(I,J)+DIF(I,J)
          410 DO 407 I=1,N
          ADDC(I)=ADDM(I)*YZ(MODE)
*         AM(I)=(AMM(I)+ADDM(I))/G
          AM(I)=(AMM(I)+ADDC(I))/G
407       CONTINUE
          BM=AM(1)
          S=0.0
          BI=0.0
          DO 30 I=2,N
          BM=BM+AM(I)
          S=S+AM(I)*XX(I)
          BI=BI+AM(I)*XX(I)**2.0
          K=I-1
          X(K)=XX(I)
          AM(K)=AM(I)
  30      CONTINUE
          N=N-1
          P=1.0/(BM*BI-S**2.0)
          DO 31 I=1,N
          A1(I)=0.0
          B1(I)=0.0
          DO 32 J=1,N
          W=AM(J)*AC(J,I)
          A1(I)=A1(I)+W
          B1(I)=B1(I)+W*X(J)
  32      CONTINUE
          AA(I)=P*(S*A1(I)*AM(I)-BM*B1(I)*AM(I))
          BB(I)=P*(S*B1(I)*AM(I)-BI*A1(I)*AM(I))
  31      CONTINUE
          DO 101 I=1,N
          DO 101 J=1,N
          A(I,J)=AC(I,J)*AM(J)
  101     CONTINUE
          DO 34 I=1,N
          DO 34 J=1,N
          A(I,J)=A(I,J)+X(I)*AA(J)+BB(J)
  34      CONTINUE
          CALL GENSS(N,1)
          DO 805 I=1,N
  805     A(I,I)=ABS(A(I,I))
          FREQ=1.0/SQRT(A(MODE,MODE))
          Z=0.0
          DO 906 I=1,N
906       Z=Z+BB(I)*T(I,MODE)
          Z=Z*FREQ**2.0
          FREQ=9.545*FREQ
          WRITE(12,420)MODE,FREQ
  420     FORMAT(/'NATURAL FREQUENCY OF MODE ',i4,' IS : ',G12.4)
  430     FORMAT(/' MODE SHAPES ',/(5G12.4))
```

```
      WRITE(12,430)(T(I,MODE),I=1,N)
      N=N+1
      IF(MODE-MOD)417,419,419
  417 MODE=MODE+1
      GO TO 410
  419 END
       ***************************************************************************
      SUBROUTINE INCOF(AI,X,N,I,S,B)
*     Subroutine to determine the influence coefficients
      DIMENSION B(50),X(50),C(50),CG(50),AI(50)
      M=N-1
      DO 30 J=1,N
      B(J)=0.
      C(J)=0.
      CG(J)=0.
   30 CONTINUE
      L=I-1
      DO 31 J=1,L
      Y1=(X(I)-X(J))/AI(J)
      Y2=(X(I)-X(J+1))/AI(J+1)
      C(J)=C(J)+(Y1+Y2)*S/2.0
      CG(J)=CG(J)+((2.0*Y1+Y2)*S)/(3.0*(Y1+Y2))
   31 CONTINUE
      DO 22 J=1,L
      DO 22 K=1,M
      IF(J-K)5,6,22
    6 B(K)=B(K)+C(J)*CG(J)
      GO TO 22
    5 B(K)=B(K)+C(J)*(CG(J)+X(K+1)-X(J+1))
   22 CONTINUE
      RETURN
      END
       ***************************************************************************
      SUBROUTINE GENSS(N,ISW)
*     Subroutine to solve eigenvalue problem by QR Transformation Method
*     The subroutine deals with solid and unsymmetric matrices
      DIMENSION AA(50,50),A(50,50),T(50,50)
      COMMON A,T
      IF(ISW)100,106,100
  100 DO 104 I=1,N
      DO 102 J=1,N
  102 T(I,J)=0.0
  104 T(I,I)=1.0
C     SET UP CONSTANTS, CALCULATE EN
  106 YR=0.1E-07
      EPS=50.0*YR
      EP=YR*1.0E-02
      MARK=0
      NM1=N-1
C     ITM LOOP
      DO 166 IT=1,100
      IF(MARK)168,108,168
  108 DO 114 I=1,NM1
      IP1=I+1
      DO 114 J=IP1,N
      ENO=ABS(A(I,J)+A(J,I))
      IF(ENO-EPS)110,110,116
  110 ENO=ABS(A(I,J)-A(J,I))
      IF(ENO-EPS)114,114,112
  112 ENO=ABS(A(I,I)-A(J,J))
      IF(ENO-EPS*10.0)114,114,116
  114 CONTINUE
      GO TO 168
  116 MARK=1
```

```
      DO 164 K=1,NM1
      KP1=K+1
      DO 164 M=KP1,N
      H=0.0
      G=0.0
      HJ=0.0
      YH=0.0
      DO 122 I=1,N
      TE=A(I,K)*A(I,K)
      TEE=A(I,M)*A(I,M)
      YH=YH+TE-TEE
      IF(I-K) 118,122,118
118   IF(I-M)120,122,120
120   AIK=A(I,K)
      AKI=A(K,I)
      AIM=A(I,M)
      AMI=A(M,I)
      H=H+AKI*AMI-AIK*AIM
      TEP=AMI*AMI+TE
      TEM=AKI*AKI+TEE
      G=G+TEP+TEM
      HJ=HJ-TEP+TEM
122   CONTINUE
      H=2.0*H
      D=A(K,K)-A(M,M)
      TEP=A(K,M)
      TEM=A(M,K)
      C=TEP+TEM
      E=TEP-TEM
      IF(ABS(C)-EP)124,124,126
124   CC=1.0
      SS=0.0
      GO TO 134
126   BY=D/C
      IF(BY) 128,130,130
128   SIG=-1.0
      GO TO 132
130   SIG=1.0
132   COT=BY+(SIG*SQRT(BY*BY+1.0))
      SS=SIG/SQRT(COT*COT+1.0)
      CC=SS*COT
134   IF(YH)136,136,138
136   TEM=CC
      CC=SS
      SS=-TEM
138   TEP=CC*CC-SS*SS
      TEM=2.0*SS*CC

      D=D*TEP+C*TEM
      H=H*TEP-HJ*TEM
      ED=2.0*E*D
      EDH=ED-H
      DEN=G+2.0*(E*E+D*D)
      IF(DEN-EP*1.E-05)142,142,140
140   TEE=EDH/(DEN+DEN)
      IF(ABS(TEE)-EP)142,142,144
142   CH=1.0
      SH=0.0
      GO TO 146
144   CH=1.0/SQRT(1.0-TEE*TEE)
      SH=CH*TEE
146   C1=CH*CC-SH*SS
      C2=CH*CC+SH*SS
      S1=CH*SS+SH*CC
```

```
      S2=-CH*SS+SH*CC
      IF(ABS(S1)-EP) 148,148,150
148   IF(ABS(S2)-EP) 164,164,150
150   MARK=0
      DO 152 J=1,N
      TEP=A(K,J)
      TEM=A(M,J)
      A(K,J)=C1*TEP+S1*TEM
152   A(M,J)=S2*TEP+C2*TEM
      DO 154 J=1,N
      TEP=A(J,K)
      TEM=A(J,M)
      A(J,K)=C2*TEP-S2*TEM
154   A(J,M)=-S1*TEP+C1*TEM
      IF(ISW)156,164,160
156   DO 158 J=1,N
      TEP=T(K,J)
      TEM=T(M,J)
      T(K,J)=C1*TEP+S1*TEM
158   T(M,J)=S2*TEP+C2*TEM
      GO TO 164
160   DO 162 J=1,N
      TEP=T(J,K)
      TEM=T(J,M)
      T(J,K)=C2*TEP-S2*TEM
162   T(J,M)=-S1*TEP+C1*TEM
164   CONTINUE
166   CONTINUE
      WRITE(*,1)
1     FORMAT(30X,'EIGEN-MAXIMUM ITERATIONS REACHED')
168   RETURN
      END
```

Table 16.7 Input Data of a ship

```
4,20,11.17
.7310,.6820,.6320,.5900
47.3,145.0,200.0,200.0,182.8
192.5,192.5,192.5,192.5,192.5
192.5,192.5,192.5,192.5,192.5
192.5,172.3,145.2,109.0,72.8
431.,1085.,1234.,1109.,986.,6019.,5702.,
4296.,4296.,4304.,3511.,2731.,3050.,3346.,
3773.,4025.,4002.,3781.,1654.,261.
88.,696.,1974.,3459.,4841.,5921.,6490.,6677.,
6677.,6677.,6677.,6677.,6591.,6489.,6177.,5554.,4370.,
2631.,734.,111.
0.249,0.765,1.054,1.054,0.962,
1.013,1.013,1.013,1.013,1.013
1.013,1.013,1.013,1.013,1.013
1.013,0.908,0.765,0.574,0.383
```

Table 16.8 Output of the ship data

```
NATURAL  FREQUENCY  OF  MODE  1  IS  :  46.59

     MODE   SHAPES
  1.112        0.8802     0.6435     0.4041       0.1732
-0.4306E-01 -0.2284    -0.3689    -0.4534      -0.4753
-0.4347      -0.3365    -0.1885    -0.2177E-02  0.2085
```

```
   0.4325       0.6576      0.8729     1.079

NATURAL   FREQUENCY   OF   MODE   2   IS   :   100.2

       MODE   SHAPES
  -0.9742      -0.6705    -0.3526   -0.3563E-01  0.2321
   0.3984       0.4270     0.3261    0.1282      -0.1103
  -0.3244      -0.4607    -0.4806   -0.3716      -0.1511
   0.1474       0.4740     0.7738    1.036

NATURAL   FREQUENCY   OF   MODE   3   IS   :   158.2

         MODE   SHAPES
   0.7599        0.4383        0.9867E-01 -0.2159 -0.4110
  -0.3569       -0.9099E-01    0.2342       0.4568  0.4666
   0.2724       -0.2896E-01   -0.3077      -0.4369 -0.3498
  -0.6285E-01    0.3321        0.6946       0.9761

NATURAL   FREQUENCY   OF   MODE   4   IS   :   221.5

          MODE   SHAPES
  -0.4757      -0.2006      0.8190E-010.3000       0.3459
   0.6881E-01 -0.2788     -0.4093    -0.2157       0.1510
   0.4104       0.3974      0.1151    -0.2390      -0.4031
  -0.2434       0.1777      0.6040     0.8955
```

The reader may draw the mode shapes given in Table 16.8. It may be mentioned that as the three - dimensional correction factor varies with modes, the mass matrix changes from mode to mode. As such in the computer program, the complete eigenvalue problem is to be solved each time we need to determine a particular natural frequency.

16.6 COMPUTER PROGRAM FOR FINITE ELEMENT FREE VIBRATION ANALYSIS OF PLATES

The computer program for the free vibration analysis of plate flexure is given in Table 16.9. The problem has been solved by the finite element method using isoparametric quadratic bending element. The element used being isoparametric, plate of any shape can be analysed with it.

The computer program consists of a number of subroutines. The element stiffness matrix STIFF and the element mass matrix ELMAS are evaluated by Gauss's numerical integration. Two - point reduced integration is to be used for isoparametric quadratic element - the Gauss point values are obtained in the subroutine GACUV. The element mass matrix is diagonal and contains the diagonal elements based on the consistent mass matrix which has been proportionately altered on the basis of the total plate element mass. The elements of the mass matrix corresponding to rotation terms are not considered. Subroutine BOUND takes care of the boundary conditions at each edge. The overall stiffness matrix GSTIF and the overall mass matrix AMASO are stored in a single array. The size of the GSTIF matrix is further reduced with the help of the subroutine CRUNCH2 where the zero elements lying in the extreme ends of the band in all rows (where applicable) are eliminated. It has been mentionad earlier (in sec 16.4) that the diagonal elements of GSTIF matrix are labelled with the help of CRUNCH2, which is indicated by KIND. Subroutine R8USIV is used for the solution of the eigenvalue program.

16.6.1 A Plate Problem

A rectangular orthotropic plate shown in Fig. 16.2 has been analysed by using the computer program of Table 16.9. All edges of the plate are simply supported. The plate has been divided into a 6´6 mesh. The input data for the problem is given in Table 16.10. The mesh division has been automatically programed. Nodes are numbered from the left hand top corner of the edge parallel to *x*-axis and then moved down - the numbering at each row starts from left hand. In all there are 36 elements, 133 nodes and 399 degrees of freedom. Only the output of natural frequencies has been given. *W* - matrix gives the eigen-vectors which have not been presented here for brevity. If required, by using suitable 'write' statements, output of eigen-vectors can be obtained.

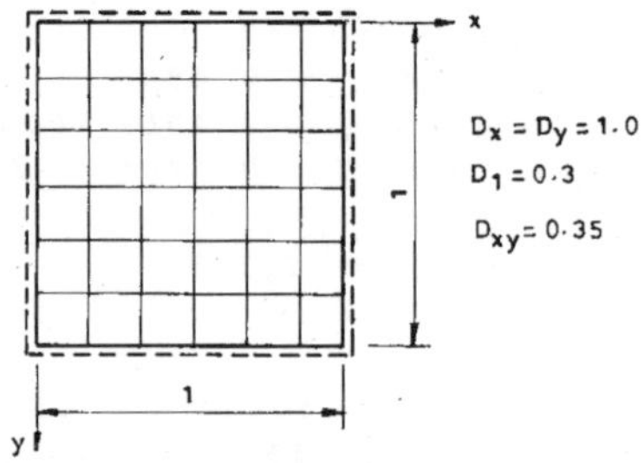

Fig. 16.2 A rectangular plate

Table 16.9 Computer program for free vibration analysis of orthotropic plates using isoparametric quadratic element

```
c*********************************************************************
c
c      FREE VIBRATION ANALYSIS OF ORTHOTROPIC PLATES USING ISOPARAMETRIC
C      QUADRATIC  ELEMENT
c
c*********************************************************************
       dimension  gstif(3300000),stifo(24,24),asat(24,24),rgd(5,5)
     *  ,sfd(8),sfde(8),sfdz(8),sfder(8),sfdzr(8),xg(8),yg(8),zx(8),
     *  zy(8),stiff(24,24),amaso(3300000),elmas(24),nod(40,8)
       dimension  kind(6000),mind(6000),bb(5,24)
c      sin(y)=dsin(y)
c      cos(y)=dcos(y)
c      abs(yy)=dabs(yy)
c      sqrt(zz)=dsqrt(zz)
c*********************************************************************
c      a        length  along  x-axis
c      b        length  along  y-axis
c      t        thickness  of  the  plate
c      dx,dy,
c      d1,dxy   flexural  properties  of  the  plate
c      sx,sy    shear  rigidities  of  the  plate
c      pr       Poisson's  ratio
c      exl,eyl  sizes  of  the  finite  element  in  the  x  and  y-directions
c      phai     angle  between  sides  in  degrees  (=90  for  rectangle)
c      nx,ny    mesh  divisions  in  the  x  and  y-directions
c      nbon     a  code  given  for  typical  boundary  conditions
c      nq       no.  of  Gauss  points  in  the  evaluation  of  stiff
c      nn       no.  of  nodes  in  the  element
```

```
c     nfree    degrees  of  freedom  per  node
c     nt       total  degrees  of  freedom  of  the  entire  plate
c     nelm     total  number  of  elements
c     nbh      half-band  width  of  the  overall  stiffness  matrix
c     nenp     total  degrees  of  freedom  per  element
c     dxz      partial  derivative  of  x  with  respect  to  zi
c     dyz      partial  derivative  of  y  with  respect  to  zi
c     dxe      partial  derivative  of  x  with  respect  to  eta
c     dye      partial  derivative  of  y  with  respect  to  eta
c     zac      the  value  of  the  jacobian
c     dnx      derivative  of  the  shape  function  with  respect  to  x
c     dny      derivative  of  the  shape  function  with  respect  to  y
c     sfd      shape  function
c     sfdz     partial  derivative  of  the  shape  function  with
respect  to  zi
c     sfde     partial  derivative  of  sfd  with  respect  to  eta
c     nod      global  number  of  nodes  of  the  element
c     rgd      rigidity  matrix
c     asat     part  of  the  element  stiffness  matrix
c     stiff    element  stiffness  matrix  in  global  coordinates
c     gstif    overall  stiffness  matrix
c     amaso    overall  mass  matrix
c     elmas    element  mass  matrix
c     xg       s-coordinate  of  the  element
c     yg       y-coordinate  of  the  element
c     zx       normalised  x-coordinate
c     zy       normalised  y-coordinate
c     rho      mass  density  of  the  material
c***********************************************************************
c     nrqd     no.  of  eigenvalue  required
c***********************************************************************
      open  (7,file='isofreein')
      open  (8,file='isofreeout')
 100  format  (2x,  6g13.6)
 101   format(16i5)
 c**********************************************************************
 c    input  data  file  and  their  out  starts
 c**********************************************************************

 105  read  (7,*)  a,b,phai,rho,t
      read  (7,*)  nx,ny,nn,nfree,nt,nbh,nq,nrqd
      read(7,*)  nbon1,nbon2,nbon3,nbon4,nsymx,nsymy
      read  (7,*)  dx,dy,d1,dxy,sx,sy,pr
      nenp=nn*nfree
      nelm=nx*ny
c***********************************************************************
c***********************************************************************

      write  (8,*)  'FREE  VIBRATION  ANALYSIS  OF  ORTHOTROPIC  PLATES'
      write  (8,*)  'USING  ISOPARAMETRIC  QUADRATIC  BENDING  ELEMENT'
      write  (8,*)  '  A  B  PHAI  RHO  THICK'
      write  (8,100)  a,b,phai,rho,t
      write  (8,*)  '  nx  ny  nn  nfree  nt  nbh  nq  nelm  nenp
    *  nrqd  '
      write  (8,101)  nx,ny,nn,nfree,nt,nbh,nq,nelm,nenp,nrqd
      write  (8,*)'nbon1  nbon2  nbon3  nbon4  nsymx  nsymy'
      write  (8,101)  nbon1,nbon2,nbon3,nbon4,nsymx,nsymy
      write  (8,*)  '  dx  dy  d1  dxy  sx  sy'
      write  (8,100)  dx,dy,d1,dxy,sx,sy
c***********************************************************************
```

```
c***************************************************************************
c     output  of  the  given  data  over
c***************************************************************************
      nt=  nfree*(((nn/2-1)*nx+nn/4)*ny+(nn/4)*nx+1)
      nbh=  nfree*((nn/2-1)*nx+(nn/2+1))
      ntot=  nbh*nt-nbh*(nbh-1)/2
      phai=  3.1415927*phai/90.0
c***************************************************************************
c     automatic  mesh  generation  starts
c***************************************************************************
c***************************************************************************
      nx21=2*nx+1
      nx1=nx+1
      do  10  ii=1,nelm
      nxi=(ii-1)/nx
      do  14  i=1,3
 14   nod(ii,i)  =  nx21*nxi+nx1*nxi+2*(ii-nx*nxi)+i-2
      nod(ii,4)  =  nx21*(nxi+1)+nx1*nxi+(ii-nx*nxi)+1
      nod(ii,8)  =nod(ii,4)  -  1
      do  15  i=5,7
      k=i-4
 15   nod(ii,i)=  (nx21+nx1)*(nxi+1)+2*(ii-nx*nxi)+2-k
 10   continue
      write(8,*)  '  node  numbers'
      do  50  ii=1,nelm
      write  (8,*)  '  The  element  number.',  ii
      write  (8,101)  (nod(ii,i),i=1,nn)
 50   continue

c***************************************************************************
c***************************************************************************
c     automatic  mesh  generation  over
c***************************************************************************
C     generation  of  rgd  matrix  starts
c***************************************************************************

      do  23  i=1,5
      do  23  j=1,5
 23   rgd(i,j)=0.0
      rgd(1,1)  =  dx
      rgd  (1,2)=  d1
      rgd(2,1)  =  d1
      rgd(2,2)  =dy
      rgd(3,3)=  dxy
      rgd(4,4)  =sx
      rgd(5,5)  =  sy
c***************************************************************************
c     generation  of  the  rgd  matrix  over
c***************************************************************************
c***************************************************************************
      do  12  i=1,nenp
      do  12  j=1,nenp
 12   stiff(i,j)=0.0
c***************************************************************************
c     element  coordinate  information  is  prescribed
c***************************************************************************
      divx=nx
      divy=ny
      exl=a/divx
      eyl=b/divy
```

```
      xg(1)  =  eyl*cos(phai)
      xg(2)  =  exl/2.0+xg(1)
      yg(1)  =  eyl*sin(phai)
      yg(2)  =  yg(1)
      xg(3)  =  exl+xg(1)
      yg(3)  =  yg(1)
      xg(4)  =  exl+xg(1)/2.0
      yg(4)  =  yg(1)/2.0
      xg(5)  =  exl
      yg(5)  =  0.0
      xg(6)  =  exl/2.0
      yg(6)  =  0.0
      xg(7)  =  0.0
      yg(7)  =  0.0
      xg(8)  =  xg(1)/2.0
      yg(8)  =  yg(1)/2.0
      zx(1)=-1.0
      zy(1)=1.0
      zx(2)=0.0
      zy(2)=1.0
      zx(3)=1.0
      zy(3)=1.0
      zx(4)=1.0
      zy(4)=0.0
      zx(5)=1.0
      zy(5)=-1.0
      zx(6)=0.0
      zy(6)=-1.0
      zx(7)=-1.0
      zy(7)=-1.0
      zx(8)=-1.0
      zy(8)=0.0

c***********************************************************************
c     element  stiffness  matrix  generation  starts
c***********************************************************************
      do  16  i=1,nq
      do  16  j=1,nq
      call  gacuv(i,j,zi,eta,hi,hj,nq)
c***********************************************************************
c     gacuv  is  a  subroutine  to  fetch  gauss  point  values  based  on  the
c     number
c     of  integration  points  required
c***********************************************************************
      call  shape  (zi,eta,zx,zy,sfd,sfdr,sfdz,sfdzr,sfde,sfder,
     * bb,zac,xg,yg,nn,nenp,nfree,elmas,rho)
      do  24  m=1,nenp
      do  24  n=1,nenp
      if  (n-m)  24,611,611
 611  asat(m,n)  =  0.0
      do  24  l=1,5
      do  24  k=1,5
      asat(m,n)  =  asat(m,n)+bb(l,m)*rgd(l,k)*bb(k,n)*zac
 24   continue
 16   continue
      do  620  i=1,nenp
      do  620  j=1,nenp
 620  stiff(j,i)  =  stiff(i,j)

c***********************************************************************
```

```
c
c     element  stiffness  matrix  generation  over
c
c***********************************************************************
c***********************************************************************
c     assembly  of  overall  stiffness  matrix  in  single  array  starts
c
c***********************************************************************

      do  120  i  =  1,ntot
  120 gstif(i)  =  0.0
      do  200  iim  =  1,nelm
      call  assem(iim,nbh,stiff,nod,gstif,nenp,nelm,nn,nfree)
  200 continue
c***********************************************************************
c
c     assembly  of  overall  stiffness  matrix  over
c
c***********************************************************************
      do  30  ij=1,nt
      kind(ij)=0
      mind(ij)=ij
  30  amaso(ij)=0

      do  125  iim=1,nelm
      call  omas  (iim,nn,nfree,nelm,nod,elmas,amaso)
  125 continue
      write(8,*)  '  the  overall  stiffness  matrix'
      call  crunch  (nt,nbh,ntot,gstif,kind)
c***********************************************************************
c
c     fitting  of  boundary  conditions
c
c***********************************************************************
      nd=0
      nx21=nfree*(nx+nx)+1
      call  bound  (nbon1,1,nx21,nfree,gstif,kind,nd)
      nd=1
      call  bound  (nbon2,nx21,nt,nxf,gstif,kind,nd)
      nin=nxf-nfree+1
      nst  =  nt-nx21-nfree
      call  bound  (nbon2,nin,nst,nxf,gstif,kind,nd)
      nd=0
      nst  =  nst+2
      call  bound  (nbon3,nst,nt,nfree,gstif,kind,nd)
      nd=1
      call  bound  (nbon4,1,nst,nxf,gstif,kind,nd)
      nx21  =  nx21+  nfree
      nst=nt-nxf+1
      call  bound  (nbon4,nx21,nst,nxf,gstif,kind,nd)
      call  r8usiv(nt,nrqd,gstif,kind,amaso,mind,bd,nfree,W)
      stop
      end
c**********************************************************************
c
c     main  program  ends
c
c***********************************************************************
c***********************************************************************
c     SHAPE  SUBROUTINE  IS  GIVEN  BELOW
```

```
c
c*****************************************************************************
      subroutine  shape  (zi,eta,zx,zy,sfd,sfdr,sfdz,sfdzr,sfde,sfder,
     * bb,zac,xg,yg,nn,nenp,nfree,elmas,rho)
      dimension  sfd(8),sfde(8),sfdz(8),sfder(8),sfdzr(8),sfdr(8),
     * bb(5,24),zx(8),zy(8),xg(8),yg(8),elmas(24)
      do  1  i=1,nn
 1    elmas(i)=0.0
      sum=0.0
      do  517  k=1,7,2
      ziz=  zi*zx(k)
      etaz=  eta*zy(k)
      sfd(k)=   (1.0+ziz)*(1.0+etaz)*(ziz+etaz-1.0)/4.0
      sfdz(k)=   (1.0+etaz)*zx(k)*(2.0*ziz+etaz)/4.0
 517  sfde(k)=   (1.0+ziz)*zy(k)*(2.0*etaz+ziz)/4.0
      do 518  ll=2,6,4
      etaz=eta*zy(ll)
      sfd(ll)=  (1.0-zi*zi)*(1.0+etaz)/2.0
      sfdz(ll)=  -zi*(1.0+etaz)
 518   sfde(ll)=  (1.0-zi*zi)*zy(ll)/2.0
      do  519  kk=4,8,4
      ziz=  zi*zx(kk)
      sfdz(kk)=  zx(kk)*(1.0-eta*eta)/2.0
      sfde(kk)=  -eta*(1.0+ziz)
 519  sfd(kk)=   (1.0+ziz)*(1.0-eta*eta)/2.0
      do  520  ii=1,5
      do  520  jj=1,nenp
 520  bb(ii,jj)=0.0
      dxz=0.
      dyz=0.
      dxe=0.
      dye=0.
      do  521  ij=1,nn
      dxz=  dxz+sfdz(ij)*xg(ij)
      dyz=  dyz+sfdz(ij)*yg(ij)
      dxe=  dxe+sfde(ij)*xg(ij)
      dye=  dye+  sfde(ij)*yg(ij)
 521  continue
      zac=  dxz*dye-dyz*dxe
      zac=abs(zac)
      call  masmat  (sfd,zac,hi,hj,rho,sum,elmas,nn)
      do  50  ij=1,nn
 50   elmas(ij)=elmas(ij)*rho/sum
      dxzi=dye/zac
      dyzi=-dye/zac
      dxei=-dxe/zac
      dyei=dxz/zac
      do  522  ii=1,nn
      k=nfree*ii-2
      dnx=  sfdz(ii)*dxzi+  sfde(ii)*dyzi
      dny=  sfdz(ii)*dxei+  sfde(ii)*dyei
      bb(1,k+1)=  -dnx
      bb(2,k+2)=-dny
      bb(3,k+1)=-dny
      bb(3,k+2)=-dnx
      bb(4,k)=  dnx
      bb(4,k+1)=  -sfd(ii)
```

```
      bb(5,k)=   dny
 522  bb(5,k+2)   =-sfd(ii)
      return
      end
c*********************************************************************
c*********************************************************************
c              TWO   POINT   GAUSS   INTEGRATION
c*******************************************************************
      subroutine  gacuv  (i,j,zi,eta,hi,hj,nq)
      if(nq-2)  70,10,70
 10   if(i-1)  30,20,30
 20   zi   =  -0.577350269189626
      hi   =  1.0
      go  to  40
 30   zi   =  0.577350269189626
      hi   =  1.00
 40   if(j-1)  60,50,60
 50   eta  =  -0.577350269189626
      hj   =  1.0
      go  to  170
 60   eta  =  0.577350269189626
      hj   =  1.0
      go  to  170
 70   if(i-1)  90,80,90
 80   zi   =  -0.774556669241483
      hi   =  0.555555555555556
      go  to  120
 90   if(i-2)  110,100,110
 100  zi   =  0.0
      hi   =  0.888888888888889
      go  to  120
 110  zi   =  0.774596669241483
      hi   =  0.555555555555556
 120  if(j-1)  140,130,140
 130  eta  =  -0.774596669241483
      hj   =  0.555555555555556
      go  to  170
 140  if(j-2)  160,150,160
 150  eta  =  0.0
      hj   =  0.888888888888889
      go  to  170
 160  eta  =  0.774596669241483
      hj   =  0.555555555555556
 170  return
      end
c******************************************************************
c     SUBROUTINE  ASSEM
c********************************************************************
      subroutine  assem(iim,nbh,stiff,nod,stifo,nenp,nelm,nn,nfree)
      dimension  stifo(1),stiff(nenp,nenp),  nod(40,8)
      ii=iim
      do  55  i=1,nn
      iii  =  nod(ii,i)
      ii1=iii-1
      m=nfree*ii1
      mi=nfree*(i-1)
      do  55  j=1,nn
      jjj=nod(ii,j)
      jj1=jjj-1
      n=nfree*jj1
```

```
      nj=  nfree*(j-1)
      do  50  k=1,nfree
      mmi=m+k
      mmm=  mmi-1
      imi=mi+k
      do  50  l=1,nfree
      nnj=  n+l
      jnj=  nj+l
      if  (nnj.gt.mmi)  go  to  50
      if  (  mmi-nbh)  35,35,40
35    kjj=  mmm*(mmm+1)/2+nnj
        kkj=kjj
      go  to  45
40    kkj=nbh*(nbh+1)/2
      lrow   =   (mmm-nbh+1)*nbh+nnj-MMI
      kkj=  kkj+lrow
45    continue
      stifo(kkj)=  stifo(kkj)+stiff(imi,jnj)
50    continue
55    continue
      return
      end
c***************************************************************************
c     end  of  subroutine  assem
c***************************************************************************
      subroutine  masmat(sfd,zac,hi,hj,rho,sum,elmas,nn)
      dimension  sfd(1),elmas(1)

      do  10  ii=1,nn
      s=sfd(ii)*sfd(ii)*zac*hi*hj
      elmas(ii)  =  elmas(ii)+s
      sum=  sum+s
10    continue
      return
      end
c***************************************************************************
c     end  of  subroutine  masmat
c***************************************************************************
      subroutine  omas  (ii,nn,nfree,nelm,nod,elmas,amaso)
      dimension  sfd(1),elmas(1),amaso(1),nod(40,8)
      do  55  i=1,nn
      jj=(nod(ii,i)-1)*nfree+1
      amaso(jj)=amaso(jj)+elmas(i)
55    continue
      return
      end
c***************************************************************************
c     end  of  subroutine  omas
c***************************************************************************
      subroutine  output  (nd,n,nrqd,value,vector,ier,m)
      real  value(1),vector(nd,m)
      do  100  in=1,nrqd
      write(8,101)'  eigenvalue  for  mode(',in,')',value(in)
      write(8,*)'  the  eigenfvectors  are'
      write(8,*)'  w  thetax  thetay'
101   format  (i5,g15.5)
      do  20  ij=1,n,.5
      write(8,102)(vector(i,in),i=  ij,ij+2)
20    continue
      write(8,103)'  error  code(ier)  =  ',ier
```

```
 102  format(5i5)
 103  format  (2x,6g13.5)
 100  continue
      return
      end
c***************************************************************************
c***************************************************************************
c          TWO  POINT  GAUSS  INTEGRATION
c***************************************************************************
      subroutine  gacuv  (i,j,zi,eta,hi,hj,nq)
      if(nq-2)  70,10,70
 10   if(i-1)  30,20,30
 20   zi   =  -0.577350269189626
      hi   =  1.0
      go  to  40
 30   zi   =  0.577350269189626
      hi   =  1.00
 40   if(j-1)  60,50,60
 50   eta  =  -0.577350269189626
      hj   =  1.0
      go  to  170
 60   eta  =  0.577350269189626
      hj   =  1.0
      go  to  170
 70   if(i-1)  90,80,90
 80   zi   =  -0.774556669241483
      hi   =  0.555555555555556
      go  to  120
 90   if(i-2)  110,100,110
 100  zi   =  0.0
      hi   =  0.888888888888889
      go  to  120
 110  zi   =  0.774596669241483
      hi   =  0.555555555555556
 120  if(j-1)  140,130,140
 130  eta  =  -0.774596669241483
      hj   =  0.555555555555556
      go  to  170
 140  if(j-2)  160,150,160
 150  eta  =  0.0
      hj   =  0.888888888888889
      go  to  170
 160  eta  =  0.774596669241483
      hj   =  0.555555555555556
 170  return
      end
c***************************************************************************
c                ASSEMBLY  IN  SKYLINE  FORM
C***************************************************************************
      subroutine  assem1(ielm,nbh,estif,nod,gstif,nfree)
      dimension  estif(40,40),nod(1000,8),gstif(3300000)
      nnpm=8
c     nnpm=no  of  nod  per  member/element
      do  60  i=1,nnpm
      ig=(nod(ielm,i)-1)*nfree
      do  60  ii=1,nfree
      mmi=ig+ii
      mmm=mmi-1
      imi=(i-1)*nfree+ii
      do  60  j=1,nnpm
```

```
      jg=(nod(ielm,j)-1)*nfree
      do  60  jj=1,nfree
      nnj=jg+jj
      jnj=(j-1)*nfree+jj
      if(nnj.gt.mmi)go  to  55
      if(mmi-nbh)35,35,40
 35   kjj=mmm*(mmm+1)/2+nnj
      kkj=kjj
      go  to  45
 40   kkj=nbh*(nbh+1)/2
      lrow=(mmm-nbh+1)*nbh+nnj-mmi
      kkj=kkj+lrow
 45   continue
      gstif(kkj)=gstif(kkj)+estif(imi,jnj)
 55   continue
 60   continue
      cwrite(*,*)'kkj=',kkj
      return
      end
c***********************************************************************
c*****************subroutine  crunch*********************************
      subroutine  crunch(npt,nbh,ntot,gstif,kind)
      dimension  ak(3300000),gstif(3300000),kind(6000)
      ist=1
      iend=0
      do  10  jj=1,nbh
      i=(jj*(jj-1))/2
      do  2  ii=1,jj
      ak(ii)=gstif(i+ii)
 2    continue
      call  search(ak,jj,ist,iend,gstif)
      kind(jj)=iend
 10   continue
      do  20  jj=nbh+1,npt
      i=i+nbh
      do  4  ii=1,nbh
      ak(ii)=gstif(i+ii)
 4    continue
      call  search(ak,nbh,ist,iend,gstif)
      kind(jj)=iend
 20   continue
      return
      end
****************************subroutine  search*********************
      subroutine  search(ak,nb1,ist,iend,gstif)
      dimension  ak(3300000),gstif(3300000)
      do  20  ii=1,nb1
      if(ak(ii))10,20,10
 10   ibl=ii
      go  to  30
 20   continue
      ibl=nb1
 30   continue
      icount=nb1-ibl+1
      iend=iend+icount
      do  40  ii=ist,iend
      jj=ii-ist+ibl
      gstif(ii)=ak(jj)
 40   continue
      ist=ist+icount
```

```
      return
      end
c*******************************************************************
c*******************************************************************
c     subroutine  bound  starts
c
c*******************************************************************
      subroutine  bound  (nbon,i,j,k,stifo,kind,nd)
      dimension  stifo(1)
      rigid=1.0e20
      go  to  (40,30,20,10,5),  nbon+1
 5    do  7  ii=i,j,k
      stifo(kind(ii+nd))=rigid
      stifo(kind(ii+4-nd))=rigid
 7    continue
      return
 10   do  15  ii=i,j,k
      stifo(kind(ii+nd))=rigid
      stifo(kind(ii+2))=rigid
 15   continue
      go  to  40
 20   do  25  ii=i,j,k
      stifo(kind(ii))=rigid
      stifo(kind(ii+1))=rigid
      stifo(kind(ii+2))=rigid
      stifo(kind(ii+4-nd))=rigid
 25   continue
      go  to  40
 30   do  35  ii=i,j,k
      stifo(kind(ii))=rigid
      stifo(kind(ii+1+nd))=rigid
 35   continue
 40   continue
      return
      end
c************************************************************************
C     end  of  bound
c************************************************************************
```

Table 16.10 Input data of plate problem

```
Input  data  for  the  plate  problem

1.  1.  90.  100.  0.01
6  6  8  3  399  69  2  2
1  1  1  1  0  0
1.  1.  0.3  0.35  35000  35000  0.3
```

Table 16.11 Out of the plate problem

```
Output  data  for  the  plate  problem


  FREE  VIBRATION  ANALYSIS  OF  ORTHOTROPIC  PLATES
  USING  ISOPARAMETRIC  QUADRATIC  BENDING  ELEMENT

Length  and  Width  of  the  Plate  :  1.000  1.000
```

```
PHAI  =  90.00
Density  of  the  Material  :  100.0
Thickness  of  the  Plate  =  0.1000E-01


No.  of  divisions  in  x-  and  y-  directions  =  6  6
No.  of  nodes  per  element  =  8
No.  of  d.o.f.  per  node  =  3
Total  d.o.f.  of  the  plate  =  399
Half  Bandwidth  =  69
No.  of  gauss  points  =  2
No.  of  elements  =  36
No.  of      d.o.f.  per  element  =  24
No.  of  required  frequencies  =  2

Boundary  Conditions  :
  nbon1   nbon2   nbon3   nbon4   nsymx   nsymy
    1       1       1       1       0       0

Rigidity  Components  :
    dx        dy         d1          dxy          sx            sy
  1.000    1.000    0.3000     0.3500    0.3500E+05    0.3500E+05

Element  Nodal  Connectivity  :
  The  element  no.  :  1
     1    2    3   15   23   22   21   14
  The  element  no.  :  2
     3    4    5   16   25   24   23   15
  The  element  no.  :  3
     5    6    7   17   27   26   25   16
  The  element  no.  :  4
     7    8    9   18   29   28   27   17
  The  element  no.  :  5
     9   10   11   19   31   30   29   18
  The  element  no.  :  6
    11   12   13   20   33   32   31   19
  The  element  no.  :  7
    21   22   23   35   43   42   41   34
  The  element  no.  :  8
    23   24   25   36   45   44   43   35
  The  element  no.  :  9
    25   26   27   37   47   46   45   36
  The  element  no.  :  10
    27   28   29   38   49   48   47   37
  The  element  no.  :  11
    29   30   31   39   51   50   49   38
  The  element  no.  :  12
    31   32   33   40   53   52   51   39
  The  element  no.  :  13
    41   42   43   55   63   62   61   54
  The  element  no.  :  14
    43   44   45   56   65   64   63   55
  The  element  no.  :  15
    45   46   47   57   67   66   65   56
  The  element  no.  :  16
    47   48   49   58   69   68   67   57
```

```
The element no. : 17
    49  50  51  59  71  70  69  58
The element no. : 18
    51  52  53  60  73  72  71  59
The element no. : 19
    61  62  63  75  83  82  81  74
The element no. : 20
    63  64  65  76  85  84  83  75
The element no. : 21
    65  66  67  77  87  86  85  76
The element no. : 22
    67  68  69  78  89  88  87  77
The element no. : 23
    69  70  71  79  91  90  89  78
The element no. : 24
    71  72  73  80  93  92  91  79
The element no. : 25
    81  82  83  95  103 102 101 94
The element no. : 26
    83  84  85  96  105 104 103 95
The element no. : 27
    85  86  87  97  107 106 105 96
The element no. : 28
    87  88  89  98  109 108 107 97
The element no. : 29
    89  90  91  99  111 110 109 98
The element no. : 30
    91  92  93  100 113 112 111 99
The element no. : 31
    101 102 103 115 123 122 121 114
The element no. : 32
    103 104 105 116 125 124 123 115
The element no. : 33
    105 106 107 117 127 126 125 116
The element no. : 34
    107 108 109 118 129 128 127 117
The element no. : 35
    109 110 111 119 131 130 129 118
The element no. : 36
    111 112 113 120 133 132 131 119

The Eigenvalues Are :

     Mode    Eigenvalue
      1      19.727364
      2      49.346203
```

References

16.1 K.J. Bathe, Finite Element Procedures in Engineering Analysis, Prentice - Hall of India Pvt. Ltd., New Delhi, 1990.

16.2 W.H. Press, S.A. Teukolsky, W.T. Vellerling and B.P. Flannery, Numerical Recipes, 2nd Edition, Cambridge University Press, 1994.

16.3 R.B. Corr and A. Jennings, A simutaneous iteration algorithm for symmetric eigenvalue problems, International Journals for Numerical Methods in Engineering, V.10, 1976, pp. 647 - 663.

16.4 K.J. Bathe and S. Ramaswamy, An accelerated subspace iteration method, Computer Methods in Applied Mechanics and Engineering, V.23, 1980, pp. 313 - 331.

16.5 J.H. Wilkinson, QR - algorithm of local symmetric matrices with multiple eigenvalues, Computer Journal, V.8, 1965, pp. 88 - 89.

APPENDIX A

THE STIFFNESS MATRIX

A.1 STIFFNESS MATRIX

The degrees of freedom of the structure are first identified. It is referred to as the number of independent joint displacements of the structure.

The intersections or interconnections between individual elements of the structure are called joints or nodes. Displacements corresponding to each degree of freedom, referred to as nodal displacements, are numbered and positive directions are assumed.

In order to evaluate the elements of the stiffness matrix, the structure is assumed to deform corresponding to a unit value of one of the nodal displacements, while all other displacement are restrained. The forces at all the required points are evaluated. The process is repeated for a unit value for each nodal displacement. The formation of the stiffness matrix is explained with the help of an example.

The continuous beam of Fig. A.1, has two degrees of freedom numbered δ_1 and δ_2. Referring to Fig. A.1(b), joint *B* is allowed to rotate by unity in the counter clockwise direction (i.e. positive direction, see Fig. A.2) while joint *C* is kept fixed. The moment developed at *B* due to unit rotation at *B* is termed as K_{11}, and the moment induced at *C* due to this displacement is K_{21} Note that there is a relation of the labels 1 and 2 in K_{11} and K_{21} to joint displacements δ_1 and δ_2.

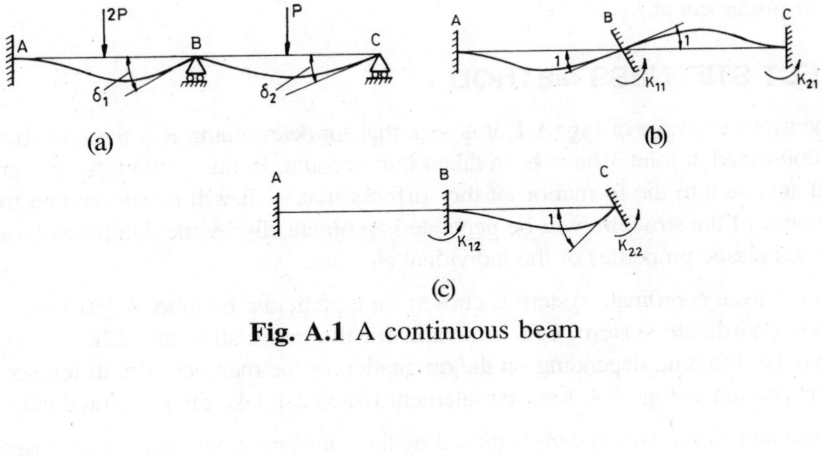

Fig. A.1 A continuous beam

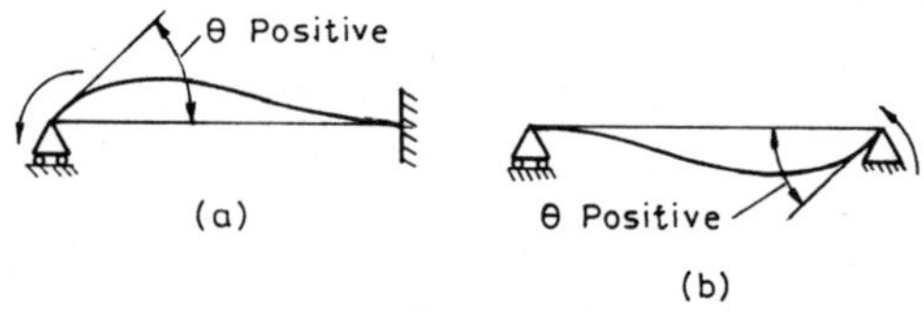

Fig. A.2 Sign convention

In order to determine K_{11} the contribution from both the members *AB* and *BC* is to be taken into account. Referring to Fig. A.3,

$$K_{11} = \frac{4EI}{L} + \frac{4EI}{L} = \frac{8EI}{L}$$

and

$$K_{21} = \frac{2EI}{L}$$

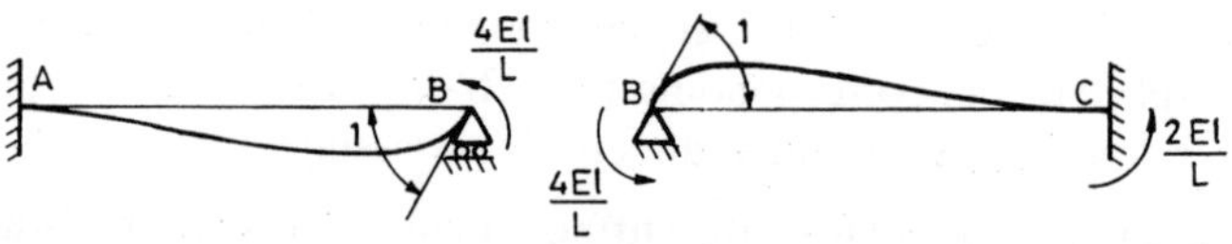

Fig. A.3 Unit rotation at joint *B*

Similarly, $K_{12} = \frac{2EI}{L}$ and $K_{22} = \frac{4EI}{L}$.

The stiffness matrix for this problem is

$$\begin{bmatrix} K_{11} & K_{12} \\ K_{21} & K_{22} \end{bmatrix} = \frac{2EI}{L}\begin{bmatrix} 4 & 1 \\ 1 & 2 \end{bmatrix}$$

The stiffness coefficient K_{ij} is defined as the force developed at point *i* corresponding to an unknown displacement at *i*, due to a unit value of the displacement at *j* corresponding to an unknown displacement at *j*.

A.2 DIRECT STIFFNESS METHOD

Even in the trivial example of Fig. A.1, it is seen that for determining K_{11}, the contribution from members connected at joint 1 have been taken into account. In this section, we present a more formalised approach to the formation of the stiffness matrix. It will be shown that the overall stiffness matrix of the structure will be generated automatically by the computer, based on the geometric and elastic properties of the individual element.

For this, a local coordinate system is chosen for a particular member. Global axis system is the common coordinate system, which is valid for the entire structure. The two-coordinate systems may be different, depending on the orientation of the member. The difference between the two is explained in Fig. A.4, for a bar element which can take up axial force only.

The local and global axes system is related by the transformation matrix. If the displacements in the local axis system for a particular member is and the $\{\delta'\}$ displacements in the corresponding global axis system is $\{\delta\}$, then

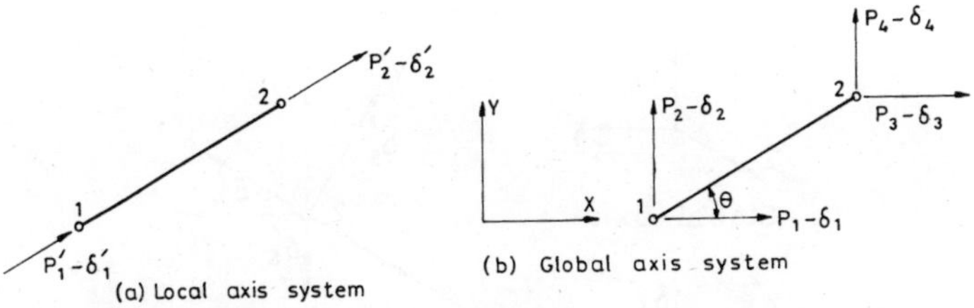

Fig. A.4 Local and global axis system

$$\{\delta'\} = [T]\{\delta'\} \tag{A.1}$$

where $[T]$ is the member transformation matrix.

For the truss, element of Fig. A.4, the stiffness matrix of the member in local coordinate is

$$[k]_e = \frac{EA}{L}\begin{bmatrix} 1 & -1 \\ -1 & 1 \end{bmatrix} \tag{A.2}$$

and
$$[T] = \begin{bmatrix} l_1 & m_1 & 0 & 0 \\ 0 & 0 & l_1 & m_1 \end{bmatrix} \tag{A-3}$$

where $l_1 = \cos\theta$ and $m_1 = \sin\theta$

It can be shown that the stiffness matrix in global coordinate system is given by

$$[K]_e = [T]^T [k]_e [T] \tag{A.4}$$

For the truss member

$$[K]_e = \frac{EA}{L}\begin{bmatrix} l_1^2 & & \text{Symmetrical} & \\ l_1 m_1 & m_1^2 & & \\ -l_1^2 & -l_1 m_1 & l_1^2 & \\ -l_1 m_1 & -m_1^2 & l_1 m_1 & m_1^2 \end{bmatrix} \tag{A.5}$$

A typical frame member is shown in Fig. A.5. The member of the frame has three degrees of freedom at each end-two translations and one rotation. X', Y', Z' is the local axis system and X, Y, Z is the global axis system. The relationship between them is given by

$$\begin{Bmatrix} \delta_1' \\ \delta_2' \\ \delta_3' \\ \cdots \\ \delta_4' \\ \delta_5' \\ \delta_6' \end{Bmatrix} = \begin{bmatrix} l_1 & m_1 & 0 & \vdots & 0 & 0 & 0 \\ -m_1 & l_1 & 0 & \vdots & 0 & 0 & 0 \\ 0 & 0 & 1 & \vdots & 0 & 0 & 0 \\ \cdots & \cdots & \cdots & \cdots & \cdots & \cdots & \cdots \\ 0 & 0 & 0 & \vdots & l_1 & m_1 & 0 \\ 0 & 0 & 0 & \vdots & -m_1 & l_1 & 0 \\ 0 & 0 & 0 & \vdots & 0 & 0 & 1 \end{bmatrix} \begin{Bmatrix} \delta_1 \\ \delta_2 \\ \delta_3 \\ \cdots \\ \delta_4 \\ \delta_5 \\ \delta_6 \end{Bmatrix} \tag{A.6}$$

or
$$\{\delta'\} = [T]\{\delta\} \tag{A.7}$$

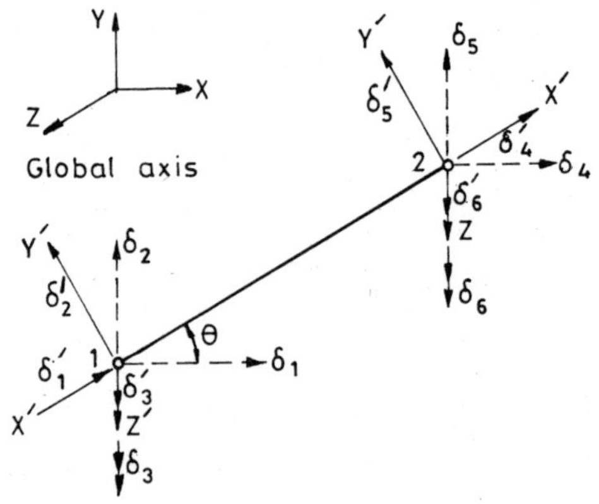

Fig. A.5 Space frame member

The stiffness matrix of the member in local axis system is

$$[k]_e = \begin{bmatrix} \frac{EA}{L} & 0 & 0 & -\frac{EA}{L} & 0 & 0 \\ & \frac{12EI}{L^3} & \frac{6EI}{L^2} & 0 & -\frac{12EI}{L^3} & \frac{6EI}{L^2} \\ & & \frac{4EI}{L} & 0 & -\frac{6EI}{L^2} & \frac{2EI}{L} \\ & & & \frac{EA}{L} & 0 & 0 \\ \text{Symmetrical} & & & & \frac{12EI}{L^3} & -\frac{6EI}{L^2} \\ & & & & & \frac{4EI}{L} \end{bmatrix} \tag{A.8}$$

The element stiffness matrix in the global coordinate system can be formed by using Eq. (A.4).

A.2.1 Overall Stiffness Matrix

The stiffness matrix of an element of the structure forms the basic component. It is now discussed here, as to how the element stiffness matrix of the individual element is assembled, to form the overall stiffness matrix of the structure.

It might have been noticed earlier, that the stiffness at a joint is obtained by adding the stiffness of all the members meeting at the joint.

The degrees of freedom of the structure are first numbered-starting with 1 and ending with *NP*, where *NP* is the total degrees of freedom. The restraints are then numbered beyond *NP*. This numbering is referred to as degrees of freedom corresponding to global numbering or global degrees of freedom. All the restraints can be given the number $(NP+1)$. This procedure will save some storage space.

We have already introduced the concept of local axis system and the global axis system. We shall henceforth deal with two different numbering in the global axis system for the displacements at member ends. The first set of numbering is referred to as local numbering, which will remain identical for every member. This numbering refers to the typical displacement labels of the member. The members meet at a joint in the structure. The same member ends will have different displacement labels, when the total structure is considered.

The continuous beam of Fig. A.6 has four degrees of freedom-rotation at four supports. In Fig. A.6(a), the numbering from left hand side corresponding to global coordinate has been indicated. In Fig. A.6(b), the local numbering of a typical member and in Fig. A.6(c), the positive direction of the displacements has been shown.

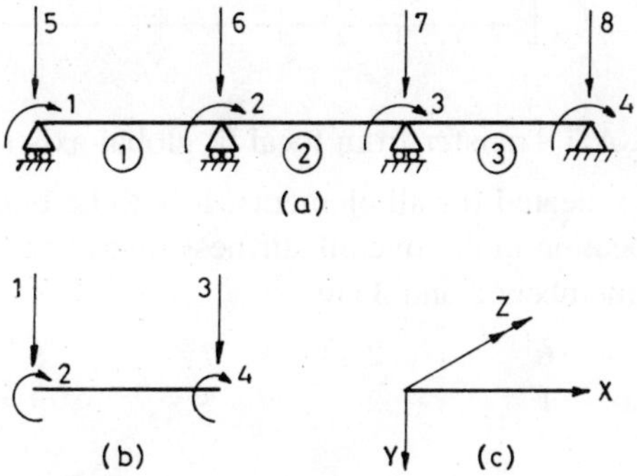

Fig. A.6 A continuous beam

Calculation of the element stiffness matrix is next to be done. This is done for a typical element on the basis of local numbering. For the problem at hand, the size of the element stiffness matrix is 4 × 4. For the first element, they are, say

Global		5	1	6	2
	Local	1	2	3	4

$$[k]_{e1} = \begin{matrix} 5 & 1 \\ 1 & 2 \\ 6 & 3 \\ 2 & 4 \end{matrix} \begin{bmatrix} a_{11} & a_{12} & a_{13} & a_{14} \\ a_{21} & a_{22} & a_{23} & a_{24} \\ a_{31} & a_{32} & a_{33} & a_{34} \\ a_{41} & a_{42} & a_{43} & a_{44} \end{bmatrix} \quad \text{(A.9)}$$

For members which are inclined and where local and global coordinates are not coincident, the element stiffness matrix is to be evaluated on the basis of transformation of Eq. (A.4). There is one-to-one correspondence between the joints of the elements and that of the assemblage. The elements of element stiffness matrix of Eq. (A.9) should now be put in their proper place, in the overall stiffness matrix. Equation (A.9) indicates that the local numbering 1 corresponds to global numbering 5, which means that a_{11} should occupy the fifth row and the fifth column (i.e. k_{55}) of the overall stiffness matrix. Local number 2 corresponds to global number 1. Therefore, a_{11}, a_{21}, a_{22} will occupy positions of k_{51}, k_{15} and k_{11} of the overall stiffness matrix. The proper location of all elements of Eq. (A.9) in the overall stiffness matrix has been shown in Fig. A.7.

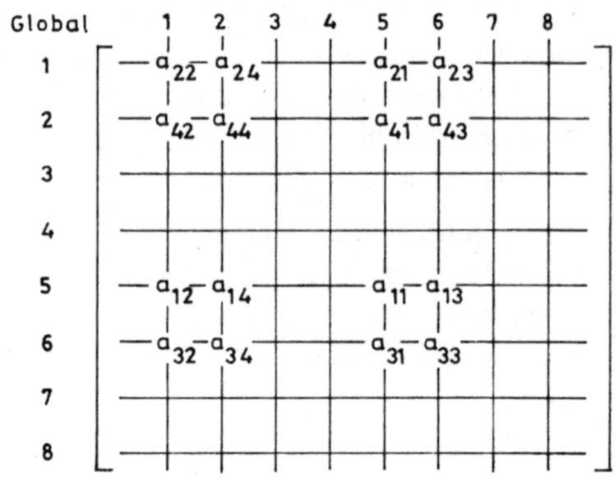

Fig. A.7 Transfer from local to global axis system

The process is to be repeated for all elements. It is to be borne in mind, that individual stiffness at a particular location in the overall stiffness matrix are to be added. If the element stiffness matrices for the members 2 and 3 are

$$\begin{array}{ccc} & \text{Global} & & 6 & 2 & 7 & 3 \\ & & \text{Local} & 1 & 2 & 3 & 4 \end{array}$$

$$[k]_{e2} = \begin{array}{cc} 6 & 1 \\ 2 & 2 \\ 7 & 3 \\ 3 & 4 \end{array} \begin{bmatrix} b_{11} & b_{12} & b_{13} & b_{14} \\ b_{21} & b_{22} & b_{23} & b_{24} \\ b_{31} & b_{32} & b_{33} & b_{34} \\ b_{41} & b_{42} & b_{43} & b_{44} \end{bmatrix} \quad \text{(A.10)}$$

and

$$\begin{array}{ccc} & \text{Global} & & 7 & 3 & 8 & 4 \\ & & \text{Local} & 1 & 2 & 3 & 4 \end{array}$$

$$[k]_{e3} = \begin{array}{cc} 7 & 1 \\ 3 & 2 \\ 8 & 3 \\ 4 & 4 \end{array} \begin{bmatrix} c_{11} & c_{12} & c_{13} & c_{14} \\ c_{21} & c_{22} & c_{23} & c_{24} \\ c_{31} & c_{32} & c_{33} & c_{34} \\ c_{41} & c_{42} & c_{43} & c_{44} \end{bmatrix}$$

then, the overall stiffness matrix is shown in Fig. A.8. The steps to be followed in the computer program is shown in the flow chart of Fig. A.9.

$$
[K] = \begin{array}{c} \\ 1 \\ 2 \\ 3 \\ 4 \\ 5 \\ 6 \\ 7 \\ 8 \end{array}
\begin{array}{c}
\begin{array}{cccccccc} 1 & 2 & 3 & 4 & 5 & 6 & 7 & 8 \end{array} \\
\begin{bmatrix}
a_{22} & a_{24} & 0 & 0 & a_{21} & a_{23} & 0 & 0 \\
a_{42} & \boxed{a_{44}+b_{22}} & b_{24} & 0 & a_{41} & \boxed{a_{43}+b_{21}} & b_{23} & 0 \\
0 & b_{42} & \boxed{b_{44}+c_{22}} & c_{24} & 0 & b_{41} & \boxed{b_{43}+c_{21}} & c_{23} \\
0 & 0 & c_{42} & c_{44} & 0 & 0 & c_{41} & c_{43} \\
a_{12} & a_{14} & 0 & 0 & a_{11} & a_{13} & 0 & 0 \\
a_{32} & \boxed{a_{34}+b_{12}} & b_{14} & 0 & a_{31} & \boxed{a_{33}+b_{11}} & b_{13} & 0 \\
0 & b_{32} & \boxed{b_{34}+c_{12}} & c_{14} & 0 & b_{31} & \boxed{b_{33}+c_{11}} & c_{13} \\
0 & 0 & c_{32} & c_{34} & 0 & 0 & c_{31} & c_{33}
\end{bmatrix}
\end{array}
$$

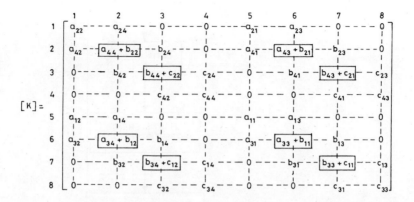

Fig. A.8 Assembly of the stiffness matrix

The overall mass matrix, overall damping matrix and the overall exciting force matrix can be similarly formed from the respective element matrices.

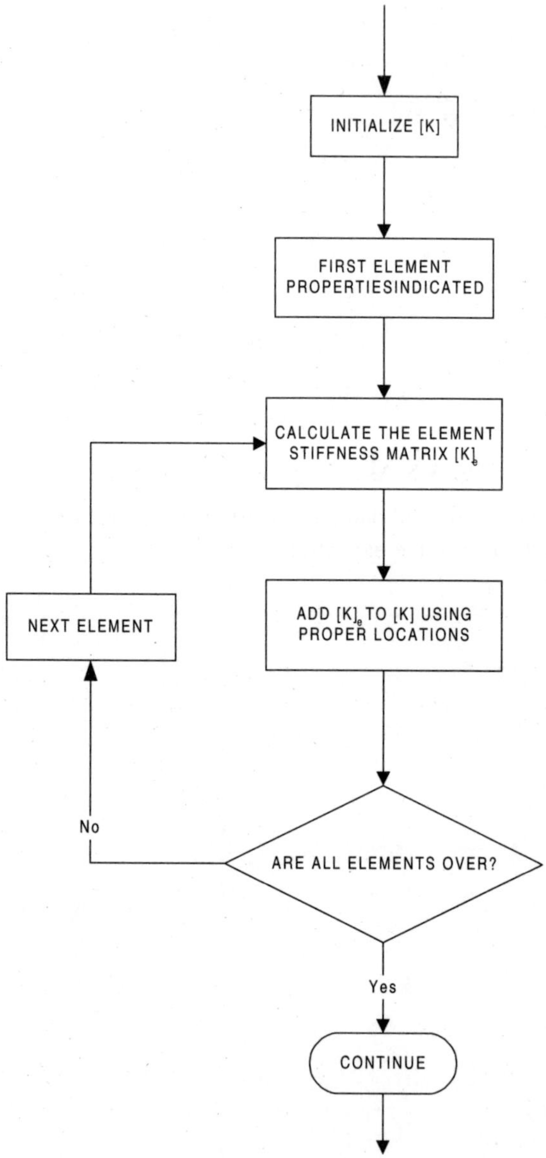

Fig. A.9

There are various ways, by which levels in the displacements are put. In the scheme that has been discussed, the active degrees of freedom are labelled (their total is *NP*) first and then the restraints are labelled. The first set of equations NP is operated upon, for obtaining the necessary solution.

APPENDIX B

TABLE OF SPRING STIFFNESS

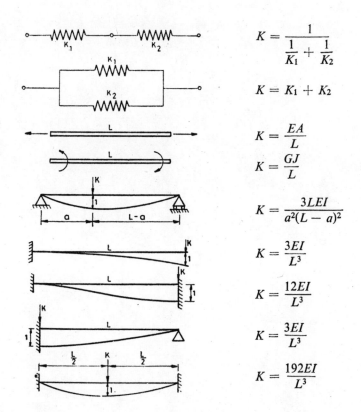

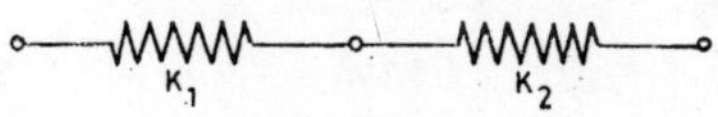

$$K = \frac{1}{\frac{1}{K_1} + \frac{1}{K_2}}$$

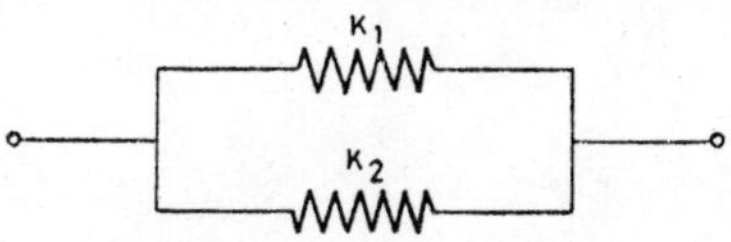

$$K = K_1 + K_2$$

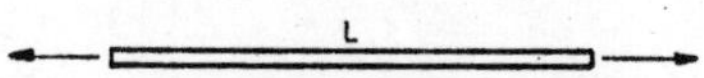

$$K = \frac{EA}{L}$$

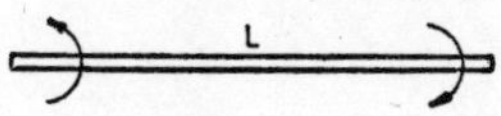

$$K = \frac{GJ}{L}$$

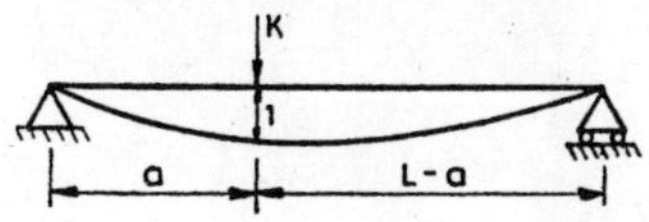

$$K = \frac{3LEI}{a^2(L - a)^2}$$

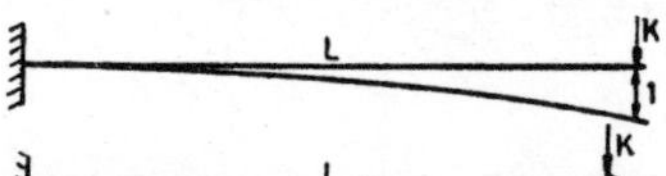

$$K = \frac{3EI}{L^3}$$

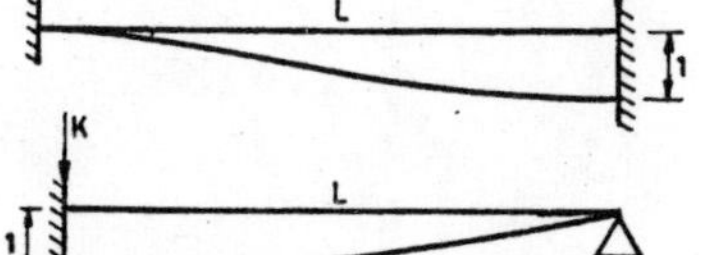

$$K = \frac{12EI}{L^3}$$

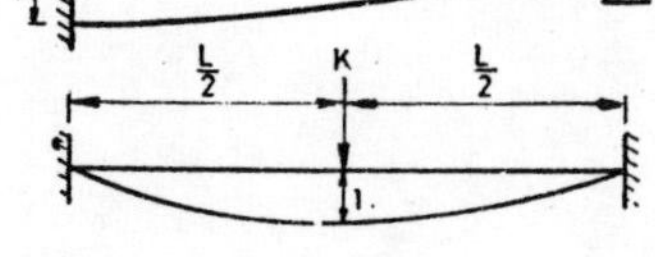

$$K = \frac{3EI}{L^3}$$

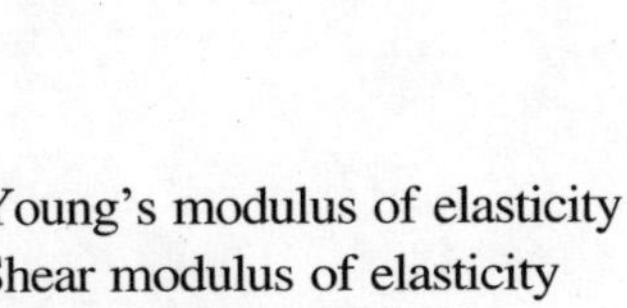

$$K = \frac{192EI}{L^3}$$

where

E	=	Young's modulus of elasticity of the material
G	=	Shear modulus of elasticity
A	=	Cross-sectional area
I	=	second moment of the area
J	=	polar moment of inertia
L	=	length of the member

Index